PROBLEMS IN REAL ANALYSIS
Second Edition

Second Edition

A Workbook with Solutions

PROBLEMS IN REAL ANALYSIS
Second Edition

A Workbook with Solutions

CHARALAMBOS D. ALIPRANTIS

Departments of Economics and Mathematics
Purdue University

and

OWEN BURKINSHAW

Departments of Mathematical Sciences
Indiana University, Purdue University, Indianapolis

ACADEMIC PRESS
An Imprint of Elsevier
San Diego London Boston
New York Sydney Tokyo Toronto

Permissions may be sought directly from Elsevier's Science and Technology Rights
Department in Oxford, UK. Phone: (44) 1865 843830, Fax: (44) 1865 853333, e-mail:
permissions@elsevier.co.uk. You may also complete your request on-line via the
Elsevier homepage: http://www.elsevier.com by selecting "Customer Support" and
then "Obtaining Permissions".

ACADEMIC PRESS
An Imprint of Elsevier
525 B Street, Suite 1900, San Diego, CA 92101-4495, USA
http://www.apnet.com

ACADEMIC PRESS LIMITED
24–28 Oval Road, London NW1 7DX, UK
http://www.hbuk.co.uk/ap/

International Standard Book Number 0-12-050253-4

This book accompanies the following title, catalogued by the Library of Congress:

Library of Congress Cataloging-in-Publication Data
Aliprantis, Charalambos D.
    Problems in real analysis / Charalambos D. Aliprantis and Owen Burkinshaw.
      p. cm.
    Includes bibliographical references and index.
    ISBN-13: 978-0-12-050253-0 ISBN-10: 0-12-050253-4 (acid-free paper)
    1. Mathematical analysis. 2. Functions of real variables. I. Burkinshaw, Owen. II. Title.
QA300.A48   1998
515—dc21   98-3955
                                                                 CIP

ISBN-13: 978-0-12-050253-0
ISBN-10: 0-12-050253-4
Transferred to Digital Printing 2008
    07   9 8 7

CONTENTS

Foreword   vii

CHAPTER 1.   FUNDAMENTALS OF REAL ANALYSIS   1

     1.   Elementary Set Theory   1
     2.   Countable and Uncountable Sets   6
     3.   The Real Numbers   11
     4.   Sequences of Real Numbers   20
     5.   The Extended Real Numbers   34
     6.   Metric Spaces   45
     7.   Compactness in Metric Spaces   54

CHAPTER 2.   TOPOLOGY AND CONTINUITY   65

     8.   Topological Spaces   65
     9.   Continuous Real-Valued Functions   73
     10.   Separation Properties of Continuous Functions   92
     11.   The Stone–Weierstrass Approximation Theorem   98

CHAPTER 3.   THE THEORY OF MEASURE   107

     12.   Semirings and Algebras of Sets   107
     13.   Measures on Semirings   112
     14.   Outer Measures and Measurable Sets   116
     15.   The Outer Measure Generated by a Measure   122
     16.   Measurable Functions   133
     17.   Simple and Step Functions   137
     18.   The Lebesgue Measure   146
     19.   Convergence in Measure   157
     20.   Abstract Measurability   160

CHAPTER 4. THE LEBESGUE INTEGRAL 171

    21. Upper Functions 171
    22. Integrable Functions 174
    23. The Riemann Integral as a Lebesgue Integral 190
    24. Applications of the Lebesgue Integral 206
    25. Approximating Integrable Functions 220
    26. Product Measures and Iterated Integrals 224

CHAPTER 5. NORMED SPACES AND L_p-SPACES 239

    27. Normed Spaces and Banach Spaces 239
    28. Operators between Banach Spaces 245
    29. Linear Functionals 251
    30. Banach Lattices 259
    31. L_p-Spaces 271

CHAPTER 6. HILBERT SPACES 297

    32. Inner Product Spaces 297
    33. Hilbert Spaces 310
    34. Orthonormal Bases 325
    35. Fourier Analysis 333

CHAPTER 7. SPECIAL TOPICS IN INTEGRATION 345

    36. Signed Measures 345
    37. Comparing Measures and the
        Radon–Nikodym Theorem 353
    38. The Riesz Representation Theorem 365
    39. Differentiation and Integration 379
    40. The Change of Variables Formula 395

FOREWORD

This book contains complete solutions to the 609 problems in the third edition of *Principles of Real Analysis*, Academic Press, 1998. The problems have been spread over forty sections which follow the format of the book.

All solutions are based on the material covered in the text with frequent references to the results in the text. For instance, a reference to Theorem 7.3 refers to Theorem 7.3 and a reference to Example 28.4 refers to Example 28.4, both in the third edition of *Principles of Real Analysis*.

This problem book will be beneficial to students only if they use it "properly." That is to say, if students look at a solution of a problem *only after* trying very hard to solve the problem. Students will do themselves great injustice by reading a solution without any prior attempt on the problem. It should be a real challenge to students to produce solutions which are different from the ones presented here.

We would like to express our most sincere thanks to all the people who made constructive recommendations and corrections regarding the text and the problems. Special thanks are due to Professor Yuri Abramovich for his contributions and suggestions during the writing of this problem book.

C. D. ALIPRANTIS AND O. BURKINSHAW
West Lafayette, Indiana
July, 1998

CHAPTER **1**_____

FUNDAMENTALS OF REAL ANALYSIS

1. ELEMENTARY SET THEORY

Problem 1.1. *Establish the following set theoretic relations:*

1. $(A \cup B) \cap C = (A \cap C) \cup (B \cap C)$ *and*
   $(A \cap B) \cup C = (A \cup C) \cap (B \cup C)$;
2. $(A \cup B) \setminus C = (A \setminus C) \cup (B \setminus C)$ *and*
   $(A \cap B) \setminus C = (A \setminus C) \cap (B \setminus C)$;
3. $A \setminus B = A \cap B^c$;
4. $A \subseteq B \iff B^c \subseteq A^c$; *and*
5. $(A \cup B)^c = A^c \cap B^c$ *and* $(A \cap B)^c = A^c \cup B^c$.

Also, for an arbitrary function $f: X \to Y$, establish the following claims:

6. $f\left(\bigcup_{i \in I} A_i\right) = \bigcup_{i \in I} f(A_i)$;
7. $f\left(\bigcap_{i \in I} A_i\right) \subseteq \bigcap_{i \in I} f(A_i)$;
8. $f^{-1}\left(\bigcup_{i \in I} B_i\right) = \bigcup_{i \in I} f^{-1}(B_i)$;
9. $f^{-1}\left(\bigcap_{i \in I} B_i\right) = \bigcap_{i \in I} f^{-1}(B_i)$; *and*
10. $f^{-1}(B^c) = \left[f^{-1}(B)\right]^c$

Solution. (1) We establish the first formula only. We have

$$x \in (A \cup B) \cap C \iff x \in A \cup B \text{ and } x \in C$$
$$\iff [\, x \in A \text{ or } x \in B \,] \text{ and } x \in C$$
$$\iff [\, x \in A \text{ and } x \in C \,] \text{ or } [\, x \in B \text{ and } x \in C \,]$$
$$\iff x \in A \cap C \text{ or } x \in B \cap C$$
$$\iff x \in (A \cap C) \cup (B \cap C).$$

(2) Again, we establish the first formula only. Observe that

$$x \in (A \cup B) \setminus C \iff x \in A \cup B \text{ and } x \notin C$$

$$\Longleftrightarrow \; [\, x \in A \text{ or } x \in B \,] \text{ and } x \notin C$$
$$\Longleftrightarrow \; [\, x \in A \text{ and } x \notin C \,] \text{ or } [\, x \in B \text{ and } x \notin C \,]$$
$$\Longleftrightarrow \; x \in A \setminus C \text{ or } x \in B \setminus C$$
$$\Longleftrightarrow \; x \in (A \setminus C) \cup (B \setminus C).$$

(3) Note that

$$x \in A \setminus B \iff x \in A \text{ and } x \notin B$$
$$\iff x \in A \text{ and } x \in B^c \iff x \in A \cap B^c.$$

(4) Let $A \subseteq B$. Then, $x \in B^c$ implies $x \notin B$ and so $x \notin A$ (i.e., $x \in A^c$) so that $B^c \subseteq A^c$. On the other hand, if $B^c \subseteq A^c$ holds, then (by the preceding case) we have $A = (A^c)^c \subseteq (B^c)^c = B$.

(5) Note that

$$x \in (A \cap B)^c \iff x \notin A \cap B \iff x \notin A \text{ or } x \notin B$$
$$\iff x \in A^c \text{ or } x \in B^c \iff x \in A^c \cup B^c.$$

Moreover,

$$x \in (A \cup B)^c \iff x \notin A \cup B \iff x \notin A \text{ and } x \notin B$$
$$\iff x \in A^c \text{ and } x \in B^c \iff x \in A^c \cap B^c.$$

(6) We have

$$y \in f\left(\bigcup_{i \in I} A_i\right) \iff \exists\, x \in \bigcup_{i \in I} A_i \text{ with } y = f(x)$$
$$\iff \exists\, i \in I \text{ with } x \in A_i \text{ and } y = f(x)$$
$$\iff \exists\, i \in I \text{ with } y \in f(A_i) \iff y \in \bigcup_{i \in I} f(A_i).$$

(7) From the inclusion $f\left(\bigcap_{i \in I} A_i\right) \subseteq f(A_j)$ for each j, we see that

$$f\left(\bigcap_{i \in I} A_i\right) \subseteq \bigcap_{i \in I} f(A_i).$$

(8) We have

$$x \in f^{-1}\left(\bigcup_{i \in I} B_i\right) \iff f(x) \in \bigcup_{i \in I} B_i \iff \exists\, i \in I \text{ with } f(x) \in B_i$$
$$\iff \exists\, i \in I \text{ with } x \in f^{-1}(B_i) \iff x \in \bigcup_{i \in I} f^{-1}(B_i).$$

(9) Note that

$$x \in f^{-1}\left(\bigcap_{i \in I} B_i\right) \iff f(x) \in \bigcap_{i \in I} B_i \iff f(x) \in B_i \text{ for each } i \in I$$

$$\iff x \in f^{-1}(B_i) \text{ for each } i \in I \iff x \in \bigcap_{i \in I} f^{-1}(B_i).$$

(10) Observe that

$$x \in f^{-1}(B^c) \iff f(x) \in B^c \iff f(x) \notin B$$

$$\iff x \notin f^{-1}(B) \iff x \in \left[f^{-1}(B)\right]^c.$$

Problem 1.2. *For two sets A and B show that the following statements are equivalent:*

a. $A \subseteq B$;
b. $A \cup B = B$;
c. $A \cap B = A$.

Solution.   (a) $\Longrightarrow$ (b) Clearly, $B \subseteq A \cup B$ holds. On the other hand, if $x \in A \cup B$, then $x \in A$ or $x \in B$, and so in either case, $x \in B$. This means $A \cup B \subseteq B$, and hence, $A \cup B = B$.

(b) $\Longrightarrow$ (c) By part (1) of the preceding problem, we have

$$A \cap B = A \cap (A \cup B) = (A \cap A) \cup (A \cap B) = A \cup (A \cap B) = A.$$

(c) $\Longrightarrow$ (a) Clearly, $A = A \cap B \subseteq B$.

Problem 1.3. *Show that $(A \triangle B) \triangle C = A \triangle (B \triangle C)$ holds for every triplet of sets $A, B,$ and C.*

Solution.   Note first that for any three sets X, Y, and Z we have

$$X \triangle Y \setminus Z = [X \setminus (Y \cup Z)] \cup [Y \setminus (X \cup Z)]$$

and

$$Z \setminus (X \triangle Y) = [Z \setminus (X \cup Y)] \cup [X \cap Y \cap Z].$$

For instance, to verify the first identity, note that

$$x \in X \triangle Y \setminus Z \iff [x \in X \setminus Y \text{ or } x \in Y \setminus X] \text{ and } x \notin Z$$

$$\iff [x \in X, x \notin Y, \text{ and } x \notin Z] \text{ or } [x \in Y, x \notin X, \text{ and } x \notin Z]$$

$$\iff x \in [X \setminus (Y \cup Z)] \cup [Y \setminus (X \cup Z)].$$

Thus,

$$
\begin{aligned}
(A\Delta B)\Delta C &= [(A\Delta B)\setminus C]\cup[C\setminus(A\Delta B)]\\
&= [A\setminus(B\cup C)]\cup[B\setminus(A\cup C)]\cup[C\setminus(A\cup B)]\cup[A\cap B\cap C]\\
&= \{[A\setminus(B\cup C)]\cup(A\cap B\cap C)\}\cup\{[B\setminus(C\cup A)]\cup[C\setminus(B\cup A)]\}\\
&= [A\setminus(B\Delta C)]\cup[(B\Delta C)\setminus A]\\
&= A\Delta(B\Delta C).
\end{aligned}
$$

Problem 1.4.   *Give an example of a function* $f:X \to Y$ *and two subsets* A *and* B *of* X *such that* $f(A\cap B)\neq f(A)\cap f(B)$.

Solution.   Define $f:\{0,1\}\to\{0,1\}$ by $f(0)=f(1)=0$. If $A=\{0\}$ and $B=\{1\}$, then $f(A\cap B)=\emptyset\neq\{0\}=f(A)\cap f(B)$.

Problem 1.5.   *For a function* $f:X\to Y$, *show that the following three statements are equivalent:*

   a. f *is one-to-one.*
   b. $f(A\cap B)=f(A)\cap f(B)$ *holds for all* $A,\,B\in\mathcal{P}(X)$.
   c.   *For every pair of disjoint subsets* A *and* B *of* X, *we have* $f(A)\cap f(B)=\emptyset$.

Solution. $(a)\implies(b)$ If $y\in f(A)\cap f(B)$, then there exist $a\in A$ and $b\in B$ with $y=f(a)=f(b)$. Since f is one-to-one, $a=b\in A\cap B$, and so $y\in f(A\cap B)$. Thus, $f(A)\cap f(B)\subseteq f(A\cap B)\subseteq f(A)\cap f(B)$.
$(b)\implies(c)$ Obvious.
$(c)\implies(a)$ Let $f(a)=f(b)$. If $a\neq b$, then the two sets $A=\{a\}$ and $B=\{b\}$ satisfy $A\cap B=\emptyset$, while $f(A)\cap f(B)=\{f(a)\}\neq\emptyset$.

Problem 1.6.   *Let* $f:X\to Y$ *be a function. Show that* $f(f^{-1}(A))\subseteq A$ *for all* $A\subseteq Y$, *and* $B\subseteq f^{-1}(f(B))$ *for all* $B\subseteq X$.

Solution.   Clearly, $x\in f^{-1}(A)$ if and only if $f(x)\in A$. Thus, $f(f^{-1}(A))\subseteq A$. Similarly, $x\in f^{-1}(f(B))$ if and only if $f(x)\in f(B)$, and so $B\subseteq f^{-1}(f(B))$ holds.

Problem 1.7.   *Show that a function* $f:X\to Y$ *is onto if and only if* $f(f^{-1}(B))=B$ *holds for all* $B\subseteq Y$.

Solution.   Assume that f is onto and $B\subseteq Y$. If $b\in B$, then there exists some $a\in X$ with $f(a)=b$; clearly, $a\in f^{-1}(B)$. Thus, $b=f(a)\in f(f^{-1}(B))$, and so $B\subseteq f(f^{-1}(B))\subseteq B$ holds.

For the converse, note that the relation $f(f^{-1}(\{b\})) = \{b\}$ implies $f^{-1}(\{b\}) \neq \emptyset$ for each $b \in Y$ so that f is onto.

Problem 1.8. *Let* $X \xrightarrow{f} Y \xrightarrow{g} Z$. *If* $A \subseteq Z$, *show that*

$$(g \circ f)^{-1}(A) = f^{-1}(g^{-1}(A)).$$

Solution. Note that

$$x \in (g \circ f)^{-1}(A) \Longleftrightarrow g(f(x)) \in A \Longleftrightarrow f(x) \in g^{-1}(A) \Longleftrightarrow x \in f^{-1}(g^{-1}(A)).$$

Problem 1.9. *Show that the composition of functions satisfies the associative law. That is, show that if* $X \xrightarrow{f} Y \xrightarrow{g} Z \xrightarrow{h} V$, *then* $(h \circ g) \circ f = h \circ (g \circ f)$.

Solution. Observe that for each $x \in X$ we have

$$[(h \circ g) \circ f](x) = h \circ g(f(x)) = h(g(f(x))) = h((g \circ f)(x)) = [h \circ (g \circ f)](x).$$

Therefore, $(h \circ g) \circ f = h \circ (g \circ f)$.

Problem 1.10. *Let* $f : X \to Y$. *Show that the relation* $\mathcal{R}$ *on* X, *defined by* $x_1 \mathcal{R} x_2$ *whenever* $f(x_1) = f(x_2)$, *is an equivalence relation.*

Solution. We must show that the relation $\mathcal{R}$ is reflexive, symmetric, and transitive.

Reflexivity: Note that $f(x) = f(x)$ implies $x \mathcal{R} x$ for each $x \in X$.

Symmetry: Let $x_1 \mathcal{R} x_2$. Then, $f(x_1) = f(x_2)$ or $f(x_2) = f(x_1)$, so that $x_2 \mathcal{R} x_1$.

Transitivity: If $x_1 \mathcal{R} x_2$ and $x_2 \mathcal{R} x_3$, then $f(x_1) = f(x_2)$ and $f(x_2) = f(x_3)$ both hold. It follows that $f(x_1) = f(x_3)$, and so $x_1 \mathcal{R} x_3$.

Problem 1.11. *If* X *and* Y *are sets, then show that*

$$\mathcal{P}(X) \cap \mathcal{P}(Y) = \mathcal{P}(X \cap Y) \quad and \quad \mathcal{P}(X) \cup \mathcal{P}(Y) \subseteq \mathcal{P}(X \cup Y).$$

Solution. (a) Note that

$$A \in \mathcal{P}(X) \cap \mathcal{P}(Y) \Longleftrightarrow A \subseteq X \text{ and } A \subseteq Y$$
$$\Longleftrightarrow A \subseteq X \cap Y \Longleftrightarrow A \in \mathcal{P}(X \cap Y).$$

(b) Clearly,

$$A \in \mathcal{P}(X) \cup \mathcal{P}(Y) \Longrightarrow A \subseteq X \text{ or } A \subseteq Y \Longrightarrow A \subseteq X \cup Y \Longrightarrow A \in \mathcal{P}(X \cup Y).$$

If X and Y are two nonempty disjoint sets, then $X \cup Y \notin \mathcal{P}(X) \cup \mathcal{P}(Y)$, and so equality is seldom valid.

2.   COUNTABLE AND UNCOUNTABLE SETS

Problem 2.1.   *Show that the set of all rational numbers is countable.*

Solution.   Let $\mathcal{Q}$ be the set of rational numbers and let $\mathcal{Q}^+ = \{r \in \mathcal{Q}: r > 0\}$. Then the function $f: \mathbb{N} \times \mathbb{N} \to \mathcal{Q}^+$ defined by $f(m, n) = \frac{m}{n}$ is onto. The conclusion now follows from Theorems 2.7 and 2.5.

Problem 2.2.   *Show that the set of all finite subsets of a countable set is countable.*

Solution.   We can assume that $A = \{p_1, p_2, \ldots\}$ is the set of all prime numbers. Let $\mathcal{F}$ denote the collection of all finite subsets of A. Define $f: \mathcal{F} \to \mathbb{N}$ by $f(F) =$ the product of the elements of F, for each $F \in \mathcal{F}$. Then f is one-to-one, and the conclusion follows from Theorem 2.5.

Problem 2.3.   *Show that a union of an at-most countable collection of sets, each of which is finite, is an at-most countable set.*

Solution.   This follows immediately from Theorem 2.6.

Problem 2.4.   *Let A be an uncountable set and let B be a countable subset of A. Show that A is equivalent to $A \setminus B$.*

Solution.   Let $B = \{b_1, b_2, \ldots\}$. Since A is uncountable, the set $A \setminus B$ is also uncountable. Let $C = \{c_1, c_2, \ldots\}$ be a countable subset of $A \setminus B$. Now define $f: A \setminus B \to A$ by

$$f(x) = \begin{cases} x, & \text{if } x \notin C; \\ c_{n+1}, & \text{if } x = c_{2n+1} \ (n = 0, 1, 2, \ldots); \\ b_n, & \text{if } x = c_{2n} \ (n = 1, 2, \ldots). \end{cases}$$

Then f is one-to-one and onto, proving that $A \approx A \setminus B$.

Problem 2.5.   *Assume that $f: A \to B$ is a surjective (onto) function between two sets. Establish the following:*

   a.   card $B \leq$ card A.
   b.   *If A is countable, then B is at-most countable.*

Solution. (a) Consider the family $\{f^{-1}(b)\colon b \in B\}$. Clearly, this is a family of disjoint subsets of A. By the Axiom of Choice there exists a subset C of A such that $C \cap f^{-1}(b)$ consists precisely of one element of A for each $b \in B$. The conclusion now follows by observing that $f\colon C \to B$ is one-to-one and onto.

(b) This follows immediately from part (a).

Problem 2.6. *Show that two nonempty sets A and B are equivalent if and only if there exists a function from A onto B and a function from B onto A.*

Solution. If A and B are equivalent, then there exists a function $f\colon A \to B$ which is one-to-one and onto. Clearly, $f^{-1}\colon B \to A$ is a surjective function.

For the converse, assume that there exists a function from A onto B and a function from B onto A. A glance at the preceding problem guarantees that card $B \le$ card A and card $A \le$ card B. Now, use the Schröder–Bernstein theorem to conclude that A and B are equivalent sets.

Problem 2.7. *Show that if a finite set X has n elements, then its power set $\mathcal{P}(X)$ has 2^n elements.*

Solution. We shall use induction on n. Assume that $\{1, \ldots, n\}$ has 2^n subsets. Then the subsets of the set $\{1, \ldots, n, n+1\}$ consist of:

a. The subsets of $\{1, \ldots, n\}$, which are 2^n altogether; and
b. The subsets of the form $A \cup \{n+1\}$, where A is a subset of $\{1, \ldots, n\}$, again 2^n altogether.

Thus, the number of subsets of $\{1, \ldots, n, n+1\}$ is $2^n + 2^n = 2^{n+1}$.

A direct proof goes as follows. Notice that the number of subsets of $\{1, 2, \ldots, n\}$ having k elements (where $0 \le k \le n$) is precisely $\binom{n}{k}$. So, the total number of subsets of $\{1, 2, \ldots, n\}$ is

$$\binom{n}{0} + \binom{n}{1} + \binom{n}{2} + \cdots + \binom{n}{n} = (1+1)^n = 2^n,$$

where the last equality holds true by virtue of the binomial theorem.

Problem 2.8. *Show that the set of all sequences with values 0 or 1 is uncountable.*

Solution. For each subset A of $\mathbb{N}$ define the sequence $f(A) = \{x_n\}$ by $x_n = 1$ if $n \in A$ and $x_n = 0$ if $n \notin A$. Then f defines a function from $\mathcal{P}(\mathbb{N})$ onto the sequences with values 0 and 1. Since f is clearly one-to-one and onto, the conclusion follows from Theorem 2.8.

Problem 2.9. *If* $2 = \{0, 1\}$, *then show that* $2^X \approx \mathcal{P}(X)$ *for every set* X.

Solution. Define $f: \mathcal{P}(X) \to 2^X$ by $A \longmapsto f_A$, where $f_A(x) = 1$ if $x \in A$ and $f_A(x) = 0$ if $x \notin A$. Note that f is one-to-one and onto. Therefore, $2^X \approx \mathcal{P}(X)$.

Problem 2.10. *Any complex number that is a root of a (nonzero) polynomial with integer coefficients is called an* **algebraic number**. *Show that the set of all algebraic numbers is countable.*

Solution. Let $\mathcal{Z} = \{\ldots, -2, -1, 0, 1, 2, \ldots\}$. Fix $n \geq 1$. Since every polynomial $p(x) = a_0 + a_1 x + \cdots + a_n x^n$ is determined uniquely by $(a_0, a_1, \ldots, a_n)$, it is easy to see that the nonzero polynomials of degree $\leq n$ with integer coefficients are in one-to-one correspondence with the countable set $\mathcal{Z}^{n+1} \backslash \{(0, 0, \ldots, 0)\}$. Let $\{p_1, p_2, \ldots\}$ be an enumeration of all these polynomials. By the Fundamental Theorem of Algebra, the set $A_k = \{x \in \mathbb{C}: \ p_k(x) = 0\}$ is a finite set. Thus, the set of all zeros of the polynomials $\{p_1, p_2, \ldots\}$ of degree $\leq n$ is precisely the set $R_n = \bigcup_{k=1}^{\infty} A_k$, which (by Theorem 2.6) is a countable set. Now, note that the set of all algebraic numbers is $\bigcup_{n=1}^{\infty} R_n$, which—as a countable union of countable sets—is itself countable; see Theorem 2.6.

Problem 2.11. *For an arbitrary function* $f: \mathbb{R} \to \mathbb{R}$ *show that the set*

$$A = \Big\{ a \in \mathbb{R}: \ \lim_{x \to a} f(x) \ \text{exists and} \ \lim_{x \to a} f(x) \neq f(a) \Big\}$$

is at-most countable.

Solution. Let $\mathcal{I}$ denote the set of all open subintervals of $\mathbb{R}$ with rational endpoints and note that $\mathcal{I}$ is a countable set. Also, let $\mathcal{Q}$ denote the countable set of all rational numbers of $\mathbb{R}$.

For each rational real number r, let

$$A_r = \Big\{ a \in A: \ \text{Either} \ f(a) < r < \lim_{x \to a} f(x) \ \text{or} \ \lim_{x \to a} f(x) < r < f(a) \Big\}.$$

Clearly, $A = \bigcup_{r \in \mathcal{Q}} A_r$ holds. Thus, in order to establish that A is at most countable, it suffices to show that each A_r is at-most countable.

So, fix some $r \in \mathcal{Q}$ and $a \in A_r$ and assume (without loss of generality) that $f(a) < r < \lim_{x \to a} f(x)$. Then there exists as $\delta > 0$ such that $a - \delta < y < a + \delta$ and $y \neq a$ imply $f(y) > r$. Next, pick an open interval I_a with rational endpoints (i.e., $I_a \in \mathcal{I}$) such that $a \in I_a$ and $I_a \subseteq (a - \delta, a + \delta)$. Since $f(y) > r$

holds for each $y \in I_a$ with $y \neq a$, we see that $y \notin A_r$ for each $y \in I_a \backslash \{a\}$. In particular, note that $A_r \cap I_a = \{a\}$.

Thus, we have established a mapping $a \longmapsto I_a$ from A_r into $\mathcal{I}$ (which in view of $A_r \cap I_a = \{a\}$ for each $a \in A_r$) is also one-to-one. This implies that A_r is at-most countable, and hence, A is likewise at-most countable.

Problem 2.12. *Show that the set of real numbers is uncountable by proving the following:*

a) $(0, 1) \approx \mathbb{R}$; *and*
b) $(0, 1)$ *is uncountable.*

Solution. (a) The function $f : (0, 1) \rightarrow \mathbb{R}$ defined by the formula $f(x) = \tan\left(\pi x - \frac{\pi}{2}\right)$ is one-to-one and onto.

(b) If $(0, 1)$ is countable, then let $\{x_1, x_2, \ldots\}$ be one enumeration of $(0, 1)$. For each n write $x_n = 0.d_{n1}d_{n2} \cdots$ in its decimal expansion, where each d_{ij} is $0, 1, \ldots,$ or 9. Now, consider the real number y of $(0, 1)$ whose decimal expansion $y = 0.y_1 y_2 \cdots$ satisfies $y_n = 1$ if $d_{nn} \neq 1$ and $y_n = 2$ if $d_{nn} = 1$. An easy argument now shows (how?) that $y \neq x_n$ for each n, which is a contradiction. Hence, the interval $(0, 1)$ is an uncountable set.

Problem 2.13. *Using mathematical induction prove the following:*

   a.   *If $a \geq -1$, then $(1+a)^n \geq 1+na$ for $n = 1, 2, \ldots$ (Bernoulli's inequality).*
   b.   *If $0 < a < 1$, then $1 + 3^n a > (1 + a)^n$ for $n = 1, 2, \ldots$.*
   c. $\cos(n\pi) = (-1)^n$ *for $n = 1, 2, \ldots$.*

Solution. (a) Let $a \geq -1$. For $n = 1$ the inequality is trivially true; in fact, it is an equality. For the induction step, assume that $(1 + a)^n \geq 1 + na$ holds true for some n. Since $1 + a > 0$ is assumed to be true, it follows that

$$(1 + a)^{n+1} = (1 + a)(1 + a)^n \geq (1 + a)(1 + na) = 1 + na + a + na^2$$
$$= 1 + (n + 1)a + na^2 \geq 1 + (n + 1)a,$$

which is the desired inequality when n takes the value $n + 1$. This completes the induction.

(b) Assume $0 < a < 1$. Since $1 + 3a > 1 + a$, the desired inequality is true for $n = 1$. For the inductive step assume $1 + 3^n a > (1 + a)^n$. Then, taking into account that $0 < a < 1$, we see that

$$(1 + a)^{n+1} = (1 + a)(1 + a)^n < (1 + a)(1 + 3^n a)$$
$$= 1 + 3^n a + a + 3^n a^2 = 1 + (3^n + 3^n a + 1)a$$
$$< 1 + (3^n + 3^n + 3^n)a = 1 + 3 \cdot 3^n a = 1 + 3^{n+1}a,$$

which is the desired inequality valid when n is replaced by $n + 1$. By the Principle of Mathematical Induction, the inequality is true for every natural number n.

(c) For $n = 1$, we have $\cos(1 \cdot \pi) = \cos \pi = -1 = (-1)^1$. Now, assume that $\cos(n\pi) = (-1)^n$. Then, using the trigonometric formula $\cos(x + y) = \cos x \cos y - \sin x \sin y$, we see that

$$\cos\big[(n + 1)\pi\big] = \cos(n\pi + \pi) = \cos(n\pi) \cos \pi - \sin(n\pi) \sin \pi$$
$$= (-1)^n(-1) - \sin(n\pi) \cdot 0 = (-1)^{n+1},$$

and the induction is complete.

Problem 2.14. *Show that the Well-Ordering Principle implies the Principle of Mathematical Induction.*

Solution. Let $S \subseteq \mathbb{N}$ satisfy

a. $1 \in S$, and
b. $n + 1 \in S$ whenever $n \in S$.

We must show that $S = \mathbb{N}$, or equivalently that $\mathbb{N} \setminus S = \emptyset$.

To this end, assume by way of contradiction that we have $\mathbb{N} \setminus S \neq \emptyset$. Then, by the Well Ordering Principle, $n = \min(\mathbb{N} \setminus S)$ exists. Clearly, $1 < n \in \mathbb{N} \setminus S$. Thus, $n - 1 \in S$, and consequently $n = (n - 1) + 1 \in S$, a contradiction. Therefore, $\mathbb{N} \setminus S = \emptyset$ or $S = \mathbb{N}$.

Problem 2.15. *Show that the Principle of Mathematical Induction implies the Well-Ordering Principle.*

Solution. Assume that the Principle of Mathematical Induction is true. Consider the subset S of $\mathbb{N}$ consisting of all natural numbers n with the property: whenever a nonempty subset A of $\mathbb{N}$ contains a natural number $m \leq n$, then A has a least element. To establish the Well-Ordering Principle, we need to show that $S = \mathbb{N}$.

To this end, note that $1 \in S$. Now assume that $n \in S$. Also, assume that a nonempty subset A of $\mathbb{N}$ contains some natural number $m \leq n + 1$. If A contains a natural number $k < n + 1$, then A also contains a natural number (namely k itself) less than or equal to n, and so, in view of $n \in S$, A must have a least element. On the other hand, if A does not contain any natural number strictly less that $n + 1$, it follows that $n + 1 \in A$, in which case $n + 1$ is the least element of A. Therefore, $n + 1 \in S$, and so by the validity of the Principle of Mathematical Induction, we infer that $S = \mathbb{N}$.

3. THE REAL NUMBERS

Problem 3.1. *If $a \vee b = \max\{a, b\}$ and $a \wedge b = \min\{a, b\}$, then show that*

$$a \vee b = \tfrac{1}{2}(a + b + |a - b|) \quad and \quad a \wedge b = \tfrac{1}{2}(a + b - |a - b|).$$

Solution. Since all expressions do not change their values if we interchange a and b, we can assume $a \geq b$. Thus,

$$\tfrac{1}{2}(a + b + |a - b|) = \tfrac{1}{2}(a + b + a - b) = a = a \vee b,$$

and

$$\tfrac{1}{2}(a + b - |a - b|) = \tfrac{1}{2}[a + b - (a - b)] = b = a \wedge b.$$

Problem 3.2. *Show that $\big||a| - |b|\big| \leq |a + b| \leq |a| + |b|$ for all $a, b \in \mathbf{R}$.*

Solution. From $-|a| \leq a \leq |a|$ and $-|b| \leq b \leq |b|$, it follows that

$$-(|a| + |b|) \leq a + b \leq |a| + |b|.$$

So, $|a + b| \leq |a| + |b|$.

Substituting $a - b$ in the place of a, we get $|a| \leq |a - b| + |b|$ so that $|a| - |b| \leq |a - b|$. Interchanging a and b yields $-(|a| - |b|) \leq |a - b|$, and so $\big||a| - |b|\big| \leq |a - b|$ also holds.

Problem 3.3. *Show that the real numbers $\sqrt{2}$ and $\sqrt{2} + \sqrt{3}$ are irrational numbers.*

Solution. Assume by way of contradiction that $\sqrt{2} = \frac{m}{n}$ with $m, n \in \mathbf{N}$. We can suppose that m and n have no common positive divisors other than 1. Squaring, we get $m^2 = 2n^2$. This implies that m is even, i.e., $m = 2k$ for some $k \in \mathbf{N}$ (otherwise $m = 2k + 1$ implies that m^2 is odd, a contradiction). It follows that $4k^2 = 2n^2$, or $n^2 = 2k^2$, which in turn implies that n is even, i.e., $n = 2\ell$ for some $\ell \in \mathbf{N}$. But then, m and n have the common factor 2, which is a contradiction. Hence, $\sqrt{2}$ is not a rational number. (This simple proof is due to Eudoxus.)

With a different and more elegant proof one can establish the following general result:

* *The square root $\sqrt{k}$ of a natural number k is a rational number if and only if k is a complete square, i.e., $k = p^2$ for some $p \in \mathbf{N}$.*

If $k = p^2$, then clearly $\sqrt{k} = p \in \mathbb{N}$. On the other hand, if $\sqrt{k}$ is a rational number, then $\sqrt{k}$ is a rational root of the polynomial $p(x) = x^2 - k$. But the positive rational roots of this polynomial are of the form $\frac{m}{n}$, where $m \in \mathbb{N}$ is a divisor of k and $n \in \mathbb{N}$ is a divisor of 1. Thus, $\sqrt{k} = m \in \mathbb{N}$, and so $k = m^2$.

To see that $\sqrt{2} + \sqrt{3}$ is not a rational number, assume by way of contradiction that $\sqrt{2} + \sqrt{3} = r > 0$ is a rational number. Then $\sqrt{3} = r - \sqrt{2}$ and by squaring, we get $3 = r^2 - 2r\sqrt{2} + 2$. This implies $\sqrt{2} = \frac{r^2 - 1}{2r}$, a rational number, contrary to our previous conclusion. Hence, $\sqrt{2} + \sqrt{3}$ is an irrational number.

Problem 3.4. *Show that between any two distinct real numbers there is an irrational number.*

Solution. Let $a < b$. Choose a rational number r with $a < r < b$, and then select some n so that $0 < \frac{\sqrt{2}}{n} < b - r$. Note that the irrational number $x = r + \frac{\sqrt{2}}{n}$ satisfies $a < x < b$.

Alternatively: Note that the open interval (a, b) is uncountable, while the set of all rational numbers is countable.

Problem 3.5. *This problem will introduce (by steps) the familiar process of subtraction in the framework of the axiomatic foundation of real numbers.*

    a. *Show that the element 0 is uniquely determined, i.e., show that if $x + 0^* = x$ for all $x \in \mathbb{R}$ and some $0^* \in \mathbb{R}$, then $0^* = 0$.*

    b. *Show that the* **cancellation law of addition** *is valid, i.e., show that $x + a = x + b$ implies $a = b$.*

    c. *Use the cancellation law of addition to show that $0 \cdot a = 0$ for all $a \in \mathbb{R}$.*

    d. *Show that for each real number a the real number $-a$ is the unique real number that satisfies the equation $a + x = 0$. (The real number $-a$ is called the* **negative** *of a.)*

    e. *Show that for any two given real numbers a and b, the equation $a + x = b$ has a unique solution, namely $x = b + (-a)$. The* **subtraction** *operation $-$ of $\mathbb{R}$ is now defined by $a - b = a + (-b)$; the real number $a - b$ is also called the* **difference** *of b from a.*

    f. *For any real numbers a and b show that $-(-a) = a$ and $-(a+b) = -a - b$.*

Solution. (a) Assume than another element $0^* \in \mathbb{R}$ satisfies $0^* + x = x + 0^* = x$ for all $x \in \mathbb{R}$. Letting $x = 0$, we get $0^* + 0 = 0$. Now, recalling that $0 + y = y + 0 = y$ also holds for all $y \in \mathbb{R}$, letting $y = 0^*$ yields $0^* = 0^* + 0 = 0$.

(b) Let $x + a = x + b$. By Axiom 5 there exists some $z \in \mathbb{R}$ such that $z + x = x + z = 0$. So,

$$a = 0 + a = (z + x) + a = z + (x + a) = z + (x + b) = (z + x) + b = 0 + b = b.$$

(c) Clearly,

$$0 \cdot a + 0 = 0 \cdot a = (0 + 0) \cdot a = 0 \cdot a + 0 \cdot a,$$

and so by the cancellation law of addition, $0 \cdot a = 0$ for each $a \in \mathbf{R}$.

(d) Assume $a + x = 0$. Since $a + (-a) = 0$, we see that $a + x = a + (-a)$, and so, by the cancellation law we have established in (b) above, $x = -a$; the negative of a.

(e) If $a + z = a + y = b$, then by the cancellation law, we get $z = y$. Thus, given a and b, the equation $a + x = b$ has at-most one solution $x \in \mathbf{R}$. Since

$$a + \big[b + (-a)\big] = (a+b) + (-a) = (-a) + (a+b) = \big[(-a) + a\big] + b = 0 + b = b,$$

we see that the only solution of the equation $a + x = b$ is $x = b + (-a)$. We denote this number by $b - a$ and call it the subtraction of a from b.

(f) A close look at the equation $a + (-a) = (-a) + a = 0$ guarantees immediately that $-(-a) = a$. Moreover, from

$$a + b + (-a-b) = a + b + \big[-a + (-b)\big] = \big[(a+b) + (-a)\big] + (-b) = b + (-b) = 0,$$

we easily infer that $-(a + b) = -a - b$.

Problem 3.6.   *This problem introduces (by steps) the familiar process of division in the framework of the axiomatic foundation of real numbers.*

  a.   *Show that the element 1 is uniquely determined, i.e., show that if $1^* \cdot x = x$ for all $x \in \mathbf{R}$ and some $1^* \in \mathbf{R}$, then $1^* = 1$.*
  b.   *Show that the* **cancellation law of multiplication** *is valid, i.e., show that $x \cdot a = x \cdot b$ with $x \neq 0$ implies $a = b$.*
  c.   *Show that for each real number $a \neq 0$ the real number a^{-1} is the unique real number that satisfies the equation $x \cdot a = 1$. The real number $x = a^{-1}$ is called the* **inverse** *(or the* **reciprocal**) *of a.*
  d.   *Show that for any two given real numbers a and b with $a \neq 0$, the equation $ax = b$ has a unique solution, namely $x = a^{-1}b$. The* **division operation** $\div$ *(or $/$) of $\mathbf{R}$ is now defined by $b \div a = a^{-1}b$; as usual, the real number $b \div a$ is also denoted by b/a or $\frac{b}{a}$.*
  e.   *For any two nonzero $a, b \in \mathbf{R}$ show that $(a^{-1})^{-1} = a$ and $(ab)^{-1} = a^{-1}b^{-1}$.*
  f.   *Show that $\frac{a}{1} = a$ for each a, $\frac{0}{b} = 0$ for each $b \neq 0$, and $\frac{a}{a} = 1$ for each $a \neq 0$.*

Solution.   (a) Assume that some real number 1^* satisfies $1^* \cdot x = x \cdot 1^* = x$ for each $x \in \mathbf{R}$. In particular, letting $x = 1$, we get $1^* \cdot 1 = 1$. Since $y \cdot 1 = y$ for all $y \in \mathbf{R}$, letting $y = 1^*$ yields $1^* = 1^* \cdot 1 = 1$. So, 1 is the only real number r which satisfies $r \cdot x = x$ for each $x \in \mathbf{R}$.

(b) Assume $x \cdot a = x \cdot b$ with $x \neq 0$. By Axiom 7 there exists a real number $y \in \mathbf{R}$ such that $y \cdot x = 1$. Now, observe that

$$a = 1 \cdot a = (yx)a = y(xa) = y(xb) = (yx)b = 1 \cdot b = b.$$

(c) If $ax = ay = 1$ with $a \neq 0$, then by (b), we must have $x = y$. This shows that the reciprocal a^{-1} of a is uniquely determined.

(d) To see that the equation $ax = b$ with $a \neq 0$ has at-most one solution x, notice that if $ax = ay = b$, then by the cancellation law of multiplication, we have $x = y$. Moreover, notice that

$$a \cdot (a^{-1}b) = (a \cdot a^{-1})b = 1 \cdot b = b.$$

The above show that the equation $ax = b$ with $a \neq 0$ has the unique solution $x = a^{-1}b$.

(e) If $a \neq 0$, then the equation $a \cdot a^{-1} = 1$ readily says that $(a^{-1})^{-1} = a$. In addition, from

$$(ab) \cdot (b^{-1}a^{-1}) = a(b \cdot b^{-1})a^{-1} = a \cdot 1 \cdot a^{-1} = 1,$$

we easily obtain $(ab)^{-1} = b^{-1}a^{-1}$.

(f) Since $1 \cdot a = a$, we obtain $\frac{a}{1} = a$ for each $a \in \mathbf{R}$. The equation $b \cdot 0 = 0$ also implies that $\frac{0}{b} = 0$ for each $b \neq 0$. From $a \cdot 1 = a$, we get immediately $\frac{a}{a} = 1$ for all $a \neq 0$.

Problem 3.7.   *Establish the following familiar properties of real numbers using the axioms of the real numbers together with the properties established in the previous two problems.*

   i.   **The zero product rule:** *$ab = 0$ if and only if either $a = 0$ or $b = 0$.*

   ii. **The multiplication rule of signs:** *$(-a)b = a(-b) = -(ab)$ and $(-a)(-b) = ab$ for all $a, b \in \mathbf{R}$.*

  iii. **The multiplication rule for fractions:** *For $b, d \neq 0$ and arbitrary real numbers a, c we have*

$$\frac{a}{b} \cdot \frac{c}{d} = \frac{ac}{bd}.$$

       In particular, if $\frac{a}{b} \neq 0$, then $\left(\frac{a}{b}\right)^{-1} = \frac{b}{a}$.

  iv. **The cancellation law of division:** *If $a \neq 0$ and $x \neq 0$, then $\frac{bx}{ax} = \frac{b}{a}$ for each real number b.*

   v. **The division rule for fractions:** *Division by a fraction is the same as* *multiplication by the reciprocal of the fraction, i.e., whenever the fraction* $\frac{a}{b} \div \frac{c}{d}$ *is defined, we have*

$$\frac{a}{b} \div \frac{c}{d} = \frac{a}{b} \cdot \frac{d}{c} = \frac{ad}{bc}.$$

Solution. (i) We already know from the previous problem that $0 \cdot b = 0$ for each $b \in \mathbf{R}$. On the other hand, if $ab = 0 (= a \cdot 0)$ and $a \neq 0$, the cancellation law of multiplication shows that $b = 0$.

(ii) Clearly,

$$ab + (-a)b = [a + (-a)]b = 0 \cdot b = 0 \quad \text{and} \quad ab + a(-b) = a[b + (-b)] = a \cdot 0 = 0,$$

and so $-(ab) = (-a)b = a(-b)$. This implies

$$(-a)(-b) = -[a(-b)] = -[-(ab)] = ab.$$

(iii) If $b, d \neq 0$, then

$$bd\left(\frac{a}{b} \cdot \frac{c}{d}\right) = \left(b \cdot \frac{a}{b}\right) \cdot \left(d \cdot \frac{c}{d}\right) = ac,$$

and this shows that $\frac{ac}{bd} = \frac{a}{b} \cdot \frac{c}{d}$. Since $\frac{a}{b} \cdot \frac{b}{a} = \frac{ab}{ab} = 1$, we see that $\left(\frac{a}{b}\right)^{-1} = \frac{b}{a}$.

(iv) If $c = \frac{b}{a}$, then $ac = b$ and so $(ax)c = bx$ for each $x \neq 0$, which shows that $c = \frac{b}{a} = \frac{bx}{ax}$.

(v) Notice that the identity

$$\frac{c}{d} \cdot \frac{ad}{bc} = \frac{adc}{dbc} = \frac{a}{b}$$

guarantees $\frac{a}{b} \div \frac{c}{d} = \frac{ad}{bc} = \frac{a}{b} \cdot \frac{d}{c}$.

Problem 3.8. *This problem establishes that there exists essentially one set of real numbers that satisfies the eleven axioms stated in Section 3. To see this, let* $\mathbf{R}$ *be a set of real numbers (i.e., a collection of objects that satisfies all eleven axioms stated in Section 3 of the text).*

a. *Show that $1 > 0$.*

b. *A real number a satisfies $a = -a$ if and only if $a = 0$.*

c. *If $n = 1 + 1 + \cdots + 1$ (where the sum has "n summands" all equal to 1), then show that these elements are all distinct; as usual, we shall call the collection $\mathbf{N}$ of all these numbers the natural numbers of $\mathbf{R}$.*

d. *Let Z consist of $\mathbf{N}$ together with its negative elements and zero; we shall call Z, of course, the set of integers of $\mathbf{R}$. Show that Z consists of distinct elements and that it is closed under addition and multiplication.*

e. *Define the set Q of rational numbers by $Q = \{\frac{m}{n}: m, n \in Z \text{ and } n \neq 0\}$. Show that Q satisfies itself axioms 1 through 10 and that*

$$a = \sup\{r \in Q: r \leq a\} = \inf\{s \in Q: a \leq s\}$$

holds for each $a \in \mathbb{R}$.

f. *Now, let $\mathbb{R}'$ be another set of real numbers and let Q' denote its rational numbers. If $1'$ denotes the unit element of $\mathbb{R}'$, then we write*

$$n' = 1' + 1' + \cdots + 1'$$

for the sum having "n-summands" all equal to $1'$. Now, define the function $f: Q \to Q'$ by

$$f(\tfrac{m}{n}) = \tfrac{m'}{n'}$$

and extend it to all of $\mathbb{R}$ via the formula

$$f(a) = \sup\{f(r): r \leq a\}.$$

Show that $\mathbb{R}$ and $\mathbb{R}'$ essentially coincide by establishing the following:

   i.   *$a \leq b$ holds in $\mathbb{R}$ if and only if $f(a) \leq f(b)$ holds in $\mathbb{R}'$.*
   ii. *f is one-to-one and onto.*
   iii. *$f(a+b) = f(a) + f(b) \quad and \quad f(ab) = f(a)f(b)$ for all $a, b \in \mathbb{R}$.*

Solution.   (a) Since $1 \neq 0$, we have two possibilities: either $1 > 0$ or $0 > 1$. If $0 > 1$, then (by Axiom 9) we have $0 + (-1) > 1 + (-1) = 0$ or $-1 > 0$, i.e., -1 is a positive number. Now, using Axiom 10, we infer that $0 \cdot (-1) \geq 1 \cdot (-1)$, or $0 \geq -1$, contrary to $-1 > 0$. Hence, $1 > 0$.

(b) Since $0 + 0 = 0$, we know that $-0 = 0$. Conversely, assume that a real number a satisfies $a = -a$. This implies $a + a = (1 + 1)a = 0$. However, since $1 > 0$, we have $1 + 1 \geq 1 + 0 = 1 > 0$, and so $1 + 1 \neq 0$. Consequently, from the zero product rule, $(1 + 1)a = 0$ implies $a = 0$.

(c) As shown in part (b) above, $1 + 1 \neq 0$ and in fact $1 + 1 \neq 1$; otherwise $1 + 1 = 1 = 1 + 0$ implies (in view of the cancellation law) $1 = 0$, which is impossible. Now, by induction, assume that

$$0 < 1 < 1 + 1 < 1 + 1 + 1 < \cdots < \underbrace{1 + 1 + \cdots + 1}_{n\text{-summands}}.$$

We claim that the real number $n + 1 = 1 + 1 + \cdots + 1 + 1$ (a sum of $n + 1$ summands) satisfies $n + 1 > n = 1 + 1 + \cdots + 1$ (where the last sum has n summands). Indeed, if $n + 1 \leq n$, then $(n + 1) + (-n) \leq n + (-n)$ or $1 \leq 0$, which is a contradiction. Hence, $n + 1 > n$ and the induction is complete.

(d) By part (c) we know that the natural numbers together with zero are all distinct real numbers. If $-m = -n$ with $m, n \in \mathbb{N}$, then $m = n$, which shows that distinct natural numbers have distinct negatives. If $m = -n$ with $m, n \in \mathbb{N}$, then

$m + n = 0$ contradicting (c), and so no natural number can be equal to a negative integer. It now follows that Z consists of distinct elements.

(e) Observe that if $\frac{m}{n}$ and $\frac{p}{q}$ are two rational numbers, then

$$\frac{m}{n} + \frac{p}{q} = \frac{mq + np}{nq} \quad \text{and} \quad \frac{m}{n} \cdot \frac{p}{q} = \frac{mp}{nq},$$

and if $\frac{m}{n} \neq 0$, then $\left(\frac{m}{n}\right)^{-1} = \frac{n}{m}$. That is, $\mathcal{Q}$ is closed under addition, multiplication and inverses. Since all real numbers satisfy axioms 1 through 10, it follows that $\mathcal{Q}$ itself satisfies axioms 1 through 10 in its own right.

For the second part, fix $a \in \mathbb{R}$ and let $A = \{r \in \mathcal{Q}: r \leq a\}$. Since there exists a rational number between $a - 1$ and a (see Theorem 3.4), A is nonempty, and clearly A is bounded from above by a. By the Completeness Axiom (Axiom 11), $\sup A$ exists in $\mathbb{R}$ and satisfies $\sup A \leq a$.

Now, let $\epsilon > 0$. By Theorem 3.4, there exists some rational number r such that $a - \epsilon < r < a$. Clearly, $r \in A$, and so $a - \epsilon < \sup A$, or $a < \sup A + \epsilon$, holds for all $\epsilon > 0$. This implies $a \leq \sup A$, and hence $a = \sup A$. The equality, $a = \inf\{s \in \mathcal{Q}: a \leq s\}$ can be proven in a similar manner.

(f) Notice that the mapping is well defined. That is, if $\frac{m}{n} = \frac{p}{q}$ in $\mathcal{Q}$, then $f\left(\frac{m}{n}\right) = f\left(\frac{p}{q}\right)$. Indeed, since $\frac{m}{n} = \frac{p}{q}$ is equivalent to $mq = np$, we see that $m'q' = n'q'$ or $\frac{m'}{n'} = \frac{p'}{q'}$. Now, let us verify properties (i), (ii), and (iii).

i. $a \leq b$ holds in $\mathbb{R}$ if and only if $f(a) \leq f(b)$ holds in $\mathbb{R}'$.

Note first that two rational numbers $r, s \in \mathcal{Q}$ satisfy $r \leq s$ if and only if $r' \leq s'$. Indeed, to see this it suffices to assume that r and s are positive rational numbers (why?). We have

$$r = \frac{m}{n} \leq s = \frac{p}{q} \iff mq \leq np \iff m'q' \leq n'q' \iff r' = \frac{m'}{n'} \leq \frac{p'}{q'} = s'.$$

Now, let $a \leq b$. Then $\{r \in \mathcal{Q}: r \leq a\} \subseteq \{s \in \mathcal{Q}: s \leq b\}$, and from this it easily follows that $f(a) \leq f(b)$. For the converse, assume that $f(a) \leq f(b)$. If $a \leq b$ is not true, then we must have $b < a$. But then, by Theorem 3.4 there exist two rational numbers $r, s \in \mathcal{Q}$ such that $b < r < s < a$. This implies $f(b) \leq r' < s' \leq f(a)$, a contradiction.

ii. f is one-to-one and onto.

To see that f is onto, let $a' \in \mathbb{R}'$. Then by Theorem 3.4,

$$a' = \sup\{t \in \mathcal{Q}': t \leq a'\}.$$

If, we let $S = \{r \in \mathcal{Q}: r' \leq a'\}$, then this set is bounded above in $\mathbb{R}$ (why?) and so $a = \sup S$ exists in $\mathbb{R}$. Moreover, notice that

$$\{t \in \mathcal{Q}': t \leq a'\} = \{r': r \in \mathcal{Q} \text{ and } r \leq a\}.$$

Now, it is easy to see that $f(a) = a'$.

To verify that f is one-to-one assume $f(a) = f(b)$. Then by part (i) we have $a \le b$ and $b \le a$, i.e., $a = b$.

iii. $f(a + b) = f(a) + f(b)$ and $f(ab) = f(a)f(b)$ for all $a, b \in \mathbf{R}$.

We verify the additivity property only and leave the multiplicative property for the reader. Clearly, $f(r + s) = f(r) + f(s)$ holds for all rational numbers r, s. Now, fix $a, b \in \mathbf{R}$ and assume $r, s \in \mathcal{Q}$ satisfy $r \le a$ and $s \le b$. Then $f(r) = r' \le f(a)$ and $f(s) = s' \le f(b)$. Since $r + s \in \mathcal{Q}$ and $r + s \le a + b$, we see that $f(r) + f(s) = r' + s' = f(r + s) \le f(a + b)$. This easily implies

$$f(a) + f(b) \le f(a + b).$$

For the reverse inequality, let $\epsilon' > 0$ in $\mathbf{R}'$. Then there exist rational numbers $r, s \in \mathcal{Q}$ with $r \le a$ and $s \le b$ such that $f(a) - \epsilon' < f(r)$ and $f(b) - \epsilon' < f(s)$. Since $r + s \le a + b$, it follows that $f(a) + f(b) - 2\epsilon' < f(r) + f(s) = f(r + s) \le f(a + b)$ for each $\epsilon' > 0$. This guarantees $f(a) + f(b) \le f(a + b)$, and therefore $f(a + b) = f(a) + f(b)$.

Problem 3.9.   *Consider a two-point set $R = \{0, 1\}$ equipped with the following operations:*

  a.   *Addition* $(+) : 0 + 0 = 0, 0 + 1 = 1 + 0 = 1$ *and* $1 + 1 = 0$,
  b.   *Multiplication* $(\cdot) : 0 \cdot 1 = 1 \cdot 0 = 0$ *and* $1 \cdot 1 = 1$, *and*
  c.   *Ordering:* $0 \ge 0, 1 \ge 1$ *and* $1 \ge 0$.

Does R with the above operations satisfy all eleven axioms defining the real numbers? Explain your answer.

Solution.   It satisfies all axioms except Axiom 9, which states that:

  • *If $x \ge y$ and $z \ge 0$, then $x + z \ge y + z$.*

To see this, assume that Axiom 9 is valid. We distinguish two cases.

CASE I: $1 > 0$.

In this case, we must have $0 = 1 + 1 \ge 0 + 1 = 1$, which contradicts $1 > 0$.

CASE II: $0 > 1$.

This implies $1 = 0 + 1 \ge 1 + 1 = 0$, which again contradicts $0 > 1$. Thus, Axiom 9 does not hold in this case.

It should be noticed that Axiom 9 is the one that guarantees that $1 + 1$ (i.e, the number 2) is distinct from 0 and 1; and, of course, it is the axiom that establishes (as we saw in part (b) of Problem 3.8) the existence of the set of integers.

Problem 3.10.   *Consider the set of rational numbers $\mathcal{Q}$ equipped with the usual operations of addition, multiplication, and ordering. Why doesn't $\mathcal{Q}$ coincide with the set of real numbers?*

Solution. The set of rational numbers satisfies all the axioms of real numbers except the completeness axiom. This was proven in part (e) of Problem 3.8. To see that $\mathcal{Q}$ does not satisfy the completeness axiom, assume by way of contradiction that it does. Consider the set

$$S = \left\{ 0 \le r \in \mathcal{Q}: r^2 \le 2 \right\}.$$

Then S is nonempty and bounded from above in $\mathcal{Q}$ (why?), and so $b = \sup S$ exists in $\mathcal{Q}$. Now, repeat the proof of Theorem 3.5 to conclude that $b^2 = 2$, i.e., that $b = \sqrt{2}$. However, we proved in Problem 3.3 that $\sqrt{2}$ is not a rational number, and we have reached a contradiction. Hence, $\mathcal{Q}$ does not satisfy the completeness axiom and it cannot coincide with the set of real numbers.

Problem 3.11. *This problem establishes the familiar rules of "exponents" based on the axiomatic foundation of real numbers. To avoid unnecessary notation, we shall assume that all real numbers encountered here are positive—and so by Theorem 3.5, all non-negative real numbers have unique roots. As usual, the "integer" powers are defined by*

$$a^n = \underbrace{a \cdot a \cdots a}_{n\text{-}factors}, \quad a^0 = 1, \quad a^1 = a, \quad \text{and} \quad a^{-n} = \frac{1}{a^n}.$$

Extending this to rational numbers, for each $m, n \in \mathbb{N}$ we define

$$a^{\frac{m}{n}} = \sqrt[n]{a^m} \quad \text{and} \quad a^{-\frac{m}{n}} = \frac{1}{a^{\frac{m}{n}}} = \frac{1}{\sqrt[n]{a^m}}.$$

Establish the following properties:
  a. $a^{\frac{m}{n}} = (\sqrt[n]{a})^m$ *for all $m, n \in \mathbb{N}$.*
  b. *If $m, n, p, q \in \mathbb{N}$ satisfy $\frac{m}{n} = \frac{p}{q}$, then $a^{\frac{m}{n}} = a^{\frac{p}{q}}$.*
  c. *If r and s are rational numbers, then:*
   i. $a^r a^s = a^{r+s}$ *and* $\frac{a^r}{a^s} = a^{r-s}$,
   ii. $(ab)^r = a^r b^r$ *and* $(\frac{a}{b})^r = \frac{a^r}{b^r}$, *and*
   iii. $(a^r)^s = a^{rs}$.

Solution. It should be noticed first that $(a^n)^m = (a^m)^n = a^{mn}$ for all $a \in \mathbb{R}$ and all natural numbers $m, n \in \mathbb{N}$.

(a) Notice that

$$\left[(\sqrt[n]{a})^m \right]^n = \underbrace{(\sqrt[n]{a})^m \cdot (\sqrt[n]{a})^m \cdots (\sqrt[n]{a})^m}_{n\text{-}factors} = \underbrace{\sqrt[n]{a} \cdot \sqrt[n]{a} \cdots \sqrt[n]{a}}_{mn\text{-}factors}$$

$$= \underbrace{(\sqrt[n]{a})^n \cdot (\sqrt[n]{a})^n \cdots (\sqrt[n]{a})^n}_{m\text{-}factors} = \underbrace{a \cdot a \cdots a}_{m\text{-}factors} = a^m.$$

Since the nth-roots are unique (Theorem 3.5), we infer that $(\sqrt[n]{a})^m = \sqrt[n]{a^m} = a^{\frac{m}{n}}$.

(b) Assume $m, n, p, q \in \mathbb{N}$ satisfy $\frac{m}{n} = \frac{p}{q}$, or $pn = mq$. Using part (a), we see that

$$\left(a^{\frac{p}{q}}\right)^n = \left(\sqrt[q]{a^p}\right)^n = \left[(\sqrt[q]{a})^p\right]^n = \left(\sqrt[q]{a}\right)^{pn} = \left(\sqrt[q]{a}\right)^{mq} = \left[(\sqrt[q]{a})^q\right]^m = a^m,$$

and this shows that $a^{\frac{p}{q}} = a^{\frac{m}{n}}$.

(c) The formulas can be established easily if r and s are integers. Now, let r and s be rational numbers. We shall assume that r and s are also positive and leave the "negative case" for the reader. By part (b), we can also suppose that $r = \frac{m}{n}$ and $s = \frac{p}{n}$. Since $(\sqrt[n]{a} \cdot \sqrt[n]{b})^n = (\sqrt[n]{a})^n \cdot (\sqrt[n]{b})^n = ab$, we see that $\sqrt[n]{ab} = \sqrt[n]{a}\sqrt[n]{b}$. Now note that

i. $a^r a^s = \sqrt[n]{a^m} \cdot \sqrt[n]{a^p} = \sqrt[n]{a^m a^p} = \sqrt[n]{a^{m+p}} = a^{\frac{m+p}{n}} = a^{r+s}$.

ii. $(ab)^r = \sqrt[n]{(ab)^m} = \sqrt[n]{a^m b^m} = (\sqrt[n]{a^m})(\sqrt[n]{b^m}) = a^r b^r$.

iii. $(a^r)^s = \sqrt[n]{(\sqrt[n]{a^m})^p} = \sqrt[n]{\sqrt[n]{a^{pm}}} = \sqrt[n^2]{a^{pm}} = a^{\frac{pm}{n^2}} = a^{rs}$.

We leave the remaining cases for the reader.

4.   SEQUENCES OF REAL NUMBERS

Problem 4.1.   *Show that if* $|x| < 1$, *then* $\lim x^n = 0$.

Solution.   Let $x_n = |x|^n$ for each n. Then $x_{n+1} = |x|x_n$ holds for each n, and the assumption $|x| < 1$ implies $0 \leq x_{n+1} \leq x_n$. By Theorem 4.3, $a = \lim x_n$ exists. It follows that $a = a|x|$ (or $(1 - |x|)a = 0$) must hold, and from this that $a = 0$.

A direct way of proving that $\lim x^n = 0$ goes as follows. Observe first that we can suppose $0 < x < 1$. Now, if $\epsilon > 0$ is given, then note that

$$x^n < \epsilon \iff \ln(x^n) = n \ln x < \ln \epsilon \iff n > \frac{\ln \epsilon}{\ln x}.$$

Problem 4.2.   *Show that* $\lim x_n = x$ *holds if and only if every subsequence of* $\{x_n\}$ *has a subsequence that converges to* x.

Solution.   If $\lim x_n = x$, then every subsequence must converge to x. So, every subsequence of a subsequence (as being itself a subsequence of $\{x_n\}$) must converge to x.

For the converse, assume that each subsequence of $\{x_n\}$ has a subsequence that converges to x. Now, suppose by way of contradiction that $\{x_n\}$ does not converge to x. Then for some $\varepsilon > 0$ we must have $|x - x_n| \geq \varepsilon$ for an infinite number of n. So, there exists a subsequence $\{y_n\}$ of $\{x_n\}$ such that $|x - y_n| \geq \varepsilon$ for each n. However, the latter contradicts the fact that $\{y_n\}$ has a subsequence that converges to x. Therefore, $\lim x_n = x$.

Problem 4.3. *Consider two sequences* $\{k_n\}$ *and* $\{m_n\}$ *of strictly increasing natural numbers such that for some* $\ell \in \mathbb{N}$ *we have*

$$\{\ell, \ell + 1, \ell + 2, \ldots\} \subseteq \{k_1, k_2, \ldots\} \cup \{m_1, m_2, \ldots\}.$$

Show that a sequence of real numbers $\{x_n\}$ *converges in* $\mathbb{R}$ *if and only if both subsequences* $\{x_{k_n}\}$ *and* $\{x_{m_n}\}$ *of* $\{x_n\}$ *converge in* $\mathbb{R}$ *and they satisfy* $\lim x_{k_n} = \lim x_{m_n}$ *(in which case the common limit is also the limit of the sequence).*

In particular, show that a sequence of real numbers $\{x_n\}$ *converges in* $\mathbb{R}$ *if and only if the "even" and "odd" subsequences* $\{x_{2n}\}$ *and* $\{x_{2n-1}\}$ *both converge in* $\mathbb{R}$ *and they satisfy* $\lim x_{2n} = \lim x_{2n-1}$.

Solution. If $x_n \to x$, then clearly $x_{k_n} \to x$ and $x_{m_n} \to x$. For the converse, assume that $x_{k_n} \to x$ and $x_{m_n} \to x$. Let $\epsilon > 0$. Choose some $n_0 \in \mathbb{N}$ such that

$$\left|x_{k_n} - x\right| < \epsilon \quad \text{and} \quad \left|x_{m_n} - x\right| < \epsilon \quad \text{for all} \ n \geq n_0. \tag{$\star$}$$

Put $\ell_0 = \max\{\ell, k_{n_0}, m_{n_0}\}$, and we claim that

$$\left|x_n - x\right| < \epsilon \quad \text{for all} \ n \geq \ell_0.$$

To see this, let $n \geq \ell_0$. Then the assumption

$$\{\ell, \ell + 1, \ell + 2, \ldots\} \subseteq \{k_1, k_2, \ldots\} \cup \{m_1, m_2, \ldots\}$$

guarantees the existence of some $r \in \mathbb{N}$ such that $k_r = n$ or $m_r = n$. Since $r < n_0$ implies $k_r < k_{n_0} \leq \ell_0$ and $m_r < m_{n_0} \leq \ell_0$, we see that $r \geq n_0$. Hence, either $x_n = x_{k_r}$ of $x_n = x_{m_r}$ (with $r \geq n_0$), and so from $(\star)$ it follows that $|x_n - x| < \epsilon$. This shows that $x_n \to x$.

The last part should be immediate from the above conclusion.

Problem 4.4. *Find the* $\limsup$ *and* $\liminf$ *for the sequence* $\{(-1)^n\}$.

Solution. We have $\liminf(-1)^n = -1$ and $\limsup(-1)^n = 1$.

Problem 4.5. *Find the* $\limsup$ *and* $\liminf$ *of the sequence* $\{x_n\}$ *defined by*

$$x_1 = \tfrac{1}{3}, \quad x_{2n} = \tfrac{1}{3}x_{2n-1}, \quad \text{and} \quad x_{2n+1} = \tfrac{1}{3} + x_{2n} \ \text{for} \ n = 1, 2, \ldots.$$

Solution. We claim that

$$x_{2n} = \tfrac{1}{3^2} \sum_{k=0}^{n-1} \tfrac{1}{3^k} \quad \text{and} \quad x_{2n+1} = \tfrac{1}{3} \sum_{k=0}^{n} \tfrac{1}{3^k}$$

hold for $n = 1, 2, \ldots$. The validity of the identities can be established by induction. We shall establish the validity of the second identity and leave the verification of the first to the reader. For $n = 1$, we have

$$x_3 = x_{2 \cdot 1 + 1} = \tfrac{1}{3} + x_2 = \tfrac{1}{3} + \tfrac{1}{3}x_1 = \tfrac{1}{3} + \tfrac{1}{9} = \tfrac{1}{3}\left(1 + \tfrac{1}{3}\right) = \tfrac{1}{3}\sum_{k=0}^{1} \tfrac{1}{3^k}.$$

Now, assume that $x_{2n+1} = \tfrac{1}{3}\sum_{k=0}^{n} \tfrac{1}{3^k}$ holds for some n. Then,

$$x_{2(n+1)+1} = \tfrac{1}{3} + x_{2(n+1)} = \tfrac{1}{3} + \tfrac{1}{3}x_{2n+1} = \tfrac{1}{3} + \tfrac{1}{9}\sum_{k=0}^{n} \tfrac{1}{3^k} = \tfrac{1}{3}\sum_{k=0}^{n+1} \tfrac{1}{3^k},$$

and the induction is complete. Consequently,

$$\lim_{n\to\infty} x_{2n} = \tfrac{1}{3^2}\sum_{k=0}^{\infty} \tfrac{1}{3^k} = \tfrac{1}{3^2} \cdot \tfrac{3}{2} = \tfrac{1}{6} \quad \text{and} \quad \lim_{n\to\infty} x_{2n+1} = \tfrac{1}{3}\sum_{k=0}^{\infty} \tfrac{1}{3^k} = \tfrac{1}{2}.$$

Now, we claim that $\tfrac{1}{6}$ and $\tfrac{1}{2}$ are the only limit points of $\{x_n\}$. To see this, let a be a real number different from $\tfrac{1}{6}$ and $\tfrac{1}{2}$. Pick some $\varepsilon > 0$ such that

$$(a - \varepsilon, a + \varepsilon) \cap (\tfrac{1}{6} - \varepsilon, \tfrac{1}{6} + \varepsilon) = \emptyset \quad \text{and} \quad (a - \varepsilon, a + \varepsilon) \cap (\tfrac{1}{2} - \varepsilon, \tfrac{1}{2} + \varepsilon) = \emptyset.$$

Next, note that there exists some k such that $x_{2n} \in (\tfrac{1}{6} - \varepsilon, \tfrac{1}{6} + \varepsilon)$ and $x_{2n+1} \in (\tfrac{1}{2} - \varepsilon, \tfrac{1}{2} + \varepsilon)$ hold for all $n \geq k$. Therefore, $|x_n - a| \geq \varepsilon$ holds for all $n \geq k$, and this shows that a cannot be a limit point of the sequence $\{x_n\}$. Consequently,

$$\liminf x_n = \tfrac{1}{6} \quad \text{and} \quad \limsup x_n = \tfrac{1}{2}.$$

Problem 4.6. *Let $\{x_n\}$ be a bounded sequence. Show that*

$$\limsup(-x_n) = -\liminf x_n \quad \text{and} \quad \liminf(-x_n) = -\limsup x_n.$$

Solution. We shall use the fact that $\limsup x_n$ and $\liminf x_n$ are the largest and smallest limit points of $\{x_n\}$, respectively. We shall establish the first formula.

Choose two subsequences $\{y_n\}$ and $\{z_n\}$ of $\{x_n\}$ such that $\lim y_n = \liminf x_n$ and $\lim(-z_n) = \limsup(-x_n)$. Then

$$
\begin{aligned}
-\liminf x_n &= \lim(-y_n) \\
&\leq \limsup(-x_n) = \lim(-z_n) = -\lim z_n \\
&\leq -\liminf x_n,
\end{aligned}
$$

and so $\limsup(-x_n) = -\liminf x_n$.

Problem 4.7. *If $\{x_n\}$ and $\{y_n\}$ are two bounded sequences, then show that*

a. $\limsup(x_n + y_n) \le \limsup x_n + \limsup y_n$, *and*
b. $\liminf(x_n + y_n) \ge \liminf x_n + \liminf y_n$.

Moreover, show that if one of the sequences converges, then equality holds in both (a) and (b).

Solution. (a) By passing to a subsequence, we can assume that $\lim(x_n + y_n) = \limsup(x_n + y_n)$. Since $\{x_n\}$ is a bounded sequence, there exists a subsequence $\{x_{k_n}\}$ that converges. Let $x = \lim x_{k_n}$. By the same reasoning, there exists a subsequence of $\{y_{k_n}\}$ that converges to some y. Thus, there exists a strictly increasing sequence $\{m_n\}$ of natural numbers such that $x = \lim x_{m_n}$ and $y = \lim y_{m_n}$. Hence,

$$\limsup(x_n + y_n) = x + y = \lim x_{m_n} + \lim y_{m_n} \le \limsup x_n + \limsup y_n.$$

Finally, if $x = \lim x_n$ holds, then pick a subsequence $\{y_{k_n}\}$ of $\{y_n\}$ such that $\lim y_{k_n} = \limsup y_n$, and note that

$$\begin{aligned} \limsup x_n + \limsup y_n &= x + \lim y_{k_n} \\ &= \lim(x_{k_n} + y_{k_n}) \le \limsup(x_n + y_n). \end{aligned}$$

(b) It follows from (a) by using the preceding problem.

Problem 4.8. *Prove that the $\limsup$ and $\liminf$ processes "preserve inequalities." That is, show that if two bounded sequences $\{x_n\}$ and $\{y_n\}$ of real numbers satisfy $x_n \le y_n$ for all $n \ge n_0$, then*

$$\liminf x_n \le \liminf y_n \quad \text{and} \quad \limsup x_n \le \limsup y_n.$$

Solution. First, we shall show that if two sequences of real numbers $\{s_n\}$ and $\{t_n\}$ converge in $\mathbb{R}$ (say $s_n \to s$ and $t_n \to t$) and $s_n \le t_n$ for each $n \ge n_0$, then $s \le t$.

Indeed if, $s > t$ is true, then let $\epsilon = \frac{s-t}{2} > 0$ and note that for all n sufficiently large, we must have

$$s_n \in (s - \epsilon, s + \epsilon) = \left(\tfrac{s+t}{2}, \tfrac{3s-t}{2}\right) \quad \text{and} \quad t_n \in (t - \epsilon, t + \epsilon) = \left(\tfrac{3t-s}{2}, \tfrac{s+t}{2}\right).$$

That is, $t_n < \frac{s+t}{2} < s_n$ must hold for all n sufficiently large, which is impossible. Hence, $s \le t$.

Now, assume that two bounded sequences of real numbers $\{x_n\}$ and $\{y_n\}$ satisfy $x_n \leq y_n$ for all $n \geq n_0$. Put

$$s_n = \inf_{k \geq n} x_k \quad \text{and} \quad t_n = \inf_{k \geq n} y_k.$$

If $n \geq n_0$, then notice that for each $r \geq n$ we have $s_n = \inf_{k \geq n_0} x_k \leq x_r \leq y_r$, and so $s_n \leq \inf_{r \geq n} y_r = t_n$ for each $n \geq n_0$. By the discussion of the first part, we infer that

$$\liminf x_n = \lim s_n \leq \lim t_n = \liminf y_n.$$

The lim sup case can be established in a similar manner, or by using the formula $\limsup x_n = -\liminf(-x_n)$.

Problem 4.9.   *Show that* $\lim \sqrt[n]{n} = 1$ *(and conclude from this that* $\lim \sqrt[n]{a} = 1$ *for each* $a > 0$*).*

Solution.   Note that $\sqrt[n]{n} = \left(\sqrt[n]{\sqrt{n}}\right)^2$. An easy inductive argument shows that $\sqrt[n]{\sqrt{n}} > 1$ holds for each n. Thus, we can write $\sqrt[n]{\sqrt{n}} = 1 + x_n$ with $x_n > 0$. Since $(1 + a)^n \geq 1 + na$ holds for each n and each $a \geq 0$ (see Problem 2.13), we get

$$\sqrt{n} = \left(\sqrt[n]{\sqrt{n}}\right)^n = (1 + x_n)^n \geq 1 + nx_n,$$

and so $0 < x_n \leq \frac{1}{\sqrt{n}} - \frac{1}{n}$. This implies $\lim x_n = 0$. Therefore,

$$\sqrt[n]{n} = \left(\sqrt[n]{\sqrt{n}}\right)^2 = (1 + x_n)^2 \longrightarrow 1.$$

An alternate proof goes as follows: By L'Hôpital's Rule, we have $\lim_{x \to \infty} \frac{\ln x}{x} = 0$, and so $\lim_{n \to \infty} \frac{\ln n}{n} = 0$. Therefore, using that the exponential function is continuous, we infer that

$$\lim_{n \to \infty} \sqrt[n]{n} = \lim_{n \to \infty} e^{\frac{\ln n}{n}} = e^0 = 1.$$

For the parenthetical part, assume first $a > 1$. Then it is easy to see that $1 \leq \sqrt[n]{a} \leq \sqrt[n]{n}$ holds true for all $n > a$. Consequently, by the "Sandwich Theorem," we see that $\lim \sqrt[n]{a} = 1$. If $0 < a < 1$, then $\frac{1}{a} > 1$, and so $\lim \sqrt[n]{\frac{1}{a}} = \lim \frac{1}{\sqrt[n]{a}} = 1$, from which it follows that $\lim \sqrt[n]{a} = 1$ holds true in this case, too.

Problem 4.10. *If $\{x_n\}$ is a sequence of strictly positive real numbers, then show that*

$$\liminf_{n\to\infty} \frac{x_{n+1}}{x_n} \leq \liminf_{n\to\infty} \sqrt[n]{x_n} \leq \limsup_{n\to\infty} \sqrt[n]{x_n} \leq \limsup_{n\to\infty} \frac{x_{n+1}}{x_n}.$$

Conclude from this that if $\lim \frac{x_{n+1}}{x_n}$ exists in $\mathbf{R}$, then $\lim \sqrt[n]{x_n}$ also exists and $\lim \sqrt[n]{x_n} = \lim \frac{x_{n+1}}{x_n}$.

Solution. Let $\{x_n\}$ be a sequence of real numbers such that $x_n > 0$ holds for each n. We shall establish $\limsup \sqrt[n]{x_n} \leq \limsup \frac{x_{n+1}}{x_n}$ and leave the similar proof of the other inequality for the reader. Put

$$x = \limsup \frac{x_{n+1}}{x_n} = \bigwedge_{n=1}^{\infty} \bigvee_{k=n}^{\infty} \frac{x_{k+1}}{x_k},$$

and note that if $x = \infty$, then there is nothing to prove. So, we can assume $x < \infty$.

Let $\varepsilon > 0$ be fixed. Then there exists some k such that $\frac{x_{n+1}}{x_n} < x + \varepsilon$ holds for all $n \geq k$. Now, for $n > k$ we have

$$x_n = \frac{x_n}{x_{n-1}} \cdot \frac{x_{n-1}}{x_{n-2}} \cdots \frac{x_{k+1}}{x_k} \cdot x_k \leq (x+\varepsilon)^{n-k} x_k = (x+\varepsilon)^n c,$$

where $c = x_k(x+\varepsilon)^{-k}$ is a constant. Therefore, $\sqrt[n]{x_n} \leq (x+\varepsilon)\sqrt[n]{c}$ holds for each $n \geq k$ and so, in view of $\lim \sqrt[n]{c} = 1$ (see Problem 4.9) and Problem 4.8, we infer that

$$\limsup_{n\to\infty} \sqrt[n]{x_n} \leq \limsup_{n\to\infty}(x+\varepsilon)\sqrt[n]{c} = (x+\varepsilon) \lim_{n\to\infty} \sqrt[n]{c} = x + \varepsilon.$$

Since $\varepsilon > 0$ is arbitrary, we infer that $\limsup \sqrt[n]{x_n} \leq x = \limsup \frac{x_{n+1}}{x_n}$.

Problem 4.11. *The **sequence of averages** of a sequence of real numbers $\{x_n\}$ is the sequence $\{a_n\}$ defined by $a_n = \frac{x_1 + x_2 + \cdots + x_n}{n}$. If $\{x_n\}$ is a bounded sequence of real numbers, then show that*

$$\liminf x_n \leq \liminf a_n \leq \limsup a_n \leq \limsup x_n.$$

In particular, if $x_n \to x$, then show that $a_n \to x$. Does the convergence of $\{a_n\}$ imply the convergence of $\{x_n\}$?

Solution. The solution will be based upon the following properties of lim sup and lim inf:

- *If $\{u_n\}$ is a bounded sequence of real numbers, then for each $\epsilon > 0$ the inequalities*

$$u_k \geq \limsup u_n + \epsilon \quad and \quad u_m \leq \liminf u_n - \epsilon$$

  hold for finitely many k and finitely many m.

To see this, assume by way of contradiction that $u_k \geq \limsup u_n + \epsilon$ holds true for infinitely many k. Then there exists a subsequence $\{v_n\}$ of $\{u_n\}$ satisfying $v_n \geq \limsup u_n + \epsilon$ for each n. Since $\{v_n\}$ is a bounded sequence, there exists a subsequence $\{w_n\}$ of $\{v_n\}$ (and hence of $\{u_n\}$) satisfying $w_n \rightarrow w \in \mathbf{R}$. By Problem 4.8, we know that $w \geq \limsup u_n + \epsilon$, i.e., w is a limit point of $\{u_n\}$ which is greater than the largest limit point ($\limsup u_n$) of $\{u_n\}$, a contradiction.

Now, let $\{x_n\}$ be a bounded sequence of real numbers and fix $\epsilon > 0$. Put $\ell = \limsup x_n$ and let $K = \{k \in \mathbf{N}: x_k \geq \ell + \epsilon\}$. By the above discussion, K is a finite set. Put

$$S_n = \{i \in \mathbf{N}: i \in K \text{ and } i \leq n\} \quad and \quad T_n = \{i \in \mathbf{N}: i \notin K \text{ and } i \leq n\},$$

and define the sequences $\{s_n\}$ and $\{t_n\}$ by

$$s_n = \sum_{i \in S_n} x_i \quad and \quad t_n = \sum_{i \in T_n} x_i.$$

Clearly, $\{s_n\}$ is an eventually constant sequence, $t_n \leq n(\ell + \epsilon)$ holds for each n and $a_n = \frac{s_n}{n} + \frac{t_n}{n}$. Since $s_n/n \rightarrow 0$ and $t_n/n \leq \ell + \epsilon$ for each n, it follows from Problems 4.7 and 4.8 that

$$\limsup a_n = \limsup\left(\tfrac{s_n}{n} + \tfrac{t_n}{n}\right) = \lim \tfrac{s_n}{n} + \limsup \tfrac{t_n}{n} = \limsup \tfrac{t_n}{n} \leq \ell + \epsilon.$$

Since $\epsilon > 0$ is arbitrary, we get $\limsup a_n \leq \ell = \limsup x_n$. Similarly, $\liminf x_n \leq \liminf a_n$. If $x_n \rightarrow x$, then $x = \liminf x_n = \limsup x_n$, and so $x = \liminf a_n = \limsup a_n$. This implies $a_n \rightarrow x$.

The convergence of the sequence $\{a_n\}$ of averages does not imply the convergence of $\{x_n\}$. For instance, if $x_n = (-1)^n$, then $a_n \rightarrow 0$ while $\{x_n\}$ fails to converge.

Problem 4.12. *For a sequence of real numbers $\{x_n\}$ establish the following:*

a. *If $x_{n+1} - x_n \rightarrow x$ in $\mathbf{R}$, then $x_n/n \rightarrow x$.*

b. *If $\{x_n\}$ is bounded and $2x_n \le x_{n+1} + x_{n-1}$ holds for all $n = 2, 3, \ldots$, then $x_{n+1} - x_n \uparrow 0$.*

Solution. (a) Assume that $x_{n+1} - x_n \to x$ in $\mathbb{R}$. Notice that $\sum_{i=1}^{n}(x_{i+1} - x_i) = x_{n+1} - x_1$ for each n. By Problem 4.11, we have

$$\frac{1}{n}\sum_{i=1}^{n}(x_{i+1} - x_i) = \frac{1}{n}\left(x_{n+1} - x_1\right) \to x.$$

Since $x_1/n \to 0$, it follows that $x_{n+1}/n \to x$. Now note that

$$\frac{x_n}{n} = \frac{x_n}{n-1} \cdot \frac{n-1}{n} \longrightarrow x \cdot 1 = x.$$

(b) The condition $2x_n \le x_{n+1} + x_{n-1}$ can be rewritten as $x_n - x_{n-1} \le x_{n+1} - x_n$ for each $n = 2, 3, \ldots$, which implies that the bounded sequence $\{x_{n+1} - x_n\}$ is an increasing sequence, and hence convergent. Let $x_{n+1} - x_n \uparrow x$ in $\mathbb{R}$. By part (a), we have $x_n/n \to x$. But, since $\{x_n\}$ is a bounded sequence, $x_n/n \to 0$. Therefore, $x = 0$, and so $x_{n+1} - x_n \uparrow 0$.

Problem 4.13. *Consider the sequence $\{x_n\}$ defined by $0 < x_1 < 1$ and $x_{n+1} = 1 - \sqrt{1 - x_n}$ for $n = 1, 2, \ldots$. Show that $x_n \downarrow 0$. Also, show that $\frac{x_{n+1}}{x_n} \to \frac{1}{2}$.*

Solution. We claim that

$$0 < x_{n+1} < x_n < 1 \tag{$\star$}$$

holds for each $n = 1, 2, \ldots$. To verify this claim, we use induction. Since $0 < x_1 < 1$, we have $0 < 1 - x_1 < 1$, and so $0 < 1 - x_1 < \sqrt{1 - x_1} < 1$. Hence, $0 < 1 - \sqrt{1 - x_1} = x_2 < x_1 < 1$. That is, $(\star)$ is true for $n = 1$.

For the inductive argument, assume that $(\star)$ is true for some n. This implies $0 < 1 - x_n < 1 - x_{n+1} < 1$, and so $0 < \sqrt{1 - x_n} < \sqrt{1 - x_{n+1}} < 1$, from which it follows that

$$0 < x_{n+2} = 1 - \sqrt{1 - x_{n+1}} < 1 - \sqrt{1 - x_n} = x_{n+1} < 1,$$

which shows that $(\star)$ is true for $n + 1$. This completes the induction and guarantees that $(\star)$ is true for each n.

Now, since $\{x_n\}$ is decreasing and bounded from below, it converges, say to $x \in \mathbb{R}$. Clearly, $0 \le x < 1$. Moreover, we have

$$x = \lim_{n\to\infty} x_{n+1} = \lim_{n\to\infty}\left(1 - \sqrt{1 - x_n}\right) = 1 - \sqrt{1 - x}.$$

In other words, x is the non-negative solution of the equation $x = 1 - \sqrt{1 - x}$. Solving the equation yields $x = 0$ or $x = 1$. Hence, $x = 0$, and so $x_n \downarrow 0$.

For the last part, notice that

$$\frac{x_{n+1}}{x_n} = \frac{1 - \sqrt{1 - x_n}}{x_n} = \frac{1}{1 + \sqrt{1 - x_n}} \longrightarrow \frac{1}{2},$$

and the solution is complete.

Problem 4.14. *Show that the sequence $\{x_n\}$ defined by*

$$x_n = \left(1 + \frac{1}{n}\right)^n$$

is a convergent sequence.

Solution. From the binomial expansion:

$$x_n = \left(1 + \tfrac{1}{n}\right)^n = \sum_{i=0}^{n} \binom{n}{i} \tfrac{1}{n^i} = 1 + \sum_{i=1}^{n} \binom{n}{i} \tfrac{1}{n^i}$$

$$= 1 + \sum_{i=1}^{n} \tfrac{n(n-1)\cdots(n-i+1)}{i!} \cdot \tfrac{1}{n^i}$$

$$= 1 + \sum_{i=1}^{n} \tfrac{1}{i!}\left(1 - \tfrac{1}{n}\right)\left(1 - \tfrac{2}{n}\right) \cdots \left(1 - \tfrac{i-1}{n}\right)$$

$$\leq 1 + \sum_{i=1}^{n+1} \tfrac{1}{i!}\left(1 - \tfrac{1}{n+1}\right)\left(1 - \tfrac{2}{n+1}\right) \cdots \left(1 - \tfrac{i-1}{n+1}\right)$$

$$= \left(1 + \tfrac{1}{n+1}\right)^{n+1} = x_{n+1}.$$

Thus, $x_n \uparrow$ holds. Also, note that for $n \geq 2$ we have

$$x_n = 1 + \sum_{i=1}^{n} \tfrac{1}{i!}\left(1 - \tfrac{1}{n}\right) \cdots \left(1 - \tfrac{i-1}{n}\right) \leq 2 + \sum_{i=2}^{n} \tfrac{1}{i!} \leq 2 + \sum_{i=2}^{n} \tfrac{1}{2^{i-1}} \leq 3.$$

By Theorem 4.3, $\{x_n\}$ converges. (Of course, $\lim x_n = e = 2.718 \cdots$.)

Problem 4.15. *Assume that a sequence $\{x_n\}$ satisfies*

$$|x_{n+1} - x_n| \leq \alpha |x_n - x_{n-1}|$$

for $n = 2, 3, \ldots$ and some fixed $0 < \alpha < 1$. Show that $\{x_n\}$ is a convergent sequence.

Solution. Let $c = |x_2 - x_1|$. An easy inductive argument shows that for each n we have $|x_{n+1} - x_n| \le c\alpha^{n-1}$. Thus,

$$\left|x_{n+p} - x_n\right| \le \sum_{i=1}^{p}\left|x_{n+i} - x_{n+i-1}\right| \le c\sum_{i=1}^{p}\alpha^{n+i-2} \le \tfrac{c}{1-\alpha}\alpha^{n-1}$$

holds for all n and all p. Since $\lim \alpha^n = 0$, it follows that $\{x_n\}$ is a Cauchy sequence, and hence, a convergent sequence.

Problem 4.16. *Show that the sequence $\{x_n\}$, defined by*

$$x_1 = 1 \ \ and \ \ x_{n+1} = \frac{1}{3 + x_n} \ for \ n = 1, 2, \ldots,$$

converges and determine its limit.

Solution. Clearly, $x_n > 0$ holds for each n. Now, note that

$$|x_{n+1} - x_n| = \left|\tfrac{1}{3+x_n} - \tfrac{1}{3+x_{n-1}}\right| = \tfrac{|x_n - x_{n-1}|}{(3+x_n)(3+x_{n-1})} \le \tfrac{1}{9}|x_n - x_{n-1}|$$

holds for $n = 2, 3, \ldots$. By Problem 4.15, the sequence $\{x_n\}$ converges. If $\lim x_n = x$, then $x \ge 0$ and

$$x = \lim x_{n+1} = \frac{1}{3 + \lim x_n} = \frac{1}{3 + x}.$$

Solving the equation, we get $x = \frac{-3+\sqrt{13}}{2}$.

Problem 4.17. *Consider the sequence $\{x_n\}$ of real numbers defined by $x_1 = 1$ and $x_{n+1} = 1 + \frac{1}{1+x_n}$ for $n = 1, 2, \ldots$. Show that $\{x_n\}$ is a convergent sequence and that $\lim x_n = \sqrt{2}$.*

Solution. An easy inductive argument shows that $x_n > 0$ for each n. This implies that, in fact, we have $1 \le x_n \le 2$ for each n. Now, note that

$$\begin{aligned}
|x_{n+1} - x_n| &= \left|\frac{1}{1 + x_n} - \frac{1}{1 + x_{n-1}}\right| \\
&= \frac{|x_n - x_{n-1}|}{(1 + x_n)(1 + x_{n-1})} \\
&\le \frac{|x_n - x_{n-1}|}{(1 + 1)(1 + 1)} = \tfrac{1}{4}|x_n - x_{n-1}|
\end{aligned}$$

for each $n = 2, 3, \ldots$. By Problem 4.15, the sequence $\{x_n\}$ converges. Let

$x_n \to x$. Since $x_n \geq 1$ for each n, we see that $x \geq 1$. Then

$$x = \lim_{n \to \infty} x_{n+1} = \lim_{n \to \infty} \left(1 + \frac{1}{1 + x_n}\right) = 1 + \frac{1}{1 + x}.$$

That is, x is the positive solution of the equation $x = 1 + \frac{1}{1+x}$, or $x^2 + x = 1 + x + 1$. This implies $x^2 = 2$, and so $x = \sqrt{2}$.

Problem 4.18. *Define the sequence* $\{x_n\}$ *by* $x_1 = 1$ *and*

$$x_{n+1} = \frac{1}{2}\left(x_n + \frac{2}{x_n}\right), \quad n = 1, 2, \ldots.$$

Show that $\{x_n\}$ *converges and that* $\lim x_n = \sqrt{2}$.

Solution. Clearly, $x_n > 0$ holds for each n. (Use induction to prove this!) Also,

$$x_{n+1}^2 - 2 = \tfrac{1}{4}\left(x_n - \tfrac{2}{x_n}\right)^2 \geq 0$$

holds for each n. Thus, if $n \geq 2$, then

$$x_{n+1} - x_n = \tfrac{1}{2}\left(x_n + \tfrac{2}{x_n}\right) - x_n = \tfrac{2 - x_n^2}{2x_n} \leq 0,$$

and so $0 < x_{n+1} < x_n$ holds for each $n \geq 2$. By Theorem 4.3, $x = \lim x_n$ exists. Since $x_n^2 \geq 2$ holds for each $n \geq 2$, we see that $x > 0$. From the recursive formula, it follows that $2x = x + \frac{2}{x}$, or $x^2 = 2$. (Note also that the limit is independent of the initial choice $x_1 > 0$.)

Problem 4.19. *Define the sequence* $x_n = \sum_{k=1}^{n} \frac{1}{k}$ *for* $n = 1, 2, \ldots$. *Show that* $\{x_n\}$ *does not converge in* $\mathbb{R}$. *(See also Problem 5.10.)*

Solution. The inequality

$$x_{2n} - x_n = \tfrac{1}{n+1} + \tfrac{1}{n+2} + \cdots + \tfrac{1}{n+n}$$
$$\geq \tfrac{1}{2n} + \tfrac{1}{2n} + \cdots + \tfrac{1}{2n} = n \cdot \tfrac{1}{2n} = \tfrac{1}{2}$$

shows that $\{x_n\}$ is not a Cauchy sequence, and hence, is not convergent in $\mathbb{R}$.

Problem 4.20. *Let* $-\infty < a < b < \infty$ *and* $0 < \lambda < 1$. *Define the sequence*

$\{x_n\}$ *by* $x_1 = a$, $x_2 = b$ *and*

$$x_{n+2} = \lambda x_n + (1 - \lambda)x_{n+1} \ \textit{for} \ n = 1, 2, \ldots.$$

Show that $\{x_n\}$ *converges in* $\mathbf{R}$ *and find its limit.*

Solution. Rewriting $x_{n+2} = \lambda x_n + (1 - \lambda)x_{n+1} = \lambda x_n + x_{n+1} - \lambda x_{n+1}$ in the form $x_{n+2} - x_{n+1} = \lambda(x_n - x_{n+1})$, we see that

$$\left|x_{n+2} - x_{n+1}\right| = \lambda \left|x_{n+1} - x_n\right|$$

holds for each n. Now, a glance at Problem 4.15 guarantees that $\{x_n\}$ is a convergent sequence. However, we cannot get the limit of the sequence $\{x_n\}$ by taking limits in both sides of the recursive formula $x_{n+2} = \lambda x_n + (1 - \lambda)x_{n+1}$. We shall compute the limit of the sequence $\{x_n\}$ using a different method.

For simplicity put $\mu = 1 - \lambda$. First, we shall verify that

$$x_1 < x_3 < \cdots < x_{2n+1} < x_{2n} < x_{2n-2} < \cdots < x_2$$

holds for each n.

The proof is by induction. For $n = 1$, the inequalities reduce to $x_1 < x_2$ which is obviously true. So, for the inductive step, assume $x_{2n-1} < x_{2n}$ for some n. Then

$$x_{2n+1} = \lambda x_{2n-1} + \mu x_{2n} = x_{2n-1} + \mu(x_{2n} - x_{2n-1}) > x_{2n-1}$$

and

$$x_{2n+1} = \lambda x_{2n-1} + \mu x_{2n} = x_{2n} - \lambda(x_{2n} - x_{2n-1}) < x_{2n}.$$

Now, note that

$$x_{2n+1} < \lambda x_{2n-1} + (1 - \lambda)x_{2n} = x_{2n+2} < x_{2n}.$$

Next, if we let $d_n = x_{2n} - x_{2n-1}$, then it is easily to verify (see Figure 1.1) that

$$d_{n+1} = \lambda \mu d_n, \tag{1}$$

$$x_{2n+1} = x_{2n-1} + \mu d_n, \quad \text{and} \tag{2}$$

$$x_{2n+2} = x_{2n} - \lambda^2 d_n. \tag{3}$$

From (1) it follows that

$$d_n = (\lambda \mu)^{n-1} d_1 = (\lambda \mu)^{n-1}(b - a),$$

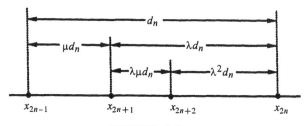

FIGURE 1.1.

and so from (2) and (3), we obtain

$$x_{2n+1} = x_1 + \sum_{i=1}^{n}(x_{2i+1} - x_{2i-1}) = x_1 + \sum_{i=1}^{n}\mu d_i$$

$$= x_1 + \mu\sum_{i=1}^{n}(\lambda\mu)^{i-1}d_1 = x_1 + \frac{\mu d_1[1 - (\lambda\mu)^n]}{1 - \lambda\mu}$$

$$= a + \frac{\mu(b-a)[1 - (\lambda\mu)^n]}{1 - \lambda\mu},$$

and

$$x_{2n+2} = x_2 - \sum_{i=1}^{n}(x_{2i} - x_{2i+2}) = x_2 - \lambda^2\sum_{i=1}^{n}(\lambda\mu)^{i-1}d_i$$

$$= x_2 - \frac{\lambda^2[1 - (\lambda\mu)^n]d_1}{1 - \lambda\mu} = b - \frac{\lambda^2(b-a)[1 - (\lambda\mu)^n]}{1 - \lambda\mu}.$$

Therefore,

$$x_{2n+1} \uparrow a + \frac{\mu(b-a)}{1 - \lambda\mu} = \frac{\lambda^2 a + \mu b}{1 - \lambda\mu} \quad \text{and} \quad x_{2n} \downarrow b - \frac{\lambda^2(b-a)}{1 - \lambda\mu} = \frac{\lambda^2 a + \mu b}{1 - \lambda\mu},$$

and consequently, $\lim x_n = \frac{\lambda^2 a + \mu b}{1 - \lambda\mu}$.

Problem 4.21. *Let G be a nonempty subset of* **R**, *which is a group under addition (i.e., if $x, y \in G$, then $x + y \in G$ and $-x \in G$). Show that between any two distinct real numbers there exists an element of G or else there exists $a \in$* **R** *such that $G = \{na: n = 0, \pm1, \pm2, \ldots\}$.*

Solution. Assume $G \neq \{0\}$. Let $a = \inf G \cap (0, \infty)$. We distinguish two cases. (1) $a > 0$. *In this case, we shall show that $G = \{na: n = 0, \pm1, \ldots\}$.*

To see this, note first that $a \in G$. Indeed, if $a \notin G$, then there exist $x, y \in G$ with $a < x < y < \frac{3a}{2}$. Then, the element $z = y - x \in G$ satisfies $0 < z < \frac{a}{2}$, contradicting the definition of a. Now, if $x \in G$, then $na \le x < (n+1)a$ must hold for some integer n. However, $x = na$ must also hold, since otherwise the element $x - na \in G$ satisfies $0 < x - na < a$, which is again a contradiction.

(2) $a = 0$. *In this case, we claim that between any two distinct real numbers there is an element of G.*

To see this, we only need to consider $0 < x < y$. Let $\delta = \min\{x, y - x\} > 0$. Choose some element $z \in G$ with $0 < z < \delta$. By the Archimedean property, the set $A = \{n \in \mathbb{N} : nz \ge y\}$ is nonempty, and by the Well Ordering Principle the element $k = \min A$ exists. Now, note that the element $b = (k - 1)z \in G$ satisfies $x < b < y$.

Problem 4.22. *Determine the limit points of the sequence $\{\cos n\}$.*

Solution. We claim that the set of limit points of the sequence $\{\cos n\}$ is $[-1, 1]$. To prove this, we shall need two facts from elementary calculus.

a) The Intermediate Value Theorem; and

b) The inequality $|\cos x - \cos y| \le |x - y|$ for all $x, y \in \mathbb{R}$.

Let $G = \{n + 2m\pi : n, m \text{ integers}\}$. Clearly, G is a group under addition, and since π is an irrational number, it is easy to see that the group G is not of the form $\{na : n = 0, \pm 1, \pm 2, \dots\}$. Now, let $x \in [-1, 1]$ and let $\varepsilon > 0$. By the Intermediate Value Theorem, there exists some $y \in \mathbb{R}$ satisfying $\cos y = x$. The preceding Problem 4.21 shows that there exist two integers n and m satisfying $y < n + 2m\pi < y + \varepsilon$. Thus,

$$|x - \cos n| = |\cos y - \cos(n + 2m\pi)| \le n + 2m\pi - y < \varepsilon.$$

The above arguments show that given $x \in [-1, 1]$ and $\varepsilon > 0$, there exists some non-negative integer n with $|x - \cos n| < \varepsilon$. From this, it easily follows (how?) that every point of $[-1, 1]$ must be a limit point of $\{\cos n\}$.

Problem 4.23. *For each n define $f_n : [-1, 1] \to \mathbb{R}$ by $f_n(x) = x^n$. Determine $\limsup f_n$ and $\liminf f_n$.*

Solution. We have

$$\limsup f_n(x) = \begin{cases} 1, & \text{if } x = -1 \\ 0, & \text{if } |x| < 1 \\ 1, & \text{if } x = 1 \end{cases}$$

and

$$\liminf f_n(x) = \begin{cases} -1, & \text{if } x = -1 \\ 0, & \text{if } |x| < 1 \\ 1, & \text{if } x = 1. \end{cases}$$

Problem 4.24. *Show that every sequence of real numbers has a monotone subsequence. Use this conclusion to provide an alternate proof of the Bolzano–Weierstrass property of the real numbers:* Every bounded sequence has a convergent subsequence. (*See Corollary 4.7.*)

Solution. Let $\{x_n\}$ be a sequence of real numbers. We consider the set of natural numbers

$$S = \{k \in \mathbb{N} : x_k \leq x_m \text{ for all } m \geq k\},$$

and distinguish two cases.

1. *S is infinite.*

In this case, we can write $S = \{k_1, k_2, \ldots\}$ with $k_1 < k_2 < \cdots$. Now, it should be clear that the subsequence $\{x_{k_n}\}$ of $\{x_n\}$ is increasing.

2. *S is finite (and possibly empty).*

In this case, if we put $k_1 = 1 + \max S$ (let $\max S = 0$ if $S = \emptyset$), then for each $k \geq k_1$ there exists some $m > k$ such that $x_m < x_k$. So, by induction, if k_n has been chosen, then we can select some natural number k_{n+1} with $k_{n+1} > k_n$ and $x_{k_{n+1}} < x_{k_n}$. This implies that $\{x_{k_n}\}$ is a strictly decreasing subsequence of $\{x_n\}$, and the claim is established.

For the Bolzano–Weierstrass property, notice that if $\{x_n\}$ is a bounded sequence, then, by the above, $\{x_n\}$ has a monotone subsequence which (by Theorem 4.3) must be convergent in $\mathbb{R}$. (We remark that this result shows that not only a bounded sequence has a convergent subsequence but it also has a monotone convergent subsequence.)

5. THE EXTENDED REAL NUMBERS

Problem 5.1. *Let $\{x_n\}$ be a sequence of $\mathbb{R}^*$. Define a limit point of $\{x_n\}$ in $\mathbb{R}^*$ to be any element x of $\mathbb{R}^*$ for which there exists a subsequence of $\{x_n\}$ that converges to x.*

Show that

$$\limsup x_n = \inf_n \left[\sup_{k \geq n} x_k\right] \quad and \quad \liminf x_n = \sup_n \left[\inf_{k \geq n} x_k\right]$$

are the largest and smallest limit points of $\{x_n\}$ in $\mathbb{R}^*$.

Solution. The lim sup case is established. Let $x = \limsup x_n \in \mathbb{R}^*$. Then three cases arise:

a) $x \in \mathbb{R}$.

In this case, repeat the proof of Theorem 4.6.

b) $x = \infty$.

In this case, we have only to show that x is a limit point of $\{x_n\}$. Note that $\bigvee_{i=n}^{\infty} x_i = \infty$ for each n. Choose some $k_1 \geq 1$ such that $x_{k_1} > 1$. Now, by induction: If k_n has been selected so that $x_{k_n} > n$, then use $\bigvee_{i>k_n} x_i = \infty$ to choose some $k_{n+1} \geq k_n + 1 > k_n$ so that $x_{k_{n+1}} > n + 1$. Clearly, $\{x_{k_n}\}$ is a subsequence of $\{x_n\}$ satisfying $\lim x_{k_n} = \infty$.

c) $x = -\infty$.

In this case, we shall show that $\lim x_n = -\infty$. Let $0 < M < \infty$. From $\bigvee_{i=n}^{\infty} x_i \downarrow -\infty$, it follows that $\bigvee_{i=n}^{\infty} x_i < -M$ for some n, and so $x_i < -M$ for all $i \geq n$. That is, $\lim x_n = -\infty$.

Problem 5.2. *Let $\{x_n\}$ be a sequence of positive real numbers such that $\ell = \lim \frac{x_{n+1}}{x_n}$ exists in $\mathbb{R}$. Show that:*

   a.   *if $\ell < 1$, then $\lim x_n = 0$, and*

   b.   *if $\ell > 1$, then $\lim x_n = \infty$.*

Solution. (a) Assume $\ell < 1$ and fix some δ such that $\ell < \delta < 1$; for instance, let $\delta = \frac{1+\ell}{2}$. Since $\lim \frac{x_{n+1}}{x_n} = \ell$, there exists some $k > 1$ such that $\frac{x_{n+1}}{x_n} < \delta$ holds for all $n \geq k$. Now, if $n > k$, then note that

$$x_n = \frac{x_n}{x_{n-1}} \cdot \frac{x_{n-1}}{x_{n-2}} \cdots \frac{x_{k+1}}{x_k} \cdot x_k$$

$$< \underbrace{\delta \cdot \delta \cdots \delta}_{(n-k)\text{-}terms} \cdot x_k = \delta^{n-k} x_k = \left(\frac{x_k}{\delta^k}\right) \delta^n,$$

and so if $c = \frac{x_k}{\delta^k}$, then

$$0 < x_n < c\delta^n$$

holds for all $n > k$. Since (in view of $0 < \delta < 1$) $\delta^n \to 0$, we easily infer that $x_n \to 0$.

(b) Assume now $\ell > 1$ and choose some δ such that $1 < \delta < \ell$. Since $\lim \frac{x_{n+1}}{x_n} = \ell$, there exists some $k > 1$ such that $\frac{x_{n+1}}{x_n} > \delta$ holds for all $n \geq k$. Then, as in the preceding case, there exists some constant $C > 0$ satisfying $x_n > C\delta^n$ for all $n > k$. From $\delta^n \to \infty$, it easily follows that $x_n \to \infty$.

Problem 5.3. *Let $0 \leq a_{n,m} \leq \infty$ for all m, n, and let $\sigma : \mathbb{N} \times \mathbb{N} \to \mathbb{N} \times \mathbb{N}$ be* *one-to-one and onto. Show that*

$$\sum_{n=1}^{\infty}\sum_{m=1}^{\infty} a_{n,m} = \sum_{n=1}^{\infty}\sum_{m=1}^{\infty} a_{\sigma(n,m)}.$$

Solution.   It follows immediately from Theorem 5.4.

Problem 5.4.   *Show that*

$$\sum_{n=1}^{\infty}\sum_{m=1}^{\infty} \frac{1}{n^2 + m^2} = \infty.$$

Solution.   The convergence or divergence of the series is according to the convergence or divergence of the double integral $\int_1^{\infty}\int_1^{\infty} \frac{dx\,dy}{x^2+y^2}$. Now, note that

$$\int_1^{\infty}\left[\int_1^{\infty} \frac{dx}{x^2 + y^2}\right] dy \geq \frac{\pi}{2} \int_1^{\infty} \frac{dy}{y} = \infty.$$

An alternate solution goes as follows. Note first that the inequality

$$\frac{1}{n^2 + m^2} > \frac{1}{(n + m)^2} > \frac{1}{(n + m)(n + m + 1)} = \frac{1}{n + m} - \frac{1}{n + m + 1}$$

implies $\sum_{m=1}^{\infty} \frac{1}{n^2+m^2} \geq \sum_{m=1}^{\infty}\left(\frac{1}{n+m} - \frac{1}{n+m+1}\right) = \frac{1}{n+1}$. Therefore,

$$\sum_{n=1}^{\infty}\sum_{m=1}^{\infty} \frac{1}{n^2 + m^2} \geq \sum_{n=1}^{\infty} \frac{1}{n + 1} = \infty.$$

Problem 5.5.   *This problem describes the p-***adic representation** *of a real number in* $(0, 1)$. *We assume that* p *is a natural number such that* $p \geq 2$ *and* $x \in (0, 1)$.

   a.   *Divide the interval* $[0, 1)$ *into the* p *closed-open intervals*

$$\left[0, \tfrac{1}{p}\right), \left[\tfrac{1}{p}, \tfrac{2}{p}\right), \ldots, \left[\tfrac{p-1}{p}, 1\right),$$

   and number them consecutively from 0 to $p - 1$. *Then* x *belongs precisely to one of these intervals, say* k_1 $(0 \leq k_1 < p)$. *Next, divide the interval* $\left[\tfrac{k_1}{p}, \tfrac{k_1+1}{p}\right)$ *into* p *closed-open intervals (of the same length), number them consecutively from 0 to* $p - 1$, *and let* k_2 *be the subinterval to which* x *belongs. Proceeding this way, we construct a sequence* $\{k_n\}$ *of non-negative* *integers such that $0 \le k_n < p$ for each n. Show that*

$$x = \sum_{n=1}^{\infty} \frac{k_n}{p^n}.$$

b. *Apply the same process as in (a) by subdividing each interval now into p open-closed intervals. For example, start with $(0, 1]$ and subdivide it into the open-closed intervals $(0, \frac{1}{p}], (\frac{1}{p}, \frac{2}{p}], \dots, (\frac{p-1}{p}, 1]$.*
    As in (a), construct a sequence $\{m_n\}$ of non-negative integers such that $0 \le m_n < p$ for each n. Show that

$$x = \sum_{n=1}^{\infty} \frac{m_n}{p^n}.$$

c. *Show by an example that the two sequences constructed in (a) and (b) may be different.*

In order to make the p-adic representation of a number unique, we shall agree to take the one determined by (a) above. As usual, it will be written as $x = 0.k_1 k_2 \cdots$.

Solution.   For (a) and (b) note that $\left| x - \sum_{i=1}^{n} \frac{k_i}{p^i} \right| \le \frac{1}{p^n}$ holds for all n.

For part (c) take, for instance, $p = 2$ and note that for $x = \frac{1}{2}$ we have $k_1 = 1$ and $k_n = 0$ if $n > 1$, while $m_1 = 0$ and $m_n = 1$ for $n > 1$.

Problem 5.6.   *Show that $\mathcal{P}(\mathbb{N}) \approx \mathbb{R}$ by establishing the following:*

  i.   *If A is an infinite set, and $f \colon A \to B$ is one-to-one such that $B \setminus f(A)$ is at-most countable, then show that $A \approx B$.*
 ii.   *Show that the set of numbers of $(0, 1)$ for which the dyadic (i.e., $p = 2$) representation determined by (a) and (b) of the preceding exercise are different is a countable set.*
iii.   *For each $x \in (0, 1)$, let $x = 0.k_1 k_2 \cdots$ be the dyadic representation determined by part (a) of the preceding exercise; clearly, each k_i is either 0 or 1. Let $f(x) = \{n \in \mathbb{N} \colon k_n = 1\}$. Show that $f \colon (0, 1) \to \mathcal{P}(\mathbb{N})$ is one-to-one such that $\mathcal{P}(\mathbb{N}) \setminus f((0, 1))$ is countable, and conclude from part (i) that $(0, 1) \approx \mathcal{P}(\mathbb{N})$.*

Solution.   (i) Let $S = \{a_1, a_2, \dots\}$ be a countable subset of A.

(a) Assume $B \setminus f(A) = \{b_1, \dots, b_n\}$ is a finite set. Then $g \colon A \to B$ defined by $g(x) = f(x)$ if $x \notin S$, $g(a_i) = b_i$ for $1 \le i \le n$, and $g(a_{i+n}) = f(a_i)$ for $i = 1, 2, \dots$ is one-to-one and onto.

(b) Assume $B \setminus f(A) = \{b_1, b_2, \dots\}$ is countable. Then $g \colon A \to B$ defined by $g(x) = f(x)$ if $x \notin S$, $g(a_{2n+1}) = f(a_n)$ and $g(a_{2n}) = b_n$ for each n is one-to-one and onto.

(ii) Let D be the set of all numbers of $(0, 1)$ for which the two sequences $\{k_n\}$ and $\{m_n\}$ determined by the preceding problem are different. Assume $x = 0.k_1 k_2 \cdots = 0.m_1 m_2 \cdots \in D$ and define the natural number $r = \min\{n: k_n \neq m_n\}$. We can assume $k_r = 1$ and $m_r = 0$. Then the inequalities

$$x - \sum_{n=1}^{r-1} \frac{m_n}{2^n} = \sum_{n=r}^{\infty} \frac{m_n}{2^n} \leq \sum_{n=r+1}^{\infty} \frac{1}{2^n}$$

$$= \frac{1}{2^r}$$

$$\leq \frac{1}{2^r} + \sum_{n=r+1}^{\infty} \frac{k_n}{2^n} = x - \sum_{n=1}^{r-1} \frac{k_n}{2^n}$$

$$= x - \sum_{n=1}^{r-1} \frac{m_n}{2^n},$$

guarantee that $x = \frac{k_1}{2} + \frac{k_2}{2^2} + \cdots + \frac{k_{r-1}}{2^{r-1}} + \frac{1}{2^r}$. In particular, note that $m_n = 1$ and $k_n = 0$ hold for each $n > r$.

On the other hand, it is not difficult to see that every x of the above type belongs to D. It is now a routine matter to verify that D is a countable set. (It is also interesting to observe that D consists precisely of the endpoints of the subintervals appearing during the construction of the expansions.)

(iii) Let $A \in \mathcal{P}(\mathbb{N})$. Define the sequence $\{m_n\}$ of $\{0, 1\}$ by $m_n = 1$ if $n \in A$ and $m_n = 0$ if $n \notin A$, and then set

$$x = \sum_{n=1}^{\infty} \frac{m_n}{2^n}.$$

Note that $A \notin f\big((0, 1)\big)$ if and only if $x \in D$. Thus, $\mathcal{P}(\mathbb{N}) \setminus f\big((0, 1)\big)$ is countable, and so by part (i) and the fact that f is one-to-one, $(0, 1) \approx \mathcal{P}(\mathbb{N})$ holds.

Problem 5.7. *For a sequence $\{x_n\}$ of real numbers show that the following conditions are equivalent:*

a. *The series $\sum_{n=1}^{\infty} x_n$ is rearrangement invariant in $\mathbb{R}$.*
b. *For every permutation σ of $\mathbb{N}$ the series $\sum_{n=1}^{\infty} x_{\sigma_n}$ converges in $\mathbb{R}$.*
c. *The series $\sum_{n=1}^{\infty} |x_n|$ converges in $\mathbb{R}$.*
d. *For every sequence $\{s_n\}$ of $\{-1, 1\}$, the series $\sum_{n=1}^{\infty} s_n x_n$ converges in $\mathbb{R}$.*
e. *For every subsequence $\{x_{k_n}\}$ of $\{x_n\}$, the series $\sum_{n=1}^{\infty} x_{k_n}$ converges in $\mathbb{R}$.*
f. *For every $\epsilon > 0$, there exists an integer k (depending on ϵ) such that for every finite subset S of $\mathbb{N}$ with $\min S \geq k$, we have $|\sum_{n \in S} x_n| < \epsilon$.*

(Any series $\sum_{n=1}^{\infty} x_n$ satisfying any one of the above conditions is also referred to as an **unconditionally convergent series.***)*

Solution. (a)$\Longrightarrow$(b) Obvious.

(b)$\Longrightarrow$(c) Assume $\sum_{n=1}^{\infty} |x_n| = \infty$. From our hypothesis it follows that $x_n > 0$ and $x_n < 0$ both hold for infinitely many n. Split $\{x_n\}$ into two subsequences $\{y_n\}$ and $\{z_n\}$ such that $y_n \geq 0$ and $z_n < 0$ hold for all n. We can assume that $\sum_{n=1}^{\infty} y_n = \infty$.

Now, use induction to construct a strictly increasing sequence of natural numbers $\{k_n\}$ such that

1. $k_1 = 1$ and $z_1 + \sum_{i=1}^{k_1} y_i > 1$; and
2. $z_n + \sum_{i=k_n+1}^{k_{n+1}} y_i > 1$ for $n = 1, 2, \ldots$.

Then note that

$$y_1, \ldots, y_{k_1}, z_1, y_{k_1+1}, \ldots, y_{k_2}, z_2, y_{k_2+1}, \cdots$$

is a permutation of $\{x_n\}$ whose series is not convergent, contrary to our hypothesis.

(c)$\Longrightarrow$(d) Obvious.

(d)$\Longrightarrow$(e) Let $\{x_{k_n}\}$ be a subsequence of $\{x_n\}$. Put $s_i = -1$ if $i \neq k_n$ for each n, and $s_{k_n} = 1$. Then

$$\sum_{n=1}^{\infty} x_{k_n} = \tfrac{1}{2}\left[\sum_{n=1}^{\infty} x_n + \sum_{n=1}^{\infty} s_n x_n\right]$$

is a convergent series.

(e)$\Longrightarrow$(f) If (f) is false, then there exists some $\varepsilon > 0$ and a sequence $\{S_n\}$ of finite subsets of natural numbers such that $\max S_n < \min S_{n+1}$ and $\left|\sum_{i \in S_n} x_i\right| \geq \varepsilon$ hold for all n. Let

$$\bigcup_{n=1}^{\infty} S_n = \{k_1, k_2, \ldots\},$$

where $k_n \uparrow$. Then, it is easy to see that the series $\sum_{n=1}^{\infty} x_{k_n}$ does not converge in $\mathbb{R}$, contradicting (e).

(f)$\Longrightarrow$(a) Let $\sigma: \mathbb{N} \longrightarrow \mathbb{N}$ be a permutation. By our hypothesis, it is readily seen that the partial sums of both series $\sum_{n=1}^{\infty} x_n$ and $\sum_{n=1}^{\infty} x_{\sigma_n}$ form Cauchy sequences, and hence, both series converge in $\mathbb{R}$. Let $x = \sum_{n=1}^{\infty} x_n$ and $y = \sum_{n=1}^{\infty} x_{\sigma_n}$.

Now, if $\varepsilon > 0$ is given, then choose k so large such that

$$\left|x - \sum_{n=1}^{r} x_n\right| < \varepsilon, \quad \left|y - \sum_{n=1}^{r} x_{\sigma_n}\right| < \varepsilon, \quad \text{and} \quad \left|\sum_{i \in S} x_i\right| < \varepsilon$$

hold for all $r \geq k$ and all finite subsets S of $\mathbb{N}$ with $\min S \geq k$. Fix some $r > k$

such that for each $1 \leq i \leq k$ there exists $1 \leq j \leq r$ with $x_i = x_{\sigma_j}$, and note that

$$\left| x - y \right| \leq \left| x - \sum_{n=1}^{k} x_n \right| + \left| \sum_{n=1}^{k} x_n - \sum_{n=1}^{r} x_{\sigma_n} \right| + \left| \sum_{n=1}^{r} x_{\sigma_n} - y \right| < \varepsilon + \varepsilon + \varepsilon = 3\varepsilon$$

holds for all $\varepsilon > 0$, and so $x = y$. In other words, the series $\sum_{n=1}^{\infty} x_n$ is rearrangement invariant.

Problem 5.8.   *A series of the form $\sum_{n=1}^{\infty}(-1)^{n-1}x_n$, where $x_n > 0$ for each n, is called an* **alternating series**. *Assume that a sequence $\{x_n\}$ of strictly positive real numbers satisfies $x_n \downarrow 0$. Then establish the following:*
   a.   *The alternating series $\sum_{n=1}^{\infty}(-1)^{n-1}x_n$ converges in $\mathbf{R}$.*
   b.   *If $\sum_{n=1}^{\infty} x_n = \infty$, then the alternating series $\sum_{n=1}^{\infty}(-1)^{n-1}x_n$ is not rearrangement invariant.*

Solution.   (a) Let $s_n = \sum_{k=1}^{n}(-1)^{k-1}x_k$. We claim that

$$s_2 \leq s_4 \leq \cdots \leq s_{2n-2} \leq s_{2n} \leq s_{2n-1} \leq s_{2n-3} \leq \cdots \leq s_3 \leq s_1$$

holds for each n. The proof is by induction. For $n = 1$, we have $s_2 = x_1 - x_2 < x_1 = s_1$. So, assume the inequalities to be true for some n. Then taking into account that $x_{2n} - x_{2n+1} \geq 0$ and $x_{2n+1} - x_{2n+2} \geq 0$, we see that

   1. $s_{2n} \leq s_{2n} + (x_{2n+1} - x_{2n+2}) = s_{2n+2} = s_{2(n+1)},$
   2. $s_{2(n+1)} = s_{2n+2} = s_{2n+1} - x_{2n+2} \leq s_{2n+1} = s_{2(n+1)-1},$
   3. $s_{2(n+1)-1} = s_{2n+1} = s_{2n-1} - (x_{2n} - x_{2n+1}) \leq s_{2n-1},$

and our claim is established.

   Now, if $s_{2n} \uparrow s$ and $s_{2n-1} \downarrow t$ hold in $\mathbf{R}$, then clearly $s \leq t$. Moreover, from $s_{2n} - s_{2n-1} = -x_{2n} \to 0$, we obtain $s = t$. But then, this implies that $\{s_n\}$ converges to s in $\mathbf{R}$; see Exercise 4.3 of Section 4. Consequently, the alternating series converges and $\sum_{k=1}^{\infty}(-1)^{k-1}x_k = \lim s_n = s$.

   (b) We must have either $\sum_{k=1}^{\infty} x_{2k-1} = \infty$ or $\sum_{k=1}^{\infty} x_{2k} = \infty$. Assume $\sum_{k=1}^{\infty} x_{2k-1} = \infty$; the other case can be treated in a similar manner.

   Since $\sum_{k=1}^{\infty} x_{2k-1} = \infty$, there exist integers $0 = k_0 < k_1 < k_2 < \cdots$ such that $\left[\sum_{i=k_n+1}^{k_{n+1}} x_{2i-1} \right] - x_{2n} > 1$ holds for each $n = 0, 1, \ldots$. Consider the rearrangement $\{y_n\}$ of the sequence $\{(-1)^{n-1}x_n\}$ given by

$$x_1, x_3, \ldots, x_{2k_1-1}, -x_2, x_{2k_1+1}, \ldots, x_{2k_2-1}, -x_4, x_{2k_2+1}, \ldots,$$

and note that $\sum_{n=1}^{\infty} y_n = \infty$ holds.

Problem 5.9.   *This problem describes the* **integral test** *for the convergence of series. Assume that* $f : [1, \infty) \to [0, \infty)$ *is a decreasing function. We define the sequences* $\{\sigma_n\}$ *and* $\{\tau_n\}$ *by*

$$\sigma_n = \sum_{k=1}^{n} f(k) \quad and \quad \tau_n = \int_{1}^{n} f(x)\,dx.$$

Establish the following:

   a. $0 \le \sigma_n - \tau_n \le f(1)$ *for all n.*
   b.   *the sequence* $\{\sigma_n - \tau_n\}$ *is decreasing—and hence, convergent in* $\mathbf{R}$.
   c.   *Show that the series* $\sum_{k=1}^{\infty} f(k)$ *converges in* $\mathbf{R}$ *if and only if the improper Riemann integral* $\int_{1}^{\infty} f(x)\,dx = \lim_{r \to \infty} \int_{1}^{r} f(x)\,dx$ *exists in* $\mathbf{R}$.

Solution.   Since f is decreasing, notice that for each $k \in \mathbf{N}$ we have $f(x) \ge f(k+1)$ and $f(x) \le f(k)$ for each $k \le x \le k+1$. So, integrating over $[k, k+1]$, we get:

$$\int_{k}^{k+1} f(x)\,dx \ge f(k+1), \quad \text{and} \tag{1}$$

$$\int_{k}^{k+1} f(x)\,dx \le f(k) \tag{2}$$

for each $k = 1, 2, \ldots$. (We remark that as a decreasing function, f is Riemann integrable over every closed subinterval of $[1, \infty)$; see Section 23 of the text.)
   (a) Using (1), we see that

$$\sigma_n = f(1) + \sum_{k=2}^{n} f(k) = f(1) + \sum_{k=1}^{n-1} f(k+1)$$

$$\le f(1) + \sum_{k=1}^{n-1} \int_{k}^{k+1} f(x)\,dx = f(1) + \int_{1}^{n} f(x)\,dx$$

$$= f(1) + \tau_n.$$

This implies $\sigma_n - \tau_n \le f(1)$ for each n.   On the other hand, using (2), we see that

$$\sigma_n \ge \sum_{k=1}^{n-1} f(k) \ge \sum_{k=1}^{n-1} \int_{k}^{k+1} f(x)\,dx = \int_{1}^{n} f(x)\,dx = \tau_n,$$

and so $\sigma_n - \tau_n \ge 0$ for each n.

(b) Using once more (1), we get

$$\sigma_{n+1} - \tau_{n+1} = \sigma_n + f(n+1) - \tau_n - \int_n^{n+1} f(x)\,dx$$

$$= \sigma_n - \tau_n - \left[\int_n^{n+1} f(x)\,dx - f(n+1) \right] \leq \sigma_n - \tau_n.$$

This shows that $\{\sigma_n - \tau_n\}$ is a decreasing sequence—and so $\lim(\sigma_n - \tau_n)$ exists in $\mathbf{R}$.

(c) Since $\{\sigma_n\}$ and $\{\tau_n\}$ are both increasing sequences of non-negative real numbers they both converge in $\mathbf{R}^*$, and clearly

$$\lim_{n \to \infty} \sigma_n = \sum_{k=1}^{\infty} f(k) \quad \text{and} \quad \lim_{n \to \infty} \tau_n = \int_1^{\infty} f(x)\,dx.$$

But from part (a), we have $\tau_n \leq \sigma_n \leq f(1) + \tau_n$ for each n, and therefore (by letting $n \to \infty$), we have

$$0 \leq \int_1^{\infty} f(x)\,dx \leq \sum_{k=1}^{\infty} f(k) \leq f(1) + \int_1^{\infty} f(x)\,dx.$$

This inequality shows that $\sum_{k=1}^{\infty} f(k)$ converges if and only the improper Riemann integral $\int_1^{\infty} f(x)\,dx$ exists.

Problem 5.10. *Use the preceding problem to show that the series $\sum_{n=1}^{\infty} \frac{1}{n^p}$ does not converge in $\mathbf{R}$ for $0 < p \leq 1$ and converges in $\mathbf{R}$ for all $p > 1$. The following are problems related to the* **harmonic series** *$\sum_{n=1}^{\infty} \frac{1}{n}$.*

a. *Prove with (at least) three different ways that $\sum_{n=1}^{\infty} \frac{1}{n} = \infty$.*
b. *If a computer starting at 12 midnight on December 31, 1939, adds one million terms of the harmonic series every second, what was the value (within an error of 1) of the sum at 12 midnight on December 31, 1997? (Assume that each year has 365 days.)*
c. *Show that $\displaystyle\sum_{n=1}^{\infty} \frac{(-1)^{n-1}}{n} = \lim_{n \to \infty} \left(\frac{1}{n+1} + \frac{1}{n+2} + \cdots + \frac{1}{2n} \right) = \ln 2.$*

Solution. Notice that

$$\int_1^{\infty} \frac{dx}{x^p} = \lim_{r \to \infty} \int_1^r \frac{dx}{x^p} = \begin{cases} \lim_{r \to \infty} \frac{r^{1-p}-1}{1-p} & \text{if } p \neq 1, \\ \lim_{r \to \infty} \ln r & \text{if } p = 1. \end{cases}$$

This limit is finite if $p > 1$ and infinity if $0 < p \leq 1$.

(a) We let $\sigma_n = \sum_{k=1}^{n} \frac{1}{k}$. Here are four proofs of the divergence of the harmonic series.

1. Notice that $\sigma_{2n} - \sigma_n = \frac{1}{n+1} + \frac{1}{n+2} + \cdots + \frac{1}{n+n} \geq n \cdot \frac{1}{2n} = \frac{1}{2}$ for each n. This shows that $\{\sigma_n\}$ is not a Cauchy sequence, and hence, divergent.

2. As shown at the beginning of the solution of the problem, $\int_1^{\infty} \frac{dx}{x} = \infty$, and so $\sum_{n=1}^{\infty} \frac{1}{n} = \infty$.

3. We claim that $\sigma_{2^n} \geq 1 + \frac{n}{2}$ for each n. (If this inequality is established, then clearly $\sum_{k=1}^{\infty} \frac{1}{k} = \lim \sigma_{2^n} = \infty$.) The proof of the inequality is by induction. For $n = 1$, we have $\sigma_{2^1} = \sigma_2 = 1 + \frac{1}{2}$. Now, if we assume the inequality true for some n, then

$$\sigma_{2^{n+1}} = \sigma_{2^n} + \frac{1}{2^n + 1} + \frac{1}{2^n + 2} + \cdots + \frac{1}{2^n + 2^n}$$

$$\geq 1 + \frac{n}{2} + 2^n \cdot \frac{1}{2 \cdot 2^n} = 1 + \frac{n+1}{2}.$$

4. Note that

$$\sum_{n=1}^{\infty} \frac{1}{n} = \left[1 + \frac{1}{2} + \cdots + \frac{1}{9}\right] + \left[\frac{1}{10} + \cdots + \frac{1}{99}\right] + \left[\frac{1}{100} + \cdots + \frac{1}{999}\right] + \cdots$$

$$\geq 9 \cdot \frac{1}{10} + 90 \cdot \frac{1}{100} + 900 \cdot \frac{1}{1000} + \cdots$$

$$= \frac{9}{10} + \frac{9}{10} + \frac{9}{10} + \cdots = \infty.$$

(b) From Problem 5.9, we know that the harmonic series is associated with the function $f(x) = \frac{1}{x}$ and that $0 \leq \sigma_n - \ln n \leq f(1) = 1$ for each n. So, $\ln n$ approximates σ_n within an error of one. If the computer started adding the terms of the harmonic series at 12 midnight on December 31, 1939, then up to 12 midnight on December 31, 1997, there are

$$57(\text{years}) \times 365(\text{days}) \times 24(\text{hours}) \times 60(\text{minutes}) \times 60(\text{seconds})$$

$$= 1,797,552 \times 10^3 \text{ seconds.}$$

So, if the computer adds $1,000,000 = 10^6$ terms per second of the harmonic series, the last number N added a second before midnight on December 31 of 1997 is $N = 1,797,552 \times 10^9$. Therefore,

$$\sum_{n=1}^{N} \frac{1}{n} = \sigma_N \approx \ln N = 35.12520213\ldots,$$

which shows that the harmonic series is a "very slow" divergent series.

(c) From Problem 5.8, we know that the alternating series $\sum_{k=1}^{\infty} \frac{(-1)^{k-1}}{k}$ is convergent in $\mathbb{R}$. Also, from Problem 5.9, we know that $\lim(\sigma_n - \ln n) = \gamma \in \mathbb{R}$. So, if we let $x_n = \gamma - (\sigma_n - \ln n)$, then $x_n \to 0$ and $\sigma_n = \gamma + \ln n + x_n$ for each n. Now, note that

$$
\begin{aligned}
\sum_{k=1}^{2n} \frac{(-1)^{k-1}}{k} &= 1 + \frac{1}{3} + \frac{1}{5} + \cdots + \frac{1}{2n-1} - \left(\frac{1}{2} + \frac{1}{4} + \cdots + \frac{1}{2n}\right) \\
&= 1 + \frac{1}{2} + \frac{1}{3} + \cdots + \frac{1}{2n} - 2\left(\frac{1}{2} + \frac{1}{4} + \cdots + \frac{1}{2n}\right) \\
&= \frac{1}{n+1} + \frac{1}{n+2} + \cdots + \frac{1}{2n} \\
&= \sigma_{2n} - \sigma_n = \left[\gamma + \ln(2n) + x_{2n}\right] - \left[\gamma + \ln n + x_n\right] \\
&= \ln 2 + x_{2n} - x_n,
\end{aligned}
$$

for each n. This implies $\sum_{k=1}^{\infty} \frac{(-1)^{k-1}}{k} = \ln 2$.

Problem 5.11 (Toeplitz). *Let $\{a_n\}$ be a sequence of positive real numbers (i.e., $a_n > 0$ for each n) and put $b_n = \sum_{i=1}^{n} a_i$. Assume that $b_n \uparrow \sum_{i=1}^{\infty} a_i = \infty$. If $\{x_n\}$ is a sequence of real numbers such that $x_n \to x$ in $\mathbb{R}$, then show that*

$$
\lim_{n\to\infty} \frac{1}{b_n} \sum_{i=1}^{n} a_i x_i = x.
$$

Solution. Let $\epsilon > 0$. Choose some k such that $|x_n - x| < \epsilon$ for each $n \geq k$. Put $M = \max\{|x_i - x|: i = 1, \ldots, k\}$, and then select some $\ell > k$ such that $\frac{Mb_k}{b_n} < \epsilon$ for all $n \geq \ell$. Now, notice that if $n \geq \ell$, then

$$
\begin{aligned}
\left|\frac{1}{b_n} \sum_{i=1}^{n} a_i x_i - x\right| &= \left|\frac{1}{b_n} \sum_{i=1}^{n} a_i x_i - \frac{1}{b_n} \sum_{i=1}^{n} a_i x\right| \\
&\leq \frac{1}{b_n} \sum_{i=1}^{k} a_i |x_i - x| + \frac{1}{b_n} \sum_{i=k+1}^{n} a_i |x_i - x| \\
&\leq \frac{Mb_k}{b_n} + \epsilon < \epsilon + \epsilon = 2\epsilon,
\end{aligned}
$$

and the conclusion follows. (Note that this problem is a substantial generalization of Problem 4.11.)

Problem 5.12 (Kronecker). *Assume that a sequence of positive numbers $\{b_n\}$ satisfies $0 < b_1 < b_2 < b_3 < \cdots$ and $b_n \uparrow \infty$. If a series $\sum_{n=1}^{\infty} x_n$ of real numbers* *converges in* $\mathbb{R}$, *then show that*

$$\lim_{n \to \infty} \frac{1}{b_n} \sum_{i=1}^{n} b_i x_i = 0.$$

In particular, show that if $\{y_n\}$ *is a sequence of real numbers such that the series* $\sum_{n=1}^{\infty} \frac{y_n}{n}$ *converges in* $\mathbb{R}$, *then* $\frac{y_1 + \cdots + y_n}{n} \to 0$.

Solution. Let $x = \sum_{n=1}^{\infty} x_n$. Put $b_0 = 0$, $s_0 = 0$, and $s_n = x_1 + \cdots + x_n$ for each $n \geq 1$. Now, notice that

$$\sum_{i=1}^{n} b_i x_i = \sum_{i=1}^{n} b_i (s_i - s_{i-1}) = b_n s_n - \sum_{i=2}^{n} s_{i-1}(b_i - b_{i-1}).$$

Therefore, $\frac{1}{b_n} \sum_{i=1}^{n} b_i x_i = s_n - \frac{1}{b_n} \sum_{i=2}^{n} s_{i-1}(b_i - b_{i-1})$. Since $b_i - b_{i-1} > 0$ for each i and $\sum_{i=1}^{n}(b_i - b_{i-1}) = b_n \uparrow \infty$, it follows from the preceding problem that $\frac{1}{b_n} \sum_{i=2}^{n} s_{i-1}(b_i - b_{i-1}) = x$. Hence,

$$\lim_{n \to \infty} \frac{1}{b_n} \sum_{i=1}^{n} b_i x_i = \lim_{n \to \infty} \left[s_n - \frac{1}{b_n} \sum_{i=2}^{n} s_{i-1}(b_i - b_{i-1}) \right] = x - x = 0.$$

For the second part, notice that if $\sum_{n=1}^{\infty} \frac{y_n}{n}$ is convergent in $\mathbb{R}$, then let $b_n = n$ for each n and notice that (by the above)

$$\frac{1}{b_n} \sum_{i=1}^{n} b_i \frac{y_i}{i} = \frac{y_1 + \cdots + y_n}{n} \longrightarrow 0,$$

as desired.

6. METRIC SPACES

Problem 6.1. *For subsets A and B of a metric space* (X, d) *show that:*

a. $(A \cap B)^\circ = A^\circ \cap B^\circ$.
b. $A^\circ \cup B^\circ \subseteq (A \cup B)^\circ$.
c. $\overline{A \cup B} = \overline{A} \cup \overline{B}$.
d. $\overline{A \cap B} \subseteq \overline{A} \cap \overline{B}$.
e. *If B is open, then* $\overline{A} \cap B \subseteq \overline{A \cap B}$.

Solution. (a) From $(A \cap B)^\circ \subseteq A^\circ$ and $(A \cap B)^\circ \subseteq B^\circ$, it follows that $(A \cap B)^\circ \subseteq A^\circ \cap B^\circ$. On the other hand, since $A^\circ \cap B^\circ \subseteq A \cap B$ holds and $A^\circ \cap B^\circ$ is open, it easily follows that $A^\circ \cap B^\circ \subseteq (A \cap B)^\circ$.

(b) From $A \subseteq A \cup B$, it follows that $A^{\circ} \subseteq (A \cup B)^{\circ}$. Similarly, $B^{\circ} \subseteq (A \cup B)^{\circ}$, and the desired inclusion follows.

(c) From $S \subseteq \overline{S}$ and the fact that $\overline{S}$ is a closed set for any subset S, we see that

$$\overline{A \cup B} \subseteq \overline{\overline{A} \cup B \cup \overline{A} \cup B} = \overline{\overline{A} \cup \overline{B}} \subseteq \overline{\overline{A} \cup \overline{B}} = \overline{A} \cup \overline{B}.$$

(d) Since $A \cap B \subseteq \overline{A} \cap \overline{B}$, we have $\overline{A \cap B} \subseteq \overline{\overline{A} \cap \overline{B}} = \overline{A} \cap \overline{B}$.

(e) If $x \in \overline{A} \cap B$ and $r > 0$, then choose some $0 < \delta < r$ with $B(x, \delta) \subseteq B$, and note that

$$B(x, r) \cap (A \cap B) \supseteq B(x, \delta) \cap B \cap A = B(x, \delta) \cap A \neq \emptyset.$$

That is, $x \in \overline{A \cap B}$, and so $\overline{A} \cap B \subseteq \overline{A \cap B}$ holds.

Problem 6.2.   *Show that in a Euclidean space $\mathbb{R}^{n}$ with the Euclidean distance, the closure of any open ball $B(a, r)$ is the closed ball $\{x \in \mathbb{R}^{n}: d(x, a) \leq r\}$. Give an example of a complete metric space for which the corresponding statement is false.*

Solution.   Let $C(a, r) = \{x \in \mathbb{R}^{k}: d(a, x) \leq r\}$. Since $C(a, r)$ is a closed set, it follows that $\overline{B(a, r)} \subseteq C(a, r)$.

For the other inclusion, let $x \in C(a, r)$. For each n let $x_{n} = \frac{1}{n}a + \left(1 - \frac{1}{n}\right)x$. The inequalities

$$d(a, x_{n}) = \left(1 - \tfrac{1}{n}\right)d(a, x) \leq \left(1 - \tfrac{1}{n}\right)r < r \text{ and } d(x, x_{n}) = \tfrac{1}{n}d(a, x) \leq \tfrac{r}{n}$$

imply that $\{x_{n}\} \subseteq B(a, r)$ and $x_{n} \longrightarrow x$. Consequently, $x \in \overline{B(a, r)}$, and thus $C(a, r) \subseteq \overline{B(a, r)}$ also holds.

For a counterexample, consider $X = \{0, 1\}$ with the discrete distance, and note that X is a complete metric space. Also, observe that $\overline{B(0, 1)} = \{0\}$, while $C(0, 1) = \{0, 1\}$.

Problem 6.3.   *If A is a nonempty subset of $\mathbb{R}$, then show that the set*

$$B = \left\{a \in \overline{A}: \text{ There exists some } \varepsilon > 0 \text{ with } (a, a + \varepsilon) \cap A = \emptyset\right\}$$

is at-most countable.

Solution.   For each $a \in B$ pick a rational number $r_{a} > a$ so that $(a, r_{a}) \cap A = \emptyset$. We claim that if $a, b \in B$ satisfy $a \neq b$, then $r_{a} \neq r_{b}$. Indeed, if $a < b$ and

$r_a = r_b$ hold, then—since $b \in (a, r_a)$—the open interval (a, r_a) is a neighborhood of $b \in \overline{A}$, and so $(a, r_a) \cap A \neq \emptyset$, contrary to the choice of r_a.

The above show that the mapping $a \longmapsto r_a$, from B into the set of rational numbers, is one-to-one. Consequently, the set B is at-most countable.

Problem 6.4. *Let $f: (X, d) \to (Y, \rho)$ be a function. Show that f is continuous if and only if $f^{-1}(B^\circ) \subseteq \left[f^{-1}(B) \right]^\circ$ for every subset B of Y.*

Solution. Assume f continuous, and $B \subseteq Y$. Since B° is open, the set $f^{-1}(B^\circ)$ is likewise open. Thus, in view of $B^\circ \subseteq B$, we have

$$f^{-1}(B^\circ) = \left[f^{-1}(B^\circ) \right]^\circ \subseteq \left[f^{-1}(B) \right]^\circ.$$

In the opposite direction, assume that the condition is satisfied. If $B \subseteq Y$ is open (i.e., if $B = B^\circ$ holds), then

$$\left[f^{-1}(B) \right]^\circ \subseteq f^{-1}(B) = f^{-1}(B^\circ) \subseteq \left[f^{-1}(B) \right]^\circ$$

shows that $f^{-1}(B)$ is open. Therefore, f is continuous.

Problem 6.5. *Show that the boundary of a closed or open set in a metric space is nowhere dense. Is this statement true for an arbitrary subset?*

Solution. Since $\partial A = \partial A^c = \overline{A} \cap \overline{A^c}$ holds, we can assume that A is closed. Thus,

$$(\partial A)^\circ = \left(\overline{A} \cap \overline{A^c} \right)^\circ = \left(A \cap \overline{A^c} \right)^\circ = A^\circ \cap \left(\overline{A^c} \right)^\circ$$
$$\subseteq A^\circ \cap \overline{A^c} = A^\circ \cap \left(A^\circ \right)^c = \emptyset.$$

Since ∂A is closed, this shows that ∂A is nowhere dense.

An alternate proof goes as follows: If $x \in (\partial A)^\circ$, then there exists some $r > 0$ such that $B(x, r) \subseteq \partial A = A \cap \overline{A^c} \subseteq A$. This implies $B(x, r) \cap A^c = \emptyset$, contrary to $x \in \overline{A^c}$.

The boundary of an arbitrary set need not be nowhere dense. An example: Let $X = \mathbb{R}$ with the Euclidean distance, and let $A = Q$ (the set of rational numbers). Note that $\partial A = \mathbb{R}$.

Problem 6.6. *Show that the set of irrational numbers is not a countable union of closed subsets of $\mathbb{R}$.*

Solution. Let I denote the set of all irrational numbers, and let $\{r_1, r_2, \dots\}$ be an enumeration of the rational numbers of $\mathbb{R}$.

Assume by way of contradiction that there exists a sequence of closed sets $\{A_n\}$ of $\mathbf{R}$ such that $I = \bigcup_{n=1}^{\infty} A_n$. Then

$$\mathbf{R} = \left(\bigcup_{n=1}^{\infty} A_n \right) \cup \left(\bigcup_{n=1}^{\infty} \{r_n\} \right),$$

and by the Baire Category Theorem (Theorem 6.18), we must have $(A_n)^{\circ} \neq \emptyset$ for some n. Thus, some A_n contains an interval. However, since $A_n \subseteq I$ holds and each interval contains rational numbers, this is impossible, and the conclusion follows.

Problem 6.7. *Let (X, d) be a metric space. Show that if $\{x_n\}$ and $\{y_n\}$ are Cauchy sequences of X, then $\{d(x_n, y_n)\}$ converges in $\mathbf{R}$.*

Solution. Use the inequality

$$\left| d(x_n, y_n) - d(x_m, y_m) \right| \leq d(x_n, x_m) + d(y_n, y_m).$$

(Also, see the discussion before Theorem 6.19.)

Problem 6.8. *Show that in a metric space a Cauchy sequence converges if and only if it has a convergent subsequence.*

Solution. Let $\{x_n\}$ be a Cauchy sequence in a metric space (X, d). If $x_n \to x$ holds in X, then every subsequence of $\{x_n\}$ converges to x.

For the converse, assume that there exists a subsequence $\{x_{k_n}\}$ of $\{x_n\}$ such that $x_{k_n} \to x$ holds in X. Let $\epsilon > 0$. Choose n_0 such that $d\left(x_{k_n}, x\right) < \epsilon$ and $d(x_n, x_m) < \epsilon$ for $n, m > n_0$. Now, if $n > n_0$, then $k_n \geq n > n_0$, and so

$$d(x_n, x) \leq d\left(x_n, x_{k_n}\right) + d\left(x_{k_n}, x\right) < \epsilon + \epsilon = 2\epsilon.$$

This shows that, $\lim x_n = x$ holds in X.

Problem 6.9. *Prove that the closed interval $[0, 1]$ is an uncountable set:*
  a. *by using Cantor's Theorem 6.14, and*
  b. *by using Baire's Theorem 6.17.*

Solution. (a) Assume by way of contradiction that $[0, 1]$ is a countable set, say $[0, 1] = \{x_1, x_2, \ldots\}$. We consider $[0, 1]$ equipped with the usual distance $d(x, y) = |x - y|$ so that $[0, 1]$ is a complete metric space.

Subdivide [0, 1] into three closed subintervals (as in the construction of the Cantor set) of equal length. Remove from [0, 1] the middle open subinterval and consider the remaining two closed subintervals (here the subintervals $\left[0, \frac{1}{3}\right]$ and $\left[\frac{2}{3}, 1\right]$) and then select one of them, say I_1, such that $x_1 \notin I_1$. Next, repeat this process with I_1 in place of [0, 1] and select a closed subinterval I_2 of I_1 of length equal to one-third of I_1 such that $x_2 \notin I_2$. Inductively, assume that we have chosen n closed intervals $I_1, \ldots, I_n$ such that:

1. $I_n \subseteq I_{n-1} \subseteq \cdots \subseteq I_2 \subseteq I_1$,
2. $x_k \notin I_k$ for $k = 1, \ldots, n$, and
3. the length of each I_k is $\frac{1}{3^k}$.

As above, there exists a closed subinterval I_{n+1} of I_n of length equal to one-third of I_n such that $x_{n+1} \notin I_{n+1}$.

Thus, there exists a sequence $\{I_n\}$ of closed subintevals of [0, 1] such that $I_{n+1} \subseteq I_n$, $x_n \notin I_n$ and $d(I_n) = \frac{1}{3^n}$ for each n. By Theorem 6.14, we infer that $\bigcap_{n=1}^{\infty} I_n$ consists exactly of one point. But, since $x_n \notin I_n$ for each n, we see that $\bigcap_{n=1}^{\infty} I_n = \emptyset$, a contradiction. Hence, [0, 1] must be uncountable.

(b) Again, assume by way of contradiction that $[0, 1] = \{x_1, x_2, \ldots\}$ and again we consider [0, 1] as a complete metric space. If $A_n = \{x_n\}$, then each A_n is closed and has no interior points. However, Theorem 6.17 (or Theorem 6.18) applied to the equality $[0, 1] = \bigcup_{n=1}^{\infty} A_n$ implies that some A_n must have an interior point, which is impossible. This shows that [0, 1] cannot be countable.

Problem 6.10. *Let $\{r_1, r_2, \ldots\}$ be an enumeration of all rational numbers in the interval* [0, 1] *and for each $x \in [0, 1]$ let $A_x = \{n \in \mathbb{N}: r_n \leq x\}$. Define the function $f: [0, 1] \to \mathbb{R}$ by the formula*

$$f(x) = \sum_{n \in A_x} \frac{1}{2^n}.$$

Show that f restricted to the set of irrational numbers of [0, 1] *is continuous.*

Solution. Fix an irrational number $a \in [0, 1]$ and let $\varepsilon > 0$. Pick a natural number k such that $\sum_{n=k}^{\infty} \frac{1}{2^n} < \varepsilon$ and let

$$\delta = \min\{|a - r_1|, |a - r_2|, \ldots, |a - r_k|\} > 0.$$

We claim that if $x \in [0, 1]$ is an irrational number, then $|a - x| < \delta$ implies $|f(x) - f(a)| < \varepsilon$ (which tells us that f is continuous when restricted to the irrational numbers).

To see this, let $x \in [0, 1]$ be an arbitrary irrational number satisfying $|x - a| < \delta$. Let I_x denote the half-open subinterval of [0, 1] which is open at left and closed at right having endpoints a and x. If $B_x = \{n \in \mathbb{N}: r_n \in I_x\}$, then note that $B_x \subseteq \{k, k+1, k+2, \ldots\}$ (why?), and so

$$\left| f(x) - f(a) \right| = \sum_{n \in B_x} \frac{1}{2^n} \leq \sum_{n=k}^{\infty} \frac{1}{2^n} < \varepsilon$$

holds, as claimed.

Problem 6.11. *This problem concerns connected metric spaces. A metric space (X, d) is said to be* **connected** *whenever $\emptyset$ and X are the only subsets of X that are simultaneously open and closed. A subset A of a metric space (X, d) is said to be* **connected** *whenever (A, d) is itself a connected metric space. Establish the following properties regarding connected metric spaces and connected sets.*

  a. *A metric space (X, d) is connected if and only if every continuous function $f: X \to \{0, 1\}$ is constant, where the two point set $\{0, 1\}$ is considered to be a metric space under the discrete metric.*

  b. *If in a metric space (X, d) we have $B \subseteq A \subseteq X$, then the set B is a connected subset of (A, d) if and only if B is a connected subset of (X, d).*

  c. *If $f: (X, d) \to (Y, \rho)$ is a continuous function and A is a connected subset of X, then $f(A)$ is a connected subset of Y.*

  d. *If $\{A_i\}_{i \in I}$ is a family of connected subsets of a metric space such that $\bigcap_{i \in I} A_i \neq \emptyset$, then $\bigcup_{i \in I} A_i$ is likewise a connected set.*

  e. *If A is a subset of a metric space and $a \in A$, then there exists a largest (with respect to inclusion) connected subset C_a of A that contains a. (The connected set C_a is called the* **component** *of a with respect to A.)*

  f. *If a, b belong to a subset A of a metric space and C_a and C_b are the components of a and b in A, then either $C_a = C_b$ or else $C_a \cap C_b = \emptyset$. Hence, the identity $A = \bigcup_{a \in A} C_a$ shows that A can be written as a disjoint union of connected sets.*

  g. *A nonempty subset of $\mathbb{R}$ with at least two elements is a connected set if and only if it is an interval. Use this and the conclusion of (f) to infer that every open subset of $\mathbb{R}$ can be written as an at-most countable union of disjoint open intervals.*

Solution. (a) Let (X, d) be a connected space and let $f: X \longrightarrow \{0, 1\}$ be a continuous function. Then the set $A = f^{-1}(0)$ is an open and closed subset of X. Since X is connected either $A = \emptyset$ (in which case $f(x) = 1$ holds for each $x \in X$) or $A = X$ (in which case $f(x) = 0$ holds for each $x \in X$).

For the converse, assume that every continuous function from X into $\{0, 1\}$ is constant and let A be a closed and open subset of X. Then the function

$f: X \rightarrow \mathbb{R}$ defined by

$$f(x) = \begin{cases} 1, & \text{if } x \in A; \\ 0, & \text{if } x \notin A, \end{cases}$$

is continuous (why?). By our hypothesis, f must be a constant function, and this implies that either $A = \emptyset$ or $A = X$, i.e., X is a connected metric space.

(b) It follows immediately from (a).

(c) Assume that f and A satisfy the stated properties and consider the continuous functions $(A, d) \xrightarrow{f} (f(A), \rho) \xrightarrow{g} \{0, 1\}$. By (a), the continuous function $g \circ f$ must be a constant function and from this, we see that g is also a constant function. By (a), $(f(A), \rho)$ is a connected metric space.

(d) Assume the family $\{A_i : i \in I\}$ satisfies the stated properties. Put $A = \bigcup_{i \in I} A_i$ and let $f: (A, d) \longrightarrow \{0, 1\}$ be a continuous function. Then the function $f: (A_i, d) \longrightarrow \{0, 1\}$ is a continuous function and so f restricted to each A_i is constant. Since $\bigcap_{i \in I} A_i \neq \emptyset$, we see that f is constant on A, and so—by (a)—the set A is connected.

(e) Fix $a \in A$ and let

$$\mathcal{A} = \{B \subseteq A: B \text{ is connected and } a \in B\}.$$

Note that $\{a\} \in \mathcal{A}$ and that $\bigcap_{B \in \mathcal{A}} B \neq \emptyset$. By (d), the set $C_a = \bigcup_{B \in \mathcal{A}} B$ is a connected subset of A that satisfies the desired properties.

(f) If $C_a \cap C_b \neq \emptyset$, then by (d) we infer that $C_a \cup C_b$ is a connected set containing a. Hence, $C_b \subseteq C_a \cup C_b \subseteq C_a$. Similarly, $C_a \subseteq C_b$ and so $C_a = C_b$.

(g) Let A be a connected subset of $\mathbb{R}$ and let $a, b \in A$ satisfy $a < b$. If $a < x < b$ and $x \notin A$, then the set $A \cap (-\infty, x)$ is a proper and closed subset of A (why?), a contradiction. Thus, $(a, b) \subseteq A$ holds and this shows that A is an interval.

For the converse, assume that I is an interval of $\mathbb{R}$. Assume by way of contradiction that there exists an onto continuous function $f: I \longrightarrow \{0, 1\}$. Pick $a, b \in I$ such that $f(a) = 0$ and $f(b) = 1$; we can suppose that $a < b$ (and so $[a, b] \subseteq I$). Now, let

$$c_0 = \sup\{c \in [a, b): f(c) = 0\}.$$

By the continuity of f, we see that $f(c_0) = 0$ and that $c_0 < b$. Then $f(x) = 1$ holds for all $c_0 < x < b$, and so (by the continuity of f again) $f(c_0) = 1$ must also hold, which is impossible. Therefore, every continuous function from I into $\{0, 1\}$ is constant, and so by (a) the interval I is a connected set.

Finally, note that if I is an open subset of $\mathbb{R}$, then by (f) we know that $I = \bigcup_{a \in I} C_a$, where each C_a is a connected set. It easily follows (how?) that each C_a is an open interval and that there are at-most countably many of them.

Problem 6.12.   *Show that* $\mathbb{R}^n$ *with the Euclidean distance is a connected metric space. Use this conclusion to establish that, if the intersection of two open subsets of* $\mathbb{R}^n$ *is a proper closed set, then the two open sets must be disjoint.*

Solution.   Let d denote the Euclidean distance of $\mathbb{R}^n$, i.e., let

$$d(\mathbf{x}, \mathbf{y}) = \left[\sum_{i=1}^{n} (x_i - y_i)^2 \right]^{\frac{1}{2}}.$$

For each $\mathbf{x} \in \mathbb{R}^n$, let $L_\mathbf{x}$ denote the line segment joining $\mathbf{0}$ and $\mathbf{x}$, i.e., let $L_\mathbf{x} = \{t\mathbf{x}: 0 \le t \le 1\}$. We claim that $L_\mathbf{x}$ is a connected set.

To see this, note that the function $f: [0, 1] \longrightarrow L_\mathbf{x}$, defined by $f(t) = t\mathbf{x}$, satisfies $|f(t) - f(s)| \le d(\mathbf{x}, \mathbf{0})|s - t|$, and so f is (uniformly) continuous. From parts (g) and (c) of Problem 6.11, we see that $L_\mathbf{x}$ is a connected set.

Now, use part (d) of the preceding problem and the identity $\mathbb{R}^n = \bigcup_{\mathbf{x} \in \mathbb{R}^n} L_\mathbf{x}$ to infer that $\mathbb{R}^n$ is itself a connected metric space.

For the last part of the problem, let U and V be two open subsets of $\mathbb{R}^n$ such that $K = U \cap V$ is a closed set. Then K is both open and closed (and since K is a proper subset of $\mathbb{R}^n$) it must be the empty set.

Problem 6.13.   *Let C be a nonempty closed subset of $\mathbb{R}$. Show that a function $f: C \to \mathbb{R}$ is continuous if and only if it can be extended to a continuous real-valued function on $\mathbb{R}$.*

Solution.   Let C be a nonempty closed subset of $\mathbb{R}$ and let $f: C \to \mathbb{R}$ be a function. If f can be extended to a continuous real-valued function on $\mathbb{R}$, then $f: C \to \mathbb{R}$ is obviously continuous.

For the converse, assume that $f: C \to \mathbb{R}$ is a continuous function. Start by observing that the complement C^c of C is an open set and so (by part (g) of Problem 6.11) C^c can be written as an at-most countable union of pairwise disjoint open intervals; say $C^c = \bigcup_{i \in I}(a_i, b_i)$, where I is at-most countable. Since the open intervals $\{(a_i, b_i): i \in I\}$ are pairwise disjoint, it follows that all the endpoints a_i and b_i belong to C. Hence, $f(a_i)$ and $f(b_i)$ are defined for each i. Now, extend the domain of f by defining the graph of the function f on the interval (a_i, b_i) to be the straight line segment joining the points $(a_i, f(a_i))$ and $(b_i, f(b_i))$. In other words, for each $a_i < x < b_i$ we let $f(x) = f(a_i) + \frac{f(b_i) - f(a_i)}{b_i - a_i}(x - a_i)$—in case $(a_i, b_i) = (-\infty, b_i)$ or $(a_i, b_i) = (a_i, \infty)$ let $f(x) = b_i$ or $f(x) = a_i$.

We claim that this extension of f to all of $\mathbb{R}$ is continuous. Clearly, f is continuous at every point of C° and at every point of C^c (why?). We only need to verify that f is continuous at the boundary points of C. So, let $a \in \partial C$ and let $x_n \longrightarrow a$ with $\{x_n\} \subseteq C^c$—if $\{x_n\} \subseteq C$, then $f(x_n) \longrightarrow f(a)$ is trivially true. Also, we shall assume that a is not one of the endpoints a_i or b_i. For each n pick (the unique) $i_n \in I$ with $a_{i_n} \leq x_n \leq b_{i_n}$. Note that in this case, we must have $\lim a_{i_n} = \lim b_{i_n} = a$ (why?). From

$$
\begin{aligned}
\left| f(x_n) - f(a) \right| &= \left| \left[f(a) + \tfrac{f(b_{i_n}) - f(a_{i_n})}{b_{i_n} - a_{i_n}} (x_n - a_{i_n}) \right] - f(a) \right| \\
&= \left| \tfrac{f(b_{i_n}) - f(a_{i_n})}{a_{i_n} - b_{i_n}} (x_n - a_{i_n}) \right| \leq \left| f(b_{i_n}) - f(a_{i_n}) \right| \\
&\longrightarrow |f(a) - f(a)| = 0,
\end{aligned}
$$

we see that $\lim f(x_n) = f(a)$. A similar conclusion holds true if a is one of the endpoints a_i or b_i. This shows that f is continuous at a, as claimed.

For an alternate proof see Problem 10.11.

Problem 6.14. *Show that a metric space is a Baire space if and only if the complement of every meager set is dense.*

Solution. Let X be a metric space. Assume first that X is a Baire space and let A be a meager set. Pick a sequence $\{A_n\}$ of nowhere dense sets such that $A = \bigcup_{n=1}^{\infty} A_n$. To show that A^c is dense, it suffices to show that $V \cap A^c \neq \emptyset$ for each nonempty open set V. To see this, let V be a nonempty open set, and assume by way of contradiction that $V \cap A^c = \emptyset$. This implies $V \subseteq A$, and so

$$
V = \bigcup_{n=1}^{\infty} V \cap A_n.
$$

Hence, V is a nonempty open meager set, a contradiction. Hence, A^c is a dense set.

For the converse, assume that the complement of every meager set is dense, and let V be an open meager set. Then V^c is dense. So, if V is nonempty, then $V \cap V^c \neq \emptyset$, which is impossible. Thus, the empty set is the only open meager set, and hence, X is a Baire space.

Problem 6.15. *A subset of a metric space is called **co-meager** if its complement is a meager set. For a subset A of a Baire space show that:*

a. *A is co-meager if and only if it contains a dense G_δ-set.*

b. *A is meager if and only if it is contained in an F_σ-set whose complement is dense.*

Solution. Notice that if A is a nowhere dense set in a metric space X, then from Lemma 6.8 we see that

$$
\emptyset = (\overline{A})^{\circ} = (\overline{A})^{c-c} = ([(\overline{A})^c]^{-})^c.
$$

This implies that a subset A is nowhere dense if and only if the open set $(\overline{A})^c$ is dense.

Now, assume that X is a Baire space and let A be a subset of X.

(a) Suppose first that A is a co-meager set. Then there exists a sequence $\{A_n\}$ of nowhere dense sets such that $A = \left(\bigcup_{n=1}^{\infty} A_n \right)^c$. This implies

$$A^c = \bigcap_{n=1}^{\infty} A_n \subseteq \bigcap_{n=1}^{\infty} \overline{A_n}. \qquad (\star)$$

By the above discussion, each set $(\overline{A_n})^c$ is an open dense set, and since X is a Baire space, the G_δ-set $E = \bigcap_{n=1}^{\infty} (\overline{A_n})^c$ is also dense (see Theorem 6.16). Now, a glance at $(\star)$ shows that $E \subseteq A$.

For the converse, assume that A contains a dense G_δ-set B, i.e., $B \subseteq A$. So, there exists a sequence $\{V_n\}$ of open sets such that $B = \bigcap_{n=1}^{\infty} V_n$. From $B \subseteq V_n$, we see that each V_n is also dense. This implies

$$[(V_n)^c]^\circ = [(V_n)^c]^{c-c} = (\overline{V_n})^c = X^c = \emptyset,$$

and so each $(V_n)^c$ is nowhere dense closed set. Now, use the inclusion

$$A^c \subseteq B^c = \bigcup_{n=1}^{\infty} (V_n)^c$$

to conclude that A^c is a meager set, i.e., A is a co-meager set.

(b) Assume first that A is a meager set, i.e., A^c is a co-meager set. By part (a), there exists a dense G_δ-set E such that $E \subseteq A^c$. This implies $A \subseteq E^c$, where now E is an F_σ-set whose complement $(E^c)^c = E$ is dense.

For the converse, assume that $A \subseteq F$ holds, where F is an F_σ-set with dense complement. It follows that $F^c \subseteq A^c$, where now F^c is a dense G_δ-set. By part (a), A^c is co-meager set, which means that A is a meager set.

7. COMPACTNESS IN METRIC SPACES

Problem 7.1. *Let $f: (X, d) \to (Y, \rho)$ be a function. Show that f is continuous if and only if f restricted to the compact subsets of X is continuous.*

Solution. Assume that f is continuous on every compact set. Let $x_n \to x$. Then the set $A = \{x_1, x_2, \ldots\} \cup \{x\}$ is compact (note that every open cover of A can be reduced to a finite cover), and $x_n \to x$ holds in A. Since f restricted to A is continuous, $\lim f(x_n) = f(x)$ holds, which shows that f is continuous.

Problem 7.2. *A metric space is said to be* **separable** *if it contains a countable subset that is dense in the space. Show that every compact space* (X, d) *is separable.*

Solution. For each n choose a finite subset F_n of X such that $X = \bigcup_{x \in F_n} B(x, \frac{1}{n})$. Let $F = \bigcup_{n=1}^{\infty} F_n$, and note that F is at-most countable.

Now, let $x \in X$ and $r > 0$. Pick some n with $\frac{1}{n} < r$. Then there exists some $y \in F_n$ with $d(x, y) < \frac{1}{n} < r$. Thus, $B(x, r) \cap F \neq \emptyset$, and so F is dense in X.

Problem 7.3. *Show that if* (X, d) *is a separable metric space (see the preceding exercise for the definition), then* card $X \leq \mathfrak{c}$.

Solution. Let $\{x_1, x_2, \ldots\}$ be a countable dense subset of X. Consider the collection of open balls $\{B(x_i, \frac{1}{j}): i, j = 1, 2, \ldots\}$. Clearly, this collection is countable; let $\{B_1, B_2, \ldots\}$ be one of its enumerations. Now, for each $x \in X$ define the set $S_x = \{n \in \mathbb{N}: x \in B_n\}$. Thus, a mapping $x \longmapsto S_x$ from X into $\mathcal{P}(\mathbb{N})$ has been established that is clearly one-to-one. Consequently, card $X \leq$ card $\mathcal{P}(\mathbb{N}) = \mathfrak{c}$. (See also Problem 5.6.)

Problem 7.4. *Let* $(X_1, d_1), \ldots, (X_n, d_n)$ *be arbitrary metric spaces, and let* $X = X_1 \times \cdots \times X_n$. *If* $x = (x_1, \ldots, x_n)$ *and* $y = (y_1, \ldots, y_n)$, *define*

$$D_1(x, y) = \sum_{m=1}^{n} d_m(x_m, y_m) \quad and \quad D_2(x, y) = \left(\sum_{m=1}^{n} [d_m(x_m, y_m)]^2 \right)^{\frac{1}{2}}.$$

    a. *Show that* D_1 *and* D_2 *are distances on* X.
    b. *Show that* D_1 *is equivalent to* D_2.
    c. *Show that* (X, D_1) *is complete if and only if each* (X_i, d_i) *is complete.*
    d. *Show that* (X, D_1) *is compact if and only if each* (X_i, d_i) *is compact.*

Solution. (a) Routine.
    (b) Use the inequalities

$$\tfrac{1}{n} D_1(x, y) \leq D_2(x, y) \leq n D_1(x, y).$$

    (c) Assume that each X_m $(m = 1, \ldots, n)$ is a complete metric space. Let $\{x_k\}$ be a D_1-Cauchy sequence of X, where $x_k = (x_1^k, \ldots, x_n^k)$. Clearly, each $\{x_m^k\}$ is a Cauchy sequence of X_m, and thus there exists $x_m \in X_m$ such that $\lim_{k \to \infty} d_m(x_m^k, x_m) = 0$. Hence, if $x = (x_1, \ldots, x_n) \in X$, then we have $\lim_{k \to \infty} D_1(x_k, x) = 0$, so that the metric space X is D_1-complete.

Now, let X be D_1-complete. Fix an element $(y_1, \ldots, y_n) \in X$. Let $\{x_m^k\}$ be a Cauchy sequence of X_m. If $x_k \in X$ is the element whose j^{th} component equals y_j for $j \neq m$ and equals x_m^k if $j = m$, then $\{x_k\}$ is a Cauchy sequence of X. If $x \in X$ is its limit, then it is easy to see that $\lim_{k \to \infty} d_m(x_m^k, x_m) = 0$, so that each X_m is complete.

(d) Assume first that each X_m is compact. Then following the proof of the second part of Theorem 7.4, we can see that every sequence of X has a convergent subsequence, and so X must be a compact metric space.

On the other hand, if X is a compact metric space, then the function $f_m \colon X \longrightarrow X_m$, defined by $f_m(x_1, \ldots, x_n) = x_m$, is continuous and onto for each $1 \leq m \leq n$. Hence, by Theorem 7.5, each $X_m = f_m(X)$ is compact.

Problem 7.5. *Let $\{(X_n, d_n)\}$ be a sequence of metric spaces, and let $X = \prod_{n=1}^{\infty} X_n$. For each $x = \{x_n\}$ and $y = \{y_n\}$ in X, define*

$$d(x, y) = \sum_{n=1}^{\infty} \frac{1}{2^n} \cdot \frac{d_n(x_n, y_n)}{1 + d_n(x_n, y_n)}.$$

a. *Show that d is a distance on X.*
b. *Show that (X, d) is a complete metric space if and only if each (X_n, d_n) is complete.*
c. *Show that (X, d) is a compact metric space if and only if each (X_n, d_n) is compact.*

Solution. (a) Note first that if d is a distance on a set X, then $\rho(x, y) = \frac{d(x,y)}{1+d(x,y)}$ is likewise a distance on X, which is equivalent to d. From this observation it easily follows that

$$d(x, y) = \sum_{n=1}^{\infty} \frac{1}{2^n} \cdot \frac{d_n(x_n, y_n)}{1 + d_n(x_n, y_n)}$$

is a distance on $X = \prod_{n=1}^{\infty} X_n$.

(b) Let $\{x^k\}$ be a sequence of X, where $x^k = (x_1^k, x_2^k, \ldots)$. The proof follows from the following two properties (whose verifications are straightforward).

1. $x^k \longrightarrow x$ holds in X if and only if $x_i^k \longrightarrow x_i$ holds in X_i for each i; and
2. $\{x^k\}$ is a Cauchy sequence in X if and only if $\{x_i^k\}$ is a Cauchy sequence in X_i for each i.

(c) Assume that (X, d) is a compact metric space. Then the function $f_i \colon X \longrightarrow X_i$, defined by $f(x) = x_i$ for each $x = (x_1, x_2, \ldots) \in X$, is continuous and onto. By Theorem 7.5, each X_i is a compact metric space.

For the converse, assume that each X_i is a compact metric space. By (b), X is a complete metric space, and so by Theorem 7.8 it suffices to show that X is totally bounded. To this end, let $\varepsilon > 0$. Choose n such that $2^{-n} < \varepsilon$, and note that

$$\rho_n(x, y) = \sum_{i=1}^{n} \frac{1}{2^i} \cdot \frac{d_i(x_i, y_i)}{1 + d_i(x_i, y_i)}$$

defines a distance on $\prod_{i=1}^{n} X_i$. It should be clear that ρ_n is equivalent to the distances of the preceding problem, and $\left(\prod_{i=1}^{n} X_i, \rho_n\right)$ is a compact metric space. Choose a finite subset F of $\prod_{i=1}^{n} X_i$ such that the ρ_n-balls with centers at the points of F and radii ε cover $\prod_{i=1}^{n} X_i$. Next, extend each $x \in F$ to an element of X (i.e., add to each $x \in F$ arbitrary components $x_{n+1}, x_{n+2}, \ldots$). Now, if $y = (y_1, y_2, \ldots) \in X$, then pick some $x \in F$ with $\rho_n(x, y) < \varepsilon$, and note that

$$d(x, y) = \rho_n(x, y) + \sum_{i=n+1}^{\infty} \frac{1}{2^i} \cdot \frac{d_i(x_i, y_i)}{1 + d_i(x_i, y_i)} < \varepsilon + 2^{-n} < \varepsilon + \varepsilon = 2\varepsilon.$$

Thus, $X = \bigcup_{x \in F} B(x, 2\varepsilon)$ holds, and therefore, X is totally bounded.

Problem 7.6. *A family of set $\mathcal{F}$ is said to have the* **finite intersection property** *if every finite intersection of sets of $\mathcal{F}$ is nonempty. Show that a metric space is compact if and only if every family of closed sets with the finite intersection property has a nonempty intersection.*

Solution. Let X be compact, and let $\{A_i : i \in I\}$ be a family of closed sets with the finite intersection property. If $\bigcap_{i \in I} A_i = \emptyset$, then $X = \bigcup_{i \in I} A_i^c$ holds, and by the compactness of X, there exist $i_1, \ldots, i_n \in I$ such that $X = \bigcup_{j=1}^{n} A_{i_j}^c$. Thus, $\bigcap_{j=1}^{n} A_{i_j} = \emptyset$, contrary to our hypothesis. Hence, $\bigcap_{i \in I} A_i \neq \emptyset$.

For the converse, assume that every family of closed sets with the finite intersection property has a nonempty intersection. Let $X = \bigcup_{i \in I} V_i$ be an open cover. Then $\bigcap_{i \in I} V_i^c = \emptyset$, and since $\{V_i^c : i \in I\}$ is a family of closed sets, our hypothesis guarantees the existence of a finite number of indices $i_1, \ldots, i_n$ such that $\bigcap_{j=1}^{n} V_{i_j}^c = \emptyset$. Thus, $X = \bigcup_{j=1}^{n} V_{i_j}$ holds, so that X is a compact metric space.

Problem 7.7. *Let $f : X \to X$ be a function from a set X into itself. A point $a \in X$ is called a* **fixed point** *for f if $f(a) = a$.*

Assume that (X, d) is a compact metric space and $f : X \to X$ satisfies the inequality $d(f(x), f(y)) < d(x, y)$ for $x \neq y$. Show that f has a unique fixed point.

Solution. Note first that f has at most one fixed point. Indeed, if $f(x) = x$ and $f(y) = y$ hold with $x \neq y$, then

$$d(x, y) = d(f(x), f(y)) < d(x, y)$$

must hold, which is absurd.

Now, define the function $g: X \to \mathbf{R}$ by $g(x) = d(x, f(x))$. From the inequality

$$\left|g(x) - g(y)\right| \leq d(f(x), f(y)) + d(x, y) \leq 2d(x, y)$$

(see the discussion preceding Theorem 6.19), it follows that g is continuous. Since X is compact, g attains its minimum at some point $a \in X$. If $f(a) \neq a$, then the inequality

$$g(f(a)) = d(f(a), f(f(a))) < d(a, f(a)) = g(a)$$

shows that g does not attain a minimum at a. Thus, $f(a) = a$ must hold, and so a is a (unique) fixed point for f.

Problem 7.8. *Let (X, d) be a metric space. A function $f: X \to X$ is called a* **contraction** *if there exists some $0 < \alpha < 1$ such that $d(f(x), f(y)) \leq \alpha d(x, y)$ for all $x, y \in X$; α is called a* **contraction constant**.

Show that every contraction f on a complete metric space (X, d) has a unique fixed point; that is, show that there exists a unique point $x \in X$ such that $f(x) = x$.

Solution. Note first that if $f(x) = x$ and $f(y) = y$ hold, then the inequality $d(x, y) = d(f(x), f(y)) \leq \alpha d(x, y)$ easily implies that $d(x, y) = 0$, and so $x = y$. That is, f has at-most one fixed point.

To see that f has a fixed point, choose some $a \in X$, and then define the sequence $\{x_n\}$ inductively by

$$x_1 = a \quad \text{and} \quad x_{n+1} = f(x_n) \text{ for } n = 1, 2, \ldots .$$

From our condition, it follows that

$$d(x_{n+1}, x_n) = d(f(x_n), f(x_{n-1})) \leq \alpha d(x_n, x_{n-1})$$

holds for $n = 2, 3, \ldots$. Thus, as in Problem 4.15, we can show that $\{x_n\}$ is a Cauchy sequence. Since X is complete, $\{x_n\}$ is a convergent sequence. Let $x = \lim x_n$. Now, by observing that f is (uniformly) continuous, we obtain that

$$x = \lim_{n \to \infty} x_{n+1} = \lim_{n \to \infty} f(x_n) = f(x),$$

and so x is a (unique) fixed point for f.

Problem 7.9. *A property of a metric space is called a* **topological property** *if it is preserved in a homeomorphic metric space.*

   a. *Show that compactness is a topological property.*

   b. *Show that completeness, boundedness, and total boundedness are not topological properties.*

Solution. (a) It follows from Theorem 7.5.

(b) Consider $(0, 1]$ and $[1, \infty)$ as metric spaces under the usual Euclidean distance $d(x, y) = |x - y|$. Clearly, $(0, 1]$ is not complete but it is bounded and totally bounded. Also, $[1, \infty)$ is complete (because it is a closed subset of $\mathbb{R}$), but is neither bounded nor totally bounded. On the other hand, $f : (0, 1] \to [1, \infty)$, defined by $f(x) = \frac{1}{x}$, is a homeomorphism, and the claims in (b) follow.

Problem 7.10. *Let (X, d) be a metric space. Define the distance of two nonempty subsets A and B of X by*

$$d(A, B) = \inf\{d(x, y) \colon x \in A \text{ and } y \in B\}.$$

   a. *Give an example of two closed sets A and B of some metric space with $A \cap B = \emptyset$ and such that $d(A, B) = 0$.*

   b. *If $A \cap B = \emptyset$, A is closed, and B is compact (and, of course, both are nonempty), then show that $d(A, B) > 0$.*

Solution. (a) Let $X = \mathbb{R}^2$ with the Euclidean distance, and consider the closed subsets of X

$$A = \left\{\left(x, \tfrac{1}{x}\right) \colon x \geq 1\right\} \quad \text{and} \quad B = \{(x, 0) \colon x \geq 1\}.$$

Note that $A \cap B = \emptyset$, while $d(A, B) = 0$.

(b) Let A and B be as stated in the problem. If $d(A, B) = 0$, then pick two sequences $\{x_n\} \subseteq A$ and $\{y_n\} \subseteq B$ with $d(x_n, y_n) \longrightarrow 0$. Since B is compact, by passing to a subsequence (if necessary), we can assume that $y_n \longrightarrow y$ holds for some $y \in B$. The inequality

$$d(x_n, y) \leq d(x_n, y_n) + d(y_n, y)$$

shows that $d(x_n, y) \longrightarrow 0$. Since A is closed, $y \in A$, and hence $A \cap B \neq \emptyset$, contrary to our hypothesis. Therefore, $d(A, B) > 0$ must hold.

Problem 7.11. *Let (X, d) be a compact metric space and $f : X \to X$ an isometry; that is, $d(f(x), f(y)) = d(x, y)$ holds for all $x, y \in X$. Then show that f is onto. Does the conclusion remain true if X is not assumed to be compact?*

Solution. Let $y \in X$. Define the sequence $\{x_n\}$ of $f(X)$ by

$$x_1 = f(y) \quad \text{and} \quad x_{n+1} = f(x_n) \text{ for } n = 1, 2, \dots .$$

Note that $d(x_n, x_{n+p}) = d(y, x_p)$ holds for all n and all p. Since $f(X)$ is compact, $\{x_n\}$ must have a limit point in $f(X)$. Let a be a limit point of $\{x_n\}$.

Now, let $\varepsilon > 0$. Pick $n > 1$ and p such that $d(x_n, a) < \varepsilon$ and $d(x_{n+p}, a) < \varepsilon$. Then

$$d\big(y, f(X)\big) \le d(y, x_p) = d(x_n, x_{n+p}) \le d(x_n, a) + d(x_{n+p}, a) < 2\varepsilon$$

holds for all $\varepsilon > 0$, and so $d\big(y, f(X)\big) = 0$. Thus, $y \in \overline{f(X)} = f(X)$, so that $f(X) = X$ holds.

If X is not supposed to be compact, then the conclusion is no longer true. A counterexample: Take $X = \mathbb{N}$ with $d(n, m) = |n - m|$ and consider the function $f: \mathbb{N} \to \mathbb{N}$ defined by $f(n) = n + 1$.

Problem 7.12. *Show that a metric space X is compact if and only if every continuous real-valued function on X attains its maximum value.*

Solution. Let (X, d) be a metric space. Assume that X is compact and that $f: X \to \mathbb{R}$ is a continuous function. By Theorem 7.5, we know that $f(X)$ is a compact subset of $\mathbb{R}$, and so (by Theorem 7.4) $f(X)$ is closed and bounded. The maximum of $f(X)$ is the maximum value of f on X.

For the converse, assume that every continuous real-valued function on X attains a maximum value. Clearly, every continuous real-valued function on X attains also a minimum value.

We shall establish first that X is a complete metric space. Let $(\hat{X}, \hat{d})$ denote the completion of (X, d) and let $\hat{x} \in \hat{X}$. The function $f: X \longrightarrow \mathbb{R}$, defined by $f(x) = \hat{d}(\hat{x}, x)$, satisfies $\inf\{f(x): x \in X\} = 0$. So, there exists some $x_0 \in X$ satisfying $f(x_0) = \hat{d}(\hat{x}, x_0) = 0$. It follows that $\hat{x} = x_0 \in X$ and so $\hat{X} = X$. This means that X is a complete metric space.

Next, we shall show that X is totally bounded. To establish this, assume by way of contradiction that X is not totally bounded. Then an easy inductive argument shows that there exist some $\varepsilon > 0$ and a sequence $\{x_n\}$ of X such that $d(x_n, x_m) \ge 3\varepsilon$ holds for $n \ne m$. For each n consider the nonempty closed set

$$C_n = [B(x_n, \varepsilon)]^c = \{x \in X: d(x, x_n) \ge \varepsilon\},$$

and then define the function $f_n: X \to \mathbb{R}$ by

$$f_n(x) = d(x, C_n) = \inf\{d(x, y): y \in C_n\}.$$

So, f_n is a bounded function, $f_n(x) = 0$ holds for each $x \in C_n$ and $f_n(x_n) > 0$. Multiplying by a constant c_n, we can assume that $\sup\{f_n(x): x \in X\} > n$ holds for each n. Now, define the function $f: X \longrightarrow \mathbf{R}$ by

$$f(x) = \begin{cases} f_n(x), & \text{if } x \in B(x_n, \varepsilon) \\ 0, & \text{if } x \notin \bigcup_{n=1}^{\infty} B(x_n, \varepsilon), \end{cases}$$

and we claim that f is a continuous function. Clearly, f is continuous at the points of the balls $B(x_n, \varepsilon)$. If $x_0 \notin \bigcup_{n=1}^{\infty} B(x_n, \varepsilon)$, note that $B(x_0, \frac{\varepsilon}{2}) \cap B(x_n, \varepsilon) \neq \emptyset$ holds for at-most one n (why?). If $B(x_0, \frac{\varepsilon}{2}) \cap B(x_n, \varepsilon) = \emptyset$ for each n, then $f(x) = 0$ for each x in $B(x_0, \frac{\varepsilon}{2})$, and so f is continuous at x_0. Thus, we can assume that $B(x_0, \frac{\varepsilon}{2}) \cap B(x_n, \varepsilon) \neq \emptyset$ for some n. We distinguish two cases.

CASE I: $d(x_0, x_n) > \varepsilon$.

In this case, there exists some $0 < r < \frac{\varepsilon}{2}$ such that $B(x_0, r) \cap B(x_n, \varepsilon) = \emptyset$. Clearly, $f(x) = 0$ holds for each $x \in B(x_0, r)$, and from this we see that f is continuous at x_0.

CASE II: $d(x_0, x_n) = \varepsilon$.

Let $\{z_k\}$ be a sequence of X satisfying $z_k \longrightarrow x_0$; we can assume that z_k belongs to $B(x_0, \frac{\varepsilon}{2})$ for each k. Note that if $z_k \notin B(x_n, \varepsilon)$, then $f(z_k) = 0$. On the other hand, if $z_k \in B(x_n, \varepsilon)$, then

$$0 \leq f(z_k) = c_n d(z_k, C_n) \leq c_n d(z_k, x_0).$$

Thus, $0 \leq f(z_k) \leq c_n d(z_k, x_0)$ holds for each k. In view of

$$\lim_{k \to \infty} d(z_k, x_0) = 0,$$

we see that $\lim f(z_k) = 0 = f(x_0)$ and so f is continuous at x_0 in this case too.

To contradict our hypothesis, note that f does not attain a maximum value. Thus, X must also be totally bounded. By Theorem 7.8, we see that X is a compact metric
space.

Problem 7.13. *This exercise presents a converse of Theorem 7.7. Assume that (X, d) is a metric space such that every real-valued continuous function on X is uniformly continuous.*

   a. *Show that X is a complete metric space.*
   b. *Give an example of a noncompact metric space with the above property.*
   c. *If X has a finite number of isolated points (an element $a \in X$ is said to be an **isolated point** whenever there exists some positive $r > 0$ such that $B(a, r) \cap (X \setminus \{a\}) = \emptyset$), then show that X is a compact metric space.*

Solution. Let (X, d) be a metric space such that every continuous real-valued function on X is uniformly continuous.

(a) If $\hat{x} \in \hat{X}$ (the completion of X) is an element that does not belong to X, then the function $f: X \to \mathbf{R}$ defined by $f(x) = \frac{1}{d(\hat{x},x)}$, $x \in X$, is a continuous real-valued function on X that fails to be uniformly continuous (why?), a contradiction. Hence, $\hat{X} = X$ holds, which means that X is a complete metric space.

(b) Let $X = \{1, 2, \ldots\}$ equipped with the discrete distance d. Then every set is open and so every real-valued function f on X is continuous. Since $d(x, y) < 1$ implies $x = y$ (and so $f(x) - f(y) = 0$), we see that every real-valued function on X is uniformly continuous. Now, note that X is not a compact metric space.

(c) In view of (a), we need to establish that X is totally bounded. To this end, assume that X is not totally bounded. Then, there exist some $\varepsilon > 0$ and a sequence of elements $\{x_n\}$ of X such that $d(x_n, x_m) > 3\varepsilon$ for $n \neq m$. From our hypothesis, we can suppose that each x_n is an accumulation point of X. For each n pick an element y_n such that $0 < d(x_n, y_n) < \frac{\varepsilon}{n}$ and let $r_n = d(x_n, y_n)$. Put

$$C_n = \{x \in X: d(x, x_n) \geq r_n\}$$

and define the functions f_n and f as in the solution of Problem 7.12 (the open ball $B(x_n, \varepsilon)$ is now replaced by $B(x_n, r_n)$). Then f is a continuous function and satisfies $f(y_n) = 0$ for each n. Pick $z_n \in B(x_n, r_n)$ such that $f(x_n) > n$, and note that

$$\left| f(y_n) - f(z_n) \right| > n \quad \text{and} \quad \lim_{n \to \infty} d(y_n, z_n) = 0.$$

This shows that the continuous function f is not uniformly continuous, contrary to our hypothesis. Hence, X is totally bounded, as desired.

Problem 7.14. *Consider a function $f: (X, d) \to (Y, \rho)$ between two metric spaces. The graph G of f is the subset of $X \times Y$ defined by*

$$G = \{(x, y) \in X \times Y: y = f(x)\}.$$

If (Y, ρ) is a compact metric space, then show that f is continuous if and only if G is a closed subset of $X \times Y$, where $X \times Y$ is considered to be a metric space under the distance $D((x, y), (u, v)) = d(x, u) + \rho(y, v)$; see Problem 7.4. Does the result hold true if (Y, ρ) is not assumed to be compact?

Solution. Observe that an arbitrary sequence $\{(x_n, y_n)\}$ of $X \times Y$ satisfies $(x_n, y_n) \to (x, y)$ in $X \times Y$ if and only if $x_n \to x$ and $y_n \to y$ both hold.

Assume (Y, ρ) compact and G closed. If f is not continuous, then there exists a sequence $\{x_n\}$ of X and some $\varepsilon > 0$ such that $x_n \to x$ and $\rho\big(f(x_n), f(x)\big) \geq \varepsilon$ for all n (why?). Since (Y, ρ) is compact, by passing to a subsequence, we can assume that $f(y_n) \to y$ holds in Y. Now, observe that $\big(x_n, f(x_n)\big) \in G$ holds for each n and $\big(x_n, f(x_n)\big) \to (x, y)$ holds in $X \times Y$. Since G is closed, it follows that $(x, y) \in G$ and so $y = f(x)$. This implies

$$\rho\big(f(x_n), f(x)\big) \to \rho\big(f(x), f(x)\big) = 0,$$

which contradicts $\rho\big(f(x_n), f(x)\big) \geq \varepsilon$ for all n. Hence, f is a continuous function.

If (Y, ρ) is not compact, then a function with closed graph need not be continuous. For an example, consider the function $f : \mathbf{R} \to \mathbf{R}$ defined by

$$f(x) = \begin{cases} \frac{1}{x} & \text{if } x \neq 0; \\ 0 & \text{if } x = 0 \ . \end{cases}$$

Problem 7.15. *A cover $\{V_i\}_{i \in I}$ of a set X is said to be a* **pointwise finite cover** *whenever each $x \in X$ belongs at-most to a finite number of the V_i.*

Show that a metric space is compact if and only if every pointwise finite open cover of the space contains a finite subcover.

Solution. Clearly, if X is compact, then every pointwise finite open cover of X contains a finite subcover. For the converse, assume that every pointwise finite open cover of X contains a finite subcover. To establish that the metric space X is compact, it suffices to show that every sequence in X contains a convergence subsequence.

Let $\{x_n\}$ be a sequence in X. We can suppose (why?) that the sequence consists of distinct elements. Suppose by way of contradiction that $\{x_n\}$ has no convergence subsequence. Then x_1 is not in the closure of the set $\{x_n : n \neq 1\}$ and thus, there exists an open ball $V_1 = B(x_1, \delta_1)$ about x_1 with radius $0 < \delta_1 < 1$ and satisfying $x_n \notin V_1$ for all $n \neq 1$. Also, x_2 is not in the closure of the set $\{x_n : n \neq 2\}$ and thus, there exists an open ball $V_2 = B(x_2, \delta_2)$ about x_2 with radius $0 < \delta_2 < \frac{1}{2}$ and such that $x_n \notin V_2$ for all $n \neq 2$. Proceeding inductively, we see that for each k there exists an open ball $V_k = B(x_k, \delta_k)$ with radius $0 < \delta_k < \frac{1}{2^k}$ satisfying $x_n \notin V_k$ for all $n \neq k$.

Since the set $F = \{x_1, x_2, \ldots\}$ contains no convergent subsequences, the set F must contain all of its closure points. Thus, F is a closed set, and hence, the set $G = X \setminus F$ is an open set. Then, the collection $\mathcal{C} = \{G, V_1, V_2, \ldots\}$ is an open cover of X. In fact, the collection $\mathcal{C}$ is a pointwise finite open cover of X because if a point x belongs to an infinite number of sets in $\mathcal{C}$, then x belongs to an infinite number of the sets V_n. However, this would imply that a subsequence of $\{x_n\}$ converges to the point x. Since the sequence $\{x_n\}$ contains no convergent subsequences, we infer that $\mathcal{C}$ is a pointwise finite open cover.

Therefore, $\mathcal{C}$ contains a finite subcover of X, say $V_1, \ldots, V_m, G$. Since G does not intersect $\{x_1, x_2, \ldots\}$, it follows that $\{x_1, x_2, \ldots\} \subseteq \bigcup_{i=1}^{m} V_i$. However, this contradicts the fact $x_n \notin V_k$ for $n \neq k$. Conclusion: The sequence $\{x_n\}$ must have a convergent subsequence—and hence, the metric space X is compact.

CHAPTER 2

TOPOLOGY AND CONTINUITY

8. TOPOLOGICAL SPACES

Problem 8.1. *For any subset A of a topological space show the following:*

a. $A^\circ = (\overline{A^c})^c$.
b. $\partial A = \overline{A} \setminus A^\circ$.
c. $(A \setminus A^\circ)^\circ = \emptyset$.

Solution. (a) Note that

$$x \in A^\circ \iff \text{there exists a neighborhood } V \text{ of } x \text{ with } V \subseteq A$$
$$\iff \text{there exists a neighborhood } V \text{ of } x \text{ with } V \cap A^c = \emptyset$$
$$\iff x \notin \overline{A^c} \iff x \in (\overline{A^c})^c.$$

(b) Using (a), we see that $\partial A = \overline{A} \cap \overline{A^c} = \overline{A} \setminus (\overline{A^c})^c = \overline{A} \setminus A^\circ$.
(c) If $x \in (A \setminus A^\circ)^\circ$, then for some open set V we have

$$x \in V \subseteq A \setminus A^\circ \subseteq A.$$

This implies $x \in A \setminus A^\circ$ and $x \in A^\circ$, a contradiction. Hence, $(A \setminus A^\circ)^\circ = \emptyset$.

Problem 8.2. *If A and B are two arbitrary subsets of a topological space, then show the following:*

a. $\overline{A \cup B} = \overline{A} \cup \overline{B}$.
b. $(A \cup B)' = A' \cup B'$.

Solution. (a) See Problem 6.1.
(b) Clearly, $A \subseteq B$ implies $A' \subseteq B'$, and so $A' \cup B' \subseteq (A \cup B)'$. For the reverse inclusion, let $x \in (A \cup B)'$. If $x \notin A' \cup B'$, then there exist two neighborhoods

V and W of x such that

$$A \cap (V \setminus \{x\}) = B \cap (W \setminus \{x\}) = \emptyset.$$

Now, note that the neighborhood $U = V \cap W$ of the point x satisfies

$$(A \cup B) \cap (U \setminus \{x\}) = \emptyset,$$

proving that $x \notin (A \cup B)'$, a contradiction.

Problem 8.3. *If A is an arbitrary subset of a Hausdorff topological space, then show that its derived set A' is a closed set.*

Solution. Let A be an arbitrary subset of a Hausdorff topological space X. We shall establish that $(A')^c$ is an open set (and this will guarantee that A' is a closed set). To this end, let $x \in (A')^c$, i.e., let $x \notin A'$. This means that there exists a neighborhood V of x such that

$$V \cap (A \setminus \{x\}) = \emptyset. \tag{$\star$}$$

We claim that $V \subseteq (A')^c$ holds. To see this, let $y \in V$ with $y \neq x$. Since X is a Hausdorff topological space, there exist neighborhoods U and W of y and x, respectively, such that $U \cap W = \emptyset$. Now, note that $V \cap U$ is a neighborhood of y with $x \notin V \cap U$ and so from $(\star)$, we see that $(V \cap U) \cap A = \emptyset$. The latter shows that $y \notin A'$. Hence, $V \subseteq (A')^c$ holds proving that every point of $(A')^c$ is an interior point, as desired.

Problem 8.4. *Let $X = \mathbf{R}$, and let τ be the topology on X defined in Example 8.4. In other words, $A \in \tau$ if and only if for each $x \in A$ there exist $\epsilon > 0$ and an at-most countable set B (both depending on x) such that $(x - \epsilon, x + \epsilon) \setminus B \subseteq A$.*

    a. *Show that τ is a topology on X.*
    b. *Verify that $0 \in \overline{(0, 1)}$.*
    c. *Show that there is no sequence $\{x_n\}$ of $(0, 1)$ with $\lim x_n = 0$.*

Solution. (a) Straightforward.

    (b) Since for each $\varepsilon > 0$ and each countable set B the set $(-\varepsilon, \varepsilon) \setminus B$ is uncountable, we must have $\big((-\varepsilon, \varepsilon) \setminus B\big) \cap (0, 1) \neq \emptyset$. This easily implies that $0 \in \overline{(0, 1)}$.

    (c) If $\{x_n\}$ is a sequence of $(0, 1)$, then $V = (-1, 1) \setminus \{x_1, x_2, \ldots\}$ is a neighborhood of zero, and $x_n \notin V$ for all n. This shows that no sequence of $(0, 1)$ can converge to 0.

Problem 8.5. *If A is a dense subset of a topological space, then show that $\mathcal{O} \subseteq \overline{A \cap \mathcal{O}}$ holds for every open set $\mathcal{O}$. Generalize this conclusion as follows: If A is open, then $A \cap \overline{B} \subseteq \overline{A \cap B}$ for each set B.*

Solution. Let $x \in \mathcal{O}$ and let V be a neighborhood of x. Since $\mathcal{O}$ is open, $V \cap \mathcal{O}$ is a neighborhood of x, and so the denseness of A implies

$$V \cap (A \cap \mathcal{O}) = (V \cap \mathcal{O}) \cap A \neq \emptyset,$$

which means that $x \in \overline{A \cap \mathcal{O}}$.

For the general case, assume A is an open set and let $x \in A \cap \overline{B}$. If V is a neighborhood of x, then $V \cap A$ is also a neighborhood of x. Since $x \in \overline{B}$, it follows that $V \cap (A \cap B) = (V \cap A) \cap B \neq \emptyset$. This shows that $x \in \overline{A \cap B}$, and hence, $A \cap \overline{B} \subseteq \overline{A \cap B}$.

Problem 8.6. *If $\{\mathcal{O}_i\}_{i \in I}$ is an open cover for a topological space X, then show that a subset A of X is closed if and only if $A \cap \mathcal{O}_i$ is closed in $\mathcal{O}_i$ for each $i \in I$ (where $\mathcal{O}_i$ is considered equipped with the relative topology).*

Solution. If A is closed, then clearly $A \cap \mathcal{O}_i$ is closed in $\mathcal{O}_i$ for each i. For the converse, assume that $A \cap \mathcal{O}_i$ is closed in $\mathcal{O}_i$ for each i. Put

$$V_i = \mathcal{O}_i \setminus A \cap \mathcal{O}_i = \mathcal{O}_i \setminus A,$$

and note that—by our hypothesis—each V_i is open in $\mathcal{O}_i$. Since each $\mathcal{O}_i$ is an open subset of X, it follows that each V_i is likewise an open subset of X. Now, note that

$$A^c = X \setminus A = \left(\bigcup_{i \in I} \mathcal{O}_i \right) \setminus A = \bigcup_{i \in I} (\mathcal{O}_i \setminus A) = \bigcup_{i \in I} V_i$$

is an open subset of X, and so A is a closed set.

Problem 8.7. *If (X, τ) is a Hausdorff topological space, then show the following:*

a. *Every finite subset of X is closed.*
b. *Every sequence of X converges to at-most one point.*

Solution. (a) Let $A = \{x\}$ be a one-point set. If $y \notin A$, then (since X is a Hausdorff space) there exists a neighborhood V of y with $x \notin V$, and so $V \subseteq A^c$. Thus, A^c is open, and hence A, is closed. Now, observe that every finite set is a finite union of one-point sets.

(b) If $x \neq y$, then there exist neighborhoods V_x and V_y of x and y respectively, such that $V_x \cap V_y = \emptyset$. Now, a sequence of X cannot converge to x and y at the same time simply because its terms cannot be eventually in both V_x and V_y.

Problem 8.8.   *For a function $f : (X, \tau) \to (Y, \tau_1)$ show the following:*

  a.   *If τ is the discrete topology, then f is continuous.*
  b.   *If τ is the indiscrete topology and τ_1 is a Hausdorff topology, then f is continuous if and only if f is a constant function.*

Solution.   (a) Note that every subset of X is open. Thus, $f^{-1}(A)$ is an open set for every subset A of Y, and so f is continuous.

(b) Recall that the indiscrete topology is the topology $\tau = \{\emptyset, X\}$. If f is a constant function, then $f^{-1}(A)$ is either $\emptyset$ or X, and so f is continuous. For the converse, let f be a continuous function. If for some x, $y \in X$ we have $f(x) \neq f(y)$, then there exists a neighborhood V of $f(x)$ such that $f(y) \notin V$. Now note that $f^{-1}(V)$ is neither equal to $\emptyset$ nor equal to X, and so $f^{-1}(V)$ is not open, a contradiction. Thus, f must be a constant function.

Problem 8.9.   *Let f and g be two continuous functions from (X, τ) into a Hausdorff topological space (Y, τ_1). Assume that there exists a dense subset A of X such that $f(x) = g(x)$ for all $x \in A$. Show that $f(x) = g(x)$ holds for all $x \in X$.*

Solution.   Suppose that for some $x \in X$ we have $f(x) \neq g(x)$. Pick a neighborhood V of $f(x)$ and another W of $g(x)$ such that $V \cap W = \emptyset$. Since $f^{-1}(V) \cap g^{-1}(W)$ is a neighborhood of x and A is dense in X, there exists some $y \in f^{-1}(V) \cap g^{-1}(W) \cap A$. Now, note that $f(y) = g(y) \in V \cap W = \emptyset$ must hold, which is absurd. Thus, $f(x) = g(x)$ holds for each $x \in X$.

Problem 8.10.   *Let $f : (X, \tau) \to (Y, \tau_1)$ be a function. Show that f is continuous if and only if $f^{-1}(B^\circ) \subseteq [f^{-1}(B)]^\circ$ holds for every subset B of Y.*

Solution.   Repeat the solution of Problem 6.4.

Problem 8.11.   *If $f : (X, \tau) \to (Y, \tau_1)$ and $g : (Y, \tau_1) \to (Z, \tau_2)$ are continuous functions, show that their composition $g \circ f : (X, \tau) \to (Z, \tau_2)$ is also continuous.*

Solution.   Use the identity $(g \circ f)^{-1}(V) = f^{-1}\big(g^{-1}(V)\big)$. (See Problem 1.8.)

Problem 8.12.   *Let X be a topological space, let $a \in X$, and let $\mathcal{N}_a$ denote the collection of all neighborhoods at a. The* **oscillation** *of a function $f : X \to \mathbf{R}$ at the point a is the extended non-negative real number*

$$\omega_f(a) = \inf_{V \in \mathcal{N}_a} \Big\{ \sup_{x,y \in V} |f(x) - f(y)| \Big\}.$$

Establish the following properties regarding the oscillation:

a. *The function f is continuous at a if and only if $\omega_f(a) = 0$.*
b. *If X is an open interval of $\mathbf{R}$ and $f: X \to \mathbf{R}$ is a monotone function, then*
$$\omega_f(a) = \left| \lim_{x \to a^+} f(x) - \lim_{x \to a^-} f(x) \right|.$$

Solution. (a) Assume that f is continuous at a. Fix $\epsilon > 0$. Then there exists some $W \in \mathcal{N}_a$ (i.e., some neighborhood W of a) such that $x \in W$ implies $|f(a) - f(x)| < \epsilon$. So, if $x, y \in W$, then

$$|f(x) - f(y)| \leq |f(x) - f(a)| + |f(a) - f(y)| < \epsilon + \epsilon = 2\epsilon,$$

and thus

$$0 \leq \omega_f(a) \leq \sup_{x,y \in W} |f(x) - f(y)| \leq 2\epsilon$$

for each $\epsilon > 0$. This implies $\omega_f(a) = 0$.

For the converse, assume $\omega_f(a) = 0$. Let $\epsilon > 0$. Then from the definition of the oscillation, we see that there exists some neighborhood V of a such that $\sup_{x,y \in V} |f(x) - f(y)| < \epsilon$. In particular, we have $|f(x) - f(a)| < \epsilon$ for all $x \in V$, and this shows that f is continuous at a.

(b) We can assume that f is an increasing function. Note that we can consider neighborhoods of a of the form (c, d) with $a \in (c, d)$. Consider first a neighborhood (c, d) of a and assume that $x, y \in (c, d)$ satisfy $x < a < y$. Since f is increasing, it follows that $0 \leq \lim_{t \to a^+} f(t) - \lim_{t \to a^-} f(t) \leq f(y) - f(x)$, and from this, we infer that

$$\lim_{t \to a^+} f(t) - \lim_{t \to a^-} f(t) \leq \omega_f(a).$$

On the other hand, if $\epsilon > 0$ is given, then there exists some $\delta > 0$ such that the open interval $J = (a - \delta, a + \delta)$ satisfies $J \subseteq X$ and

$$\omega_f(a) \leq \sup_{x,y \in J} |f(x) - f(y)| < \left[\lim_{t \to a^+} f(t) - \lim_{t \to a^-} f(t) \right] + \epsilon.$$

This implies $\omega_f(a) \leq \lim_{t \to a^+} f(t) - \lim_{t \to a^-} f(t)$, and so

$$\omega_f(a) = \lim_{t \to a^+} f(t) - \lim_{t \to a^-} f(t)$$

holds true.

Problem 8.13. *Show that a finite union of nowhere dense sets is again a nowhere dense set. Is this statement true for a countable union of nowhere dense sets?*

Solution.   Let A and B be two nowhere dense sets. Using the identity $S^\circ = S^{c-c}$ (see Problem 8.1), we have

$$\left(\overline{A \cup B}\right)^\circ = \left(\overline{A} \cup \overline{B}\right)^{c-c} = \left(\overline{A}^{-c} \cap \overline{B}^{-c}\right)^{-c} \subseteq \left(A^{-c-} \cap B^{-c-}\right)^c$$
$$= A^{-c-c} \cup B^{-c-c} = \left(\overline{A}\right)^\circ \cup \left(\overline{B}\right)^\circ = \emptyset \cup \emptyset = \emptyset.$$

An easy induction argument can now complete the proof.

The countable union of nowhere dense sets need not be nowhere dense. An example: Take $X = \mathbf{R}$, and let $E_n = \{r_n\}$, where $\{r_1, r_2, \ldots\}$ is an enumeration of the rational numbers. Clearly, each E_n is nowhere dense, while $\bigcup_{n=1}^\infty E_n = \{r_1, r_2, \ldots\}$ is not nowhere dense.

Problem 8.14.   *Show that the boundary of an open or closed set is nowhere dense.*

Solution.   Repeat the solution of Problem 6.5.

Problem 8.15.   *Let $f:(X, \tau) \to \mathbf{R}$, and let D be the set of all points of X where f is discontinuous. If D^c is dense in X, then show that D is a meager set.*

Solution.   From $\overline{D^c} = X$, it follows that $D^\circ = \left(\overline{D^c}\right)^c = \emptyset$. Now, the proof can be completed by observing that D is an F_σ-set (Theorem 8.10).

Problem 8.16.   *Show that there is no function $f: \mathbf{R} \to \mathbf{R}$ having the irrational numbers as the set of its discontinuities.*

Solution.   Let I denote the set of all irrational numbers of $\mathbf{R}$. If I is the set of discontinuities of a function $f: \mathbf{R} \longrightarrow \mathbf{R}$, then (by Theorem 8.10) I is an F_σ-set. However, this is impossible by Problem 6.6.

Problem 8.17.   *Show that every closed subset of a metric space is a G_δ-set and every open set is an F_σ-set.*

Solution.   Let A be a nonempty closed subset of a metric space X. Then the function $f: X \longrightarrow \mathbf{R}$, defined by

$$f(x) = d(x, A) = \inf\{d(x, y): \ y \in A\},$$

is continuous (see the proof of Lemma 10.4) and satisfies

$$A = f^{-1}(\{0\}) = f^{-1}\left(\bigcap_{n=1}^\infty \left(-\tfrac{1}{n}, \tfrac{1}{n}\right)\right) = \bigcap_{n=1}^\infty f^{-1}\left(\left(-\tfrac{1}{n}, \tfrac{1}{n}\right)\right).$$

(See the discussion at the end of Section 6 of the text.) Thus, A is a G_δ-set. By Theorem 8.9 every open set is an F_σ-set.

Problem 8.18. *Let $\mathcal{B}$ be a collection of open sets in a topological space (X, τ). If for each x in an arbitrary open set V there exists some $B \in \mathcal{B}$ with $x \in B \subseteq V$, then $\mathcal{B}$ is called a* **base** *for τ. In general, a collection $\mathcal{B}$ of subsets of a nonempty set X is said to be a* **base** *if*

   1. $\bigcup_{B \in \mathcal{B}} B = X$, *and*
   2.   *for every pair $A, B \in \mathcal{B}$ and $x \in A \cap B$, there exists some $C \in \mathcal{B}$ with $x \in C \subseteq A \cap B$.*

Show that if $\mathcal{B}$ is a base for a set X, then the collection

$$\tau = \{V \subseteq X \colon \forall\, x \in V \text{ there exists } B \in \mathcal{B} \text{ with } x \in B \subseteq V\}$$

is a topology on X having $\mathcal{B}$ as a base.

Solution. Obviously, $\mathcal{B} \subseteq \tau$ holds. Clearly, $\emptyset \in \tau$, and from condition (1) it follows that $X \in \tau$. Also, it should be clear that τ is closed under arbitrary unions.

   Now, let $V, W \in \tau$ and $x \in V \cap W$. Choose two sets $A, B \in \mathcal{B}$ with $x \in A \subseteq V$ and $x \in B \subseteq W$. By condition (2), there exists some $C \in \mathcal{B}$ with $x \in C \subseteq A \cap B \subseteq V \cap W$, that is, $V \cap W \in \tau$. Thus, τ is a topology.

   The verification that $\mathcal{B}$ is a base for τ is straightforward.

Problem 8.19. *Let (X, τ) be a topological space, and let $\mathcal{B}$ be a base for the topology τ (see the preceding exercise for the definition). Show that there exists a dense subset A of X such that $\operatorname{card} A \leq \operatorname{card} \mathcal{B}$.*

Solution. If $B \in \mathcal{B}$ and $B \neq \emptyset$, then fix some $x_B \in B$ and consider the set $A = \{x_B \colon B \in \mathcal{B} \setminus \{\emptyset\}\}$. We claim that:

   1. A is dense in X, and
   2. $\operatorname{card} A \leq \operatorname{card} \mathcal{B}$.

To see (1) let V be a nonempty open set. If $x \in V$, then there exists some $B \in \mathcal{B}$ with $x \in B \subseteq V$. It follows that $x_B \in V$, and so $V \cap A \neq \emptyset$. This shows that A is dense in X.

   For (2) note that the function $f \colon \mathcal{B} \setminus \{\emptyset\} \longrightarrow A$, defined by $f(B) = x_B$, is onto. By the Axiom of Choice there exists a subset $\mathcal{C}$ of $\mathcal{B}$ such that $\mathcal{C} \cap f^{-1}(\{x\})$ consists precisely of one point for each $x \in A$. Then $f \colon \mathcal{C} \longrightarrow A$ is one-to-one and onto, proving that $\operatorname{card} A = \operatorname{card} \mathcal{C} \leq \operatorname{card} \mathcal{B}$.

Problem 8.20. *Let* $f: X \to Y$ *be a function. If* τ *is a topology on* X*, then the* **quotient topology** τ_f *determined by* f *on* Y *is defined by* $\tau_f = \{\mathcal{O} \subseteq Y: f^{-1}(\mathcal{O}) \in \tau\}$*.*

   a. *Show that* τ_f *is indeed a topology on* Y *and that* $f:(X, \tau) \to (Y, \tau_f)$ *is continuous.*

   b. *If* $g:(Y, \tau_f) \to (Z, \tau_1)$ *is a function, then show that the composition function* $g \circ f:(X, \tau) \to (Z, \tau_1)$ *is continuous if and only if* g *is continuous.*

   c. *Assume that* $f: X \to Y$ *is onto and that* τ^* *is a topology on* Y *such that* $f:(X, \tau) \to (Y, \tau^*)$ *is an open mapping (i.e., it carries open sets of* X *onto open sets of* Y*) and continuous. Show that* $\tau^* = \tau_f$*.*

Solution. a. (1) Since $f^{-1}(\emptyset) = \emptyset \in \tau$ and $f^{-1}(Y) = X \in \tau$, we see that $\emptyset, Y \in \tau_f$.

(2) If $V, W \in \tau_f$, then the identity $f^{-1}(V \cap W) = f^{-1}(V) \cap f^{-1}(W)$ implies that $V \cap W \in \tau_f$.

(3) If $\{V_i: i \in I\}$ is a family of τ_f, then in view of the identity $f^{-1}(\bigcup V_i) = \bigcup f^{-1}(V_i)$, we see that $\bigcup V_i \in \tau_f$.

b. Assume $g \circ f$ is continuous. If V is an open subset of Z, then $f^{-1}(g^{-1}(V)) = (g \circ f)^{-1}(V) \in \tau$ shows that $g^{-1}(V) \in \tau_f$. That is, g is continuous.

c. Since f is continuous, it is easy to see that $\tau^* \subseteq \tau_f$ holds. On the other hand, let $V \in \tau_f$. Then $f^{-1}(V) \in \tau$, and moreover, since f is an open mapping and onto, we have $V = f(f^{-1}(V)) \in \tau^*$ (see Problem 1.7). That is, $\tau_f \subseteq \tau^*$ also holds, and so $\tau_f = \tau^*$.

Problem 8.21. *This exercise presents an example of a compact set whose closure is not compact. Start by considering the interval* $[0, 1]$ *with the topology* τ *generated by the metric* $d(x, y) = |x - y|$*. It should be clear that* $([0, 1], \tau)$ *is a compact topological space. Next, put* $X = [0, 1] \cup \mathbb{N} = [0, 1] \cup \{2, 3, 4, \ldots\}$*, and define*

$$\tau^* = \tau \cup \{[0, 1] \cup A: A \subseteq \mathbb{N}\}.$$

   a. *Show that* τ^* *is a non-Hausdorff topology on* X *and that* τ^* *induces* τ *on* $[0, 1]$*.*

   b. *Show that* (X, τ^*) *is not a compact topological space.*

   c. *Show that* $[0, 1]$ *is a compact subset of* (X, τ^*)*.*

   d. *Show that* $[0, 1]$ *is dense in* X *(and hence, its closure is not compact.*

   e. *Why doesn't this contradict Theorem 8.12(1)?*

Solution. a. (1) Clearly, $\emptyset, X \in \tau^*$.

(2) Let $V, W \in \tau^*$. Then we have the following cases:

   CASE I. $V, W \in \tau$. In this case, $V \cap W \in \tau \subseteq \tau^*$.

   CASE II. $V \in \tau$ and $W \notin \tau$ (and vice versa). Note that $V \cap W = V \in \tau \subseteq \tau^*$.

CASE III. $V \notin \tau$ and $W \notin \tau$. In this case, we have $V \cap W = [0, 1] \cup A$ for some $A \subseteq \mathcal{N}$. That is, $V \cap W \in \tau^*$.

(3) Let $\{V_i : i \in I\}$ be a family of τ^*. If $V_i \in \tau$ holds for each i, then clearly $\bigcup V_i \in \tau \subseteq \tau^*$ holds. On the other hand, if some V_i is of the form $[0, 1] \cup A$, then $\bigcup V_i$ is of the same type, and hence, it belongs to τ^*.

Thus, τ^* is a topology on X that induces τ on $[0, 1]$.

b. The cover $X = \bigcup_{n=2}^{\infty} ([0, 1] \cup \{n\})$ cannot be reduced to a finite cover.

c. Since $([0, 1], \tau)$ is a compact topological space and τ^* induces τ on $[0, 1]$, it follows that $[0, 1]$ is a compact subset of X.

d. If $x \in X \setminus [0, 1]$, then every neighborhood V of x is of the form $V = [0, 1] \cup A$ for some $A \subseteq \mathcal{N}$. Thus, $V \cap [0, 1] = [0, 1] \neq \emptyset$ holds for every neighborhood V of x. Therefore, $\overline{[0, 1]} = X$ holds.

e. This does not contradict Theorem 8.12(1) because (X, τ^*) is not a Hausdorff topological space.

Problem 8.22. *A topological space (X, τ) is said to be* **connected** *if a subset of X that is simultaneously closed and open (called a* **clopen set***) is either empty or else equal to X.*

   a. *Show that (X, τ) is connected if and only if the only continuous functions from (X, τ) into $\{0, 1\}$ (with the discrete topology) are the constant ones.*

   b. *Let $f : (X, \tau) \to (Y, \tau^*)$ be onto and continuous. If (X, τ) is connected, then show that (Y, τ^*) is also connected.*

Solution. (a) If $f : X \longrightarrow \{0, 1\}$ is a nonconstant continuous function, then $f^{-1}(\{0\})$ is a nonempty clopen set which is different from X, and so X is not connected.

For the converse, assume that every continuous function from X into $\{0, 1\}$ is constant. If A is a clopen subset of X different from $\emptyset$ and X, then the function $f : X \longrightarrow \{0, 1\}$, defined by $f(x) = 1$ if $x \in A$ and $f(x) = 0$ if $x \notin A$, is a nonconstant continuous function, a contradiction. Thus, X is a connected topological space.

(b) Let A be a clopen subset of Y. By the continuity of f, the set $f^{-1}(A)$ is a clopen subset of X. Since X is connected, $f^{-1}(A) = \emptyset$ or $f^{-1}(A) = X$. Also, since f is onto, $f(f^{-1}(A)) = A$ holds (Problem 1.7). Thus, $A = \emptyset$ or $A = Y$, proving that Y is a connected topological space.

9. CONTINUOUS REAL-VALUED FUNCTIONS

Problem 9.1. *If u, v, and w are vectors in a vector lattice, then establish the following identities:*

   a. $u \vee v + u \wedge v = u + v$;

b. $u - v \vee w = (u - v) \wedge (u - w)$;
c. $u - v \wedge w = (u - v) \vee (u - w)$;
d. $\alpha(u \wedge v) = (\alpha u) \wedge (\alpha v)$ if $\alpha \geq 0$;
e. $|u - v| = u \vee v - u \wedge v$;
f. $u \vee v = \frac{1}{2}(u + v + |u - v|)$;
g. $u \wedge v = \frac{1}{2}(u + v - |u - v|)$.

Solution.   We use the identities (a), (b), and (d) in Section 9 of the text.
   (a) Replace w by $-(u + v)$ in $u \wedge v + w = (u + w) \wedge (v + w)$ to get

$$u \wedge v - (u + v) = (-v) \wedge (-u) = -u \vee v.$$

   (b) $u - v \vee w = u + (-v) \wedge (-w) = (u - v) \wedge (u - w)$.
   (c) $u - v \wedge w = u + (-v) \vee (-w) = (u - v) \vee (u - w)$.
   (d) If $\alpha \geq 0$, then

$$\begin{aligned}
\alpha(u \wedge v) &= \alpha[-(-u) \vee (-v)] = -\alpha[(-u) \vee (-v)] \\
&= -(-\alpha u) \vee (-\alpha v) = (\alpha u) \wedge (\alpha v).
\end{aligned}$$

   (e) Using (a), we see that

$$\begin{aligned}
u \vee v - u \wedge v &= u \vee v + \left[u \vee v - (u + v)\right] = 2(u \vee v) - (u + v) \\
&= (2u) \vee (2v) - (u + v) = (u - v) \vee (v - u) = |u - v|.
\end{aligned}$$

   (f) Using (e) and (a), we get

$$u + v + |u - v| = \left(u \vee v + u \wedge v\right) + \left(u \vee v - u \wedge v\right) = 2(u \vee v).$$

   (g) As in (f), we get

$$u + v - |u - v| = u \vee v + u \wedge v - (u \vee v - u \wedge v) = 2(u \wedge v).$$

Problem 9.2.   *If u and v are elements in a vector lattice, then show that:*
   a. $|u + v| \vee |u - v| = |u| + |v|$, *and*
   b. $|u + v| \wedge |u - v| = ||u| - |v||$.

Solution.   (a) Note that

$$\begin{aligned}
|u + v| \vee |u - v| &= [(u + v) \vee (-u - v)] \vee [(u - v) \vee (-u + v)] \\
&= [(u + v) \vee (-u + v)] \vee [(-u - v) \vee (u - v)] \\
&= [u \vee (-u) + v] \vee [(-u) \vee u - v] \\
&= [u \vee (-u)] + [v \vee (-v)] = |u| + |v|.
\end{aligned}$$

(b) Using the distributive law, we see that

$$|u + v| \wedge |u - v| = [(u + v) \vee (-u - v)] \wedge [(u - v) \vee (-u + v)]$$
$$= [(u+v) \wedge (u-v)] \vee [(-u-v) \wedge (u-v)] \vee [(u+v) \wedge (-u+v)] \vee \cdots$$
$$\cdots \vee [(-u - v) \wedge (-u + v)]$$
$$= [u + v \wedge (-v)] \vee [(-u) \wedge u - v] \vee [u \wedge (-u) + v] \vee [v \wedge (-v) - u]$$
$$= (u - |v|) \vee (-u - |v|) \vee (v - |u|) \vee (-v - |u|)$$
$$= [u \vee (-u) - |v|] \vee [v \vee (-v) - |u|] = (|u| - |v|) \vee (|v| - |u|)$$
$$= \big| |u| - |v| \big|.$$

Problem 9.3. *Show that* $|u| \wedge |v| = 0$ *holds if and only if* $|u + v| = |u - v|$ *holds.*

Solution. If $|u| \wedge |v| = 0$, then using parts (a) and (b) and part (e) of Problem 9.1, we get

$$|u + v| \wedge |u - v| = \big| |u| - |v| \big| = |u| \vee |v| - |u| \wedge |v| = |u| \vee |v|$$
$$= |u| + |v| - |u| \wedge |v| = |u| + |v| = |u + v| \vee |u - v|.$$

This easily implies that $|u + v| = |u - v|$ holds.

For the converse, assume that $|u + v| = |u - v|$. Then by parts (a) and (b) of Problem 9.2, we have

$$|u| + |v| = \big| |u| - |v| \big| = |u| \vee |v| - |u| \wedge |v|$$
$$= (|u| + |v| - |u| \wedge |v|) - |u| \wedge |v| = |u| + |v| - 2(|u| \wedge |v|),$$

from which it follows that $|u| \wedge |v| = 0$.

Problem 9.4. *Show that the vector space consisting of all polynomials (with real coefficients) on* **R** *is not a function space. Prove a similar result for the vector space of all real-valued differentiable functions on* **R**.

Solution. If p is the polynomial defined by $p(x) = x$, then $|p|(x) = |p(x)| = |x|$ holds. Clearly, $|p|$ is not differentiable (and hence, it is not a polynomial either).

Problem 9.5. *Let* X *be a topological space. Consider the collection* L *of all real-valued functions on* X *defined by*

$$L = \{f \in \mathbf{R}^X : \exists \{f_n\} \subseteq C(X) \text{ such that } \lim f_n(x) = f(x) \; \forall \, x \in X\}.$$

Show that L *is a function space.*

Solution. Clearly, L is a vector space. Now, let $f, g \in L$. Choose two sequences $\{f_n\}$ and $\{g_n\}$ of $C(X)$ with $\lim f_n(x) = f(x)$ and $\lim g_n(x) = g(x)$ for all x. Then $f_n \vee g_n \in C(X)$ for each n and

$$\lim f_n \vee g_n(x) = \lim \frac{1}{2}\big[f_n(x) + g_n(x) + |f_n(x) - g_n(x)|\big]$$

$$= \frac{1}{2}\big[f(x) + g(x) + |f(x) - g(x)|\big] = f \vee g(x),$$

so that $f \vee g \in L$. Similarly, $f \wedge g \in L$, so that L is a function space.

Problem 9.6. *Let L be a vector space of real-valued functions defined on a set X. If for every function $f \in L$ the function $|f|$ (defined by $|f|(x) = |f(x)|$ for each $x \in X$) belongs to L, then show that L is a function space.*

Solution. Use the identities

$$f \vee g = \frac{1}{2}(f + g + |f - g|) \quad \text{and} \quad f \wedge g = \frac{1}{2}(f + g - |f - g|).$$

Problem 9.7. *Consider each rational number written in the form $\frac{m}{n}$, where $n > 0$, and m and n are integers without any common factors other than ± 1. Clearly, such a representation is unique. Now, define $f: \mathbb{R} \to \mathbb{R}$ by $f(x) = 0$ if x is irrational and $f(x) = \frac{1}{n}$ if $x = \frac{m}{n}$ as above. Show that f is continuous at every irrational number and discontinuous at every rational number.*

Solution. The proof will be based upon the following property: *Let $\{r_n\}$ be a bounded sequence of distinct rational numbers. If $r_n = \frac{m_n}{k_n}$ (where $k_n > 0$, and m_n and k_n do not have common factors), then $\lim k_n = \infty$.*

To see this, pick some number $M > 0$ such that $|r_n| \leq M$ for each n, and so $|m_n| \leq M k_n$. Now, if for some $C > 0$ we have $|k_n| \leq C$ for infinitely many n, then $|m_n| \leq MC$ must also hold for the same infinitely many n. However, this contradicts the fact that there is a finite number of rational numbers $\frac{m}{n}$ with $|m| \leq MC$ and $|n| < C$.

Now, let x be an irrational number. If $\{x_n\}$ is a sequence of irrational numbers with $x_n \longrightarrow x$, then $0 = f(x_n) \longrightarrow 0 = f(x)$. Thus, if f is not continuous at x, then there exists a sequence $\{r_n\}$ of rational numbers with $r_n \longrightarrow x$ and $\lim f(r_n) \neq 0$. Since x is irrational, we can assume $r_n \neq r_m$ whenever $n \neq m$. Write $r_n = \frac{m_n}{k_n}$, and note that $f(r_n) = \frac{1}{k_n} \not\to 0$ implies $k_n \not\to \infty$, a contradiction. Therefore, f is continuous at every irrational number.

Now, let r be a rational number. Choose a sequence $\{r_n\}$ of distinct rational numbers with $r_n = \frac{m_n}{k_n} \longrightarrow r$. Now, note that $\lim f(r_n) = \lim \frac{1}{k_n} = 0 \neq f(r)$ holds, which shows that f is not continuous at r. That is, f is discontinuous at every rational number.

Problem 9.8.   *Let $f:[a, b] \rightarrow \mathbb{R}$ be increasing, i.e., $x < y$ implies $f(x) \leq f(y)$. Show that the set of points where f is discontinuous is at-most countable.*

Solution.   Let $f:[a, b] \longrightarrow \mathbb{R}$ be increasing, and let D be the set of discontinuities of f. For each $x \in D$ choose a rational number r_x such that $\lim_{t \uparrow x} f(t) < r_x < \lim_{t \downarrow x} f(t)$. Since $x, y \in D$ with $x < y$ implies

$$r_x < \lim_{t \downarrow x} f(t) < \lim_{t \uparrow y} f(t) < r_y,$$

it follows that $r_x \neq r_y$ whenever $x \neq y$. Thus, $x \longmapsto r_x$ is a one-to-one function from D into the set of rational numbers, and so D is at-most countable.

Problem 9.9.   *Give an example of a strictly increasing function $f:[0, 1] \rightarrow \mathbb{R}$ which is continuous at every irrational number and discontinuous at every rational number.*

Solution.   For each $t \in [0, 1]$, let $f_t:[0, 1] \rightarrow [0, 1]$ be a strictly increasing function which is continuous everywhere except at $x = t$. For instance, for $0 < t \leq 1$ let

$$f_t(x) = \begin{cases} 0.5x & \text{if } 0 \leq x < t, \\ x & \text{if } t \leq x \leq 1, \end{cases}$$

and $f_0(x) = 0.5 + 0.5x$ if $0 < x \leq 1$ and $f_0(0) = 0$.

If $\{r_1, r_2, \ldots\}$ is an enumeration of the rational numbers of $[0, 1]$, then define the function $f:[0, 1] \rightarrow [0, 1]$ by

$$f(x) = \sum_{n=1}^{\infty} \frac{1}{2^n} f_{r_n}(x),$$

and note that f satisfies the desired properties.

Problem 9.10.   *Recall that a function $f:(X, \tau) \rightarrow (Y, \tau_1)$ is called an open mapping if $f(V)$ is open whenever V is open. Prove that if $f:\mathbb{R} \rightarrow \mathbb{R}$ is a continuous open mapping, then f is a strictly monotone function—and hence, a homeomorphism.*

Solution.   Let (a, b) be a finite open interval of $\mathbb{R}$. Since f attains a maximum value on $[a, b]$ and $f((a, b))$ is an open set, it is easy to see that the extrema of f on $[a, b]$ take place at the end points. In particular, this implies $f(a) \neq f(b)$. (If $f(a) = f(b)$, then $f((a, b))$ must be a one-point set, contradicting the fact that f is an open mapping.) Next, we claim that f is strictly monotone on (a, b). To see this, assume $f(a) < f(b)$, and $a < x < y < b$. Then note first that

$f(a) < f(x) < f(b)$ must hold. Indeed, if $f(x) \leq f(a)$ holds, then f attains its minimum on $[a, b]$ at some interior point. Similarly, if $f(x) \geq f(b)$ holds, then f attains its maximum value on $[a, b]$ at some interior point. However, (since f is an open mapping) both cases are impossible, and so $f(a) < f(x) < f(b)$ holds. By the same arguments, $f(x) < f(y) < f(b)$. Thus, f is strictly increasing on (a, b). Similarly, if $f(a) > f(b)$ holds, then f is strictly decreasing on (a, b).

Now, assume that f is strictly increasing on $(0, 1)$, and let $x < y$. Choose some n with $(0, 1) \subseteq (-n, n)$ and $x, y \in (-n, n)$. Since f is strictly monotone on $(-n, n)$, and strictly increasing on $(0, 1)$, it is easy to see that f must be strictly increasing on $(-n, n)$. Thus, $f(x) < f(y)$ holds, and this shows that f is strictly increasing on $\mathbf{R}$. (We remark that the function f need not be onto. However, the mapping $f: \mathbf{R} \longrightarrow f(\mathbf{R})$ is a homeomorphism.)

Problem 9.11. *Let X be a nonempty set, and for any two functions $f, g \in \mathbf{R}^X$ let*

$$d(f, g) = \sup_{x \in X} \frac{|f(x) - g(x)|}{1 + |f(x) - g(x)|}.$$

Establish the following:

  a. *($\mathbf{R}^X$, d) is a metric space.*
  b. *A sequence $\{f_n\} \subseteq \mathbf{R}^X$ satisfies $d(f_n, f) \to 0$ for some $f \in \mathbf{R}^X$ if and only if $\{f_n\}$ converges uniformly to f.*

Solution. (a) Clearly, $d(f, g) \geq 0$ for all $f, g \in \mathbf{R}^X$ and $d(f, g) = 0$ if and only if $f = g$. Moreover, it should be clear that $d(f, g) = d(g, f)$ for all $f, g \in \mathbf{R}^X$. What needs verification is the triangle inequality. To do this, we need the following two properties:

  1. $0 \leq x \leq y$ implies $\frac{x}{1+x} \leq \frac{y}{1+y}$, and
  2. $\frac{x+y}{1+x+y} \leq \frac{x}{1+x} + \frac{y}{1+y}$ for all $x, y \geq 0$.

Property (1) follows from the fact that the function $f(t) = \frac{t}{1+t}$ $(t \geq 0)$ is strictly increasing on $[0, \infty)$; notice that $f'(t) = (1+t)^{-2} > 0$ for each $t > -1$. For (2) fix $x, y \geq 0$, and note that

$$(x + y)(1 + x)(1 + y) = x(1 + x)(1 + y) + y(1 + x)(1 + y)$$
$$\leq \left[x(1 + x)(1 + y) + xy(1 + y) \right] + \left[y(1 + x)(1 + y) + xy(1 + x) \right]$$
$$= x(1 + y)(1 + x + y) + y(1 + x)(1 + x + y).$$

Dividing across by $(1+x)(1+y)(1+x+y)$, the validity of (2) can be established.

Now, let $f, g, h \in \mathbf{R}^X$ and $x \in X$. From

$$|f(x) - g(x)| \le |f(x) - h(x)| + |h(x) - g(x)|$$

and (1) and (2), we get

$$
\begin{aligned}
\frac{|f(x) - g(x)|}{1 + |f(x) - g(x)|} &\le \frac{|f(x) - h(x)| + |h(x) - g(x)|}{1 + |f(x) - h(x)| + |h(x) - g(x)|} \\
&\le \frac{|f(x) - h(x)|}{1 + |f(x) - h(x)|} + \frac{|h(x) - g(x)|}{1 + |h(x) - g(x)|} \\
&\le d(f, h) + d(h, g),
\end{aligned}
$$

for all $x \in X$. This implies

$$d(f, g) = \sup_{x \in X} \frac{|f(x) - g(x)|}{1 + |f(x) - g(x)|} \le d(f, h) + d(h, g).$$

(b) Let $\{f_n\} \subseteq \mathbf{R}^X$. Assume first that $\{f_n\}$ converges uniformly to some function $f \in \mathbf{R}^X$, and let $\epsilon > 0$. So, there exists n_0 such that $|f_n(x) - f(x)| < \epsilon$ holds for all $n \ge n_0$ and all $x \in X$, and hence, $\frac{|f_n(x) - f(x)|}{1 + |f_n(x) - f(x)|} \le |f_n(x) - f(x)| < \epsilon$ for all $n \ge n_0$ and all $x \in X$. It follows that

$$d(f_n, f) = \sup_{x \in X} \frac{|f_n(x) - f(x)|}{1 + |f_n(x) - f(x)|} \le \epsilon$$

for all $n \ge n_0$. This shows that $d(f_n, f) \to 0$.

For the converse, assume $d(f_n, f) \to 0$, and let $\epsilon > 0$. Then there exists some n_0 such that

$$d(f_n, f) = \sup_{x \in X} \frac{|f_n(x) - f(x)|}{1 + |f_n(x) - f(x)|} < \frac{\epsilon}{1 + \epsilon}$$

for all $n \ge n_0$, and hence, $\frac{|f_n(x) - f(x)|}{1 + |f_n(x) - f(x)|} < \frac{\epsilon}{1 + \epsilon}$ for all $n \ge n_0$ and all $x \in X$. This implies $|f_n(x) - f(x)| < \epsilon$ for all $n \ge n_0$ and all $x \in X$, which means that $\{f_n\}$ converges uniformly to f.

Problem 9.12. *Let $f, f_1, f_2, \ldots$ be real-valued functions defined on a compact metric space (X, d) such that $x_n \to x$ in X implies $f_n(x_n) \to f(x)$ in $\mathbf{R}$. If f is continuous, then show that the sequence of functions $\{f_n\}$ converges uniformly to f.*

Solution. Assume that the functions $f, f_1, f_2, \ldots$ satisfy the stated properties and that the function $f : X \longrightarrow \mathbf{R}$ is continuous. Also, assume by way of contradiction that the sequence $\{f_n\}$ does not converge uniformly to f. Then an easy argument shows (how?) that there exist $\varepsilon > 0$, a subsequence $\{g_n\}$ of $\{f_n\}$, and a sequence $\{x_n\}$ of X such that

$$\left|g_n(x_n) - f(x_n)\right| \geq \varepsilon \quad \text{for each } n. \tag{$\star$}$$

Since X is compact, the sequence $\{x_n\}$ has a convergent subsequence in X, say $x_{k_n} \longrightarrow x$. By the continuity of f, we see that $f(x_{k_n}) \longrightarrow f(x)$. Also, from our hypothesis, it follows that $g_{k_n}(x_{k_n}) \longrightarrow f(x)$, and so

$$\left|g_{k_n}(x_{k_n}) - f(x_{k_n})\right| \longrightarrow \left|f(x) - f(x)\right| = 0,$$

contrary to $(\star)$. Therefore, the sequence $\{f_n\}$ converges uniformly to f.

Problem 9.13. *For a sequence $\{f_n\}$ of real-valued functions defined on a topological space X that converges uniformly to a real-valued function f on X establish the following.*

   a. *If $x_n \to x$ and f is continuous at x, then $f_n(x_n) \to f(x)$.*
   b. *If each f_n is continuous at some point $x_0 \in X$, then f is also continuous at the point x_0 and*

$$\lim_{x \to x_0} \lim_{n \to \infty} f_n(x) = \lim_{n \to \infty} \lim_{x \to x_0} f_n(x) = f(x_0).$$

Solution. (a) Assume f is continuous at x, $x_n \to x$ and let $\varepsilon > 0$. Choose some k with $|f_n(y) - f(y)| < \varepsilon$ for all $n > k$ and all $y \in X$. By the continuity of f at x, there exists some $m > k$ with $|f(x_n) - f(x)| < \varepsilon$ for all $n > m$. Thus,

$$\left|f_n(x_n) - f(x)\right| \leq \left|f_n(x_n) - f(x_n)\right| + \left|f(x_n) - f(x)\right| < 2\varepsilon$$

holds for all $n > m$, so that $\lim f_n(x_n) = f(x)$.

   (b) Assume that each f_n is continuous at $x_0 \in X$ and let $\epsilon > 0$. Since $\{f_n\}$ converges uniformly to f on X, there exists some k satisfying $|f_k(x) - f(x)| < \epsilon$ for all $x \in X$. Now, the continuity of f_k at x_0 guarantees the existence of a neighborhood V of x_0 such that $|f_k(x) - f_k(x_0)| < \epsilon$ for all $x \in V$. Then

$$
\begin{aligned}
|f(x) - f(x_0)| &\leq |f(x) - f_k(x)| + |f_k(x) - f_k(x_0)| + |f_k(x_0) - f(x_0)| \\
&< \epsilon + \epsilon + \epsilon = 3\epsilon
\end{aligned}
$$

holds for all $x \in V$, which shows that f is continuous at x_0. For the equality, note that

$$\lim_{x \to x_0} \lim_{n \to \infty} f_n(x) = \lim_{x \to x_0} f(x) = f(x_0),$$

while

$$\lim_{n \to \infty} \lim_{x \to x_0} f_n(x) = \lim_{n \to \infty} f_n(x_0) = f(x_0).$$

Problem 9.14. *Let $f_n: [0, 1] \to \mathbb{R}$ be defined by $f_n(x) = x^n$ for $x \in [0, 1]$. Show that $\{f_n\}$ converges pointwise and find its limit function. Is the convergence uniform?*

Solution. Clearly,

$$f_n(x) \longrightarrow f(x) = \begin{cases} 0, & \text{if } 0 \le x < 1; \\ 1, & \text{if } x = 1. \end{cases}$$

Since f is not continuous, the convergence cannot be uniform; see Theorem 9.2.

Problem 9.15. *Let $g: [0, 1] \to \mathbb{R}$ be a continuous function with $g(1) = 0$. Show that the sequence of functions $\{f_n\}$ defined by $f_n(x) = x^n g(x)$ for $x \in [0, 1]$, converges uniformly to the constant zero function.*

Solution. Let $\varepsilon > 0$. Choose some $0 < \delta < 1$ with $|g(x)| < \varepsilon$ whenever $\delta < x \le 1$. Now, pick some $M > 0$ with $|g(x)| \le M$ for all $x \in [0, 1]$, and then select some k with $M\delta^n < \varepsilon$ whenever $n > k$. Thus, for each $n > k$ we have $|x^n g(x)| \le M\delta^n < \varepsilon$ for $0 \le x \le \delta$ and $|x^n g(x)| \le |g(x)| < \varepsilon$ for all $\delta < x \le 1$. That is, the sequence $\{f_n\}$ converges uniformly to the constant zero function.

Problem 9.16. *Let $\{f_n\}$ be a sequence of continuous real-valued functions defined on $[a, b]$, and let $\{a_n\}$ and $\{b_n\}$ be two sequences of $[a, b]$ such that $\lim a_n = a$ and $\lim b_n = b$. If $\{f_n\}$ converges uniformly to f on $[a, b]$, then show that*

$$\lim_{n \to \infty} \int_{a_n}^{b_n} f_n(x) \, dx = \int_a^b f(x) \, dx.$$

Solution. Let $\varepsilon > 0$. Pick some k such that for all $n > k$ we have:

1. $a_n - a < \varepsilon$ and $b - b_n < \varepsilon$; and
2. $|f_n(x) - f(x)| < \varepsilon$ for all $x \in [a, b]$.

Also, since f is continuous (Theorem 9.2), there exists some $M > 0$ satisfying $|f(x)| \leq M$ for all $x \in [a, b]$. Thus,

$$\left| \int_{a_n}^{b_n} f_n(x)\,dx - \int_a^b f(x)\,dx \right|$$
$$= \left| \int_{a_n}^{b_n} f_n(x)\,dx - \int_{a_n}^{b_n} f(x)\,dx - \int_a^{a_n} f(x)\,dx - \int_{b_n}^b f(x)\,dx \right|$$
$$\leq \int_a^b \left| f_n(x) - f(x) \right| dx + \int_a^{a_n} |f(x)|\,dx + \int_{b_n}^b |f(x)|\,dx$$
$$\leq \varepsilon(b-a) + M(a_n - a) + M(b - b_n) < \varepsilon(2M + b - a)$$

holds for all $n > k$, and our conclusion follows.

Problem 9.17. *Let $\{f_n\}$ be a sequence of continuous real-valued functions on a metric space X such that $\{f_n\}$ converges uniformly to some function f on every compact subset of X. Show that f is a continuous function.*

Solution. Let $x_n \longrightarrow x$ in X. Put $K = \{x_1, x_2, \ldots\} \cup \{x\}$, and note that K is a compact set—every open cover of K can be reduced to a finite cover. Since $\{f_n\}$ is a sequence of continuous functions that converges uniformly to f on K, it follows from Theorem 9.2 that f is continuous on K. Since $x_n \longrightarrow x$ holds in K, we get $f(x_n) \longrightarrow f(x)$. That is, f is a continuous function.

Problem 9.18. *Let $\{f_n\}$ and $\{g_n\}$ be two uniformly bounded sequences of real-valued functions on a set X. If both $\{f_n\}$ and $\{g_n\}$ converge uniformly on X, then show that $\{f_n g_n\}$ also converges uniformly on X.*

Solution. Assume that $\{f_n\}$ and $\{g_n\}$ converge uniformly to f and g, respectively. Let $\varepsilon > 0$. Choose some k with $|f_n(x) - f(x)| < \varepsilon$ and $|g_n(x) - g(x)| < \varepsilon$ for all $n > k$ and all $x \in X$. Also, pick some $M > 0$ so that $|f_n(x)| \leq M$ and $|g_n(x)| \leq M$ hold for all n and all x. Now, note that

$$\left| f_n(x)g_n(x) - f(x)g(x) \right|$$
$$\leq \left| f_n(x) \right| \cdot \left| g_n(x) - g(x) \right| + \left| g(x) \right| \cdot \left| f_n(x) - f(x) \right| < 2M\varepsilon$$

holds for all $n > k$ and all $x \in X$.

Problem 9.19. *Suppose that $\{f_n\}$ is a sequence of monotone real-valued functions defined on $[a, b]$ and not necessarily all increasing or decreasing. Show* *that if* $\{f_n\}$ *converges pointwise to a continuous function* f *on* $[a, b]$, *then* $\{f_n\}$ *converges uniformly to* f *on* $[a, b]$.

Solution. Let $\varepsilon > 0$. Since f is uniformly continuous (Theorem 7.7), there exists some $\delta > 0$ so that $|f(x) - f(y)| < \varepsilon$ holds whenever $|x - y| < \delta$. Fix a finite number of points $a = x_0 < x_1 < \cdots < x_k = b$ with $x_i - x_{i-1} < \delta$ for $1 \le i \le k$, and then pick some m such that $|f_n(x_i) - f(x_i)| < \varepsilon$ holds for each $0 \le i \le k$ and all $n > m$.

Now, let $n > m$. Assume that f_n is decreasing. If $x \in [a, b]$, then $x_{i-1} \le x \le x_i$ holds for some $1 \le i \le k$, and so

$$
\begin{aligned}
\left| f_n(x) - f_n(x_i) \right| &= f_n(x) - f_n(x_i) \le f_n(x_{i-1}) - f_n(x_i) \\
&= [f_n(x_{i-1}) - f(x_{i-1})] + [f(x_{i-1}) - f(x_i)] + [f(x_i) - f_n(x_i)] \\
&< \varepsilon + \varepsilon + \varepsilon < 3\varepsilon.
\end{aligned}
$$

A similar inequality holds true if f_n is increasing. Therefore,

$$
\begin{aligned}
\left| f_n(x) - f(x) \right| &\le \left| f_n(x) - f_n(x_i) \right| + \left| f_n(x_i) - f(x_i) \right| + \left| f(x_i) - f(x) \right| \\
&< 3\varepsilon + \varepsilon + \varepsilon = 5\varepsilon
\end{aligned}
$$

holds for all $x \in [a, b]$ and all $n > m$. That is, $\{f_n\}$ converges uniformly to f.

Problem 9.20. *Let X be a topological space and let $\{f_n\}$ be a sequence of real-valued continuous functions defined on X. Suppose that there is a function $f : X \to \mathbf{R}$ such that $f(x) = \lim f_n(x)$ holds for all $x \in X$. Show that f is continuous at a point a if and only if for each $\epsilon > 0$ and each m there exist a neighborhood V of a and some $k > m$ such that $|f(x) - f_k(x)| < \epsilon$ holds for all $x \in V$.*

Solution. Assume that f is continuous at some point a. Let $\varepsilon > 0$ and an integer m be given. Pick a neighborhood U of a such that $|f(x) - f(a)| < \varepsilon$ holds for all $x \in U$. Since $\lim f_n(a) = f(a)$ holds, there exists an integer $r > m$ such that $|f(a) - f_n(a)| < \varepsilon$ holds for all $n > r$. Fix any integer $k > r$ and note that $k > m$. Since f_k is a continuous function, there exists a neighborhood W of a such that $|f_k(a) - f_k(x)| < \varepsilon$ holds for all $x \in W$. Now, note that if $x \in V = U \cap W$, then

$$
\left| f(x) - f_k(x) \right| \le \left| f(x) - f(a) \right| + \left| f(a) - f_k(a) \right| + \left| f_k(a) - f_k(x) \right| < 3\varepsilon.
$$

For the converse, assume that f satisfies the stated condition at the point a and let $\varepsilon > 0$. Since $f(a) = \lim f_n(a)$ holds, there exists an integer m such that $|f(a) - f_n(a)| < \varepsilon$ holds for all $n > m$. By the hypothesis, there exist a neighborhood V of a and an integer $k > m$ such that $|f(x) - f_k(x)| < \varepsilon$ holds for all $x \in V$. By the continuity of f_k, there exists another neighborhood U of a such that $|f_k(a) - f_k(x)| < \varepsilon$ holds for all $x \in U$. Now, note that $x \in U \cap V$ implies

$$\left| f(a) - f(x) \right| \le \left| f(a) - f_k(a) \right| + \left| f_k(a) - f_k(x) \right| + \left| f_k(x) - f(x) \right| < 3\varepsilon,$$

which shows that f is continuous at the point a.

Problem 9.21. *Let $\{f_n\}$ be a uniformly bounded sequence of continuous real-valued functions on a closed interval $[a, b]$. Show that the sequence of functions $\{\phi_n\}$ defined by $\phi_n(x) = \int_a^x f_n(t)\,dt$ for each $x \in [a, b]$, contains a uniformly convergent subsequence on $[a, b]$.*

Solution. Since the sequence $\{f_n\}$ is uniformly bounded, there is some $M > 0$ such that $|f_n(x)| < M$ holds for all $x \in [a, b]$ and all n. Clearly,

$$\left| \phi_n(x) \right| = \left| \int_a^x f_n(t)\,dt \right| \le M(b - a)$$

holds for all $x \in [a, b]$ and all n. So, the sequence $\{\phi_n\}$ is uniformly bounded and we claim that it is an equicontinuous sequence.

To see this, let $\varepsilon > 0$ and put $\delta = \varepsilon/M$. Now, note that $x, y \in [a, b]$ and $|x - y| < \delta$ imply

$$\left| \phi_n(x) - \phi_n(y) \right| = \left| \int_a^x f_n(t)\,dt - \int_a^y f_n(t)\,dt \right|$$

$$\le \left| \int_x^y |f_n(t)|\,dt \right| \le \left| \int_x^y M\,dt \right| = M|x - y| < \varepsilon.$$

Thus, the set $A = \{\phi_1, \phi_2, , \ldots\}$ is equicontinuous. If $\overline{A}$ denotes the (uniform) closure of A in $C[a, b]$, then $\overline{A}$ is bounded, closed, and equicontinuous (why?). By the Ascoli–Arzelà theorem (Theorem 9.10), the set $\overline{A}$ is a compact set. Since $\{\phi_n\}$ is a sequence of $\overline{A}$, it follows that $\{\phi_n\}$ has a subsequence that converges uniformly on $[a, b]$.

Problem 9.22. *For each n let $f_n: \mathbf{R} \to \mathbf{R}$ be a monotone (either increasing or decreasing) function. If there exists a dense subset A of $\mathbf{R}$ such that $\lim f_n(x)$ exists in $\mathbf{R}$ for each $x \in A$, then show that $\lim f_n(x)$ exists in $\mathbf{R}$ at-most for all but countably many x.*

Solution.   Assume that the functions f_n and the dense subset A of $\mathbb{R}$ satisfy the properties of the problem. Also, assume at the beginning that all but a finite number of the f_n are increasing functions.

Define the function $f: \mathbb{R} \longrightarrow \mathbb{R}$ by

$$f(x) = \limsup f_n(x), \quad x \in \mathbb{R}.$$

Note that $f(x)$ is a real number for each $x \in \mathbb{R}$. Indeed, if $x \in \mathbb{R}$, then there exist $a, b \in A$ with $a < x < b$, and so $f_n(a) \leq f_n(x) \leq f_n(b)$ holds for all sufficiently large n. Consequently,

$$\begin{aligned}
-\infty < \lim f_n(a) = \limsup f_n(a) \\
\leq \limsup f_n(x) = f(x) \\
\leq \limsup f_n(b) = \lim f_n(b) < \infty
\end{aligned}$$

Clearly, $f(x) = \lim f_n(x)$ holds for each $x \in A$. Next, note that f is an increasing function. Indeed, if $x < y$ holds, then from $f_n(x) \leq f_n(y)$ for all sufficiently large n, we see that $f(x) = \limsup f_n(x) \leq \limsup f_n(y) = f(y)$. By Problem 9.8, we know that f has at-most countably many discontinuities in every closed subinterval of $\mathbb{R}$. Hence, f has at-most countably many discontinuities (why?). Now, we claim that

$$\lim f_n(x) = f(x)$$

holds at every point of continuity of f. To see this, let x_0 be a point of continuity of f and let $\varepsilon > 0$. Pick some $\delta > 0$ such that $x_0 - \delta < x < x_0 + \delta$ implies $|f(x) - f(x_0)| < \varepsilon$, and then choose $a, b \in A$ with $x_0 - \delta < a < x_0 < b < x_0 + \delta$. Also, pick some n_0 such that for each $n \geq n_0$ the function f_n is increasing and satisfies

$$\left| f_n(b) - f(b) \right| < \varepsilon \quad \text{and} \quad \left| f_n(a) - f(a) \right| < \varepsilon.$$

Now, note that for $n \geq n_0$, we have

$$\begin{aligned}
f(x_0) - f_n(x_0) &\leq f(x_0) - f_n(a) \\
&= [f(x_0) - f(a)] + [f(a) - f_n(a)] < \varepsilon + \varepsilon = 2\varepsilon
\end{aligned}$$

and

$$\begin{aligned}
f_n(x_0) - f(x_0) &\leq f_n(b) - f(x_0) \\
&= [f_n(b) - f(b)] + [f(b) - f(x_0)] < \varepsilon + \varepsilon = 2\varepsilon.
\end{aligned}$$

Thus, $|f_n(x_0) - f(x_0)| < 2\varepsilon$ holds for all $n \geq n_0$, proving that $\lim f_n(x_0) = f(x_0)$.

For the general case, assume that there are infinitely many increasing and in-finitely many decreasing f_n. Split the sequence $\{f_n\}$ into two subsequences $\{g_n\}$ and $\{h_n\}$ such that each g_n is increasing and each h_n is decreasing. Put

$$g(x) = \limsup g_n(x) \quad \text{and} \quad h(x) = \liminf h_n(x) = -\limsup[-h_n(x)],$$

and note that $g(a) = h(a)$ holds for each $a \in A$. By the above conclusion, g and h are continuous except possibly at the points of an at-most countable subset C of $\mathbf{R}$, and for each point $x \notin C$ we have

$$\lim g_n(x) = g(x) \quad \text{and} \quad \lim h_n(x) = h(x).$$

Now, let $c \notin C$ and fix $\varepsilon > 0$. Pick some $\delta > 0$ such that $|x - c| < \delta$ implies $|g(x) - g(c)| < \varepsilon$ and $|h(x) - h(c)| < \varepsilon$. Pick $a \in A$ with $|a - c| < \delta$ and note that from $g(a) = h(a)$, it follows that

$$\left| g(c) - h(c) \right| \leq \left| g(c) - g(a) \right| + \left| h(a) - h(c) \right| < \varepsilon + \varepsilon = 2\varepsilon.$$

Since $\varepsilon > 0$ is arbitrary, we see that $g(c) = h(c)$ holds for each $c \notin C$. This implies (how?) that $\lim f_n(c)$ exists in $\mathbf{R}$ for each $c \notin C$.

Problem 9.23.   *Consider a continuous function $f: [0, \infty) \to \mathbf{R}$. For each n define the continuous function $f_n: [0, \infty) \to \mathbf{R}$ by $f_n(x) = f(x^n)$. Show that the set of continuous functions $\{f_1, f_2, \ldots\}$ is equicontinuous at $x = 1$ if and only if f is a constant function.*

Solution.   Let $f \in C[0, \infty)$, let $f_n: [0, \infty) \longrightarrow \mathbf{R}$ be defined by $f_n(x) = f(x^n)$, and let $E = \{f_1, f_2, \ldots\}$. If f is a constant function, then it should be clear that the set E is equicontinuous at $x = 1$.

For the converse, assume that the set E is equicontinuous at $x = 1$. Fix $a > 0$ and let $\varepsilon > 0$. The equicontinuity of E at $x = 1$ guarantees the existence of some $0 < \delta < 1$ such that $|x - 1| < \delta$ implies $|f_n(x) - f_n(1)| < \varepsilon$ for each n. From $\lim \sqrt[n]{a} = 1$ (why?), we see that there exists some n_0 such that $|\sqrt[n]{a} - 1| < \delta$ holds for each $n \geq n_0$. Thus, if $n \geq n_0$, then we have

$$|f(a) - f(1)| = \left| f\left((\sqrt[n]{a})^n \right) - f(1^n) \right| = |f_n(\sqrt[n]{a}) - f_n(1)| < \varepsilon.$$

Since $\varepsilon > 0$ is arbitrary, it follows that $f(a) = f(1)$ holds for each $a > 0$. By continuity, we see that $f(a) = f(1)$ for each $a \geq 0$, and so f is a constant function.

Problem 9.24. *Let* (X, d) *be a compact metric space and let* $\mathcal{A}$ *be an equicontinuous subset of* $C(X)$. *Show that* $\mathcal{A}$ *is uniformly equicontinuous, i.e., show that for each* $\epsilon > 0$ *there exists some* $\delta > 0$ *such that* $x, y \in X$ *and* $d(x, y) < \delta$ *imply* $|f(x) - f(y)| < \epsilon$ *for all* $f \in \mathcal{A}$.

Solution. Let (X, d) be a compact metric space, let $\mathcal{A}$ be an equicontinuous subset of $C(X)$, and let $\varepsilon > 0$. For each $x \in X$ there exists (by the equicontinuity of $\mathcal{A}$) some $\delta_x > 0$ such that $d(x, y) < \delta_x$ implies $|f(x) - f(y)| < \varepsilon$ for all $f \in \mathcal{A}$. From $X = \bigcup_{x \in X} B(x, \frac{\delta_x}{2})$ and the compactness of X, we see that there exist $x_1, \ldots, x_n \in X$ such that $X = \bigcup_{i=1}^{n} B(x_i, \frac{\delta_{x_i}}{2})$.

Let $\delta = \frac{1}{2} \min\{\delta_{x_1}, \ldots, \delta_{x_n}\} > 0$ and let $x, y \in X$ satisfy $d(x, y) < \delta$. Now, pick some $1 \le i \le n$ with $d(x, x_i) < \frac{\delta_{x_i}}{2}$ and observe that $|f(x) - f(x_i)| < \varepsilon$ for all $f \in \mathcal{A}$. In addition, from

$$ d(y, x_i) \le d(y, x) + d(x, x_i) < \frac{\delta_{x_i}}{2} + \frac{\delta_{x_i}}{2} = \delta_{x_i}, $$

we see that $|f(y) - f(x_i)| < \varepsilon$ holds for all $f \in \mathcal{A}$. Therefore, from the above, if $d(x, y) < \delta$, then

$$ \left| f(x) - f(y) \right| \le \left| f(x) - f(x_i) \right| + \left| f(x_i) - f(y) \right| < \varepsilon + \varepsilon = 2\varepsilon $$

holds for all $f \in \mathcal{A}$. That is, $\mathcal{A}$ is a uniformly equicontinuous subset of $C(X)$.

Problem 9.25. *Let* X *be a connected topological space (see Problem 8.22 of Section 8 for the definition) and let* $\mathcal{A}$ *be an equicontinuous subset of* $C(X)$. *If for some* $x_0 \in X$, *the set of real numbers* $\{f(x_0): f \in \mathcal{A}\}$ *is bounded, then show that* $\{f(x): f \in \mathcal{A}\}$ *is also bounded for each* $x \in X$.

Solution. Let X be a connected topological space, let $\mathcal{A}$ be an equicontinuous subset of $C(X)$, and let $x_0 \in X$ be a point such that the collection of real numbers $\{f(x_0): f \in \mathcal{A}\}$ is bounded. Consider the set

$$ E = \left\{x \in X: \text{The set } \{f(x): f \in \mathcal{A}\} \text{ is bounded}\right\}. $$

Since $x_0 \in E$, we see that E is nonempty. We claim that E is both open and closed. If this is the case, then by the connectedness of X we must have $E = X$, and the desired conclusion follows.

We shall show first that E is a closed set. To this end, let $y \in \overline{E}$. By the equicontinuity of $\mathcal{A}$, there exists a neighborhood V of y such that

$$ |f(x) - f(y)| < 1 $$

holds for all $x \in V$ and all $f \in \mathcal{A}$. From $y \in \overline{E}$, we see that $V \cap E \ne \emptyset$. Fix some $z \in V \cap E$, and then pick some $M > 0$ such that $|f(z)| \le M$ holds for each $f \in \mathcal{A}$. In particular, we have

$$\left| f(y) \right| \le \left| f(y) - f(z) \right| + \left| f(z) \right| < 1 + M$$

for all $f \in \mathcal{A}$. This means that $y \in E$, and so $\overline{E} = E$, i.e., E is a closed set.

Next, we shall establish that E is an open set. To this end, let $y \in E$. Pick some $C > 0$ such that $|f(y)| \le C$ holds for each $f \in \mathcal{A}$. By the equicontinuity of $\mathcal{A}$, there exists a neighborhood W of y such that $|f(x) - f(y)| < 1$ holds for each $x \in W$ and all $f \in \mathcal{A}$. In particular, if $x \in W$, then

$$\left| f(x) \right| \le \left| f(x) - f(y) \right| + \left| f(y) \right| < 1 + C$$

holds for all $f \in \mathcal{A}$, and so $x \in E$. That is, $W \subseteq E$ holds, which shows that y is an interior point of E. Therefore, E is also an open set.

Problem 9.26. *Let $\{f_n\}$ be an equicontinuous sequence in $C(X)$, where X is not necessarily compact. If for some function $f: X \to \mathbf{R}$ we have $\lim f_n(x) = f(x)$ for each $x \in X$, then show that $f \in C(X)$.*

Solution. Let $x \in X$ and let $\varepsilon > 0$. Since $\{f_n\}$ is an equicontinuous sequence, there exists a neighborhood V of the point x such that $|f_n(y) - f_n(x)| < \varepsilon$ holds for all n and each $y \in V$.

Now, let $y \in V$. Pick some k with $|f_k(x) - f(x)| < \varepsilon$ and $|f_k(y) - f(y)| < \varepsilon$, and note that

$$\left| f(x) - f(y) \right| \le \left| f(x) - f_k(x) \right| + \left| f_k(x) - f_k(y) \right| + \left| f_k(y) - f(y) \right| < 3\varepsilon.$$

That is, f is continuous at the arbitrary point x.

Problem 9.27. *Let X be a compact topological space, and let $\{f_n\}$ be an equicontinuous sequence of $C(X)$. Assume that there exists some $f \in C(X)$ and some dense subset A of X such that $\lim f_n(x) = f(x)$ holds for each $x \in A$. Then show that $\{f_n\}$ converges uniformly to f.*

Solution. Let $\varepsilon > 0$. By the equicontinuity of $\{f_n\}$ and the continuity of f, for each $x \in X$, there exists some neighborhood V_x of x such that

1. $|f_n(y) - f_n(x)| < \varepsilon$ holds for all $y \in V_x$ and all n; and
2. $|f(y) - f(x)| < \varepsilon$ holds for all $y \in V_x$.

By the compactness of X, there exist $x_1, \ldots, x_k \in X$ such that $X = \bigcup_{i=1}^k V_{x_i}$.

Now, let $y \in V_{x_i}$. Choose some $x \in A \cap V_{x_i}$, and then pick some m_i with $|f_n(x) - f(x)| < \varepsilon$ for all $n > m_i$. Clearly, $|f(x) - f(y)| < 2\varepsilon$. Thus,

$$\left| f_n(y) - f(y) \right|$$
$$\leq \left| f_n(y) - f_n(x_i) \right| + \left| f_n(x_i) - f_n(x) \right| + \left| f_n(x) - f(x) \right| + \left| f(x) - f(y) \right|$$
$$< 5\varepsilon$$

holds for all $y \in V_{x_i}$ and all $n > m_i$.

Finally, put $m = \max\{m_i: 1 \leq i \leq k\}$, and note that $|f_n(y) - f(y)| < 5\varepsilon$ for all $y \in X$ and all $n > m$.

Problem 9.28.   *Show that for any fixed integer $n > 1$ the set of functions f in $C[0, 1]$ such that there is some $x \in [0, 1 - \frac{1}{n}]$ for which*

$$|f(x + h) - f(x)| \leq nh \quad whenever \quad 0 < h < \frac{1}{n},$$

is nowhere dense in $C[0, 1]$ (with the uniform metric).

Use the above conclusion and Baire's theorem to prove that there exists a continuous real-valued function defined on $[0, 1]$ that is not differentiable at any point of $[0, 1]$.

Solution.   Let $D(f, g) = \|f - g\|_\infty = \sup\{|f(x) - g(x)|: x \in [0, 1]\}$. For $n \geq 2$ define

$$A_n = \left\{ f \in C[0, 1]: \exists \, x \in [0, 1 - \tfrac{1}{n}] \text{ with } \left| f(x + h) - f(x) \right| \leq nh \right.$$
$$\left. \text{whenever } 0 < h < \tfrac{1}{n} \right\}.$$

We claim that:

1.   Each A_n is closed; and
2.   Each A_n is nowhere dense in $C[0, 1]$ (i.e., $(A_n)^\circ = \emptyset$).

To see that each A_n is closed, let $\{f_k\} \subseteq A_n$ satisfy $\lim D(f_k, f) = 0$ (i.e., $\{f_k\}$ converges uniformly to f on $[0, 1]$). For each k choose some $x_k \in [0, 1 - \frac{1}{n}]$ with $|f_k(x_k + h) - f_k(x_k)| \leq nh$ for all $0 < h < \frac{1}{n}$. Since $[0, 1 - \frac{1}{n}]$ is compact, there exists a subsequence of $\{x_k\}$ that converges to some $x \in [0, 1 - \frac{1}{n}]$. We can assume that $\lim x_k = x$. By Problem 9.13, $\lim f_k(x_k + h) = f(x + h)$ and $\lim f_k(x_k) = f(x)$, and so $|f(x + h) - f(x)| \leq nh$ holds for all $0 < h < \frac{1}{n}$. Thus, $f \in A_n$, and hence, A_n is a closed subset of $C[0, 1]$.

Now, let $f \in A_n$ and let $\varepsilon > 0$. Consider the function $g \in C[0, 1]$ whose graph is shown in Figure 2.1. Note that for each $x \in [0, 1)$ we have $|g(x + h) - g(x)| = 3nh$ for all sufficiently small $h > 0$. Put $f_1 = f + g$, and note that

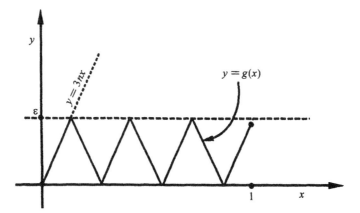

FIGURE 2.1. The Construction of a Nowhere Differentiable Function

$D(f, f_1) = \|g\|_\infty = \varepsilon$. On the other hand, if $x \in [0, 1)$ is fixed, then for all sufficiently small $h > 0$ we have

$$nh < 2nh = 3nh - nh \leq |g(x + h) - g(x)| - |f(x + h) - f(x)|$$
$$\leq |g(x + h) - g(x) - [f(x) - f(x + h)]| = |f_1(x + h) - f_1(x)|.$$

Thus, $f_1 \notin A_n$, and so $B(f, 2\varepsilon) \not\subseteq A_n$ for all $\varepsilon > 0$. This shows that $(A_n)^\circ = \emptyset$. Now, for each $n \geq 2$ let

$$B_n = \left\{ f \in C[0, 1] : \exists\, x \in [\tfrac{1}{n}, 1] \text{ with } |f(x - h) - f(x)| \leq nh \right.$$
$$\left. \text{whenever } 0 < h < \tfrac{1}{n} \right\}.$$

By the same arguments, each B_n is closed and nowhere dense. Consequently, from Baire's Theorem 6.17, we have

$$C[0, 1] \neq \left(\bigcup_{n=2}^{\infty} A_n \right) \cup \left(\bigcup_{n=2}^{\infty} B_n \right).$$

In particular, note that every $f \in C[0, 1] \setminus \left(\bigcup_{n=2}^{\infty} A_n \right) \cup \left(\bigcup_{n=2}^{\infty} B_n \right)$ does not have any one-sided derivative at any point of $[0, 1]$.

Problem 9.29. *Establish the following result regarding differentiability and uniform convergence. Let $\{f_n\}$ be a sequence of differentiable real-valued functions defined on a bounded open interval (a, b) such that:*

    a. *for some $x_0 \in (a, b)$ the sequence of real numbers $\{f_n(x_0)\}$ converges in $\mathbf{R}$, and*

b. *the sequence of derivatives $\{f_n'\}$ converges uniformly to a function $g:(a, b) \to \mathbb{R}$.*

Then the sequence $\{f_n\}$ converges uniformly to a function $f:(a, b) \to \mathbb{R}$ which is differentiable at x_0 and satisfies $f'(x_0) = g(x_0)$.

Solution. First, we shall show that $\{f_n\}$ is a uniformly Cauchy sequence. To this end, let $\epsilon > 0$ and pick some $M > 0$ such that $|x - x_0| \leq M$ for each $x \in (a, b)$. Next, choose some k such that

$$\left| f_n'(x) - f_m'(x) \right| < \epsilon \quad \text{for all } m, n \geq k \text{ and all } x \in (a, b) \qquad (\star)$$

and

$$\left| f_n(x_0) - f_m(x_0) \right| < \epsilon \quad \text{for all } m, n \geq k. \qquad (\star\star)$$

Using the Mean Value Theorem, $(\star)$ and $(\star\star)$, we see that for each $x \in (a, b)$ and each pair $n, m \geq k$ there exists some $t \in (a, b)$ such that

$$\begin{aligned}
\left| f_n(x) - f_m(x) \right| &\leq \left| \left[f_n(x) - f_m(x) \right] - \left[f_n(x_0) - f_m(x_0) \right] \right| + \left| f_n(x_0) - f_m(x_0) \right| \\
&= \left| f_n'(t) - f_m'(t) \right| \cdot |x - x_0| + \left| f_n(x_0) - f_m(x_0) \right| \\
&\leq M\epsilon + \epsilon = (1 + M)\epsilon.
\end{aligned}$$

This shows that $\{f_n\}$ is a uniformly Cauchy sequence, and hence, $\{f_n\}$ converges uniformly to a function $f:(a, b) \to \mathbb{R}$.

Next, for each n we consider the continuous function $\phi_n:(a, b) \to \mathbb{R}$ defined by $\phi_n(x) = \frac{f_n(x) - f_n(x_0)}{x - x_0}$ if $x \neq x_0$ and $\phi_n(x_0) = f_n'(x_0)$. Using the Mean Value Theorem and $(\star)$, we see that for each $x \in (a, b)$ there exists some $c_x \in (a, b)$ such that

$$\left| \phi_n(x) - \phi_m(x) \right| = \left| \frac{[f_n(x) - f_m(x)] - [f_n(x_0) - f_m(x_0)]}{x - x_0} \right| = \left| f_n'(c_x) - f_m'(c_x) \right| < \epsilon,$$

for all $n, m \geq k$. This shows that $\{\phi_n\}$ is a uniformly Cauchy sequence, and hence, it converges uniformly to the function $\phi:(a, b) \to \mathbb{R}$ defined by $\phi(x) = \frac{f(x) - f(x_0)}{x - x_0}$ if $x \neq x_0$ and $\phi(x_0) = g(x_0)$.

Finally, from Problem 9.13, we obtain

$$\begin{aligned}
g(x_0) &= \lim_{n \to \infty} f_n'(x_0) = \lim_{n \to \infty} \lim_{x \to x_0} \phi_n(x) = \lim_{x \to x_0} \lim_{n \to \infty} \phi_n(x) \\
&= \lim_{x \to x_0} \lim_{n \to \infty} \frac{f_n(x) - f(x)}{x - x_0} = \lim_{x \to x_0} \frac{f(x) - f(x_0)}{x - x_0}.
\end{aligned}$$

This shows that f is differentiable at x_0 and that $f'(x_0) = g(x_0)$.

10.   SEPARATION PROPERTIES OF CONTINUOUS FUNCTIONS

Problem 10.1.   *Let (X, d) be a metric space and let A be a nonempty subset of X. The **distance function** of A is the function $d(\cdot, A): X \to \mathbb{R}$ defined by*

$$d(x, A) = \inf\{d(x, a): a \in A\}.$$

Show that $d(x, A) = 0$ if and only if $x \in \overline{A}$.

Solution.   Clearly, $d(x, A) \geq 0$ for each $x \in X$. Assume that $x \in \overline{A}$ and let $\epsilon > 0$. Then $B(x, \epsilon) \cap A \neq \emptyset$, and so there exists some $y \in A$ such that $d(x, y) < \epsilon$. From the definition of the distance function, we see that $0 \leq d(x, A) \leq d(x, y) < \epsilon$. Since $\epsilon > 0$ is arbitrary, this implies $d(x, A) = 0$.

   For the converse, assume that $d(x, A) = 0$. If $\epsilon > 0$, then it follows from $d(x, A) = \inf\{d(x, a): a \in A\} < \epsilon$ that there exists some $a \in A$ with $d(x, a) < \epsilon$. Hence, $B(x, \epsilon) \cap A \neq \emptyset$ for each $\epsilon > 0$, and this implies that $x \in \overline{A}$.

Problem 10.2.   *Let (X, d) be a metric space, let A and B be two nonempty disjoint closed sets and consider the function $f: X \to [0, 1]$ defined by $f(x) = \frac{d(x,A)}{d(x,A)+d(x,B)}$. Show that:*
   a.   *f is a continuous function,*
   b.   *$f^{-1}(\{0\}) = A$ and $f^{-1}(\{1\}) = B$, and*
   c.   *if $d(A, B) = \inf\{d(a, b): a \in A$ and $b \in B\} > 0$, then f is uniformly continuous.*

Solution.   Let C be an arbitrary nonempty subset of X. We shall show first that the function $x \mapsto d(x, C)$ is uniformly continuous. To see this, fix $x, y \in X$. Choosing some $c \in C$, we see that

$$d(x, C) \leq d(x, c) \leq d(x, y) + d(y, c) \leq d(x, y) + d(y, C),$$

or $d(x, C) - d(y, C) \leq d(x, y)$. Exchanging the roles of x and y in the last inequality, we get $d(y, C) - d(x, C) \leq d(x, y)$. Therefore,

$$\left| d(x, C) - d(y, C) \right| \leq d(x, y),$$

and the uniform continuity of $x \mapsto d(x, C)$ follows.

   (a) Observe that since A and B are disjoint closed sets, it follows from the Problem 10.1 that $d(x, A) + d(x, B) > 0$ for each $x \in X$. This, in connection with the (uniform) continuity of the functions $d(\cdot, A)$ and $d(\cdot, B)$, guarantees that f is a continuous function.

   (b) Note that $f(x) = 0$ if and only if $d(x, A) = 0$. Now, by Problem 10.1, we have $d(x, A) = 0$ if and only if $x \in \overline{A} = A$. In other words, we have $f(x) = 0$ if and only if $x \in A$. This means $f^{-1}(\{0\}) = A$.

Similarly, notice that $f(x) = \frac{d(x,A)}{d(x,A)+d(x,B)} = 1$ if and only if $d(x, B) = 0$. As above, this shows that $f^{-1}(\{1\}) = B$.

(c) Fix some $\epsilon > 0$ such that $d(u, v) \geq \epsilon$ for all $u \in A$ and $v \in B$. If $a \in A$ and $b \in B$ are arbitrary, then for each $z \in X$ we have

$$\epsilon \leq d(a, b) \leq d(z, a) + d(z, b) \leq d(z, A) + d(z, B).$$

Now, if $x, y \in X$, then the inequalities

$$\begin{aligned}
|f(x) - f(y)| &= \left| \frac{d(x, A)}{d(x, A) + d(x, B)} - \frac{d(y, A)}{d(y, A) + d(y, B)} \right| \\
&= \frac{|[d(y, A) + d(y, B)]d(x, A) - [d(x, A) + d(x, B)]d(y, A)|}{[d(x, A) + d(x, B)][d(y, A) + d(y, B)]} \\
&= \frac{|[d(x, A) - d(y, A)]d(x, B) + [d(y, B) - d(x, B)]d(x, A)|}{[d(x, A) + d(x, B)][d(y, A) + d(y, B)]} \\
&\leq \frac{[d(x, B) + d(x, A)]d(x, y)}{[d(x, A) + d(x, B)][d(y, A) + d(y, B)]} \\
&\leq \frac{d(x, y)}{\epsilon}
\end{aligned}$$

guarantee that f is uniformly continuous.

Problem 10.3. *Let A and B be two nonempty subsets of a metric space X such that $A \cap \overline{B} = \overline{A} \cap B = \emptyset$. Show that there exist two open disjoint set U and V such that $A \subseteq U$ and $B \subseteq V$.*

Solution. From the solution of Problem 10.2, we know that for each nonempty subset C of X the function $x \mapsto d(x, C)$ is (uniformly) continuous. Now, consider the function $f: X \to \mathbb{R}$ defined by

$$f(x) = d(x, A) - d(x, B).$$

By the above, f is a continuous function. From $A \cap \overline{B} = \emptyset$ and Problem 10.1, we see that $f(x) = -d(x, B) < 0$ holds for each $x \in A$. Similarly, $f(x) > 0$ holds for each $x \in B$. Consequently, the two disjoint open sets $U = f^{-1}\big((-\infty, 0)\big)$ and $V = f^{-1}\big((0, \infty)\big)$ satisfy $A \subseteq U$ and $B \subseteq V$.

Problem 10.4. *Show that a closed set of a normal space is itself a normal space.*

Solution. Let C be a closed subset of a normal space X. We consider C equipped with the topology induced by X. Now, assume that A and B are two disjoint closed subsets of C. Since C is closed, it is easy to see that A and B are also closed subsets of X. Pick two open subsets V_1 and W_1 of X satisfying $A \subseteq V_1$, $B \subseteq W_1$ and $V_1 \cap W_1 = \emptyset$. Now, if $V = C \cap V_1$ and $W = C \cap W_1$, then V and W are two disjoint open subsets of C satisfying $A \subseteq V$ and $B \subseteq W$. This shows that C equipped with the relative topology is a normal space.

Problem 10.5. *Let X be a normal space and let A and B be two disjoint closed subsets of X. Show that there exist open sets V and W such that $A \subseteq V$, $B \subseteq W$ and $\overline{V} \cap \overline{W} = \emptyset$.*

Solution. Assume that A and B are two disjoint closed subsets of a normal space X. Pick two disjoint open sets V and W_1 satisfying $A \subseteq V$ and $B \subseteq W_1$. We claim that $\overline{V} \cap W_1 = \emptyset$. Indeed, if $x \in \overline{V} \cap W_1$, then on one hand W_1 is a neighborhood of x, and on the other hand, x belongs to the closure of V, which imply $W_1 \cap V \neq \emptyset$, a contradiction.

Now, since $\overline{V} \cap B = \emptyset$ and X is normal, there exist two disjoint open sets V_1 and W such that $\overline{V} \subseteq V_1$ and $B \subseteq W$. As before, $V_1 \cap \overline{W} = \emptyset$, and clearly the open sets V and W satisfy the desired properties.

Alternatively: If a continuous function $f \colon X \to [0, 1]$ satisfies $A \subseteq f^{-1}(\{0\})$ and $B \subseteq f^{-1}(\{1\})$, then the open sets $V = f^{-1}\big([0, \frac{1}{2})\big)$ and $W = f^{-1}\big((\frac{3}{4}, 1]\big)$ satisfy $A \subseteq \overline{V}$, $B \subseteq \overline{W}$, and $\overline{V} \cap \overline{W} = \emptyset$.

Problem 10.6. *Show that a topological space is normal if and only if for each closed set A and each open set V with $A \subseteq V$, there exists an open set W such that $A \subseteq W \subseteq \overline{W} \subseteq V$.*

Solution. Let X be a topological space. Assume first that X is a normal space and let a closed set A and an open set V satisfy $A \subseteq V$. Then $A \cap V^c = \emptyset$ and V^c is a closed set. Pick two disjoint open sets W and U such that $A \subseteq W$ and $V^c \subseteq U$. In particular, $\overline{W} \cap U = \emptyset$. This implies $\overline{W} \cap V^c = \emptyset$, and so $\overline{W} \subseteq V$.

For the converse, assume that the property is satisfied and let A and B be two disjoint nonempty closed sets. If $V = B^c$, then V is an open set such that $A \subseteq V$. By our hypothesis, there exists an open set W such that $A \subseteq W \subseteq \overline{W} \subseteq V = B^c$. If $U = \overline{W}^c$, then U is an open set disjoint from W and satisfies $B \subseteq U$. This shows that X is a normal space.

Problem 10.7. *For a closed subset A of a normal topological space X, establish the following:*

  a. *There exists a continuous function $f \colon X \to [0, 1]$ satisfying $f^{-1}(\{0\}) = A$ if and only if A is a G_δ-set.*

b.   *If A is a G_δ-set and B is another closed set satisfying $A \cap B = \emptyset$, then there exists a continuous function $g: X \to [0, 1]$ such that $g^{-1}(\{0\}) = A$ and $g(b) = 1$ for each $b \in B$.*

Solution.   Let A be a closed subset of a normal topological space X.

(a) If there exists a continuous function $f: X \to [0, 1]$ such that $f^{-1}(\{0\}) = A$, then the identity

$$A = f^{-1}(\{0\}) = f^{-1}\left(\bigcap_{n=1}^{\infty}[0, \tfrac{1}{n})\right) = \bigcap_{n=1}^{\infty} f^{-1}\left([0, \tfrac{1}{n})\right)$$

shows that A is a G_δ-set.

For the converse, assume that A is a G_δ-set. Pick a sequence $\{V_n\}$ of open sets such that $A = \bigcap_{n=1}^{\infty} V_n$. Since $A \cap V_n^c = \emptyset$, it follows from Uryson's lemma that there exists a continuous function $f_n: X \to [0, 1]$ satisfying $f_n(a) = 0$ for each $a \in A$ and $f_n(x) = 1$ for all $x \in V_n^c$. Now, consider the function $f: X \to [0, 1]$ defined by

$$f(x) = \sum_{n=1}^{\infty} \tfrac{1}{2^n} f_n(x).$$

From the Weierstrass' M-test (Theorem 9.5) and Theorem 9.2, it is easy to see that f is a continuous function, and we claim that $f^{-1}(\{0\}) = A$. Clearly, $f(x) = 0$ for each $x \in A$. Now, assume $f(x) = 0$. Then $f_n(x) = 0$ for all n, and so (in view of $f_n(v) = 1$ for each $v \in V_n^c$) we have $x \in V_n$ for each n, i.e., $x \in \bigcap_{n=1}^{\infty} V_n = A$. Therefore, $f^{-1}(\{0\}) = A$.

(b) Assume now that A is a closed G_δ-set and B is another closed set such that $A \cap B = \emptyset$. So, there exist two disjoint open set V and W such that $A \subseteq V$ and $B \subseteq W$. This implies that the sequence $\{V_n\}$ introduced in part (a) can be assumed to satisfy $V_n \subseteq V$ for each n. In particular, each f_n satisfies $f_n(b) = 1$ for each $b \in B$. Now, it is easy to see that the continuous function f constructed in the preceding part satisfies the desired property.

Problem 10.8.   *Show that a compact subset A of a Hausdorff locally compact topological space is a G_δ-set if and only if there exists a continuous function $f: X \to [0, 1]$ such that $A = f^{-1}(\{0\})$.*

Solution.   If $A = f^{-1}(\{0\})$, then—as in the solution of part (a) of the preceding problem—the set A is a G_δ-set. For the converse, assume that $A = \bigcap_{n=1}^{\infty} V_n$, where each V_n is an open set. By Theorem 10.8, for each n there exists a continuous function $f_n: X \to [0, 1]$ such that $f_n(x) = 1$ for each $x \in A$ and $f_n(x) = 0$ for each $x \notin V_n$. Now, as in the the solution of part (a) of the preceding problem, notice that the function $f: X \to [0, 1]$ defined by $f(x) = \sum_{n=1}^{\infty} \frac{1}{2^n} f_n(x)$ satisfies the desired properties.

Problem 10.9. *A topological space X is said to be **perfectly normal** if for every pair of disjoint closed sets A and B, there is a continuous function $f: X \to [0, 1]$ such that $A = f^{-1}(\{0\})$ and $B = f^{-1}(\{1\})$. (Part (b) of Problem 10.2 shows that every metric space is perfectly normal.)*

Show that a Hausdorff normal topological space is perfectly normal if and only if every closed set is a G_δ-set.

Solution. Let X be a Hausdorff normal topological space. Assume first that X is perfectly normal and let A be a proper closed subset of X. If $a \in X$ satisfies $a \notin A$, then $A \cap \{a\} = \emptyset$ and $\{a\}$ is a closed set. So, there exists a continuous function $f: X \to [0, 1]$ with $f^{-1}(\{0\}) = A$. This implies (as in the solution of Problem 10.7), that A is a G_δ-set.

For the converse, assume that every closed set is a G_δ-set. Let A and B be two closed disjoint sets. By Problem 10.7 there exist two continuous functions $g, h: X \to [0, 1]$ such that:

  i. $g^{-1}(\{0\}) = A$ and $g(b) = 1$ for each $b \in B$, and
  ii. $h^{-1}(\{0\}) = B$ and $h(a) = 1$ for each $a \in A$.

Now, let $f = \frac{1}{2}g + \frac{1}{2}(1 - h)$, and note that $f: X \to [0, 1]$, $A = f^{-1}(\{0\})$ and $B = f^{-1}(\{1\})$.

Problem 10.10. *Show that a nonempty connected normal space is either a singleton or uncountable.*

Solution. Let X be a (nonempty) Hausdorff connected normal space. If X is not a singleton, then there exist $a, b \in X$ with $a \neq b$. Since X is Hausdorff, singletons are closed sets, and we have $\{a\} \cap \{b\} = \emptyset$. Now, pick a continuous function $f: X \to [0, 1]$ such that $f(a) = 0$ and $f(b) = 1$. The assumption that X is connected guarantees (according to Problem 6.11(g)) that $f(X)$ is an interval and so $f(X) = [0, 1]$. This easily implies that X is uncountable—in fact, it has cardinality greater than or equal to the cardinality of the continuum.

Problem 10.11. *Let X be a normal space, let C be a closed subset of X, and let I be a nonempty interval—with the possibility $I = (-\infty, \infty)$. If $f: C \to I$ is a continuous function, then show that f has a continuous extension to all of X with values in I.*

Solution.   Assume that C is a closed subset of a normal space X and that $f: X \rightarrow I$ is a continuous function, where I is an interval. The interval I must be one of the following type: (a, b), $[a, b]$, $[a, b)$, $(a, b]$. So, we shall establish the continuous extension of f by steps.

STEP I: *I is either of the form $[a, b)$ or $(b, a]$.*

In this case, there exists a homeomorphism $h: I \rightarrow [0, 1)$. For instance, if $-\infty < a < b < \infty$, then $h(x) = \frac{b-x}{b-a}$ is a homeomorphism between $(a, b]$ and $[0, 1)$. Likewise, if $a \in \mathbb{R}$, then $h(x) = \frac{a-x}{1+a-x}$ defines a homeomorphism between $(-\infty, a]$ and $[0, 1)$.

Fix a homeomorphism $h: I \rightarrow [0, 1)$ and consider the continuous (composition) function $h \circ f: C \rightarrow [0, 1) \subseteq [0, 1]$. By Tietze's extension theorem, there exists a continuous function $g: X \rightarrow [0, 1]$ satisfying $g(x) = h(f(x))$ for all $x \in C$. The continuity of g guarantees that the set $A = g^{-1}(\{1\})$ is a closed subset of X. Also, since for each $x \in C$, we have $g(x) = h(f(x)) \in [0, 1)$, we see that $C \cap A = \emptyset$. By Uryson's lemma, there exists a continuous function $\theta: X \rightarrow [0, 1]$ such that $\theta(a) = 0$ for each $a \in A$ and $\theta(c) = 1$ for each $c \in C$.

Now, consider the function $\phi: X \rightarrow [0, 1]$ defined by $\phi(x) = \theta(x)g(x)$. We claim that $\phi(X) \subseteq [0, 1)$. To see this, let $x \in X$. If $x \in A$, then $\phi(x) = \theta(x)g(x) = 0 \cdot 1 = 0$, and if $x \notin A$, then $0 \le g(x) < 1$ and so $\phi(x) = \theta(x)g(x) < 1$ is also true. Next, define the function $\hat{f}: X \rightarrow I$ by

$$\hat{f}(x) = (h^{-1} \circ \phi)(x) = h^{-1}(\theta(x)g(x)).$$

If $x \in C$, then $\theta(x)g(x) = g(x) = h(f(x))$, and hence,

$$\hat{f}(x) = h^{-1}(h(f(x))) = f(x).$$

This shows that $\hat{f}: X \rightarrow I$ is a continuous extension of f to all of X.

STEP II: *$I = [a, b]$ with $-\infty < a < b < \infty$.*

The function $h: [a, b] \rightarrow [0, 1]$, defined by $h(x) = \frac{x-a}{b-a}$, is a homeomorphism. By Tietze's extension theorem, there exists a continuous function $g: X \rightarrow [0, 1]$ satisfying $g(x) = (h \circ f)(x)$ for each $x \in C$. Then the continuous function $\hat{f} = h^{-1} \circ g: X \rightarrow [a, b]$ satisfies $\hat{f}(c) = f(c)$ for each $c \in C$.

STEP III: *Assume $I = (a, b)$ with $-\infty \le a < b \le \infty$.*

In this case, there exists a homeomorphism $h: (a, b) \rightarrow (-1, 1)$. (For instance, for $-\infty < a < b < \infty$ let $h(x) = \frac{2(x-a)}{b-a} - 1$ and if $(a, b) = (-\infty, \infty)$ take $h(x) = \frac{2}{\pi} \arctan x$.) Now, consider the continuous function $h \circ f: C \rightarrow (-1, 1) \subseteq [-1, 1]$ and note that by STEP II there exists a continuous function $g: X \rightarrow [-1, 1]$ satisfying $g(c) = (h \circ f)(c)$ for each $c \in C$.

Next, let $B = g^{-1}(\{-1, 1\})$. Then B is closed and $B \cap C = \emptyset$. By Uryson's lemma, there exists a continuous function $\theta \colon X \to [0, 1]$ satisfying $\theta(b) = 0$ for each $b \in B$ and $\theta(c) = 1$ for each $c \in C$. As before, define the continuous function $\phi \colon X \to [-1, 1]$ by $\phi(x) = \theta(x)g(x)$. Then it is easy to see that $\phi(X) \subseteq (-1, 1)$ and the function $\hat{f} \colon X \to (a, b)$, defined by $\hat{f} = h^{-1} \circ \phi$, is a continuous extension of f.

11. THE STONE–WEIERSTRASS APPROXIMATION THEOREM

Problem 11.1. *Let X be a compact topological space. For a subset L of $C(X)$, let $\overline{L}$ denote the uniform closure of L in $C(X)$. Show the following:*
   a. *If L is a function space, then so is $\overline{L}$.*
   b. *If L is an algebra, then so is $\overline{L}$.*

Solution. Let $f, g \in \overline{L}$. Pick two sequences $\{f_n\}$ and $\{g_n\}$ of L that converge uniformly to f and g, respectively. Also, pick some $M > 0$ so that $\|f_n\|_\infty \leq M$ and $\|g_n\|_\infty \leq M$ hold for all n.
(a) The inequality $\big||f_n| - |f|\big| \leq |f_n - f|$ shows that $\{|f_n|\}$ converges uniformly to $|f|$. Since $|f_n| \in L$ for each n, it follows that $|f| \in \overline{L}$. This implies that $\overline{L}$ is a function space.
(b) From the inequalities

$$\|f_n g_n - fg\|_\infty \leq \|g\|_\infty \cdot \|f_n - f\|_\infty + \|f_n\|_\infty \cdot \|g_n - g\|_\infty$$
$$\leq M\big(\|f_n - f\|_\infty + \|g_n - g\|_\infty\big),$$

it follows that the sequence $\{f_n g_n\}$ of L converges uniformly to fg. Thus, $fg \in \overline{L}$, and so $\overline{L}$ is an algebra.

Problem 11.2. *Let L be the collection of all continuous piecewise linear functions defined on $[0, 1]$. That is, $f \in L$ if and only if $f \in C[0, 1]$ and there exists a finite number of points $0 = x_0 < x_1 < \cdots < x_n = 1$ (depending on f) such that f is linear on each interval $[x_{m-1}, x_m]$. Show that L is a function space but not an algebra. Moreover, show that L is dense in $C[0, 1]$ with respect to the uniform metric.*

Solution. The verification that L is a function space is routine. Since the function $f(x) = x$ satisfies $f \in L$ and $f^2 \notin L$, it follows that L is not an algebra of functions.
   To see that L is dense, let $f \in C[0, 1]$ and let $\varepsilon > 0$. By the uniform continuity of f, there exists some $\delta > 0$ such that $|x - y| < \delta$ implies $|f(x) - f(y)| < \varepsilon$. Let $0 = x_0 < x_1 < \cdots < x_n = 1$ be a finite collection of points with $x_i - x_{i-1} < \delta$

for each $1 \leq i \leq n$. The function g, defined on each subinterval $[x_{i-1}, x_i]$ by

$$g(t) = f(x_{i-1}) + \frac{f(x_i) - f(x_{i-1})}{x_i - x_{i-1}}(t - x_{i-1}),$$

belongs to L and satisfies $\|f - g\|_\infty < \varepsilon$.

An alternate way of proving the denseness of L is the following: Note that $1 \in L$ and L separates the points of $[0, 1]$ (why?). Thus, by the Stone–Weierstrass theorem, L is dense in $C[0, 1]$.

Problem 11.3. *Show that a continuous function $f: (0, 1) \to \mathbb{R}$ is the uniform limit of a sequence of polynomials on $(0, 1)$ if and only if it admits a continuous extension to $[0, 1]$.*

Solution. Let $f: (0, 1) \longrightarrow \mathbb{R}$ be a continuous function. Assume first that f has a continuous extension to $[0, 1]$—which we denote by $\hat{f}$. Then, by Corollary 11.6, the function $\hat{f}$ is the uniform limit of a sequence of polynomials on $[0, 1]$, and consequently $f: (0, 1) \longrightarrow \mathbb{R}$ is likewise the uniform limit of a sequence of polynomials on $(0, 1)$.

For the converse, assume that there exists a sequence of polynomials $\{p_n\}$ that converges uniformly to f on $(0, 1)$. Let $\varepsilon > 0$ and then pick some n_0 such that $|p_n(x) - f(x)| < \varepsilon$ holds for all $x \in (0, 1)$ and all $n \geq n_0$. From the triangle inequality, we see that

$$\left| p_n(x) - p_m(x) \right| \leq \left| p_n(x) - f(x) \right| + \left| p_m(x) - f(x) \right| < \varepsilon + \varepsilon = 2\varepsilon$$

for all $x \in (0, 1)$ and all $n \geq n_0$. By continuity, we infer that

$$|p_n(x) - p_m(x)| \leq 2\varepsilon$$

holds for all $x \in [0, 1]$ and all $n \geq n_0$. The above show that $\{p_n\}$ is a Cauchy sequence of $C[0, 1]$, and so (by Theorem 9.3) the sequence $\{p_n\}$ converges in $C[0, 1]$, say to $g \in C[0, 1]$. It follows that $f(x) = g(x)$ for all $x \in (0, 1)$, and so g is a continuous extension of $f: (0, 1) \longrightarrow \mathbb{R}$ to $[0, 1]$.

Problem 11.4. *If f is a continuous function on $[0, 1]$ such that $\int_0^1 x^n f(x)\, dx = 0$ for $n = 0, 1, \ldots$, then show that $f(x) = 0$ for all $x \in [0, 1]$.*

Solution. By Corollary 11.6, there exists a sequence of polynomials $\{p_n\}$ that converges uniformly to f. It easily follows that $\{p_n f\}$ also converges uniformly to f^2, and by our hypothesis we see that $\int_0^1 p_n(x) f(x)\, dx = 0$ holds for each n. Now, invoke Problem 9.16 to infer that $\int_0^1 f^2(x)\, dx = \lim \int_0^1 p_n(x) f(x)\, dx = 0$. The latter easily implies that $f(x) = 0$ holds for each $x \in [0, 1]$.

Problem 11.5. *Show that the algebra generated by the set $\{1, x^2\}$ is dense in $C[0, 1]$ but fails to be dense in $C[-1, 1]$.*

Solution. Since the function $f(x) = x^2$ separates the points of $[0, 1]$, the algebra generated by $\{1, x^2\}$ also separates the points of $[0, 1]$. Thus, by the Stone–Weierstrass, this algebra must be dense in $C[0, 1]$.

To see that the algebra generated by $\{1, x^2\}$ is not dense in $C[-1, 1]$, note that for every f in the closure of this algebra, we have $f(-1) = f(1)$. Thus, this algebra is not dense in $C[-1, 1]$.

Problem 11.6. *Let us say that a polynomial is **odd** (resp. **even**) whenever it does not contain any monomial of even (resp. odd) degree.*

Show that a continuous function $f: [0, 1] \to \mathbf{R}$ vanishes at zero (i.e., $f(0) = 0$) if and only if it is the uniform limit of a sequence of odd polynomials on $[0, 1]$.

Solution. If f is the uniform limit of a sequence of odd polynomials, then it should be clear that f vanishes at zero. For the converse, assume that $f \in C[0, 1]$ satisfies $f(0) = 0$ and let $\varepsilon > 0$. Define the function $g: [-1, 1] \to \mathbf{R}$ by

$$g(x) = \begin{cases} f(x), & \text{if } 0 \le x \le 1; \\ -f(-x), & \text{if } -1 \le x < 0, \end{cases}$$

and note that $g \in C[-1, 1]$. By the Stone–Weierstrass theorem there exists a polynomial p such that $|g(x) - p(x)| < \varepsilon$ for each $x \in [-1, 1]$.

Next, write $p = q + r$, where q is the odd polynomial consisting of the sum of all odd terms of p and r is the even polynomial consisting of the sum of all even terms (including the constant term) of p. In particular, note that $q(-x) = -q(x)$ and $r(-x) = r(x)$ hold for each x. Thus, if $0 \le x \le 1$, then

$$\big| f(x) - q(x) - r(x) \big| = \big| g(x) - p(x) \big| < \varepsilon,$$

and $g(-x) = -f(x)$ implies

$$\big| f(x) - q(x) + r(x) \big| = \big| p(-x) - g(-x) \big| < \varepsilon.$$

from which it follows that $\big| f(x) - q(x) \big| < \varepsilon$. (Here we use the elementary property: if $|a+b| < \varepsilon$ and $|a-b| < \varepsilon$, then $|a| = \big| \frac{a+b}{2} + \frac{a-b}{2} \big| \le \big| \frac{a+b}{2} \big| + \big| \frac{a-b}{2} \big| < \varepsilon$ and $|b| < \varepsilon$.) In other words, the odd polynomial q is ε-uniformly close to f on $[0, 1]$, and the desired conclusion follows.

Problem 11.7. *If $f:[0, 1] \to \mathbf{R}$ is a continuous function such that $\int_0^1 f(\sqrt[2n+1]{x})$ $dx = 0$ for $n = 0, 1, 2, \ldots$, then show that $f(x) = 0$ for all $x \in [0, 1]$. Does the same conclusion hold true if the interval $[0, 1]$ is replaced by the interval $[-1, 1]$?*

Solution. Assume that a continuous function $f \in C[0, 1]$ satisfies $\int_0^1 f(\sqrt[2n+1]{x})$ $dx = 0$ for each $n = 0, 1, 2, \ldots$. Then the change of variable $u = \sqrt[2n+1]{x}$ (or $x = u^{2n+1}$) yields

$$\int_0^1 f(\sqrt[2n+1]{x})\, dx = (2n + 1) \int_0^1 u^{2n} f(u)\, du = 0,$$

and so $\int_0^1 x^{2n} f(x)\, dx = 0$ holds for all $n = 0, 1, 2, \ldots$. The conclusion now follows immediately from Problem 11.5.

The conclusion is not valid if we replace the interval $[0, 1]$ by the interval $[-1, 1]$. For instance, if $f(x) = x$ for all $x \in [-1, 1]$, then note that $\int_{-1}^1 f(\sqrt[2n+1]{x})\, dx = 0$ holds for all $n = 0, 1, 2, \ldots$.

Problem 11.8. *Assume that a function $f:[0, \infty) \to \mathbf{R}$ is either a polynomial or else a continuous bounded function. Then show that f is identically equal to zero (i.e., show that $f = 0$) if and only if $\int_0^\infty f(x)e^{-nx}\, dx = 0$ for all $n = 1, 2, 3, \ldots$.*

Solution. Let $f:[0, \infty) \longrightarrow \mathbf{R}$ be a continuous bounded function. If $f = 0$, then clearly $\int_0^\infty f(x)e^{-nx}\, dx = 0$ holds for all $n = 1, 2, 3, \ldots$.

For the converse, assume that

$$\int_0^\infty f(x)e^{-nx}\, dx = 0 \quad \text{holds for all} \quad n = 1, 2, 3, \ldots. \tag{$\star$}$$

Using the change of variable $u = e^{-x}$, it follows from $(\star)$ that

$$\int_0^\infty f(x)e^{-nx}\, dx = \int_{0+}^1 f(-\ln u)u^{n-1}\, du = 0, \quad n = 1, 2, \ldots. \tag{$\star\star$}$$

In particular, $\int_{0+}^1 g(u)u^n\, du = 0$ holds for each $n = 0, 1, \ldots$, where $g(u) = uf(-\ln u)$. Since f is bounded, note that $\lim_{u \to 0+} g(u) = 0$ holds, and so g defines a continuous function on $[0, 1]$. From $(\star\star)$, we see that $\int_0^1 g(x)x^n\, dx = 0$ holds for all $n = 0, 1, 2, \ldots$. Problem 11.4 implies that $g = 0$, and consequently $f = 0$.

A closer look at the above arguments reveals that we have actually proven the following result.

- *Assume that $f:[0,\infty) \to \mathbb{R}$ is a continuous function such that*

$$\int_0^\infty f(x)e^{-nx}\, dx = 0 \ \ \text{for all}\ \ n = k, k+1, k+2, \ldots,$$

where k is a positive integer. If $\lim_{u \to 0^+} u^m f(-\ln u) = 0$ for some natural number m, then the function f is identically equal to zero.

Indeed, replacing n by $n + k + m + 1$ in ($\star\star$), we get

$$\int_{0^+}^1 u^{m+k} f(-\ln u) u^n\, du = 0, \ \ n = 0, 1, 2, \ldots,$$

which implies (as above) that $f = 0$. The reader can verify easily that any function f that satisfies $|f(x)| \le Ce^{\alpha x}$ for some $C > 0$ and $\alpha > 0$ and all $x \ge x_0$ also satisfies $\lim_{u \to 0^+} u^m f(-\ln u) = 0$ for some natural number m. In particular, the reader should notice that every polynomial p satisfies an estimate of the form $|p(x)| \le Ce^{\alpha x}$.

One more comment regarding the above discussion is in order. Recall that if $f:[0,\infty) \to \mathbb{R}$ is a "nice" function, then the formula

$$\mathcal{L}(f)(s) = \int_0^\infty e^{-st} f(t)\, dt$$

is called the **Laplace transform** of f. The Laplace transform is a linear operator and plays an important role in a wide range of applications. The reader should notice that in actuality property ($\bullet$) asserts that the Laplace transform is a one-to-one operator when defined on an appropriate linear space of functions. (See also Example 30 of Chapter 5 in the text.)

Problem 11.9. *Show that a continuous bounded function $f:[1,\infty) \to \mathbb{R}$ is identically equal to zero if and only if $\int_1^\infty x^{-n} f(x)\, dx = 0$ for each $n = 8, 9, 10, \ldots$.*

Solution. The "if" part only needs verification. Therefore, assume that the function $f:[1,\infty) \longrightarrow \mathbb{R}$ satisfies $\int_1^\infty x^{-n} f(x)\, dx = 0$ for each $n = 8, 9, 10, \ldots$. Using the change of variable $u = x^{-1}$, we see that

$$\int_1^\infty x^{-n} f(x)\, dx = \int_{0^+}^1 u^{n-2} f\big(\tfrac{1}{u}\big)\, du = \int_{0^+}^1 u^{n-8} g(u)\, du = 0, \qquad (\star\star\star)$$

where $g(u) = u^6 f(\tfrac{1}{u})$. Since f is bounded, we see that $\lim_{u \to 0^+} g(u) = 0$, and so g defines a continuous function on $[0, 1]$. In addition, from ($\star \star \star$), we see that

$\int_0^1 x^n g(x)\,dx = 0$ holds for each $n = 0, 1, 2, \ldots$. By Problem 11.4, it follows that $g = 0$, and consequently, $f = 0$.

Problem 11.10. *Let A be an algebra of continuous real-valued functions defined on a compact topological space X and separating the points of X. Show that the closure $\overline{A}$ of A in $C(X)$ with respect to the uniform metric is either all of $C(X)$ or else that there exists $a \in X$ such that $\overline{A} = \{f \in C(X):\ f(a) = 0\}$.*

Solution. Let $A \subseteq C(X)$ be an algebra, where X is compact. Now, consider the sequence of polynomials $\{P_n(x)\}$ on $[0, 1]$ defined by

$$P_1(x) = 0 \quad \text{and} \quad P_{n+1}(x) = P_n(x) + \tfrac{1}{2}\big[x - \big(P_n(x)\big)^2\big] \text{ for } n = 1, 2, \ldots.$$

An easy inductive argument shows that each polynomial $P_n(x)$ has a constant term equal to zero. This guarantees that if $f \in \overline{A}$, then $P_n(f) \in \overline{A}$ for each n. Also, by Lemma 11.4, we know that the sequence $\{P_n(x)\}$ converges uniformly to $\sqrt{x}$ on $[0, 1]$. Thus, if $f \in \overline{A}$ is non-zero, then put $c = \|f\|_\infty$, and note that:

1. The sequence $\left\{P_n(\frac{f^2}{c^2})\right\} \subseteq \overline{A}$ converges uniformly to $\frac{|f|}{c}$. Hence, $|f| \in \overline{A}$.
2. Since $\left\{P_n(\frac{|f|}{c})\right\} \subseteq \overline{A}$ converges uniformly to $\sqrt{\frac{|f|}{c}}$, we see that $\sqrt{\frac{|f|}{c}} \in \overline{A}$.

Thus, if $f \in \overline{A}$, then both $|f|$ and $\sqrt{|f|}$ belong to $\overline{A}$. In particular, $\overline{A}$ is an algebra and a function space.

Now, suppose that $\overline{A}$ is not of the form $\{f \in C(X):\ f(a) = 0\}$ for some $a \in X$. This implies that for each $x \in X$, there exists some $f \in \overline{A}$ with $f(x) \neq 0$. Thus, for each $x \in X$, there exists some $f_x \in \overline{A}$ and a neighborhood V_x of x with $f_x(y) \neq 0$ for all $y \in V_x$. By the compactness of X, there exist $x_1, \ldots, x_n \in X$ with $X = \bigcup_{i=1}^n V_{x_i}$. Note that the function $g = f_{x_1}^2 + \cdots + f_{x_n}^2$ of $\overline{A}$ satisfies $g(x) > 0$ for each $x \in X$. Multiplying by an appropriate constant, we can assume that $g(x) > 1$ holds for all x. Put $h_n = \sqrt[2^n]{g}$, and note that $h_n \in \overline{A}$ and that $h_n(x) \downarrow 1$ for each $x \in X$. By Dini's theorem, $\{h_n\}$ converges uniformly to the constant function $\mathbf{1}$, and so $\mathbf{1} \in \overline{A}$. Theorem 11.5 now guarantees that $\overline{A} = C(X)$ must hold.

Problem 11.11. *Let A be the vector space generated by the functions*

$$\mathbf{1}, \sin x, \sin^2 x, \sin^3 x, \ldots$$

defined on $[0, 1]$. That is, $f \in A$ if and only if there is a non-negative integer k and real numbers $\alpha_0, \alpha_1, \ldots, \alpha_k$ (all depending on f) such that $f(x) = \sum_{n=0}^k \alpha_n \sin^n x$

for each $x \in [0, 1]$. Show that $\mathcal{A}$ is an algebra and that $\mathcal{A}$ is dense in $C[0, 1]$ with respect to the uniform metric.

Solution. Clearly, $\mathcal{A}$ is an algebra of functions that contains the constant function **1**. Also, since the function $f(x) = \sin x$ separates the points of $[0, 1]$, the algebra $\mathcal{A}$ likewise separates the points of $[0, 1]$. By the Stone–Weierstrass theorem, $\mathcal{A}$ is dense in $C[0, 1]$.

Problem 11.12. *Let X be a compact subset of $\mathbb{R}$. Show that $C(X)$ is a separable metric space (with respect to the uniform metric).*

Solution. The polynomials with rational coefficients form a countable set (why?). By Corollary 11.6, this set is dense in $C(X)$.

Problem 11.13. *Generalize the previous exercise as follows: Show that if (X, d) is a compact metric space, then $C(X)$ is a separable metric space.*

Solution. By Problem 7.2, we know that X is a separable metric space. Fix a countable dense subset $\{x_1, x_2, \ldots\}$ of X and for each n let $f_n: X \longrightarrow \mathbb{R}$ be the function defined by $f_n(t) = d(t, x_n)$ for each $t \in X$.

Now, let $x, y \in X$ satisfy $x \neq y$. Put $d(x, y) = 2\delta > 0$. Choose some n with $d(x, x_n) < \delta$, and note that

$$f_n(y) = d(y, x_n) \geq d(x, y) - d(x, x_n) \geq 2\delta - \delta = \delta > d(x, x_n) = f_n(x),$$

so that $f_n(x) \neq f_n(y)$. This implies that the algebra generated by $\{\mathbf{1}, f_1, f_2, \ldots\}$ separates the points of X. By the Stone–Weierstrass theorem (Theorem 11.5), this algebra must be dense in $C(X)$.

Next, consider the collection $\mathcal{C}$ of all finite products of the countable collection $\{\mathbf{1}, f_1, f_2, \ldots\}$ and note that $\mathcal{C}$ is a countable set, say $\mathcal{C} = \{g_1, g_2, \ldots\}$. To complete the proof note that the finite linear combinations of $\{\mathbf{1}, g_1, g_2, \ldots\}$ with rational coefficients form a countable dense subset of $C(X)$.

Problem 11.14. *Let X and Y be two compact metric spaces. Consider the Cartesian product $X \times Y$ equipped with the distance D_1 given in Problem 7.4, so that $X \times Y$ is a compact metric space. Show that if $f \in C(X \times Y)$ and $\epsilon > 0$, then there exist functions $\{f_1, \ldots, f_n\} \subseteq C(X)$ and $\{g_1, \ldots, g_n\} \subseteq C(Y)$ such that*

$$\left| f(x, y) - \sum_{i=1}^{n} f_i(x) g_i(y) \right| < \epsilon$$

holds for all $(x, y) \in X \times Y$.

Solution.   Consider the set

$$A = \left\{ h \in C(X \times Y) \colon \ \exists \, \{f_1, \dots, f_n\} \subseteq C(X), \ \{g_1, \dots, g_n\} \subseteq C(Y) \right.$$

$$\left. \text{with } h(x, y) = \sum_{i=1}^{n} f_i(x) g_i(y) \ \forall \, (x, y) \in X \times Y \right\}.$$

Then, A is an algebra of functions of $C(X \times Y)$ and $\mathbf{1} \in A$. On the other hand, if $(x_1, y_1) \neq (x_2, y_2)$, then either $x_1 \neq x_2$ or $y_1 \neq y_2$. If $x_1 \neq x_2$, then select some $f \in C(X)$ with $f(x_1) \neq f(x_2)$, and let $F(x, y) = f(x)$ for all $(x, y) \in X \times Y$. If $y_1 \neq y_2$, then pick some $g \in C(Y)$ with $g(y_1) \neq g(y_2)$, and put $F(x, y) = g(y)$. In either case, $F \in A$ and $F(x_1, y_1) \neq F(x_2, y_2)$ holds, so that A separates the points of $X \times Y$. Now, by the Stone–Weierstrass theorem (Theorem 11.5), we have $\overline{A} = C(X \times Y)$, and the desired conclusion follows.

CHAPTER 3

THE THEORY OF MEASURE

12. SEMIRINGS AND ALGEBRAS OF SETS

Problem 12.1. *If X is a topological space, then show that the collection*

$$S = \{C \cap O: C \text{ closed and } O \text{ open}\} = \{C_1 \setminus C_2: C_1, C_2 \text{ closed sets}\}$$

is a semiring of subsets of X.

Solution. From $\emptyset = \emptyset \cap \emptyset$ and $X = X \cap X$, we see that $\emptyset, X \in S$. Next, notice that $C_1 \cap O_1, C_2 \cap O_2 \in S$ imply

$$(C_1 \cap O_1) \cap (C_2 \cap O_2) = (C_1 \cap C_2) \cap (O_1 \cap O_2) \in S.$$

Now, if $C_1 \cap O_1, C_2 \cap O_2 \in S$, then

$$
\begin{aligned}
C_1 \cap O_1 \setminus C_2 \cap O_2 &= (C_1 \cap O_2) \cap (C_2 \cap O_2)^c \\
&= (C_1 \cap O_1) \cap (C_2^c \cup O_2^c) \\
&= (C_1 \cap O_1) \cap [C_2^c \cup (O_2^c \cap C_2)] \\
&= [C_1 \cap (O_1 \cap C_2^c)] \cup [(C_1 \cap C_2 \cap O_2^c) \cap O_1] \\
&= A \cup B,
\end{aligned}
$$

where $A = C_1 \cap (O_1 \cap C_2^c) \in S$ and $B = (C_1 \cap C_2 \cap O_2^c) \cap O_1 \in S$ satisfy $A \cap B = \emptyset$.

Problem 12.2. *Let S be a semiring of subsets of a set X, and let $Y \subseteq X$. Show that $S_Y = \{Y \cap A: A \in S\}$ is a semiring of Y (called the **restriction semiring** of S to Y).*

Solution.   The conclusion follows from the identities:

a. $Y \cap \emptyset = \emptyset$;
b. $(Y \cap A) \cap (Y \cap B) = Y \cap (A \cap B)$; and
c. $Y \cap A \setminus Y \cap B = Y \cap (A \setminus B)$.

Problem 12.3.   *Let S be the collection of all subsets of $[0, 1)$ that can be written as finite unions of subsets of $[0, 1)$ of the form $[a, b)$. Show that S is an algebra of sets but not a σ-algebra.*

Solution.   Let $A = \bigcup_{i=1}^{n} [a_i, b_i)$ and $B = \bigcup_{j=1}^{m} [c_j, d_j)$. Then, we have

a. $A \cup B \in S$;
b. $A \cap B = \bigcup_{i=1}^{n} \bigcup_{j=1}^{m} [a_i, b_i) \cap [c_j, d_j) \in S$; and
c. $[0, 1) \setminus A = \bigcap_{i=1}^{n} \big([0, 1) \setminus [a_i, b_i) \big) \in S$, where the last membership holds since each $[0, 1) \setminus [a_i, b_i)$ can be written as a finite union of sets of the form $[a, b)$.

To see that S is not a σ-algebra note that $\bigcap_{n=1}^{\infty} [0, \frac{1}{n}) = \{0\} \notin S$.

Problem 12.4.   *Prove that the σ-sets of the semiring*

$$S = \{ [a, b): a, b \in \mathbb{R} \}$$

form a topology for the real numbers.

Solution.   Let τ be the collection of all σ-sets of S. Clearly, $\emptyset \in \tau$ and $\mathbb{R} = \bigcup_{n=1}^{\infty} [-n, n) \in \tau$. It should be clear that τ is closed under finite intersections. Thus, in order to establish that τ is a topology, we need to show that τ is closed under arbitrary unions. That is, if $\{[a_i, b_i): i \in I\}$ is a collection of nonempty members of S, then we must show that $A = \bigcup_{i \in I} [a_i, b_i)$ belongs to τ (i.e., that A is a σ-set).

To see this, let $V = \bigcup_{i \in I} (a_i, b_i)$. Then, V is an open set, and thus, there exists an at-most countable collection of pairwise disjoint open interval $\{(c_j, d_j): j \in J\}$ (see part (g) of Problem 6.11) such that $V = \bigcup_{j \in J} (c_j, d_j)$. For each $j \in J$, let $A_j = [c_j, d_j)$ if $c_j = a_i$ for some $i \in I$ and let $A_j = (c_j, d_j)$ if $c_j \neq a_i$ for all $i \in I$. Clearly, each A_j is a σ-set. Moreover, it is easy to see that $A = \bigcup_{j \in J} A_j$ holds, which shows that A is a σ-set.

Problem 12.5.   *Let S be a semiring of subsets of a nonempty set X. What additional requirements must be satisfied for S to be a base for a topology on X? (For the definition of a base see Problem 8.18.) Prove that if such is the case, then each member of S is both open and closed in this topology.*

Solution. Since S is already closed under finite intersections, it follows from the definition of a base that S will be a base if and only if $\bigcup_{A \in S} A = X$.

Now, assume that $\bigcup_{A \in S} A = X$ holds. Note first that if $A, B \in S$, then (since S is a semiring) $A \setminus B$ can be written as a finite union of (disjoint) members of S. It follows that $A \setminus B$ belongs to the topology generated by S. Thus, if $B \in S$, then the relation

$$B^c = X \setminus B = \left(\bigcup_{A \in S} A \right) \setminus B = \bigcup_{A \in S} (A \setminus B),$$

shows that B^c belongs to the topology generated by S. That is, in this case, every $B \in S$ is a closed and open set.

Problem 12.6. *Let A be a fixed subset of a set X. Determine the two σ-algebras of subsets of X generated by*

a. *$\{A\}$, and*
b. *$\{B: A \subseteq B \subseteq X\}$.*

Solution. (a) $\{\emptyset, A, A^c, X\}$ and   (b) $\{B: A \subseteq B \text{ or } A \subseteq B^c\}$.

Problem 12.7. *Let X be an uncountable set, and let*

$$S = \{E \subseteq X: E \text{ or } E^c \text{ is at-most countable}\}.$$

Show that S is the σ-algebra generated by the one-point subsets of X.

Solution. Clearly, S contains the one-point subsets of X, and every member of S must be a member of the σ-algebra generated by the one-point sets. Thus, it remains to be shown that S is a σ-algebra.

Clearly, $\emptyset, X \in S$ and S is closed under complementation. Then let $\{A_n\} \subseteq S$. If each A_n is at-most countable, then $\bigcup_{n=1}^{\infty} A_n$ is at-most countable, and consequently $\bigcup_{n=1}^{\infty} A_n \in S$. On the other hand, if some A_k is uncountable, then $(A_k)^c$ is at-most countable and the inclusion $\left(\bigcup_{n=1}^{\infty} A_n \right)^c \subseteq (A_k)^c$ shows that $\bigcup_{n=1}^{\infty} A_n \in S$.

Problem 12.8. *Characterize the metric spaces whose open sets form a σ-algebra.*

Solution. We shall show that the open sets of a metric space X form a σ-algebra if and only if X is a discrete metric space (i.e., if and only if every subset of X is open).

Let τ be the collection of all open sets. If $\tau = \mathcal{P}(X)$, then clearly τ is a σ-algebra. On the other hand, if τ is a σ-algebra and $x \in X$, then

$\{x\} = \bigcap_{n=1}^{\infty} B(x, \frac{1}{n})$ shows that $\{x\}$ is an open set. This easily implies that every subset of X is open (i.e., $\tau = \mathcal{P}(X)$ holds).

Problem 12.9. *Determine the σ-algebra generated by the nowhere dense subsets of a topological space.*

Solution. Let X be a topological space. Define

$$\mathcal{A} = \big\{ A \subseteq X \colon \ A \text{ is meager or } A^c \text{ is meager} \big\}.$$

Recall that a set is called meager if it can be written as a countable union of nowhere dense sets—a set A is nowhere dense if $(\overline{A})^\circ = \emptyset$. We claim that $\mathcal{A}$ is the σ-algebra generated by the nowhere dense sets of X. Clearly, every nowhere dense set belongs to $\mathcal{A}$, and every member of $\mathcal{A}$ belongs to the σ-algebra generated by the nowhere dense sets. So, it suffices to establish that $\mathcal{A}$ is a σ-algebra of sets.

Clearly, $\emptyset$, $X \in \mathcal{A}$. Also, it should be obvious that $\mathcal{A}$ is closed under complementation. Now, let $\{A_n\} \subseteq \mathcal{A}$. If each A_n is meager, then clearly $\bigcup_{n=1}^{\infty} A_n \in \mathcal{A}$. On the other hand, if $(A_k)^c$ is a meager set for some k, then the set inclusion $\big(\bigcup_{n=1}^{\infty} A_n\big)^c \subseteq (A_k)^c$ implies $\bigcup_{n=1}^{\infty} A_n \in \mathcal{A}$. Therefore, $\mathcal{A}$ is a σ-algebra.

Problem 12.10. *Let X be a nonempty set, and let $\mathcal{F}$ be an uncountable collection of subsets of X. Show that any element of the σ-algebra generated by $\mathcal{F}$ belongs to the σ-algebra generated by some countable subcollection of $\mathcal{F}$.*

Solution. Assume $\mathcal{F}$ to be uncountable. Let $\mathcal{A}$ be the σ-algebra generated by $\mathcal{F}$. Denote by $\big\{ \mathcal{A}_i \colon i \in I \big\}$ the family of all σ-algebras each of which is generated by a countable subset of $\mathcal{F}$. It suffices to show that $\mathcal{B} = \bigcup_{i \in I} \mathcal{A}_i$ is a σ-algebra (because if this is the case, then $\mathcal{A} = \mathcal{B}$ must hold, and the conclusion follows).

Clearly, $\emptyset \in \mathcal{B}$. Also, if $A \in \mathcal{B}$, then it is easy to see that $A^c \in \mathcal{B}$ likewise holds. Now, let $\{A_n\} \subseteq \mathcal{B}$. Since each A_n belongs to a σ-algebra generated by a countable subset of $\mathcal{F}$, it easily follows that there exists some $i \in I$ with $\{A_n\} \subseteq \mathcal{A}_i$. Thus, $\bigcup_{n=1}^{\infty} A_n \in \mathcal{A}_i \subseteq \mathcal{B}$. That is, $\mathcal{B}$ is a σ-algebra, as required.

Problem 12.11. *Show that every F_σ- and every G_δ-subset of a topological space is a Borel set.*

Solution. The Borel sets are the members of the σ-algebra generated by the open sets. So, a countable intersection of open sets (or a countable union of closed sets) is always a Borel set.

Problem 12.12. *Show that every infinite σ-algebra of sets has uncountably many sets.*

Solution. Let $\mathcal{A}$ be an infinite σ-algebra of subsets of a set X. If $\mathcal{A}$ contains a sequence $\{A_n\}$ of nonempty pairwise disjoint sets, then $\mathcal{A}$ has uncountably many members. Indeed, if this is the case, then for each subset s of natural numbers let $A_s = \bigcup_{n \in s} A_n \in \mathcal{A}$, and note that $A_s \neq A_t$ if $s \neq t$. By Problem 5.6, the collection $\{A_s : s \in \mathcal{P}(\mathbb{N})\}$ has uncountably many members, and so $\mathcal{A}$ must likewise have uncountably many members.

Next, we shall show that there exists a sequence $\{B_n\} \subseteq \mathcal{A}$ with $B_{n+1} \subseteq B_n$ and $B_{n+1} \neq B_n$ for all n. If this is done, then put $A_n = B_n \setminus B_{n+1}$, and use the above arguments to see that $\mathcal{A}$ is an uncountable set.

Using induction, we shall establish the existence of a sequence $\{B_n\}$ such that:

1. $B_{n+1} \subseteq B_n$ and $B_{n+1} \neq B_n$ for all n, and
2. $\{B_n \cap A : A \in \mathcal{A}\}$ is an infinite set.

The basic step of the induction is the following: Assume that $B_n \in \mathcal{A}$ has been chosen so that $\{B_n \cap A : A \in \mathcal{A}\}$ has infinitely many members. Choose $C \in \mathcal{A}$ so that $\emptyset \subseteq B_n \cap C \subseteq B_n$ is a proper inclusion at both ends. In view of

$$B_n \cap A = \left[(B_n \cap C) \cap A\right] \cup \left[(B_n \setminus C) \cap A\right],$$

we see that either $\{(B_n \cap C) \cap A : A \in \mathcal{A}\}$ or $\{(B_n \setminus C) \cap A : A \in \mathcal{A}\}$ is infinite. If $\{(B_n \cap C) \cap A : A \in \mathcal{A}\}$ is infinite, put $B_{n+1} = B_n \cap C$. If $\{(B_n \cap C) \cap A : A \in \mathcal{A}\}$ is finite, put $B_{n+1} = B_n \setminus C$.

Start the induction with $B_1 = X$.

Problem 12.13. *Let (X, τ) be a topological space, let $\mathcal{B}$ be the σ-algebra of its Borel sets, and let Y be an arbitrary subset of X. If Y is considered equipped with the induced topology and $\mathcal{B}_Y$ denotes the σ-algebra of Borel sets of (Y, τ), then show that*

$$\mathcal{B}_Y = \{A \cap Y : A \in \mathcal{B}\}.$$

Solution. Let (X, τ), Y, and $\mathcal{B}_Y$ be as in the problem, and let

$$\mathcal{A} = \{A \cap Y : A \in \mathcal{B}\}.$$

We have to show that $\mathcal{B}_Y \subseteq \mathcal{A}$ and $\mathcal{A} \subseteq \mathcal{B}_Y$ both hold.

Clearly, $\mathcal{A}$ is a σ-algebra of subsets of Y and $O \cap Y \in \mathcal{A}$ holds for each $O \in \tau$. Thus, $\mathcal{A}$ contains the open sets of Y, and so $\mathcal{B}_Y \subseteq \mathcal{A}$. Now, consider the collection of sets

$$\mathcal{C} = \{A \in \mathcal{B} : A \cap Y \in \mathcal{B}_Y\}.$$

It is easy to see that $\mathcal{C}$ is a σ-algebra of subsets of X satisfying $\tau \subseteq \mathcal{C}$. Hence, $\mathcal{C} = \mathcal{B}$, and this implies that $\mathcal{A} \subseteq \mathcal{B}_Y$. Therefore, $\mathcal{B}_Y = \mathcal{A}$, as claimed.

Problem 12.14. *Let $A_1, \ldots, A_n$ be sets in some semiring $\mathcal{S}$. Show that there exists a finite number of pairwise disjoint sets $B_1, \ldots, B_m$ of $\mathcal{S}$ such that each A_i can be written as a union of sets from the $B_1, \ldots, B_m$.*

Solution. We use induction on n. For $n = 1$ the result is trivial. Thus, assume that the result is true for some n, and let $A_1, \ldots, A_n, A_{n+1}$ be members of $\mathcal{S}$. Pick a finite number of pairwise disjoint members $B_1, \ldots, B_m$ of $\mathcal{S}$ such that each A_i, $1 \leq i \leq n$, can be written as a union of sets from $B_1, \ldots, B_m$. Clearly, $\bigcup_{i=1}^n A_i \subseteq \bigcup_{j=1}^m B_j$. The sets $B_1 \cap A_{n+1}, \ldots, B_m \cap A_{n+1}$ are pairwise disjoint members of $\mathcal{S}$. On the other hand, for each $1 \leq i \leq m$ there exists a finite pairwise disjoint collection $\mathcal{F}_i \subseteq \mathcal{S}$ with $B_i \setminus A_{n+1} = \bigcup_{C \in \mathcal{F}_i} C$ (by the definition of the semiring). Thus, the collection

$$\mathcal{F} = \mathcal{F}_1 \cup \cdots \cup \mathcal{F}_m \cup \{B_1 \cap A_{n+1}, \ldots, B_m \cap A_{n+1}\} \subseteq \mathcal{S}$$

is finite and pairwise disjoint. Moreover, each A_i $(1 \leq i \leq n)$ can be written as a union of members of $\mathcal{F}$. Now, observe that

$$A_{n+1} = \left(A_{n+1} \setminus \bigcup_{j=1}^m B_j \right) \cup (B_1 \cap A_{n+1}) \cup \cdots \cup (B_m \cap A_{n+1}).$$

By Theorem 12.2(1) there exist pairwise disjoint sets $D_1, \ldots, D_k$ in $\mathcal{S}$ such that $A_{n+1} \setminus \bigcup_{j=1}^m B_j = \bigcup_{r=1}^k D_r$. Finally, the collection $\mathcal{F} \cup \{D_1, \ldots, D_k\} \subseteq \mathcal{S}$ is finite and pairwise disjoint, and each set A_i $(1 \leq i \leq n + 1)$ can be written as a union from these sets.

13. MEASURES ON SEMIRINGS

Problem 13.1. *Let $\{a_n\}$ be a sequence of non-negative real numbers. Let $\mu(\emptyset) = 0$, and for every nonempty subset A of $\mathbb{N}$ put $\mu(A) = \sum_{n \in A} a_n$. Show that $\mu: \mathcal{P}(\mathbb{N}) \to [0, \infty]$ is a measure.*

Solution. If $\{A_n\}$ is a sequence of pairwise disjoint subsets of $\mathbb{N}$ and $A = \bigcup_{n=1}^\infty A_n$, then note that

$$\mu(A) = \sum_{k \in A} a_k = \sum_{n=1}^\infty \left(\sum_{k \in A_n} a_k \right) = \sum_{n=1}^\infty \mu(A_n).$$

Problem 13.2. *Let S be a semiring, and let $\mu: S \to [0, \infty]$ be a set function such that $\mu(A) < \infty$ for some $A \in S$. If μ is σ-additive, then show that μ is a measure.*

Solution. Write $A = A \cup \emptyset \cup \emptyset \cup \cdots$. Then,

$$\mu(A) = \mu(A) + \mu(\emptyset) + \mu(\emptyset) + \cdots .$$

If $\mu(\emptyset) > 0$, then $\mu(A) = \infty$, contrary to our hypothesis. Thus, $\mu(\emptyset) = 0$, and so μ is a measure.

Problem 13.3. *Let X be an uncountable set, and let the σ-algebra*

$$S = \{E \subseteq X: E \text{ or } E^c \text{ is at-most countable}\};$$

see also Problem 12.7. Show that $\mu: S \to [0, \infty)$, defined by $\mu(E) = 0$ if E is at-most countable and $\mu(E) = 1$ if E^c is at-most countable, is a measure on S.

Solution. Clearly, $\mu(\emptyset) = 0$. For the σ-additivity of μ let $\{E_n\} \subseteq S$ be a pairwise disjoint sequence. Let $E = \bigcup_{n=1}^{\infty} E_n$. If each E_n is at-most countable, then E itself is at-most countable, and so $\mu(E) = \sum_{n=1}^{\infty} \mu(E_n) = 0$ holds. On the other hand, if E_k^c is at-most countable for some k, then (in view of $E_n \cap E_k = \emptyset$ for $n \neq k$) we must have $E_n \subseteq E_k^c$ for $n \neq k$, and so E_n is at-most countable for each $n \neq k$. Thus,

$$1 = \mu(E) = \mu(E_k) = \sum_{n=1}^{\infty} \mu(E_n).$$

It is interesting to observe that if $X = [0, 1]$, then S is a σ-subalgebra of the Lebesgue measurable subsets of $[0, 1]$, and μ is the restriction of the Lebesgue measure to S.

Problem 13.4. *Let X be a nonempty set, and let $f: X \to [0, \infty]$ be a function. Define $\mu: \mathcal{P}(X) \to [0, \infty]$ by $\mu(A) = \sum_{x \in A} f(x)$ if $A \neq \emptyset$ and is at-most countable, $\mu(A) = \infty$ if A is uncountable, and $\mu(\emptyset) = 0$. Show that μ is a measure.*

Solution. For the σ-additivity of μ, let $\{A_n\}$ be a pairwise disjoint sequence of subsets of X. Let $A = \bigcup_{n=1}^{\infty} A_n$. If some A_n is uncountable, then A is likewise uncountable, and hence, in this case $\mu(A) = \sum_{n=1}^{\infty} \mu(A_n) = \infty$ holds. On the other hand, if each A_n is at-most countable, then A is also at-most countable, and so

$$\mu(A) = \sum_{x \in A} f(x) = \sum_{n=1}^{\infty} \left[\sum_{x \in A_n} f(x) \right] = \sum_{n=1}^{\infty} \mu(A_n)$$

also holds.

Problem 13.5. *Let S be a semiring, and let $\mu: S \to [0, \infty]$ be a finitely additive measure. Show that if μ is σ-subadditive, then μ is a measure.*

Solution. Let $\{A_n\} \subseteq S$ be a pairwise disjoint such that $A = \bigcup_{n=1}^{\infty} A_n \in S$. By hypothesis, $\mu(A) \leq \sum_{n=1}^{\infty} \mu(A_n)$ holds. On the other hand, if k is fixed, then there exist pairwise disjoint sets $B_1, \ldots, B_m \in S$ such that $A \setminus \bigcup_{n=1}^{k} A_n = \bigcup_{i=1}^{m} B_i$ (see Theorem 12.2). Since $A = \left[\bigcup_{n=1}^{k} A_n \right] \cup \left[\bigcup_{i=1}^{m} B_i \right]$ is a finite union of pairwise disjoint members of S, the finite additivity of μ implies

$$\sum_{n=1}^{k} \mu(A_n) \leq \sum_{n=1}^{k} \mu(A_n) + \sum_{i=1}^{m} \mu(B_i) = \mu(A).$$

Since k is arbitrary, $\sum_{n=1}^{\infty} \mu(A_n) \leq \mu(A)$ also holds, and so μ is a measure.

Problem 13.6. *Let $\{\mu_n\}$ be an increasing sequence of measures on a semiring S; that is, $\mu_n(A) \leq \mu_{n+1}(A)$ holds for all $A \in S$ and all n. Define $\mu: S \to [0, \infty]$ by $\mu(A) = \sup\{\mu_n(A)\}$ for each $A \in S$. Show that μ is a measure.*

Solution. Clearly, $\mu(\emptyset) = 0$. Now, let $\{A_n\} \subseteq S$ be a pairwise disjoint sequence such that $\bigcup_{n=1}^{\infty} A_n = A \in S$. Since each μ_i is a measure,

$$\mu_i(A) = \sum_{n=1}^{\infty} \mu_i(A_n) \leq \sum_{n=1}^{\infty} \mu(A_n)$$

holds, and so $\mu(A) \leq \sum_{n=1}^{\infty} \mu(A_n)$. On the other hand, for each k we have

$$\sum_{n=1}^{k} \mu(A_n) = \lim_{i \to \infty} \sum_{n=1}^{k} \mu_i(A_n) = \lim_{i \to \infty} \mu_i \left(\bigcup_{n=1}^{k} A_n \right) \leq \mu(A).$$

Thus, $\sum_{n=1}^{\infty} \mu(A_n) \leq \mu(A)$ also holds, which shows that the measure μ is σ-additive.

Problem 13.7. *Consider the semiring* $S = \{A \subseteq \mathbb{R}: A \text{ is at-most countable}\}$, *and define* $\mu: S \to [0, \infty]$ *by* $\mu(A) = 0$ *if* A *is finite and* $\mu(A) = \infty$ *if* A *is countable. Show that* μ *is a finitely additive measure that is not a measure.*

Solution. Let $A_1, \ldots, A_n$ be pairwise disjoint members of S. Put $A = \bigcup_{i=1}^n A_i$. If each A_i is a finite set, then A is likewise a finite set, and $\sum_{i=1}^n \mu(A_i) = \mu(A) = 0$ holds. On the other hand, if one of the A_i is countable, then A itself is also countable, and $\sum_{i=1}^n \mu(A_i) = \mu(A) = \infty$ holds. Thus, μ is a finitely additive measure.

To see that μ is not σ-additive, note that $\mathbb{N} = \bigcup_{n=1}^\infty \{n\}$, while

$$0 = \sum_{n=1}^\infty \mu(\{n\}) < \mu(\mathbb{N}) = \infty.$$

Problem 13.8. *Show that every finitely additive measure is monotone.*

Solution. Assume that $\mu: S \longrightarrow [0, \infty]$ is a finitely additive measure. Let $A, B \in S$ satisfy $A \subseteq B$. Choose a finite collection of disjoint sets $C_1, \ldots, C_n$ of S such that $B \setminus A = \bigcup_{i=1}^n C_i$. Then,

$$B = A \cup C_1 \cup \cdots \cup C_n$$

is a finite union of pairwise disjoint sets of S. Thus, by the finite additivity of μ, we have

$$\mu(A) \leq \mu(A) + \mu(C_1) + \cdots + \mu(C_n) = \mu(B).$$

Problem 13.9. *Consider the set function* μ *defined in Example 13.6. That is, consider a nondecreasing and left-continuous function* $f: \mathbb{R} \to \mathbb{R}$ *and then define the set function* $\mu: S \to [0, \infty)$ *by* $\mu([a, b)) = f(b) - f(a)$, *where* S *is the semiring* $S = \{[a, b): -\infty < a \leq b < \infty\}$. *Prove alternately the fact that* μ *is a measure.*

Solution. An alternate way of proving the σ-additivity of μ is as follows. Let $a < b$ and let $[a, b) = \bigcup_{n=1}^\infty [a_n, b_n)$ with the sequence $\{[a_n, b_n)\}$ pairwise disjoint. For each $a < x \leq b$ let

$$s_x = \sum_i \left[f(b_i) - f(a_i) \right],$$

where the sum (possibly a series) extends over all i for which $[a_i, b_i) \subseteq [a, x)$ holds; we let $s_x = 0$ if there is no such interval. Since f is nondecreasing, we have $s_x \leq f(x) - f(a)$. Next, note that the set

$$A = \{x \in (a, b]: \ s_x = f(x) - f(a)\}$$

is nonempty. Let $t = \sup A$, and note that $a < t \leq b$. Now, for $x \in A$, we have

$$f(x) - f(a) = s_x \leq s_t \leq f(t) - f(a),$$

and so, by the left-continuity of f, we get $s_t = f(t) - f(a)$. That is, $t \in A$.

Our objective is to establish that $t = b$ holds. Assume by way of contradiction that $a < t < b$. Then $a_k \leq t < b_k$ must hold for some k. Since the sequence $\{[a_n, b_n)\}$ is pairwise disjoint, observe that $[a_i, b_i) \subseteq [a, t)$ holds if and only if $[a_i, b_i) \subseteq [a, a_k)$. Thus, $s_t = s_{a_k}$ holds. In particular, the relation

$$f(t) - f(a) = s_t = s_{a_k} \leq f(a_k) - f(a) \leq f(t) - f(a)$$

guarantees that $a_k \in A$. However, this implies $b_k \in A$, which is impossible. Therefore, $t = b$ holds, which guarantees that

$$\mu\big([a, b)\big) = \sum_{n=1}^{\infty} \mu\big([a_n, b_n)\big).$$

14.   OUTER MEASURES AND MEASURABLE SETS

Problem 14.1.   *Show that a countable union of null sets is again a null set.*

Solution.   The conclusion follows from the inequality

$$\mu\left(\bigcup_{n=1}^{\infty} A_n\right) \leq \sum_{n=1}^{\infty} \mu(A_n).$$

Problem 14.2.   *If μ is an outer measure on a set X and A is a null set, then show that*

$$\mu(B) = \mu(A \cup B) = \mu(B \setminus A)$$

holds for every subset B of X.

Solution. The conclusion follows from the inequalities:

$$\mu(B) \le \mu(B \cup A) = \mu\big((B \setminus A) \cup A\big)$$
$$\le \mu(B \setminus A) + \mu(A) = \mu(B \setminus A) \le \mu(B).$$

Problem 14.3. *Let μ be an outer measure on a set X. If a sequence $\{A_n\}$ of subsets of X satisfies $\sum_{n=1}^{\infty} \mu(A_n) < \infty$, then show that the set*

$$E = \big\{x \in X: \ x \ belongs \ to \ A_n \ for \ infinitely \ many \ n \big\}$$

is a null set.

Solution. Assume that a sequence $\{A_n\}$ satisfies $\sum_{n=1}^{\infty} \mu(A_n) < \infty$. For each n let $E_n = \bigcup_{i=n}^{\infty} A_i$, and note that $E \subseteq E_n$ holds for each n. Therefore,

$$0 \le \mu(E) \le \mu(E_n) \le \sum_{i=n}^{\infty} \mu(A_i) \longrightarrow 0,$$

and hence $\mu(E) = 0$.

Problem 14.4. *If E is a measurable subset of X, then show that for every subset A of X the following equality holds:*

$$\mu(E \cap A) + \mu(E \cup A) = \mu(E) + \mu(A).$$

Solution. The measurability of E gives

$$\mu(E \cup A) = \mu\big((E \cup A) \cap E\big) + \mu\big((E \cup A) \cap E^c\big) = \mu(E) + \mu(A \cap E^c).$$

Consequently, we have

$$\mu(E \cup A) + \mu(E \cap A) = \mu(E) + \mu(A \cap E^c) + \mu(A \cap E) = \mu(E) + \mu(A).$$

Problem 14.5. *Let μ be an outer measure on a set X. If A is a nonmeasurable subset of X and E is a measurable set such that $A \subseteq E$, then show that $\mu(E \setminus A) > 0$.*

Solution. If $\mu(E \setminus A) = 0$ holds, then $E \setminus A \in \Lambda$. Thus, $A = E \setminus (E \setminus A) \in \Lambda$, which is a contradiction. Therefore, $\mu(E \setminus A) > 0$.

Problem 14.6.   *Let A be a subset of X, and let $\{E_n\}$ be a disjoint sequence of measurable sets. Show that*

$$\mu\left(\bigcup_{n=1}^{\infty} A \cap E_n\right) = \sum_{n=1}^{\infty} \mu(A \cap E_n).$$

Solution.   From the σ-subadditivity of μ, we see that

$$\mu\left(A \cap \left[\bigcup_{n=1}^{\infty} E_n\right]\right) = \mu\left(\bigcup_{n=1}^{\infty} A \cap E_n\right) \le \sum_{n=1}^{\infty} \mu(A \cap E_n).$$

On the other hand, Lemma 14.5 implies

$$\sum_{n=1}^{k} \mu(A \cap E_n) = \mu\left(A \cap \left[\bigcup_{n=1}^{k} E_n\right]\right) \le \mu\left(A \cap \left[\bigcup_{n=1}^{\infty} E_n\right]\right)$$

for each k, and so $\sum_{n=1}^{\infty} \mu(A \cap E_n) \le \mu\left(A \cap \left[\bigcup_{n=1}^{\infty} E_n\right]\right)$ also holds.

Problem 14.7.   *Let $\{A_n\}$ be a sequence of subsets of X. Assume that there exists a disjoint sequence $\{B_n\}$ of measurable sets such that $A_n \subseteq B_n$ holds for each n. Show that*

$$\mu\left(\bigcup_{n=1}^{\infty} A_n\right) = \sum_{n=1}^{\infty} \mu(A_n).$$

Solution.   Put $A = \bigcup_{n=1}^{\infty} A_n$ and note that $A \cap B_n = A_n$ holds for each n. Thus, using the preceding problem, we see that

$$\mu\left(\bigcup_{n=1}^{\infty} A_n\right) = \mu\left(A \cap \left[\bigcup_{n=1}^{\infty} B_n\right]\right) = \sum_{n=1}^{\infty} \mu(A \cap B_n) = \sum_{n=1}^{\infty} \mu(A_n).$$

Problem 14.8.   *Let μ be an outer measure on a set X. Show that a subset E of X is measurable if and only if for each $\epsilon > 0$ there exists a measurable set F such that $F \subseteq E$, and $\mu(E \setminus F) < \epsilon$.*

Solution.   If E is measurable, then $F = E$ satisfies the condition for each $\varepsilon > 0$. For the converse, assume that the condition is satisfied.

Start by choosing for each n a measurable set F_n with $F_n \subseteq E$ and $\mu(E \setminus F_n) < \frac{1}{n}$. Put $F = \bigcup_{n=1}^{\infty} F_n \subseteq E$, and note that F is measurable. Consequently, $\mu(E \setminus F) \le \mu(E \setminus F_n) < \frac{1}{n}$ for each n implies $\mu(E \setminus F) = 0$, and so $E \setminus F$ is measurable. The measurability of E now follows from the identity $E = F \cup (E \setminus F)$.

An alternate proof of the preceding part goes as follows. Let A be a subset of X with $\mu(A) < \infty$. If $\epsilon > 0$ is given, pick a measurable subset F with $F \subseteq E$ and $\mu(E \setminus F) < \epsilon$. Then

$$\mu(A \cap E) = \mu\big(A \cap [F \cup (E \setminus F)]\big)$$
$$\leq \mu(A \cap F) + \mu\big(A \cap (E \setminus F)\big) \leq \mu(A \cap F) + \epsilon$$

implies $\mu(A \cap F) - \mu(A \cap E) > -\epsilon$, and so

$$\mu(A) = \mu(A \cap F) + \mu(A \cap F^c)$$
$$\geq \mu(A \cap F) + \mu(A \cap E^c)$$
$$= \mu(A \cap E) + +\mu(A \cap E^c) + \big[\mu(A \cap F) - \mu(A \cap E)\big]$$
$$> \mu(A \cap E) + \mu(A \cap E^c) - \epsilon$$

for all $\epsilon > 0$. This implies $\mu(A) \geq \mu(A \cap E) + \mu(A \cap E^c)$, which shows that E is a measurable set.

Problem 14.9. *Let μ be an outer measure on a set X. Assume that a subset E of X has the property that for each $\epsilon > 0$, there exists a measurable set F such that $\mu(E \triangle F) < \epsilon$. Show that E is a measurable set.*

Solution. Let $\varepsilon > 0$. According to the preceding problem, it suffices to show that $\mu(E \setminus G) < \varepsilon$ holds for some measurable set G with $G \subseteq E$.

For each n choose $F_n \in \Lambda$ with $\mu(E \triangle F_n) < 2^{-n}\varepsilon$. Put $F = \bigcap_{n=1}^{\infty} F_n \in \Lambda$. Since $F \setminus E \subseteq F_n \setminus E$ holds, we have

$$\mu(F \setminus E) \leq \mu(F_n \setminus E) < 2^{-n}\varepsilon$$

for each n, and so $\mu(F \setminus E) = 0$. Thus, $F \setminus E \in \Lambda$, and hence $F \cap E = F \setminus (F \setminus E)$ is also a measurable set. Now, note that $F \cap E \subseteq E$ holds and

$$\mu(E \setminus E \cap F) = \mu(E \setminus F) = \mu\left(\bigcup_{n=1}^{\infty}(E \setminus F_n)\right) \leq \sum_{n=1}^{\infty} \mu(E \setminus F_n) < \varepsilon.$$

Problem 14.10. *Let $X = \{1, 2, 3\}$, $\mathcal{F} = \{\emptyset, \{1\}, \{1, 2\}\}$ and consider the set function $\mu: \mathcal{F} \to [0, \infty]$ defined by $\mu(\emptyset) = 0$, $\mu(\{1\}) = 2$ and $\mu(\{1, 2\}) = 1$.*

   a. *Describe the outer measure μ^* generated by the set function μ.*

   b. *Describe the σ-algebra of all μ^*-measurable subsets of X (and conclude that the set $\{1\} \in \mathcal{F}$ is not a measurable set).*

Solution.   (a) The outer measure $\mu^*: \mathcal{P}(X) \to [0, \infty]$ is given by

$$\mu^*(\emptyset) = 0, \quad \mu^*(\{1\}) = \mu^*(\{2\}) = 1, \quad \mu^*(\{3\}) = \infty,$$
$$\mu^*(\{1, 2\}) = 1, \quad \mu^*(\{1, 3\}) = \mu^*(\{2, 3\}) = \mu^*(\{1, 2, 3\}) = \infty.$$

(b) The σ-algebra of all measurable sets is $\Lambda = \{\emptyset, \{3\}, \{1, 2\}, X\}$.

Problem 14.11.   *Let $v: \mathcal{P}(X) \to [0, \infty]$ be a set function. Show that v is an outer measure if and only if there exist a collection $\mathcal{F}$ of subsets of X containing the empty set and a set function $\mu: \mathcal{F} \to [0, \infty]$ with $\mu(\emptyset) = 0$ satisfying $v(A) = \mu^*(A)$ for all $A \in \mathcal{P}(X)$.*

Solution.   Assume first that $v: \mathcal{P}(X) \to [0, \infty]$ is an outer measure. Let $\mathcal{F} = \mathcal{P}(X)$ and $\mu = v$. We claim that $v(A) = \mu^*(A)$ holds for each $A \in \mathcal{P}(X)$, where

$$\mu^*(A) = \inf\left\{ \sum_{n=1}^{\infty} \mu(A_n) : \{A_n\} \subseteq \mathcal{F} \text{ and } A \subseteq \bigcup_{n=1}^{\infty} A_n \right\},$$

and $\inf \emptyset = \infty$. To see this, let $A \in \mathcal{P}(X)$. From $A = A \cup \emptyset \cup \emptyset \cup \emptyset \cdots$, we see that $\mu^*(A) \leq \mu(A) = v(A)$. On the other hand, if $A \subseteq \bigcup_{n=1}^{\infty} A_n$ holds true, then from the σ-subadditivity of v, we see that

$$v(A) \leq \sum_{n=1}^{\infty} v(A_n) = \sum_{n=1}^{\infty} \mu(A_n),$$

and so $v(A) \leq \mu^*(A)$ is also true. Hence, $v(A) = \mu^*(A)$ for each subset A of X.

For the converse, assume that the outer measure μ^* generated by a set function $\mu: \mathcal{F} \to [0, \infty]$ satisfies $\mu(\emptyset) = 0$ and $v(A) = \mu^*(A)$ for each $A \in \mathcal{P}(X)$. We shall show that v is an outer measure by verifying the three properties required to be satisfied by v in order to be a measure.

(1) From $0 \leq v(\emptyset) = \mu^*(\emptyset) \leq \mu(\emptyset) + \mu(\emptyset) + \mu(\emptyset) + \cdots = 0$, we see that $v(\emptyset) = 0$.

(2) (*Monotonicity*) Let $A \subseteq B$ and let $\{A_n\}$ be a sequence of $\mathcal{F}$ with $B \subseteq \bigcup_{n=1}^{\infty} A_n$. Then, $A \subseteq \bigcup_{n=1}^{\infty} A_n$, and so $\mu^*(A) \leq \sum_{n=1}^{\infty} \mu(A_n)$. Therefore,

$$v(A) = \mu^*(A) \leq \inf\left\{ \sum_{n=1}^{\infty} \mu(A_n) : \{A_n\} \subseteq \mathcal{F} \text{ and } B \subseteq \bigcup_{n=1}^{\infty} A_n \right\} = \mu^*(B) = v(B).$$

(If there is no sequence $\{A_n\}$ of $\mathcal{F}$ with $B \subseteq \bigcup_{n=1}^{\infty} A_n$, then $\mu^*(B) = \infty$, and $v(A) \leq v(B) = \mu^*(B)$ is trivially true.)

(3) (σ-Subadditivity) Let $\{E_n\}$ be a sequence of subsets of X and let $E = \bigcup_{n=1}^{\infty} E_n$. If $\sum_{n=1}^{\infty} \mu^*(E_n) = \infty$, then $\nu(E) = \mu^*(E) \leq \sum_{n=1}^{\infty} \mu^*(E_n) = \sum_{n=1}^{\infty} \nu(E_n)$ is trivially true. So, assume $\sum_{n=1}^{\infty} \mu^*(E_n) < \infty$ and let $\varepsilon > 0$. For each n pick a sequence $\{A_n^k\}$ of $\mathcal{F}$ with $E_n \subseteq \bigcup_{k=1}^{\infty} A_n^k$ and

$$\sum_{k=1}^{\infty} \mu(A_n^k) < \mu^*(E_n) + 2^{-n}\varepsilon = \nu(E_n) + 2^{-n}\varepsilon.$$

Clearly, $E \subseteq \bigcup_{n=1}^{\infty} \bigcup_{k=1}^{\infty} A_n^k$ holds, and so

$$\nu(E) = \mu^*(E) \leq \sum_{n=1}^{\infty} \sum_{k=1}^{\infty} \mu(A_n^k) < \sum_{n=1}^{\infty} \left[\nu(E_n) + 2^{-n}\varepsilon\right] = \sum_{n=1}^{\infty} \nu(E_n) + \varepsilon.$$

Since $\varepsilon > 0$ is arbitrary, $\nu(E) \leq \sum_{n=1}^{\infty} \nu(E_n)$, and we are done.

Problem 14.12. *Consider an outer measure μ on a set X and let $\mathcal{A}$ be the collection of all measurable subsets of X of finite measure. That is, consider the family $\mathcal{A} = \{A \in \Lambda : \mu(A) < \infty\}$.*

   a. *Show that $\mathcal{A}$ is a semiring.*
   b. *Define a relation $\simeq$ on $\mathcal{A}$ by $A \simeq B$ if $\mu(A \Delta B) = 0$. Show that $\simeq$ is an equivalence relation on $\mathcal{A}$.*
   c. *Let D denote the set of all equivalence classes of $\mathcal{A}$. For $A \in \mathcal{A}$ let $\dot{A}$ denote the equivalence class of A in D. Now, for $\dot{A}, \dot{B} \in D$ define $d(\dot{A}, \dot{B}) = \mu(A \Delta B)$. Show that d is well defined and that (D, d) is a complete metric space.*

Solution. Note that if A, B, and C are three arbitrary sets, then

$$A \Delta C \subseteq (A \Delta B) \cup (B \Delta C).$$

(a) Straightforward. (Note that in actuality $\mathcal{A}$ is a ring of sets.)
(b) If A, B, and C in $\mathcal{A}$ satisfy $A \simeq B$ and $B \simeq C$, then the relation

$$\mu(A \Delta C) \leq \mu\big((A \Delta B) \cup (B \Delta C)\big) \leq \mu(A \Delta B) + \mu(B \Delta C) = 0$$

shows that $A \simeq C$.
(c) If $A \simeq A_1$ and $B \simeq B_1$, then

$$\begin{aligned}
\mu(A \Delta B) &\leq \mu\big((A \Delta A_1) \cup (A_1 \Delta B_1) \cup (B_1 \Delta B)\big) \\
&\leq \mu(A \Delta A_1) + \mu(A_1 \Delta B_1) + \mu(B_1 \Delta B) = \mu(A_1 \Delta B_1).
\end{aligned}$$

Similarly, $\mu(A_1 \triangle B_1) \leq \mu(A \triangle B)$, and so $\mu(A \triangle B) = \mu(A_1 \triangle B_1)$. This shows that $d(\dot{A}, \dot{B}) = \mu(A \triangle B)$ is well defined.

For the triangle inequality, note that

$$d(\dot{A}, \dot{B}) = \mu(A \triangle B) \leq \mu(A \triangle C) + \mu(C \triangle B) = d(\dot{A}, \dot{C}) + d(\dot{C}, \dot{B}).$$

Thus, (D, d) is a metric space. What remains to be shown is that (D, d) is a complete metric space.

To this end, let $\{\dot{A}_n\}$ be a Cauchy sequence of D. By passing to a subsequence, we can assume that

$$d(\dot{A}_{n+1}, \dot{A}_n) = \mu(A_{n+1} \triangle A_n) < 2^{-n-1}$$

holds for each n. Set $A = \bigcap_{k=1}^{\infty} \bigcup_{i=k}^{\infty} A_i \in \Lambda$. Now, let n be fixed and note that $A \subseteq \bigcup_{i=n}^{\infty} A_i = A_n \cup \left(\bigcup_{i=n}^{\infty} (A_{i+1} \setminus A_i) \right)$ holds. Thus,

$$\mu(A) \leq \mu(A_n) + \sum_{i=n}^{\infty} \mu(A_{i+1} \setminus A_i) < \mu(A_n) + 2^{-n} < \infty,$$

and so $\dot{A} \in D$. Moreover, we have

$$\mu(A \setminus A_n) \leq \mu\left(\bigcup_{i=n}^{\infty} (A_{i+1} \setminus A_i) \right) \leq \sum_{i=n}^{\infty} \mu(A_{i+1} \setminus A_i) < 2^{-n}.$$

On the other hand, if $x \in A_n \setminus A$, then $x \in A_n$ and $x \notin A = \bigcap_{k=1}^{\infty} \bigcup_{i=k}^{\infty} A_i$. Consequently, there exists some $k \geq n$ with $x \notin A_i$ for each $i \geq k$. This implies $A_n \setminus A \subseteq \bigcup_{i=n}^{\infty} (A_i \setminus A_{i+1})$, and so $\mu(A_n \setminus A) \leq \sum_{i=n}^{\infty} \mu(A_i \setminus A_{i+1}) < 2^{-n}$ also holds. Therefore,

$$d(\dot{A}_n, \dot{A}) = \mu(A_n \triangle A) = \mu(A_n \setminus A) + \mu(A \setminus A_n) < 2^{1-n}$$

holds for each n. This shows that $\lim d(\dot{A}_n, \dot{A}) = 0$, and so (D, d) is a complete metric space. (For an alternate proof of this part, see Problem 31.3.)

15. THE OUTER MEASURE GENERATED BY A MEASURE

Problem 15.1. *Let* $(X, \mathcal{S}, \mu)$ *be a measure space, and let* E *be a measurable subset of* X. *Put* $\mathcal{S}_E = \{E \cap A: A \in \mathcal{S}\}$, *the restriction of* $\mathcal{S}$ *to* E. *Show that* $(E, \mathcal{S}_E, \mu^*)$ *is a measure space.*

Solution. Let E be a measurable subset of X and let $\{A_n\}$ be a sequence of Λ such that

a. $\{A_n \cap E\}$ is a pairwise disjoint sequence; and
b. there exists some $A \in S$ such that $A \cap E = \bigcup_{n=1}^{\infty} A_n \cap E$.

Using the fact that $\mu^* \colon \Lambda \longrightarrow [0, \infty]$ is a measure, we see that

$$\mu^*(A \cap E) = \mu^*\left(\bigcup_{n=1}^{\infty}(A_n \cap E)\right) = \sum_{n=1}^{\infty} \mu^*(A \cap E),$$

and so μ^* is a measure when restricted to the semiring S_E.

Problem 15.2. *Let (X, S, μ) be a measure space. Show that*

$$\mu^*(A) = \inf\{\mu^*(B)\colon B \text{ is a } \sigma\text{-set such that } A \subseteq B\}$$

holds for every subset A of X.

Solution. Let $\{A_n\} \subseteq S$ and let $B = \bigcup_{n=1}^{\infty} A_n$. By Theorem 12.2(3), there exists a pairwise disjoint sequence $\{B_n\}$ of S such that $B = \bigcup_{n=1}^{\infty} B_n$. Thus, for $A \subseteq X$, there exists a sequence $\{A_n\}$ of S with $A \subseteq \bigcup_{n=1}^{\infty} A_n$ if and only if there exists a σ-set B with $A \subseteq B$. The desired equality now follows from the relation

$$\mu^*(A) \leq \sum_{n=1}^{\infty} \mu(B_n) = \mu^*(B) \leq \sum_{n=1}^{\infty} \mu(A_n).$$

Problem 15.3. *Show that every interval I of $\mathbb{R}$ is Lebesgue measurable and $\lambda^*(I) = |I|$ (=the length of I).*

Solution. In Example 15.5, we established that the intervals I of the form $[a, b]$ and $[a, \infty)$ are Lebesgue measurable and that $\lambda^*(I) = |I|$ holds for these cases. We shall consider the other cases separately. Assume $-\infty < a < b < \infty$.

a. $I = (a, b]$. Choose a sequence $\{x_n\}$ of real numbers with $x_n \downarrow a$ and $a < x_n < b$ for each n. Thus, by Example 15.5, we have

$$\lambda^*\big((a, b]\big) = \lim_{n \to \infty} \lambda^*\big([x_n, b]\big) = \lim_{n \to \infty} (b - x_n) = b - a = |I|.$$

b. $I = (a, b)$. Pick $a < x_n < b$ with $x_n \downarrow a$ and observe that $[x_n, b) \uparrow (a, b)$.

c. $I = (-\infty, a)$. Note that $[a - n, a) \uparrow (-\infty, a)$ and so

$$\lambda^*\big((-\infty, a)\big) = \lim_{n \to \infty} \lambda\big([a - n, a)\big) = \lim_{n \to \infty} n = \infty = |I|.$$

d. $I = (-\infty, a]$. Note that $(-\infty, a) \subseteq (-\infty, a]$ and so from the inequality

$$\infty = |(-\infty, a)| = \lambda^*\big((-\infty, a)\big) \le \lambda^*\big((-\infty, a]\big),$$

we see that $|I| = \lambda^*(I) = \infty$.

e. $I = (a, \infty)$. The conclusion follows immediately from the obvious inclusion $[a + 1, \infty) \subseteq (a, \infty)$.

f. $I = (-\infty, \infty)$. Note that $[0, \infty) \subseteq (-\infty, \infty)$.

Problem 15.4.   *Show that every countable subset of $\mathbb{R}$ has Lebesgue measure zero.*

Solution.   Let $a \in \mathbb{R}$. Then, $\{a\} \subseteq [a - \varepsilon, a + \varepsilon)$ holds for each $\varepsilon > 0$ and so $\lambda^*(\{a\}) \le \lambda^*\big([a - \varepsilon, a + \varepsilon)\big) = 2\varepsilon$ for all $\varepsilon > 0$. Therefore, $\lambda^*(\{a\}) = 0$ holds for all $a \in \mathbb{R}$. If $A = \{a_1, a_2, \ldots\} = \bigcup_{n=1}^{\infty} \{a_n\}$ is a countable set, then note that $\lambda^*(A) \le \sum_{n=1}^{\infty} \lambda^*(\{a_n\}) = 0$ so that $\lambda^*(A) = 0$.

Problem 15.5.   *For a subset A of $\mathbb{R}$ and real numbers a and b, define the set $aA + b = \{ax + b\colon x \in A\}$. Show that*

a. $\lambda^*(aA + b) = |a|\lambda^*(A)$, *and*

b.   *if A is Lebesgue measurable, then so is $aA + b$.*

Solution.   Let $A \subseteq \mathbb{R}$ and fix two real numbers a and b. Since $A \subseteq \bigcup_{n=1}^{\infty} [a_n, b_n)$ holds if and only if $A + b \subseteq \bigcup_{n=1}^{\infty} [a_n + b, b_n + b)$ holds, it is easy to see that $\lambda^*(A + b) = \lambda^*(A)$. The identities

$$E \cap (b + A) = b + (E - b) \cap A \quad \text{and} \quad E \cap (b + A)^c = b + (E - b) \cap A^c$$

imply

$$\lambda^*\big(E \cap (b + A)\big) + \lambda^*\big(E \cap (b + A)^c\big) = \lambda^*\big((E - b) \cap A\big) + \lambda^*\big((E - b) \cap A^c\big),$$

which shows that A is measurable if and only if $b + A$ is measurable for each $b \in \mathbb{R}$.

Next, note that $\lambda^*\big(c(s, t)\big) = |c|\lambda^*\big((s, t)\big) = |c|(t - s)$ holds. On the other hand, since $A \subseteq \bigcup_{n=1}^{\infty} (a_n, b_n)$ holds if and only if $aA \subseteq \bigcup_{n=1}^{\infty} a(a_n, b_n)$ holds for each

$a \in \mathbb{R}$ and since $\lambda^*([a_n, b_n)) = \lambda^*((a_n, b_n))$, it follows that $\lambda^*(aA) = |a|\lambda^*(A)$ for each $a \in \mathbb{R}$. Now, the identities

$$E \cap aA = a((a^{-1}E) \cap A) \quad \text{and} \quad E \cap (aA)^c = a((a^{-1}E) \cap A^c) \ (a \neq 0),$$

imply

$$\lambda^*(E \cap aA) + \lambda^*(E \cap (aA)^c) = |a|[\lambda^*((a^{-1}E) \cap A) + \lambda^*((a^{-1}E) \cap A^c)],$$

which shows that A is measurable if and only if aA is measurable for each $a \in \mathbb{R}$.

Now, (a) and (b) follow from the preceding discussion.

Problem 15.6. *Let S be a semiring of subsets of a set X, and let $\mu: S \to [0, \infty]$ be a finitely additive measure that is not a measure. For each $A \subseteq X$ define (as usual)*

$$\mu^*(A) = \inf\left\{ \sum_{n=1}^{\infty} \mu(A_n) \colon \{A_n\} \subseteq S \text{ and } A \subseteq \bigcup_{n=1}^{\infty} A_n \right\}.$$

Show by a counterexample that it is possible to have $\mu \neq \mu^*$ on S. Why does this not contradict Theorem 15.1?

Solution. Consider the finitely additive measure μ of Problem 13.7. Clearly, $\mu(\mathbb{N}) = \infty$. Since $\mathbb{N} = \bigcup_{n=1}^{\infty}\{n\} \in S$, we have $\mu^*(\mathbb{N}) \leq \sum_{n=1}^{\infty} \mu(\{n\}) = 0$, and so $0 = \mu^*(\mathbb{N}) < \mu(\mathbb{N}) = \infty$.

This conclusion does not contradict Theorem 15.1, since the σ-additivity of the measure was essential for its proof.

Problem 15.7. *Let E be an arbitrary measurable subset of a measure space (X, S, μ) and consider the measure space (E, S_E, ν), where $S_E = \{E \cap A \colon A \subset S\}$ and $\nu(E \cap A) = \mu^*(E \cap A)$; see Problem 15.1. Establish the following properties regarding the measure space (E, S_E, ν):*

a.   *The outer measure ν^* is the restriction of μ^* on E, i.e., $\nu^*(B) = \mu^*(B)$ for each $B \subseteq E$.*

b.   *The ν-measurable sets of the measure space (E, S_E, ν) are precisely the sets of the form $E \cap A$ where A is a μ-measurable subset of X, i.e.,*

$$\Lambda_\nu = \{F \subseteq E \colon F \in \Lambda_\mu\}.$$

Solution. Let $(X, \mathcal{S}, \mu)$, E, and ν be as defined in the problem.

(a) Let B be an arbitrary subset of E. If $\{A_n\}$ is a sequence of $\mathcal{S}$ satisfying $B \subseteq \bigcup_{n=1}^{\infty} A_n$, then note that $B \subseteq \bigcup_{n=1}^{\infty} E \cap A_n$ and so

$$\nu^*(B) \le \sum_{n=1}^{\infty} \nu(E \cap A_n) = \sum_{n=1}^{\infty} \mu^*(E \cap A_n) \le \sum_{n=1}^{\infty} \mu(A_n).$$

This implies $\nu^*(B) \le \mu^*(B)$. On the other hand, if $\{A_n\}$ is a sequence of $\mathcal{S}$ satisfying $B \subseteq \bigcup_{n=1}^{\infty} E \cap A_n$, then we have

$$\mu^*(B) \le \sum_{n=1}^{\infty} \mu^*(E \cap A_n) = \sum_{n=1}^{\infty} \nu(E \cap A_n).$$

Thus, $\mu^*(B) \le \nu^*(B)$ also holds, and so $\nu^*(B) = \mu^*(B)$ for each subset B of E.

(b) Let F be a subset of E. Assume first that F is ν-measurable. If $A \in \mathcal{S}$, then note that

$$
\begin{aligned}
\mu(A) &= \mu^*(A \cap E) + \mu^*\big(A \cap (X \setminus E)\big) \\
&= \nu^*(A \cap E) + \mu^*\big(A \cap (X \setminus E)\big) \\
&= \nu^*\big((A \cap E) \cap F\big) + \nu^*\big((A \cap E) \cap (E \setminus F)\big) + \mu^*\big(A \cap (X \setminus E)\big) \\
&\ge \mu^*(A \cap F) + \mu^*\big([A \cap (E \setminus F)] \cup [A \cap (X \setminus E)]\big) \\
&= \mu^*(A \cap F) + \mu^*\big(A \cap (X \setminus F)\big),
\end{aligned}
$$

which shows that F is μ-measurable.

For the converse, assume that F is μ-measurable. If A is an arbitrary subset of E, then note that

$$
\begin{aligned}
\nu^*(A) = \mu^*(A) &= \mu^*(A \cap F) + \mu^*\big(A \cap (X \setminus F)\big) \\
&= \mu^*(A \cap F) + \mu^*\big(A \cap (E \setminus F)\big) \\
&= \nu^*(A \cap F) + \nu^*\big(A \cap (E \setminus F)\big),
\end{aligned}
$$

which means that F is also ν-measurable.

Problem 15.8. *Show that a subset E of a measure space $(X, \mathcal{S}, \mu)$ is measurable if and only if for each $\epsilon > 0$ there exists a measurable set A_ϵ and two subsets B_ϵ and C_ϵ satisfying*

$$E = (A_\epsilon \cup B_\epsilon) \setminus C_\epsilon, \quad \mu^*(B_\epsilon) < \epsilon, \quad \text{and} \quad \mu^*(C_\epsilon) < \epsilon.$$

Solution. Let E be a subset of a measure space (X, S, μ). If E is a measurable set and $\varepsilon > 0$ is given, then let $A_\varepsilon = E$ and $B_\varepsilon = C_\varepsilon = \emptyset$, and note that these sets satisfy the desired properties.

For the converse, assume that for each $\varepsilon > 0$ there exist a measurable set A_ε and subsets B_ε and C_ε satisfying

$$E = (A_\varepsilon \cup B_\varepsilon) \setminus C_\varepsilon, \quad \mu^*(B_\varepsilon) < \varepsilon, \quad \text{and} \quad \mu^*(C_\varepsilon) < \varepsilon. \qquad (\star)$$

Replacing C_ε by $(A_\varepsilon \cup B_\varepsilon) \cap C_\varepsilon$, we can assume that C_ε is a subset of $A_\varepsilon \cup B_\varepsilon$. From $(\star)$, we see that

$$E \cup C_\varepsilon = A_\varepsilon \cup B_\varepsilon. \qquad (\star\star)$$

Now, by Theorem 15.11, there exists a measurable set D_ε such that $B_\varepsilon \subseteq D_\varepsilon$ and $\mu^*(D_\varepsilon) = \mu^*(B_\varepsilon)$. Using $(\star\star)$, we get

$$E \cup C_\varepsilon \cup (D_\varepsilon \setminus B_\varepsilon) = A_\varepsilon \cup B_\varepsilon \cup (D_\varepsilon \setminus B_\varepsilon) = A_\varepsilon \cup D_\varepsilon.$$

Clearly, $A_\varepsilon \cup D_\varepsilon$ is a measurable set and

$$\mu^*(C_\varepsilon \cup (D_\varepsilon \setminus B_\varepsilon)) \le \mu^*(C_\varepsilon) + \mu^*(D_\varepsilon) < 2\varepsilon.$$

In other words, the preceding show that for each $\varepsilon > 0$ there exist a measurable set F_ε and a subset G_ε such that

$$E \cup G_\varepsilon = F_\varepsilon \quad \text{and} \quad \mu^*(G_\varepsilon) < \varepsilon.$$

Now, for each n pick a measurable set F_n and a subset G_n with $\mu^*(G_n) < \frac{1}{n}$ and $E \cup G_n = F_n$. Clearly, the set $F = \bigcap_{n=1}^\infty F_n$ is measurable. Also, the set $G = \bigcap_{n=1}^\infty G_n$ is a null set—and hence $G \setminus E$ is also measurable. In view of

$$E \cup G = \bigcap_{n=1}^\infty (E \cup G_n) = \bigcap_{n=1}^\infty F_n = F,$$

we see that $E \cup G$ is a measurable set. Finally, the measurability of E follows immediately from the identity

$$E = (E \cup G) \setminus (G \setminus E) = F \setminus (G \setminus E).$$

Problem 15.9. *Let (X, S, μ) be a measure space, and let A be a subset of X. Show that if there exists a measurable subset E of X such that $A \subseteq E$, $\mu^*(E) < \infty$, and $\mu^*(E) = \mu^*(A) + \mu^*(E \setminus A)$, then A is measurable.*

Solution. By Problem 15.7, we know that the outer measure generated by the measure space $(E, \mathcal{S}_E, \mu^*)$ coincides with μ^* and the σ-algebra of all measurable sets of the measure space $(E, \mathcal{S}_E, \mu^*)$ is $\{A \in \Lambda_\mu : A \subseteq E\}$.

Now, to complete the proof, assume $E \in \Lambda_\mu$ and that a subset A of E satisfies $\mu^*(A) + \mu^*(E \setminus A) = \mu^*(E)$. If $\mu^*(E) < \infty$ holds, then it follows from Theorem 15.8 that A is a measurable set for $(E, \mathcal{S}_E, \mu^*)$. Thus, by the preceding discussion, $A \in \Lambda_\mu$.

Problem 15.10. *Let A be a subset of $\mathbb{R}$ with $\lambda^*(A) > 0$. Show that there exists a nonmeasurable subset B of $\mathbb{R}$ such that $B \subseteq A$.*

Solution. If A is nonmeasurable, then there is nothing to prove. So, assume that A is measurable. Since some $[n, n+1] \cap A$ must have nonzero measure (why?), by translating appropriately (and using Problem 15.5), we can also assume that $A \subseteq [0, 1]$.

As in Example 15.13 define an equivalence relation $\simeq$ on A by saying that $x \simeq y$ whenever $x - y$ is a rational number. By the Axiom of Choice, there exists a subset B of A containing precisely one member from each equivalence class. Let $\{r_1, r_2, \ldots\}$ be an enumeration of the rationals of $[-1, 1]$ and let $B_n = r_n + B$. Then:

a. The sequence $\{B_n\}$ is pairwise disjoint;
b. $\lambda^*(B_n) = \lambda^*(B)$ holds (by Problem 15.5) for each n; and
c. $A \subseteq \bigcup_{n=1}^\infty B_n \subseteq [-1, 2]$.

Now, note that if B is a measurable set, then each B_n is likewise a measurable set (see Problem 15.5 again). Thus, from (c), it follows that

$$0 < \lambda^*(A) \le \lambda^*\left(\bigcup_{i=1}^\infty B_i\right) = \sum_{i=1}^\infty \lambda^*(B_i)$$

$$= \lim_{n \to \infty} \sum_{i=1}^n \lambda^*(B_i) = \lim_{n \to \infty} n\lambda^*(B) \le 3,$$

which is impossible. Therefore, B is a nonmeasurable subset of A.

Problem 15.11. *Give an example of a disjoint sequence $\{E_n\}$ of subsets of some measure space $(X, \mathcal{S}, \mu)$ such that*

$$\mu^*\left(\bigcup_{n=1}^\infty E_n\right) < \sum_{n=1}^\infty \mu^*(E_n).$$

Solution. Let E_n be the disjoint sequence of nonmeasurable sets described in Example 15.13, where $E_n = r_n + E$. Since E is a nonmeasurable set, $\lambda^*(E) > 0$ holds, and so $\lambda^*(E_n) = \lambda^*(E) > 0$. In particular, $\sum_{n=1}^{\infty} \lambda^*(E_n) = \infty$. On the other hand, $\bigcup_{n=1}^{\infty} E_n \subseteq [-1, 2]$ implies $\lambda^*\left(\bigcup_{n=1}^{\infty} E_n\right) \leq 3 < \infty$.

Problem 15.12. *Let $(X, \mathcal{S}, \mu)$ be a measure space, and let $\{A_n\}$ be a sequence of subsets of X such that $A_n \subseteq A_{n+1}$ holds for all n. If $A = \bigcup_{n=1}^{\infty} A_n$, then show that $\mu^*(A_n) \uparrow \mu^*(A)$.*

Solution. Choose some $E \in \Lambda$ with $A \subseteq E$ and $\mu^*(A) = \mu^*(E)$. (This is possible by Theorem 15.11.) By the same theorem, for each n there exists some $E_n \in \Lambda$ with $A_n \subseteq E_n \subseteq E$ and $\mu^*(A_n) = \mu^*(E_n)$. Now, for each n put $F_n = \bigcap_{k=n}^{\infty} E_k \in \Lambda$, and then let $F = \bigcup_{n=1}^{\infty} F_n \in \Lambda$. Then, we have:

a. $A_n \subseteq F_n$ and $\mu^*(A_n) = \mu^*(F_n)$ for each n; and
b. $F_n \uparrow F$ and $\mu^*(A) = \mu^*(F)$.

By Theorem 15.4, it follows that

$$\mu^*(A_n) = \mu^*(F_n) \uparrow \mu^*(F) = \mu^*(A).$$

Problem 15.13. *For subsets of a measure space $(X, \mathcal{S}, \mu)$ let us define the following almost everywhere (a.e.) relations:*

a. $A \subseteq B$ *a.e. if $\mu^*(A \setminus B) = 0$;*
b. $A = B$ *a.e. if $\mu^*(A \triangle B) = 0$;*
c. $A_n \uparrow A$ *a.e. if $A_n \subseteq A_{n+1}$ a.e. for all n and $A = \bigcup_{n=1}^{\infty} A_n$ a.e. (The meaning of $A_n \downarrow A$ a.e. is similar.)*

Generalize Theorem 15.4 by establishing the following properties for a sequence $\{E_n\}$ of measurable sets:

i.   *If $E_n \uparrow E$ a.e., then $\mu^*(E_n) \uparrow \mu^*(E)$.*
ii. *If $E_n \downarrow E$ a.e. and $\mu^*(E_k) < \infty$ for some k, then $\mu^*(E_n) \downarrow \mu^*(E)$.*

Is (i) true without assuming measurability for the sets E_n?

Solution. (i) Assume that $\{E_n\}$ is a sequence of measurable sets such that $E_n \uparrow E$ a.e. holds. Let

$$A = \left[\left(\bigcup_{n=1}^{\infty} E_n\right) \triangle E\right] \cup \left[\bigcup_{n=1}^{\infty}\left(E_n \setminus E_{n+1}\right)\right]$$

and note that $\mu^*(A) = 0$. Now, define $F = E \cup A$ and $F_n = E_n \cup A$ for each n.

Clearly, $\mu^*(E) = \mu^*(F)$ and $\mu^*(E_n) = \mu^*(F_n)$ for each n (see Problem 14.2) and $F_n \uparrow F$.

Now, apply Theorem 15.4(1) to get

$$\mu^*(E_n) = \mu^*(F_n) \uparrow \mu^*(F) = \mu^*(E).$$

(ii) Assume that $\{E_n\}$ is a sequence of measurable sets such that $E_n \downarrow E$ a.e. and $\mu^*(E_k) < \infty$ holds for some k. Define $B = [(\bigcap_{n=1}^{\infty} E_n) \triangle E] \cup [\bigcup_{n=1}^{\infty} (E_{n+1} \setminus E_n)]$. Clearly, $\mu^*(B) = 0$. Now, apply Theorem 15.4(2) to $E_n \cup E \cup B \downarrow E \cup B$.

Statement (i) is also true without assuming measurability for the E_n. This follows from the arguments of (i) previously and Problem 15.12.

Problem 15.14. *Give an example of a sequence $\{E_n\}$ of measurable sets of some measure space (X, S, μ) such that $E_{n+1} \subseteq E_n$ holds for all n and*

$$\lim_{n \to \infty} \mu^*(E_n) > \mu^*\left(\bigcap_{n=1}^{\infty} E_n \right).$$

Solution. Consider $\mathbf{R}$ with the Lebesgue measure, and let $E_n = (n, \infty)$ for each n. Then, $E_n \downarrow \emptyset$ holds, while $\lambda^*(E_n) = \infty$ for each n.

Problem 15.15. *For a sequence $\{A_n\}$ of subsets of a set X define*

$$\liminf A_n = \bigcup_{n=1}^{\infty} \bigcap_{i=n}^{\infty} A_i \quad and \quad \limsup A_n = \bigcap_{n=1}^{\infty} \bigcup_{i=n}^{\infty} A_i.$$

Now, let (X, S, μ) be a measure space and let $\{E_n\}$ be the sequence of measurable sets. Show the following:

a. $\mu^*(\liminf E_n) \le \liminf \mu^*(E_n)$.
b. *If $\mu^*(\bigcup_{n=1}^{\infty} E_n) < \infty$, then $\mu^*(\limsup E_n) \ge \limsup \mu^*(E_n)$.*

Solution. (a) Note that $\bigcap_{i=n}^{\infty} E_i \uparrow \liminf E_n$ and $\bigcap_{i=n}^{\infty} E_i \subseteq E_n$ holds for each n. By Theorem 15.4(1), we get

$$\mu^*(\liminf E_n) = \lim_{n \to \infty} \mu^*\left(\bigcap_{i=n}^{\infty} E_i \right) \le \liminf \mu^*(E_n).$$

(b) Use similar arguments and Theorem 15.4(2).

Problem 15.16. *Give an example of a sequence $\{A_n\}$ of subsets of some measure space $(X, \mathcal{S}, \mu)$ such that $A_{n+1} \subseteq A_n$ holds for each n, $\mu^*(A_1) < \infty$, and*

$$\lim_{n \to \infty} \mu^*(A_n) > \mu^*\left(\bigcap_{n=1}^{\infty} A_n\right).$$

Solution. Let $A_n = \bigcup_{i=n}^{\infty} E_i$, where $\{E_n\}$ is the sequence of Example 15.13. Note that $A_n \downarrow \emptyset$ holds. Indeed, if $x \in A_n$, then $x \in E_k$ for some $k \geq n$. Since $E_i \cap E_j = \emptyset$ whenever $i \neq j$, it follows that $x \notin A_{k+1}$ so that $A_n \downarrow \emptyset$ holds. Now, observe that $\lambda^*(A_n) \geq \lambda^*(E_n) = \lambda^*(E) > 0$ holds for all n.

Problem 15.17. *Let $(X, \mathcal{S}_1, \mu_1)$ and $(X, \mathcal{S}_2, \mu_2)$ be two measure spaces. Show that μ_1 and μ_2 generate the same outer measure on X if and only if $\mu_1 = \mu_2^*$ on $\mathcal{S}_1$ and $\mu_2 = \mu_1^*$ on $\mathcal{S}_2$ both hold.*

Solution. If μ_1 and μ_2 generate the same outer measure, then clearly $\mu_1 = \mu_2^*$ on $\mathcal{S}_1$ and $\mu_2 = \mu_1^*$ on $\mathcal{S}_2$ both hold.

For the converse, assume that $\mu_1 = \mu_2^*$ on $\mathcal{S}_1$ and $\mu_2 = \mu_1^*$ on $\mathcal{S}_2$ both hold. Let $A \subseteq X$. If $\mu_2^*(A) = \infty$, then $\mu_1^*(A) \leq \mu_2^*(A)$ holds. If $\mu_2^*(A) < \infty$, then given $\varepsilon > 0$ there exists a sequence $\{A_n\}$ of $\mathcal{S}_2$ such that $A \subseteq \bigcup_{n=1}^{\infty} A_n$ and $\sum_{n=1}^{\infty} \mu_2(A_n) < \mu_2^*(A) + \varepsilon$. Thus,

$$\mu_1^*(A) \leq \sum_{n=1}^{\infty} \mu_1^*(A_n) = \sum_{n=1}^{\infty} \mu_2(A_n) < \mu_2^*(A) + \varepsilon$$

holds for all $\varepsilon > 0$, and so $\mu_1^*(A) \leq \mu_2^*(A)$.

Similarly, $\mu_1^*(A) \geq \mu_2^*(A)$ holds, and therefore $\mu_1^*(A) = \mu_2^*(A)$ holds for all $A \subseteq X$.

Problem 15.18. *Let $(X, \mathcal{S}, \mu)$ be a measure space. A measurable set A is called an **atom** if $\mu^*(A) > 0$ and for every measurable subset E of A we have either $\mu^*(E) = 0$ or $\mu^*(A \setminus E) = 0$. If $(X, \mathcal{S}, \mu)$ does not have any atoms, then it is called a **nonatomic measure space**.*

  a. *Find the atoms of:*
    i. *the counting measure, and*
    ii. *the Dirac measure based at a point a.*
  b. *Show that the real line with the Lebesgue measure is a nonatomic measure space.*

Solution. a. (i) The atoms of the counting measure on a set are precisely the one-point sets.

(ii) The atoms of the Dirac measure based at a point a are precisely the sets containing the point a.

b. Let $A \subseteq \mathbb{R}$ be measurable with $\lambda^*(A) > 0$. Pick some integer n such that $\lambda^*([n, n+1] \cap A) = \delta > 0$. Subdivide $[n, n+1]$ into a finite number of subintervals all of the same length less than δ. For one of them, say I, we must have

$$\lambda^*\big([n, n+1] \cap A \cap I\big) > 0.$$

Now, note that the set $E = [n, n+1] \cap A \cap I \subseteq A$ is measurable and satisfies $0 < \lambda^*(E) < \delta \leq \lambda^*(A)$. This shows that A is not an atom, and hence $\mathbb{R}$ with the Lebesgue measure is nonatomic. (For more about this problem, see Problem 18.19.)

Problem 15.19. *This exercise presents an example of a measure that has infinitely many extensions to a measure on the σ-algebra generated by S. Fix a proper nonempty subset A of a set X (i.e., $A \neq X$) and consider the collection of subsets $S = \{\emptyset, A\}$.*

   a. *Show that S is a semiring.*
   b. *Show that the set function $\mu: S \to [0, \infty]$ defined by $\mu(\emptyset) = 0$ and $\mu(A) = 1$ is a measure.*
   c. *Describe the Carathéodory extension μ^* of μ.*
   d. *Determine the σ-algebra of measurable sets Λ_μ.*
   e. *Show that μ has uncountably many extensions to a measure on the σ-algebra generated by S. Why doesn't this contradict Theorem 15.10?*

Solution. The validity of (a) and (b) should be obvious.

(c) The Carathéodory extension of μ is given by

$$\mu^*(B) = \begin{cases} 0 & \text{if} \quad B = \emptyset; \\ 1 & \text{if} \quad B \neq \emptyset \text{ and } B \subseteq A; \\ \infty & \text{if} \quad B \not\subseteq A. \end{cases}$$

(d) The σ-algebra generated by S is

$$\mathcal{A} = \big\{\emptyset, A, A^c, X\big\}.$$

(e) If a is any non-negative extended real number, then the set function $\nu: \mathcal{A} \to [0, \infty]$, defined by

$$\nu(\emptyset) = 0, \quad \nu(A) = 1, \quad \nu(A^c) = a, \quad \text{and} \quad \nu(X) = 1 + a,$$

is a measure which is an extension of μ to all of $\mathcal{A}$. This shows that there are uncountably many extensions of μ to the σ-algebra generated by $\mathcal{S}$.

The latter conclusion does not contradict Theorem 15.10 because μ is not a σ-finite measure.

16.   MEASURABLE FUNCTIONS

Problem 16.1.   *Let $(X, \mathcal{S}, \mu)$ be a measure space. For a function $f: X \to \mathbb{R}$ show that the following statements are equivalent:*

a.   *f is a measurable function.*
b.   *$f^{-1}((-\infty, a))$ is measurable for each $a \in \mathbb{R}$.*
c.   *$f^{-1}((a, \infty))$ is measurable for each $a \in \mathbb{R}$.*

Solution.   (a)$\Longrightarrow$(b) Note that $f^{-1}((-\infty, a))$ is a measurable set simply because the interval $(-\infty, a)$ is an open set.
(b)$\Longrightarrow$(c) Observe that the identity

$$f^{-1}((-\infty, a]) = \bigcap_{n=1}^{\infty} f^{-1}((-\infty, a + \tfrac{1}{n}))$$

implies that $f^{-1}((-\infty, a])$ is a measurable set for each $a \in \mathbb{R}$. Consequently, the set $f^{-1}((a, \infty)) = X \setminus f^{-1}((-\infty, a])$ is also measurable for each $a \in \mathbb{R}$.
(c)$\Longrightarrow$(a) Clearly, $f^{-1}((-\infty, a]) = X \setminus f^{-1}((a, \infty))$ is measurable for each $a \in \mathbb{R}$. Thus, by condition (5) of Theorem 16.2, the function f is measurable.

Problem 16.2.   *Let $(X, \mathcal{S}, \mu)$ be a measure space, and let A be a dense subset of $\mathbb{R}$. Show that a function $f: X \to \mathbb{R}$ is measurable if and only if the set $\{x \in X: f(x) \geq a\}$ is measurable for each $a \in A$.*

Solution.   Only the "if" part needs proof. Let $a \in \mathbb{R}$. Since A is dense in $\mathbb{R}$, there exists a sequence $\{a_n\}$ of A with $a_n < a$ for each n and $a_n \uparrow a$. Now, note that the identity

$$f^{-1}([a, \infty)) = \bigcap_{n=1}^{\infty} f^{-1}([a_n, \infty))$$

shows that the set $f^{-1}([a, \infty))$ is measurable. Therefore, by Theorem 16.2, the function f is measurable.

Problem 16.3. *Give an example of a nonmeasurable function f such that $|f|$ is a measurable function and $f^{-1}(\{a\})$ is a measurable set for each $a \in \mathbf{R}$.*

Solution. Take a non-Lebesgue measurable subset E of $[0, 1]$ and consider the function $f:[0, 1] \longrightarrow \mathbf{R}$ defined by

$$f(x) = \begin{cases} x, & \text{if } x \in E; \\ -x, & \text{if } x \in [0, 1] \setminus E. \end{cases}$$

It is straightforward to verify that the function f satisfies the desired properties.

Problem 16.4. *Show that if $f:\mathbf{R} \to \mathbf{R}$ is continuous a.e., then f is a Lebesgue measurable function.*

Solution. Let $f:\mathbf{R} \longrightarrow \mathbf{R}$ be a function that is continuous almost everywhere. Put $E = \{x \in \mathbf{R}: f \text{ is continuous at } x\}$ and note that $\lambda^*(\mathbf{R} \setminus E) = 0$. Hence, $\mathbf{R} \setminus E$ and E are both measurable sets.

Now, let $\mathcal{O}$ be an arbitrary open subset of $\mathbf{R}$. Clearly, the set $f^{-1}(\mathcal{O}) \cap (\mathbf{R} \setminus E)$ (as a null set) is measurable. Since f restricted to E is continuous, $f^{-1}(\mathcal{O}) \cap E$ is an open set in E, and consequently there exists an open subset V of $\mathbf{R}$ such that $f^{-1}(\mathcal{O}) \cap E = V \cap E$. In particular, note that $f^{-1}(\mathcal{O}) \cap E$ is a measurable set. Therefore,

$$f^{-1}(\mathcal{O}) = \left[f^{-1}(\mathcal{O}) \cap E\right] \cup \left[f^{-1}(\mathcal{O}) \cap (\mathbf{R} \setminus E)\right]$$

is likewise measurable, so that f is a measurable function.

Problem 16.5. *Let $f:\mathbf{R} \to \mathbf{R}$ be a differentiable function. Show that f' is Lebesgue measurable.*

Solution. For each n define

$$g_n(x) = n\left[f(x + \tfrac{1}{n}) - f(x)\right] = \frac{f(x+\frac{1}{n})-f(x)}{\frac{1}{n}},$$

and note that each g_n is measurable (since it is continuous). In view of $g_n(x) \to f'(x)$ for each $x \in \mathbf{R}$, it follows from Theorem 16.6 that f' is a measurable function.

Problem 16.6. *Let $(X, \mathcal{S}, \mu)$ be a measure space and let $f: X \to \mathbf{R}$ be a measurable function. Show that:*
   a. *$|f|^p$ is a measurable function for all $p \geq 0$, and*
   b. *if $f(x) \neq 0$ for each $x \in X$, then $1/f$ is a measurable function.*

Solution. Let $f: X \longrightarrow \mathbb{R}$ be a measurable function.

(a) Assume $p > 0$. By Theorem 16.5, $|f|$ is measurable. The conclusion now follows from the identities $\{x \in X: |f|^p(x) \geq a\} = X$ if $a \leq 0$, and $\{x \in X: |f|^p(x) \geq a\} = \{x \in X: |f(x)| \geq a^{\frac{1}{p}}\}$ if $a > 0$.

(b) Assume that $f(x) \neq 0$ holds for each $x \in X$. Note that

$$\{x \in X: \tfrac{1}{f}(x) > 0\} = \{x \in X: f(x) > 0\},$$

$$\{x \in X: \tfrac{1}{f}(x) > a\} = \{x \in X: f(x) < \tfrac{1}{a}\} \text{ if } a > 0, \text{ and}$$

$$\{x \in X: \tfrac{1}{f}(x) > a\} = \{x \in X: f(x) < \tfrac{1}{a}\} \cup \{x \in X: f(x) > 0\} \text{ if } a < 0.$$

The preceding identities guarantee that $\frac{1}{f}$ is measurable.

Problem 16.7. *Let $\{f_n\}$ be a sequence of real-valued measurable functions on a measure space $(X, \mathcal{S}, \mu)$. Then show that the sets*

a. $A = \{x \in X: f_n(x) \to \infty\}$,
b. $B = \{x \in X: f_n(x) \to -\infty\}$, *and*
c. $C = \{x \in X: \lim f_n(x) \text{ exists in } \mathbb{R}\}$

are all measurable.

Solution. (a) For each m and k let $A_{m,k} = \{x \in X: f_n(x) \geq k \text{ for all } n \geq m\}$. From $A_{m,k} = \bigcap_{n=m}^{\infty}\{x \in X: f_n(x) \geq k\}$, we see that $A_{m,k} \in \Lambda_\mu$ for each m, k. Now, note that $A = \bigcap_{k=1}^{\infty} \bigcup_{m=1}^{\infty} A_{m,k}$.

(b) Put $B_{m,k} = \{x \in X: f_n(x) \leq -k \text{ for each } n \geq m\}$ and note that $B = \bigcap_{k=1}^{\infty} \bigcup_{m=1}^{\infty} B_{m,k}$.

(c) Let $Y = X \setminus (A \cup B)$ and consider the measure space $(Y, \mathcal{S}_Y, \mu^*)$. Also, consider all functions restricted to Y. In view of Problem 15.7, all functions are measurable with respect to this space. By Theorem 16.6, both functions $\liminf f_n$ and $\limsup f_n$ are measurable. The conclusion now follows from Theorem 16.4(c) by observing that

$$C = \{x \in X: \lim f_n(x) \text{ exists in } \mathbb{R}\}$$
$$= \{x \in Y: \limsup f_n(x) = \liminf f_n(x)\}.$$

Problem 16.8. *Let $(X, \mathcal{S}, \mu)$ be a measure space. Assume that $f: X \to \mathbb{R}$ is a measurable function and $g: \mathbb{R} \to \mathbb{R}$ is a continuous function. Show that $g \circ f$ is a measurable function.*

Solution. Consider the functions $X \xrightarrow{f} \mathbb{R} \xrightarrow{g} \mathbb{R}$ with f measurable and g continuous, and let $\mathcal{O}$ be an open subset of $\mathbb{R}$. Since g is continuous, we know that $g^{-1}(\mathcal{O})$ is an open set, and the conclusion follows from the identity

$$\left(g \circ f\right)^{-1}(\mathcal{O}) = f^{-1}\left(g^{-1}(\mathcal{O})\right).$$

Problem 16.9. *Let $\mathcal{F}$ be a nonempty family of continuous real-valued functions defined on $\mathbf{R}$. Assume that there exists a function $g: \mathbf{R} \to \mathbf{R}$ such that $f(x) \le g(x)$ for each $x \in \mathbf{R}$ and all $f \in \mathcal{F}$. Show that the supremum function $h: \mathbf{R} \to \mathbf{R}$, defined by $h(x) = \sup\{f(x): f \in \mathcal{F}\}$, is (Lebesgue) measurable.*

Solution. We shall show that $h^{-1}\left((a, \infty)\right)$ is an open set for each $a \in \mathbf{R}$ (and hence, a Lebesgue measurable set).

To see this, let $a \in \mathbf{R}$ and fix $x_0 \in h^{-1}\left((a, \infty)\right)$, i.e., $h(x_0) > a$. So, there exists some $f \in \mathcal{F}$ such that $f(x_0) > a$. Since f is a continuous function, there exists some neighborhood V of x_0 such that $f(x) > a$ for each $x \in V$. This implies $h(x) \ge f(x) > a$ for each $x \in V$, and so $V \subseteq h^{-1}\left((a, \infty)\right)$. This shows that x_0 is an interior point of $h^{-1}\left((a, \infty)\right)$ and consequently, $h^{-1}\left((a, \infty)\right)$ is an open set.

Note: A real-valued function $f: X \to \mathbf{R}$ defined on a topological space X is said to be *lower semicontinuous* if $f^{-1}\left((a, \infty)\right)$ is an open set for each $a \in \mathbf{R}$. The preceding arguments show that we have proven the following result: *The pointwise supremum of a family of lower semicontinuous functions is likewise lower semicontinuous.*

Problem 16.10. *Show that if $f: X \to \mathbf{R}$ is a measurable function, then either f is constant almost everywhere or else (exclusively) there exists a constant c such that*

$$\mu^*(\{x \in X: f(x) > c\}) > 0 \quad and \quad \mu^*(\{x \in X: f(x) < c\}) > 0.$$

Solution. Let $f: X \to \mathbf{R}$ be a measurable function which is not a constant almost everywhere. Assume first that $f(x) \ge 0$ holds for each $x \in X$ and let

$$c_0 = \sup\{c \in \mathbf{R}: \mu^*(\{x \in X: f(x) \le c\} = 0\}.$$

Clearly, $0 \le c_0 < \infty$ and $\mu^*\left(\{x \in X: f(x) < c_0\}\right) = 0$. Since f is not constant almost everywhere, there exists some $c > c_0$ such that $\mu^*\left(\{x \in X: f(x) > c\}\right) > 0$. Now, if k satisfies $c_0 < k < c$, then by the definition of c_0 we have $\mu^*\left(\{x \in X: f(x) < c\}\right) \ge \mu^*\left(\{x \in X: f(x) \le k\}\right) > 0$, and the desired conclusion is established in this case.

In the general case, either f^+ or f^- is not equal to a constant almost everywhere. We consider the case where f^+ is not equal to a constant almost everywhere (the other case can be treated in a similar fashion). By the preceding case, there exists $c > 0$ with $\mu^*\left(\{x \in X: f^+(x) > c\}\right) > 0$ and $\mu^*\left(\{x \in X: f^+(x) < c\}\right) > 0$.

To finish the proof, notice that

$$\{x \in X: f^+(x) > c\} = \{x \in X: f(x) > c\}$$

and $\{x \in X: f^+(x) < c\} = \{x \in X: f(x) < c\}$.

17. SIMPLE AND STEP FUNCTIONS

Problem 17.1. *For subsets* A *and* B *of a set* X, *establish the following statements:*

1. $\chi_\emptyset = 0$ *and* $\chi_X = 1$.
2. $A \subseteq B$ *if and only if* $\chi_A \leq \chi_B$.
3. $\chi_{A \cap B} = \chi_A \cdot \chi_B = \chi_A \wedge \chi_B$.
4. $\chi_{A \cup B} = \chi_A + \chi_B - \chi_{A \cap B} = \chi_A \vee \chi_B$.
5. $\chi_{A \setminus B} = \chi_A - \chi_{A \cap B}$.
6. *If* $A = \bigcup_{n=1}^{\infty} A_n$ *and* $\{A_n\}$ *is a pairwise disjoint sequence of subsets of* X, *then* $\chi_A = \sum_{n=1}^{\infty} \chi_{A_n}$.
7. $\chi_{A \times B} = \chi_A \cdot \chi_B$. *(Here the set* B *can be considered to be a subset of some other set* Y.)

Solution. The proofs of the statements are straightforward. To indicate how one can prove them, we shall establish the validity of statements (3) and (7).
(3) We have

$$
\begin{aligned}
\chi_{A \cap B}(x) &= \begin{cases} 1, & \text{if } x \in A \cap B \\ 0, & \text{if } x \notin A \cap B \end{cases} \\
&= \begin{cases} 1, & \text{if } x \in A \text{ and } x \in B \\ 0, & \text{if } x \notin A \\ 0, & \text{if } x \notin B \end{cases} \\
&= \chi_A(x) \cdot \chi_B(x) \\
&= \chi_A \cdot \chi_B(x) \\
&= \min\{\chi_A(x), \chi_B(x)\}.
\end{aligned}
$$

(7) Note that

$$
\begin{aligned}
\chi_{A \times B}(x, y) &= \begin{cases} 1, & \text{if } (x, y) \in A \times B \\ 0, & \text{if } (x, y) \notin A \times B \end{cases} \\
&= \begin{cases} 1, & \text{if } x \in A \text{ and } y \in B \\ 0, & \text{if } x \notin A \\ 0, & \text{if } y \notin B \end{cases} \\
&= \chi_A(x) \cdot \chi_B(y).
\end{aligned}
$$

Problem 17.2.   *Let ϕ be a step function and ψ a simple function such that $0 \le \psi \le \phi$ a.e. Show that ψ is a step function.*

Solution. Let $E = \{x\colon 0 \le \psi(x) \le \phi(x)\}$, and observe that $\mu^*(X \setminus E) = 0$. If $F = \{x \in X\colon \phi(x) > 0\}$, then the measurable set $A = (X \setminus E) \cup F$ satisfies $\mu^*(A) < \infty$, and $\psi(x) = 0$ for each $x \in X \setminus A$.

Problem 17.3.   *Show that if $(X, \mathcal{S}, \mu)$ is a finite measure space, then every simple function is a step function.*

Solution. If ϕ is a simple function and $E = \{x \in X\colon \phi(x) \neq 0\}$, then note that $\mu^*(E) \le \mu^*(X) < \infty$ holds.

Problem 17.4.   *Give an alternate proof of the linearity of the integral (Theorem 17.2) based on Problem 12.14.*

Solution. The linearity follows immediately from the following property.

- *If ϕ is a step function and $\phi = \sum_{j=1}^{m} b_j \chi_{B_j}$ is an arbitrary representation of ϕ, then*

$$I(\phi) = \sum_{j=1}^{m} b_j \mu^*(B_j).$$

We shall establish the preceding property below.

To this end, let $\phi = \sum_{i=1}^{n} a_i \chi_{A_i}$ be the standard representation of ϕ. Assume first that the B_j are pairwise disjoint. Since neither the function ϕ nor the sum $\sum_{j=1}^{m} b_j \mu^*(B_j)$ changes by deleting the terms with $b_j = 0$, we can assume that $b_j \neq 0$ for each j. In such a case, we have $\bigcup_{i=1}^{n} A_i = \bigcup_{j=1}^{m} B_j$. Moreover, note that $a_i \mu^*(A_i \cap B_j) = b_j \mu^*(A_i \cap B_j)$ for all i and j. Indeed, if $A_i \cap B_j = \emptyset$ the equality is obvious and if $x \in A_i \cap B_j$, then $a_i = b_j = \phi(x)$. Therefore,

$$I(\phi) = \sum_{i=1}^{n} a_i \mu^*(A_i) = \sum_{i=j}^{m} \sum_{i=1}^{n} a_i \mu^*(A_i \cap B_j)$$

$$= \sum_{j=1}^{m} \sum_{i=1}^{n} b_j \mu^*(A_i \cap B_j) = \sum_{j=1}^{m} b_j \mu^*(B_j).$$

Now, consider the general case. By Problem 12.14, there exist pairwise disjoint measurable sets $C_1, \ldots, C_k$ such that each C_i is included in some B_j and $B_j = \bigcup\{C_i\colon C_i \subseteq B_j\}$. For each i and j let $\delta_i^j = 1$ if $C_i \subseteq B_j$ and $\delta_i^j = 0$ if $C_i \not\subseteq B_j$.

Clearly, $\chi_{B_j} = \sum_{i=1}^{k} \delta_i^j \chi_{C_i}$ and $\mu(B_j) = \sum_{i=1}^{k} \delta_i^j \mu^*(C_i)$. Therefore,

$$\phi = \sum_{j=1}^{m} b_j \chi_{B_j} = \sum_{j=1}^{m} b_j \left[\sum_{i=1}^{k} \delta_i^j \chi_{C_i} \right] = \sum_{i=1}^{k} \left[\sum_{j=1}^{m} b_j \delta_i^j \right] \chi_{C_i}.$$

So, by the preceding case, we have

$$I(\phi) = \sum_{i=1}^{k} \left[\sum_{j=1}^{m} b_j \delta_i^j \right] \mu^*(C_i) = \sum_{j=1}^{m} b_j \left[\sum_{i=1}^{k} \delta_i^j \mu^*(C_i) \right] = \sum_{j=1}^{m} b_j \mu^*(B_j).$$

Problem 17.5. *Show that $|I(\phi)| \leq I(|\phi|)$ holds for every step function ϕ.*

Solution. From $-|\phi| \leq \phi \leq |\phi|$ and the monotonicity of the integral (Theorem 17.3), it follows that

$$-I(|\phi|) = I(-|\phi|) \leq I(\phi) \leq I(|\phi|),$$

and so $|I(\phi)| \leq I(|\phi|)$ holds.

Problem 17.6. *Let ϕ be a step function such that $I(|\phi|) = 0$. Show that $\phi = 0$ a.e. holds.*

Solution. Let $\phi = \sum_{i=1}^{n} a_i \chi_{A_i}$ be the standard representation of ϕ. Then, note that $|\phi| = \sum_{i=1}^{n} |a_i| \chi_{A_i}$ is a representation of $|\phi|$, and therefore

$$0 = I(|\phi|) = \sum_{i=1}^{n} |a_i| \mu^*(A_i).$$

Since $|a_i| > 0$ holds for each $1 < i < n$, it follows that $\mu^*(A_i) = 0$ for each $1 \leq i \leq n$, and so $\phi = 0$ a.e. holds.

Problem 17.7. *Let ϕ be a step function. Let $A = \{x \in X : \phi(x) \neq 0\}$ and $M = \max\{|\phi(x)| : x \in X\}$. Show that $|I(\phi)| \leq M\mu^*(A)$.*

Solution. Apply the monotonicity of the integral (Theorem 17.3) to the inequality $-M\chi_A \leq \phi \leq M\chi_A$.

Problem 17.8. *Let $\{\phi_n\}$ be a sequence of step functions. Show that if ϕ is a step function and $\phi_n \downarrow \phi$ a.e. holds, then $I(\phi_n) \downarrow I(\phi)$ also holds.*

Solution. If $\phi_n \downarrow \phi$ a.e. holds, then $\phi_n - \phi \downarrow 0$ a.e. likewise holds. Thus, by the order continuity of the integral (Theorem 17.4), $I(\phi_n) - I(\phi) = I(\phi_n - \phi) \downarrow 0$ so that $I(\phi_n) \downarrow I(\phi)$.

Problem 17.9. *Let $\{\phi_n\}$ be a sequence of step functions and ϕ a simple function such that $0 \le \phi_n \uparrow \phi$ a.e. holds. Show that if $\lim I(\phi_n) < \infty$, then ϕ is a step function.*

Solution. Assume $\phi(x) \ge 0$ for each x and let $\phi = \sum_{i=1}^{k} a_i \chi_{A_i}$ be the standard representation of ϕ. Now, let i be fixed. Then, for each n the function $\psi_n = \phi_n \wedge a_i \chi_{A_i}$ is a step function, $\psi_n \le \phi_n$ holds, and $\psi_n \uparrow_n \phi \wedge a_i \chi_{A_i} = a_i \chi_{A_i}$ a.e. By Theorem 17.6, we see that

$$0 \le a_i \mu^*(A_i) = \lim_{n \to \infty} I(\psi_n) \le \lim_{n \to \infty} I(\phi_n) < \infty,$$

and so $\mu^*(A_i) < \infty$ holds for each $1 \le i \le k$. That is, ϕ is a step function.

Problem 17.10. *Let (X, S, μ) be a measure space, and let $f: X \to \mathbf{R}$ be a function. Show that f is a measurable function if and only if there exists a sequence $\{\phi_n\}$ of simple functions such that $\lim \phi_n(x) = f(x)$ holds for all $x \in X$.*

Solution. Assume f to be measurable. Then both f^+ and f^- are measurable functions. By Theorem 17.7 there exist two sequences of simple functions $\{s_n\}$ and $\{t_n\}$ with $0 \le s_n(x) \uparrow f^+(x)$ and $0 \le t_n(x) \uparrow f^-(x)$ for each $x \in X$. Now, note that $\phi_n = s_n - t_n$ satisfies $\phi_n(x) \longrightarrow f(x)$ for all x.

For the converse, note that (by Theorem 16.6) the pointwise limit of a sequence of measurable functions is always a measurable function.

Problem 17.11. *Let (X, S, μ) be a σ-finite measure space, and let $f: X \to \mathbf{R}$ be a measurable function such that $f(x) \ge 0$ for all $x \in X$. Show that there exists a sequence $\{\phi_n\}$ of step functions such that $0 \le \phi_n \uparrow f(x)$ holds for all $x \in X$.*

Solution. By Theorem 17.7 there exists a sequence $\{\psi_n\}$ of simple functions satisfying $0 \le \psi_n(x) \uparrow f(x)$ for all $x \in X$. Now, pick a sequence $\{E_n\}$ of measurable sets with $\mu^*(E_n) < \infty$ for each n, and $E_n \uparrow X$. Let $\phi_n = \psi_n \wedge \chi_{E_n}$. Then, $\{\phi_n\}$ is a sequence of step functions satisfying $0 \le \phi_n(x) \uparrow f(x)$ for each $x \in X$.

Problem 17.12. *Give a proof of the order continuity of the integral, i.e., $\phi_n \downarrow 0$ a.e. implies $I(\phi_n) \downarrow 0$, based on Egorov's Theorem 16.7.*

Solution. Assume that $\{\phi_n\}$ is a sequence of step functions of some measure space $(X, \mathcal{S}, \mu)$ satisfying $\phi_n \downarrow 0$ a.e. Without loss of generality, we can suppose that $\phi_n(x) \downarrow 0$ for each $x \in X$. Let $E = \{x \in X : \phi_1(x) > 0\}$ and note that $\mu^*(E) < \infty$. Also, let $M = \max\{\phi_1(x) : x \in X\}$.

Now, let $\epsilon > 0$. By Egorov's Theorem 16.7 there exists a measurable set $F \subseteq E$ such that $\mu^*(F) < \epsilon$ and $\{\phi_n\}$ converges uniformly to zero on $E \setminus F$. So, there exists some k such that $0 \leq \phi_n(x) < \epsilon$ for all $x \in E \setminus F$ and all $n \geq k$. Thus, for $n \geq k$, we have

$$0 \leq \phi_n \leq \epsilon \chi_{E \setminus F} + M \chi_F \leq \epsilon \chi_E + M \chi_F,$$

and consequently, by the monotonicity of the integral

$$0 \leq I(\phi_n) \leq \epsilon \mu^*(E) + M \mu^*(F) < \left[\mu^*(E) + M \right] \epsilon$$

for all $n \geq k$. This shows that $I(\phi_n) \downarrow 0$.

Problem 17.13. *Let $(X, \mathcal{S}, \mu)$ be a measure space, and let $f : X \to [0, \infty)$ be a function. Show that f is measurable if and only if there exist non-negative constants $c_1, c_2, \ldots$ and measurable sets $E_1, E_2, \ldots$ such that*

$$f(x) = \sum_{n=1}^{\infty} c_n \chi_{E_n}(x)$$

holds for each $x \in X$.

Solution. Consider a measure space $(X, \mathcal{S}, \mu)$ and a non-negative real-valued function $f : X \to [0, \infty)$. Assume first that there exist non-negative constants $c_1, c_2, \ldots$ and measurable sets $E_1, E_2, \ldots$ such that $f(x) = \sum_{n=1}^{\infty} c_n \chi_{E_n}(x)$ holds for each $x \in X$. If we let $\phi_n = \sum_{i=1}^{n} c_i \chi_{E_i}$, then ϕ_n is a measurable function (in fact, it is a simple function) and $\phi_n(x) \longrightarrow f(x)$ holds for each $x \in X$. Now, by Theorem 16.6(1), the function f is necessarily a measurable function.

For the converse, assume that f is a measurable function. By Theorem 17.7 there exists a sequence of simple functions $\{\phi_n\}$ such that $0 \leq \phi_n(x) \uparrow f(x)$ for each $x \in X$. If we let $\phi_0 = 0$, then $f(x) = \sum_{n=1}^{\infty} \left[\phi_n(x) - \phi_{n-1}(x) \right]$ holds for each $x \in X$. For each n write $\phi_n - \phi_{n-1} = \sum_{i=1}^{k_n} c_i^n \chi_{E_i^n}$ with $c_i^n \geq 0$ for each i and n. Thus,

$$f(x) = \sum_{n=1}^{\infty} \sum_{i=1}^{k_n} c_i^n \chi_{E_i^n},$$

and by rearranging the terms of the preceding series in a single series, our conclusion follows.

Problem 17.14. *Let (X, S, μ) be a finite measure space satisfying $\mu^*(X) = 1$, and let $E_1, E_2, \ldots, E_{10}$ be ten measurable sets such that $\mu^*(E_i) = \frac{1}{3}$ holds for each i. Show that four of these sets have an intersection of positive measure. Is the conclusion true for nine measurable sets instead of ten?*

Solution. Consider the step function $\phi = \sum_{i=1}^{10} \chi_{E_i}$. Clearly, the function ϕ assumes only integer values and

$$\phi(x) = \text{the cardinality of the set } \{i \in \{1, \ldots, 10\}: x \in E_i\}.$$

If $\phi(x) \leq 3 = 3\chi_X(x)$ for almost all x, then

$$3 < \tfrac{10}{3} = \sum_{i=1}^{10} \mu^*(E_i) = \sum_{i=1}^{10} I(\chi_{E_i}) = I(\phi) \leq I(3\chi_X) = 3$$

a contradiction. Hence, the measurable set

$$A = \{x \in X: \phi(x) \geq 4\}$$

must have positive measure.

Next, let $A_1, A_2, \ldots, A_k$ denote the collection of all (nonempty) intersections of the sets E_i taken four at a time; clearly, $k \leq \binom{10}{4} = 210$. Now, an easy argument guarantees that $A \subseteq \bigcup_{j=1}^{k} A_j$, and from this it easily follows that at least one of the A_j must have positive measure.

For nine sets the conclusion is false. For a counterexample take $X = [0, 1]$, $\mu = \lambda$, $E_1 = E_2 = E_3 = (0, \frac{1}{3})$, $E_4 = E_5 = E_6 = (\frac{1}{3}, \frac{2}{3})$, and $E_7 = E_8 = E_9 = (\frac{2}{3}, 1)$.

Problem 17.15. *If $f: X \to [0, 1]$ is a measurable function, then show that either $f = \chi_A$ a.e. for some measurable set A or else (exclusively) there exists a constant $0 < c < \frac{1}{2}$ such that*

$$\mu^*(\{x \in X: c < f(x) < 1 - c\}) > 0.$$

Solution. For each n let $A_n = \{x \in X: \frac{1}{2n} < f(x) < 1 - \frac{1}{2n}\}$. If $\mu^*(A_n) > 0$ for some n, then the constant $c = \frac{1}{2n}$ satisfies $\mu^*(\{x \in X: c < f(x) < 1 - c\}) > 0$.

Now, assume that $\mu^*(A_n) = 0$ for each n. Then from

$$A_n \uparrow \{x \in X : 0 < f(x) < 1\},$$

we see that $\mu^*(\{x \in X : 0 < f(x) < 1\}) = 0$. This easily implies that $f = \chi_A$ a.e. for the measurable set $A = f^{-1}(\{1\})$.

Problem 17.16. *Let $(X, \mathcal{S}, \mu)$ be a measure space, and let $\phi : X \to \mathbb{R}$ be a simple function having the standard representation $\phi = \sum_{i=1}^{n} a_i \chi_{A_i}$. If $\phi \geq 0$ a.e., then the sum $\sum_{i=1}^{n} a_i \mu^*(A_i)$ makes sense as an extended real number (it may be infinite). Call this extended real number the* **Lebesgue integral** *of ϕ, and write $I(\phi) = \sum_{i=1}^{n} a_i \mu^*(A_i)$.*

a. *If ϕ and ψ are simple functions such that $\phi \geq 0$ a.e., then $\psi \geq 0$ a.e., then show that $I(\phi + \psi) = I(\phi) + I(\psi)$.*

b. *If ϕ and ψ are simple functions such that $0 \leq \phi \leq \psi$ a.e., then show that $I(\phi) \leq I(\psi)$.*

c. *Show that if $\{\phi_n\}$ and $\{\psi_n\}$ are two sequences of simple functions and $f : X \to \mathbb{R}^*$ such that $0 \leq \phi_n \uparrow f$ a.e. and $0 \leq \psi_n \uparrow f$ a.e., then $\lim I(\phi_n) = \lim I(\psi_n)$ holds (with the limits possibly being infinite).*

d. *Assume that $\{\phi_n\}$ is a sequence of simple functions such that $0 \leq \phi_n \uparrow \chi_A$ a.e. holds. Show that $\lim I(\phi_n) = \mu^*(A)$.*

e. *Give an example of a sequence $\{\phi_n\}$ of simple functions on some measure space such that $\phi_n \downarrow 0$ (everywhere) and $\lim I(\phi_n) \neq 0$.*

Solution. Clearly, a simple function ϕ is a step function if and only if $I(\phi) < \infty$.

(a) Note that $\phi + \psi$ is a step function if and only if both ϕ and ψ are step functions. In this case, the equality $I(\phi + \psi) = I(\phi) + I(\psi)$ follows from Theorem 17.2. On the other hand, if $\phi + \psi$ is not a step function, then either ϕ or ψ fails to be a step function and hence, in this case, $I(\phi + \psi) = I(\phi) + I(\psi) = \infty$ holds.

(b) If $I(\psi) = \infty$, then $I(\phi) \leq I(\psi)$ holds trivially. On the other hand, if $I(\psi) < \infty$, then ψ is a step function. It follows (from Problem 17.2) that ϕ is a step function, and the desired inequality follows from Theorem 17.3.

(c) If both $\{\phi_n\}$ and $\{\psi_n\}$ are sequences of step functions, then the conclusion follows from Theorem 17.5. Thus, we only need to consider the case when $\{\phi_n\}$ is a sequence of step functions and $I(\psi_k) = \infty$ holds for some k.

In view of $\phi_n \wedge \psi_k \uparrow_n f \wedge \psi_k = \psi_k$ a.e., it follows from Problem 17.9 that $\lim_{n \to \infty} I(\phi_n \wedge \psi_k) = \infty$. From $\phi_n \wedge \psi_k \leq \phi_n$, we obtain that $\lim I(\phi_n) = \infty$. Hence, $\lim I(\psi_n) = \lim I(\phi_n) = \infty$ holds in this case.

(d) We can suppose $0 \leq \phi_n(x) \uparrow \chi_A(x)$ holds for each x. If for each n we let $A_n = \{x \in X : \phi_n(x) > 0\}$, then each A_n is measurable and $A_n \uparrow A$ holds. Since

$\chi_{A_n} \uparrow \chi_A$, part (c) coupled with Theorem 15.4 gives

$$\lim_{n\to\infty} I(\phi_n) = \lim_{n\to\infty} I(\chi_{A_n}) = \lim_{n\to\infty} \mu^*(A_n) = \mu^*(A).$$

(e) Consider $\mathbb{R}$ with the Lebesgue measure, and let $\phi_n = \chi_{(n,\infty)}$.

Problem 17.17. *Let (X, Σ, μ) be a measure space with Σ being a σ-algebra. Let us say that a function $f: X \to \mathbb{R}$ is Σ-measurable if $f^{-1}(A) \in \Sigma$ for each open subset A of $\mathbb{R}$. Also, let $\mathcal{M}_\Sigma$ denote the collection of all Σ-measurable functions. Establish the following:*

   a. *$\mathcal{M}_\Sigma$ is a function space and an algebra of functions.*
   b. *$\mathcal{M}_\Sigma$ is closed under sequential pointwise limits.*
   c. *If μ is σ-finite and $f: X \to \mathbb{R}$ is a measurable function, then there exists a Σ-measurable function $g: X \to \mathbb{R}$ such that $f = g$ a.e.*

Solution.   (a) In order to show that $\mathcal{M}_\Sigma$ is closed under addition and multiplication, we need the following properties among Σ-measurable functions f and g: The sets

   1. $\{x \in X: \ f(x) > g(x)\}$,
   2. $\{x \in X: \ f(x) \geq g(x)\}$, and
   3. $\{x \in X: \ f(x) = g(x)\}$

all belong to Σ. To see (1), let $r_1, r_2, \ldots$ be an enumeration of the rational numbers of $\mathbb{R}$, and note that

$$\{x \in X: \ f(x) > g(x)\} = \bigcup_{n=1}^{\infty}\Big[\{x \in X: \ f(x) > r_n\} \cap \{x \in X: \ g(x) < r_n\} \Big],$$

which belongs to Σ, since it is a countable union of sets from the σ-algebra Σ. For (2), note that $\{x \in X: \ f(x) \geq g(x)\} = \{x \in X: \ g(x) > f(x)\}^c$, which belongs to Σ by (1). Finally, for (3), observe that

$$\{x \in X: \ f(x) = g(x)\} = \{x \in X: \ f(x) \geq g(x)\} \cap \{x \in X: \ g(x) \geq f(x)\},$$

which belongs to Σ by (2).

   To complete the proof of part (a), we shall establish that for Σ-measurable functions f and g, the following statements hold:

   i. $f + g$ is a Σ-measurable function.
   ii. fg is a Σ-measurable function.

    iii. $|f|$, f^+, and f^- are Σ-measurable functions.

    iv. $f \vee g$ and $f \wedge g$ are Σ-measurable functions.

The proofs of these claims are given below.

(i) Note first that if c is a constant number, then $c - g$ is a Σ-measurable function. [Reason: If $a \in \mathbf{R}$, then $\{x \in X: c - g(x) \geq a\} = \{x \in X: g(x) \leq c - a\} \in \Sigma$.] Now, if $a \in \mathbf{R}$, then the set

$$(f + g)^{-1}\big([a, \infty)\big) = \{x \in X: f(x) + g(x) \geq a\} = \{x \in X: f(x) \geq a - g(x)\}$$

belongs to Σ by the preceding observation and (2). This implies (how?) that $f + g$ is a Σ-measurable function.

(ii) Note first that f^2 is a Σ- measurable function. To see this, let $a \in \mathbf{R}$. Then $\{x \in X: f^2(x) \leq a\} = \emptyset$ if $a < 0$ and $\{x \in X: f^2(x) \leq a\} = f^{-1}\big([-\sqrt{a}, \sqrt{a}\,]\big)$ if $a \geq 0$. This implies that f^2 is a Σ-measurable function. Also, if c is a constant, then cf is measurable. [Reason: If $A = \{x \in X: cf(x) \geq a\}$, then $A = \{x \in X: f(x) \geq a/c\}$ for $c > 0$ and $A = \{x \in X: f(x) \leq a/c\}$ for $c < 0$.] The result now follows from the preceding observations combined with (i) and the relation

$$fg = \tfrac{1}{2}\big[(f + g)^2 - f^2 - g^2\big].$$

(iii) The Σ-measurability of $|f|$ follows from the relation

$$\{x \in X: |f(x)| \leq a\} = \emptyset \quad \text{if } a < 0,$$

and

$$\{x \in X: |f(x)| \leq a\} = \{x \in X: f(x) \leq a\} \cap \{x \in X: f(x) \geq -a\} \quad \text{if } a \geq 0.$$

For the Σ-measurability of f^+ and f^- use the identities

$$f^+ = \tfrac{1}{2}(|f| + f) \quad \text{and} \quad f^- = \tfrac{1}{2}(|f| - f).$$

(iv) The identities

$$f \vee g = \tfrac{1}{2}(f + g + |f - g|) \quad \text{and} \quad f \wedge g = \tfrac{1}{2}(f + g - |f - g|)$$

show that $f \vee g$ and $f \wedge g$ are Σ-measurable functions.

(b) Assume that $\{f_n\}$ is a sequence of Σ-measurable functions such that $f_n(x) \to f(x)$ holds for each $x \in X$. Observe that the equality

$$f^{-1}\big((a, \infty)\big) = \bigcup_{n=1}^{\infty} \bigcap_{i=n}^{\infty} f_i^{-1}\big((a + \tfrac{1}{n}, \infty)\big)$$

and the Σ-measurability of each f_i show that $f^{-1}\big((a, \infty)\big)$ belongs to Σ. This implies that f is a Σ-measurable function.

(c) We can assume $f(x) \geq 0$ for each $x \in X$ (otherwise, we apply the arguments below to f^+ and f^- separately). Assume first that $f = \chi_A$ for some $A \in \Sigma$. Since μ is σ-finite, it follows from Theorem 15.11 that there exists a μ-null set C such that $B = A \cup C \in \Sigma$. So, if $g = \chi_B$, then g is Σ-measurable and $f = g$ μ-a.e. It follows that if ϕ is a μ-simple function, then there exists a Σ-simple function ψ such that $\psi = \phi$ μ-a.e.

Now, by Theorem 17.7, there exists a sequence $\{\phi_n\}$ of simple functions such that $\phi_n(x) \uparrow f(x)$ for each $x \in X$. Replacing each ϕ_n by a Σ-simple function ψ_n (as above) we have $\psi_n(x) \uparrow f(x)$ for μ-almost all x. So, there exists a μ-measurable set E such that $\psi_n(x) \uparrow f(x)$ for each $x \notin E$. Now, use Theorem 15.11 to select a set $F \in \Sigma$ with $E \subseteq F$ and $\mu^*(F) = 0$. Clearly, $\psi_n(x)\chi_{F^c}(x) \uparrow f(x)\chi_{F^c}(x) = g(x)$ for each $x \in X$. By part (b), g is Σ-measurable and satisfies $g = f$ μ-a.e.

18.   THE LEBESGUE MEASURE

Problem 18.1.   *Let* $I = \prod_{i=1}^{n} I_i$ *be an interval of* $\mathbb{R}^n$. *Show that* I *is Lebesgue measurable and that* $\lambda(I) = \prod_{i=1}^{n} |I_i|$, *where* $|I_i|$ *denotes the length of the interval* I_i.

Solution.   The verification of the formula can be done by cases as in Problem 15.3. To show this, we establish the formula for two cases, and leave the rest for the reader.

The first case is when $I_i = [a_i, b_i]$, where $-\infty < a_i < b_i < \infty$ holds for each $1 \leq i \leq n$. Then, $\prod_{i=1}^{n}[a_i, b_i + \tfrac{1}{k}) \downarrow_k \prod_{i=1}^{n} I_i = I$. Thus, from Theorem 15.4, it follows that

$$\lambda(I) = \lim_{k \to \infty} \lambda\Big(\prod_{i=1}^{n}[a_i, b_i + \tfrac{1}{k}) \Big) = \lim_{k \to \infty} \prod_{i=1}^{n} (b_i - a_i + \tfrac{1}{k})$$

$$= \prod_{i=1}^{n} (b_i - a_i) = \prod_{i=1}^{n} |I_i|.$$

The second case is when $I = [a, \infty) \times [a_2, b_2] \times \cdots \times [a_n, b_n]$. Then, note that $[a, a+k] \times [a_2, b_2] \times \cdots \times [a_n, b_n] \uparrow_k I$. Taking into account the preceding case, it follows from Theorem 15.4 that

$$\lambda(I) = \lim_{k \to \infty} \lambda\big([a, a + k] \times [a_2, b_2] \times \cdots \times [a_n, b_n]\big)$$

$$= \lim_{k \to \infty} k \cdot (b_2 - a_2) \cdots (b_n - a_n) = \infty = \prod_{i=1}^{\infty} |I_i|.$$

Problem 18.2.   *Let $\mathcal{O}$ be an open subset of $\mathbf{R}$. Show that there exists an at-most countable collection $\{I_\alpha \colon \alpha \in A\}$ of pairwise disjoint open intervals such that $\mathcal{O} = \bigcup_{\alpha \in A} I_\alpha$. Also, show that $\lambda(\mathcal{O}) = \sum_{\alpha \in A} |I_\alpha|$.*

Solution.   Let $\mathcal{O}$ be an open subset of $\mathbf{R}$. By part (g) of Problem 6.11, we know that there exists an at-most countable collection $\{I_\alpha \colon \alpha \in A\}$ of pairwise disjoint open intervals such that $\mathcal{O} = \bigcup_{\alpha \in A} I_\alpha$.

Now, using the fact that the length of each I_α coincides with its Lebesgue measure (Problem 15.3), we see that

$$\lambda(\mathcal{O}) = \lambda\left(\bigcup_{\alpha \in A} I_\alpha\right) = \sum_{\alpha \in A} \lambda(I_\alpha) = \sum_{\alpha \in A} |I_\alpha|.$$

Problem 18.3.   *Show that the Borel sets of $\mathbf{R}^n$ are precisely the members of the σ-algebra generated by the compact sets.*

Solution.   Let $\mathcal{C}$ denote the σ-algebra generated by the compact sets. Since every compact set is closed (which is the complement of an open set), it follows that $\mathcal{C} \subseteq \mathcal{B}$. On the other hand, if C is a closed set and $C_n = \{x \in C \colon d(0, x) \le n\}$, then $\{C_n\}$ is a sequence of compact sets satisfying $C_n \uparrow C$. This implies that $\mathcal{C}$ contains all the closed sets (and hence, all the open sets). Thus, $\mathcal{B} \subseteq \mathcal{C}$ also holds, and so $\mathcal{B} = \mathcal{C}$.

Problem 18.4.   *Show that a subset E of $\mathbf{R}^n$ is Lebesgue measurable if and only if for each $\epsilon > 0$ there exists a closed subset F of $\mathbf{R}^n$ such that $F \subset E$ and $\lambda(E \setminus F) < \epsilon$.*

Solution.   Assume that E is Lebesgue measurable and let $\varepsilon > 0$. Since E^c is also Lebesgue measurable, there exists an open set V such that $E^c \subseteq V$ and $\lambda(V \setminus E^c) = \lambda(E \cap V) < \varepsilon$. Then, the closed set $C = V^c$ satisfies $C \subseteq E$ and $\lambda(E \setminus C) = \lambda(E \cap V) < \varepsilon$. For the converse, either reverse the preceding arguments and use Theorem 18.2, or else use Problem 14.8.

Problem 18.5.   *Show that if a subset E of $[0, 1]$ satisfies $\lambda(E) = 1$, then E is dense in $[0, 1]$.*

Solution. Let I be a (nonempty) subinterval of $[0, 1]$. If $I \cap E = \emptyset$, then we have $\lambda(E) + \lambda(I) = \lambda(E \cup I) \leq \lambda([0, 1]) = 1$, and hence, in this case, $\lambda(E) \leq 1 - \lambda(I) < 1$ holds, which is a contradiction. Thus, $I \cap E \neq \emptyset$ holds for each subinterval I of $[0, 1]$, and so the set E is dense in $[0, 1]$.

Problem 18.6. *If $E \subseteq \mathbb{R}^n$ satisfies $\lambda(E) = 0$, then show that $E^\circ = \emptyset$.*

Solution. If V is a nonempty open set with $V \subset E$, then note that $0 < \lambda(V) \leq \lambda(E)$ holds. Therefore, the open set E° must be empty.

Problem 18.7. *Show that if E is a Lebesgue measurable subset of $\mathbb{R}^n$, then there exist an F_σ-set A and a G_δ-set B such that $A \subseteq E \subseteq B$ and $\lambda(B \setminus A) = 0$.*

Solution. By Problem 18.4, for each k there exists a closed set C_k with $C_k \subseteq E$ and $\lambda(E \setminus C_k) < \frac{1}{k}$. Similarly, by Theorem 18.2, for every k there exists an open set V_k with $E \subseteq V_k$ and $\lambda(V_k \setminus E) < \frac{1}{k}$. Put $A = \bigcup_{k=1}^\infty C_k$ (an F_σ-set) and $B = \bigcap_{k=1}^\infty V_k$ (a G_δ-set). Clearly, $A \subseteq E \subseteq B$ holds, and in view of

$$\lambda(B \setminus A) \leq \lambda(V_k \setminus C_k) = \lambda\big((V_k \setminus E) \cup (E \setminus C_k)\big)$$
$$\leq \lambda(V_k \setminus E) + \lambda(E \setminus C_k) < \tfrac{2}{k}$$

for each k, we see that $\lambda(B \setminus A) = 0$.

Problem 18.8. *Let $\{E_n\}$ be a sequence of nonempty (Lebesgue) measurable subsets of $[0, 1]$ satisfying $\lim \lambda(E_n) = 1$.*

    a. *Show that for each $0 < \epsilon < 1$ there exists a subsequence $\{E_{k_n}\}$ of $\{E_n\}$ such that $\lambda(\bigcap_{n=1}^\infty E_{k_n}) > \epsilon$.*

    b. *Show that $\bigcap_{k=n}^\infty E_k = \emptyset$ is possible for each $n = 1, 2, \ldots$.*

Solution. Let $\{E_n\}$ be a sequence of nonempty Lebesgue measurable subsets of $[0, 1]$ satisfying $\lim \lambda(E_n) = 1$.

(a) Fix $0 < \varepsilon < 1$. From $\lim \lambda(E_n) = 1$, we see that there exists a subsequence $\{E_{k_n}\}$ of $\{E_n\}$ satisfying $\lambda(E_{k_n}) > 1 - \frac{1-\varepsilon}{2^n}$. Now, consider the measurable sets $E = \bigcap_{n=1}^\infty E_{k_n}$ and $F = [0, 1] \setminus E$. Then, we have

$$\lambda(F) = \lambda\big([0, 1] \setminus E\big) = \lambda\left(\bigcup_{n=1}^\infty ([0, 1] \setminus E_{k_n})\right)$$
$$= \sum_{n=1}^\infty \lambda\big([0, 1] \setminus E_{k_n}\big) = \sum_{n=1}^\infty \big[1 - \lambda(E_{k_n})\big] < \sum_{n=1}^\infty \tfrac{1-\varepsilon}{2^n} = 1 - \varepsilon.$$

Hence, $\lambda(E) = 1 - \lambda(F) > 1 - (1 - \varepsilon) = \varepsilon$.

(b) Let $A_k^n = [0, 1] \setminus \left[\frac{k-1}{n}, \frac{k}{n}\right]$, $1 \le k \le n$; $n \ge 2$. Clearly, $\lambda(A_k^n) = 1 - \frac{1}{n}$ holds for each $1 \le k \le n$ and $\bigcap_{k=1}^{n} A_k^n = \emptyset$ holds for each $n \ge 2$. Let E_n denote the sequence

$$A_1^2,\ A_2^2,\ A_1^3,\ A_2^3,\ A_3^3,\ \ldots,\ A_1^n,\ A_2^n,\ \ldots,\ A_n^n,\ A_1^{n+1},\ \ldots.$$

Now, note that $\lambda(E_n) \longrightarrow 1$ and $\bigcap_{k=n}^{\infty} E_k = \emptyset$ holds for each $n \ge 1$.

Problem 18.9. *Assume that a function $f : I \to \mathbb{R}$ defined on a subinterval of $\mathbb{R}$ satisfies a Lipschitz condition. That is, assume that there exists a constant $C > 0$ such that $|f(x) - f(y)| \le C|x - y|$ holds for all $x, y \in I$. Show that f carries (Lebesgue) null sets to null sets.*

In particular, if a function $f : I \to \mathbb{R}$ defined on a subinterval of $\mathbb{R}$ has a continuous derivative, then show that f carries null sets to null sets.

Solution. Assume that a function $f : I \longrightarrow \mathbb{R}$ satisfies the condition of the problem. Clearly, f is a (uniformly) continuous function. In particular, note that if J is a subinterval of I, then $f(J)$ is also a subinterval of $\mathbb{R}$ (see part (g) of Problem 6.11), and our condition implies (how?) that the length of $f(J)$ is less than or equal to C times the length of J, i.e., $\lambda^*\big(f(J)\big) \le C\lambda^*(J)$ holds.

Now, let A be a null subset of I and let $\varepsilon > 0$. Pick a sequence $\big\{[a_n, b_n)\big\}$ of half-open intervals such that

$$A \subseteq \bigcup_{n=1}^{\infty} [a_n, b_n) \quad \text{and} \quad \sum_{n=1}^{\infty} \lambda^*\big([a_n, b_n)\big) = \sum_{n=1}^{\infty}(b_n - a_n) < \varepsilon.$$

Hence, $f(A) \subseteq f\big(\bigcup_{n=1}^{\infty}[a_n, b_n) \cap I\big) = \bigcup_{n=1}^{\infty} f\big([a_n, b_n) \cap I\big)$, and so by the preceding

$$\lambda^*\big(f(A)\big) \le \lambda^*\left(\bigcup_{n-1}^{\infty} f\big([a_n, b_n) \cap I\big)\right)$$

$$\le \sum_{n=1}^{\infty} \lambda^*\big(f([a_n, b_n) \cap I)\big) \le \sum_{n=1}^{\infty} C(b_n - a_n) < C\varepsilon.$$

Since $\varepsilon > 0$ is arbitrary, we see that $\lambda^*\big(f(A)\big) = 0$ holds, as desired.

For the second part notice that if $[a, b]$ is a closed subinterval of I, then there exists some constant $M > 0$ satisfying $|f'(t)| \le M$ for all $t \in [a, b]$. Now, if $x, y \in [a, b]$, then there exists (by the Mean Value Theorem) some z between x and y satisfying $f(x) - f(y) = f'(z)(x - y)$. This implies $|f(x) - f(y)| \le M|x - y|$ for all $x, y \in [a, b]$. So, by the first part, f carries null sets of $[a, b]$ to null sets.

Now, fix a sequence $\{[a_n, b_n]\}$ of closed subintervals of I such that $I = \bigcup_{n=1}^{\infty}[a_n, b_n]$ and let A be a null subset of I. Then $A \cap [a_n, b_n]$ is a null subset of $[a_n, b_n]$, and so $f(A \cap [a_n, b_n])$ is a null subset of $\mathbf{R}$. Now, notice that the identity

$$f(A) = f\left(\bigcup_{n=1}^{\infty} A \cap [a_n, b_n]\right) = \bigcup_{n=1}^{\infty} f(A \cap [a_n, b_n])$$

guarantees that $f(A)$ is a null subset of $\mathbf{R}$.

Problem 18.10. *Show that the Lebesgue measure of a triangle in $\mathbf{R}^2$ equals its area. Also, determine the Lebesgue measure of a disk in $\mathbf{R}^2$.*

Solution.   Start by observing that every line segment has Lebesgue measure zero (why?). Thus, the Lebesgue measure of a triangle is the same with or without some of its edges. Also, every triangle is Lebesgue measurable (since without its edges it is an open set). Since λ is translation invariant, we can assume that all triangles have one of their vertices at zero. Let T be such a triangle, and let $A(T)$ denote its area. Following the graphs in Figure 3.1 (from left to right) we see that:

$$2\lambda(T) = \lambda(T) + \lambda(-T) = \lambda(T) + \lambda(T_2) = \lambda(T) + \lambda(T_1)$$
$$= \lambda(T_1 \cup T) = \lambda(P) = \lambda(Q) = A(P) = 2A(T).$$

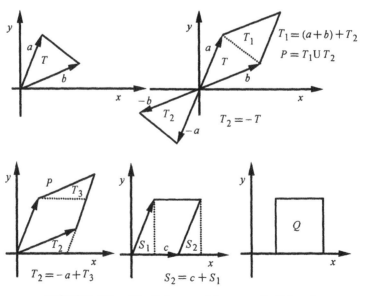

FIGURE 3.1.   The Lebesgue Measure of a Triangle

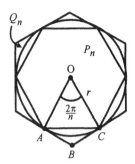

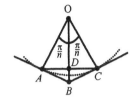

Area OAD $= \frac{1}{2} r^2 \sin\left(\frac{\pi}{n}\right) \cos\left(\frac{\pi}{n}\right)$

Area OAB $= \frac{1}{2} r^2 \tan\left(\frac{\pi}{n}\right)$

FIGURE 3.2.   The Computation of the Lebesgue Measure of a Disk

That is, $\lambda(T) = A(T)$. In particular, this implies that the Lebesgue measure of any polygon equals its area.

Now, let D be a closed disk of radius r; see Figure 3.2. To compute its Lebesgue measure, we use the Eudoxus–Archimedes *Method of Exhaustion*. For each n, let P_n and Q_n be the inscribed and circumscribed regular n-polygons, respectively. Clearly, $P_n \subseteq D \subseteq Q_n$ holds. Now, note that

$$\lambda(P_n) = \pi r^2 \left[\frac{\sin\left(\frac{\pi}{n}\right)}{\frac{\pi}{n}}\right] \cos\left(\frac{\pi}{n}\right) \leq \lambda(D) \leq \lambda(Q_n) = \pi r^2 \left[\frac{\tan\left(\frac{\pi}{n}\right)}{\frac{\pi}{n}}\right],$$

and so, by letting $n \longrightarrow \infty$, we see that

$$\lambda(D) = \pi r^2 = A(D).$$

Problem 18.11.   *If μ is a translation invariant Borel measure on $\mathbf{R}^n$, then show that there exists some $c \geq 0$ such that $\mu^*(A) = c\lambda^*(A)$ for all subset A of $\mathbf{R}^n$.*

Solution.   By Theorem 18.8, $\mu = c\lambda$ holds on $\mathcal{B}$ for some constant $c \geq 0$. Now, by Theorem 14.10, $(c\lambda)^* = c\lambda^*$ holds, and consequently $\mu^*(A) = (c\lambda)^*(A) = c\lambda^*(A)$ for each subset A of $\mathbf{R}^n$.

Problem 18.12.   *Show that an arbitrary collection of pairwise disjoint measurable subsets of $\mathbf{R}$, each of which has positive measure, is at-most countable.*

Solution.   Let $\mathcal{C}$ be a collection of pairwise disjoint measurable subsets of $\mathbf{R}$ such that $\lambda(C) > 0$ holds for each $C \in \mathcal{C}$. For each n let

$$\mathcal{C}_n = \left\{ C \in \mathcal{C}:\ \lambda(C \cap [-n, n]) \geq \tfrac{1}{n} \right\},$$

and note that $C = \bigcup_{n=1}^{\infty} C_n$. Now if $C_1, \dots, C_k \in C_n$, then we have

$$\tfrac{k}{n} \le \sum_{i=1}^{k} \lambda(C_i \cap [-n, n]) = \lambda\left(\left(\bigcup_{i=1}^{k} C_i\right) \cap [-n, n]\right) \le \lambda([-n, n]) = 2n,$$

and so $k \le 2n^2$ holds. This shows that each C_n is a finite set, and consequently C is at-most countable.

Problem 18.13. *Let G be a proper additive subgroup of $\mathbf{R}^n$. If G is a measurable set, then show that $\lambda(G) = 0$.*

Solution. If $\lambda(G) > 0$, then, by Theorem 18.13, the element zero is an interior point of $G - G$. Since G is an additive group, $G - G = G$ holds, and from this it follows that $G = \mathbf{R}^n$, which is a contradiction.

Problem 18.14. *Let $f: \mathbf{R} \to \mathbf{R}$ be additive (i.e., $f(x + y) = f(x) + f(y)$ for all $x, y \in \mathbf{R}$) and Lebesgue measurable. Show that f is continuous—and hence, of the form $f(x) = cx$.*

Solution. Assume $f \ne 0$ and let $\varepsilon > 0$. Since f is an additive function, $nf^{-1}([0, \varepsilon]) = f^{-1}([0, n\varepsilon])$ holds (why?). Thus, if $\lambda(f^{-1}([0, \varepsilon])) = 0$, then

$$\lambda(f^{-1}([-n\varepsilon, 0])) = \lambda(f^{-1}([0, n\varepsilon])) = n\lambda(f^{-1}([0, \varepsilon])) = 0$$

holds for each n, and so $\lambda(f^{-1}([-n\varepsilon, n\varepsilon])) = 0$ for all n. From

$$f^{-1}([-n\varepsilon, n\varepsilon]) \uparrow \mathbf{R},$$

it follows that $\lambda(\mathbf{R}) = 0$, which is impossible. Thus, $\lambda(f^{-1}([0, \varepsilon])) > 0$. Since f is also measurable, there exists (by Theorem 18.13) some $\delta > 0$ with

$$(-\delta, \delta) \subseteq f^{-1}([0, \varepsilon]) - f^{-1}([0, \varepsilon]) = f^{-1}([-\varepsilon, \varepsilon]).$$

That is, $-\delta < x < \delta$ implies $-\varepsilon \le f(x) \le \varepsilon$ so that f is continuous at zero. Now apply Lemma 18.7.

Problem 18.15. *Show that an arbitrary union of proper intervals of $\mathbf{R}$ is a Lebesgue measurable set.*

Solution. Let $\{I_\alpha: \alpha \in A\}$ be a family of "proper" intervals (an interval is proper whenever its endpoints a and b satisfy $a < b$) and let $E = \bigcup_{\alpha \in A} I_\alpha$. Write $E = \bigcup_{x \in E} C_x$, where C_x denotes the component of x in E. Since each x belongs to a proper subinterval of E, we see that each C_x is a proper interval; see part (g) Problem 6.11. Since the distinct components C_x are pairwise disjoint, we see that there are at-most countably many C_x and so E is the union of at-most countably many intervals. Now, use the fact that each interval is a Lebesgue measurable set to infer that E itself is a Lebesgue measurable set.

Problem 18.16. *Let C be a closed nowhere dense subset of $\mathbf{R}^n$ such that $\lambda(C) > 0$. Show that the characteristic function χ_C cannot be continuous on the complement of any Lebesgue null set of $\mathbf{R}^n$. Also, show that χ_C will be continuous on the complement of a properly chosen open set whose Lebesgue measure can be made arbitrarily small.*

Solution. Let $A \subseteq \mathbf{R}^n$ be a Lebesgue null set. Since $\lambda(C) > 0$ and $\lambda(A) = 0$, it follows that $A^c \cap C \neq \emptyset$. Fix some $a \in A^c \cap C$. We claim that $\chi_C \colon A^c \longrightarrow \mathbf{R}$ is not continuous at $x = a$.

Indeed, if $\chi_C \colon A^c \longrightarrow \mathbf{R}$ is continuous at a, then there exists some open ball $B(a, r)$ with $\chi_C(x) = 1$ for all $x \in B(a, r) \cap A^c$; i.e., $B(a, r) \cap A^c \subseteq C$ holds. Since $\lambda(A) = 0$, it follows that $B(a, r) \cap A^c$ is dense in $B(a, r)$, and therefore, $B(a, r) \subseteq C$ (since $B(a, r) \cap A^c \subseteq C$ and C is closed), contradicting the fact that C is nowhere dense.

Now, let $\varepsilon > 0$. By Theorem 18.2, there exists an open set V with $C \subseteq V$ and $\lambda(V \setminus C) < \varepsilon$. Note that the set $O = V \setminus C = V \cap C^c$ is open, and $\lambda(O) < \varepsilon$. We claim that $\chi_C \colon O^c \longrightarrow \mathbf{R}$ is continuous.

To see this, let $a \notin O = V \cap C^c$. We have two cases.
1) $a \in C$. Since $C \subseteq V$, there exists some open ball $B(a, r)$ with $B(a, r) \subseteq V$. Now, note that

$$B(a, r) \cap O^c = B(a, r) \cap [V^c \cup C] = B(a, r) \cap C \subseteq C.$$

Thus, if $x \in B(a, r) \cap O^c$, then $\chi_C(x) = 1$. This shows that the function $\chi_C \colon O^c \longrightarrow \mathbf{R}$ is continuous at $x = a$.
2) $a \in C^c$. Choose an open ball $B(a, r)$ such that $B(a, r) \subseteq C^c$. Then,

$$B(a, r) \cap O^c = B(a, r) \cap [V^c \cup C] = B(a, r) \cap V^c \subseteq V^c \subseteq C^c.$$

Thus, $x \in B(a, r) \cap O^c$ implies $\chi_C(x) = 0$, which shows that in this case $\chi_C \colon O^c \longrightarrow \mathbf{R}$ is continuous at $x = a$.

Problem 18.17. *Let $f \colon \mathbf{R}^n \to \mathbf{R}$ be a continuous function. Show that the graph*

$$G = \{(x_1, \ldots, x_n, f(x_1, \ldots, x_n)) \colon (x_1, \ldots, x_n) \in \mathbf{R}^n\}$$

of f has $(n + 1)$-dimensional Lebesgue measure zero.

Solution. Denote by λ_{n+1} and λ_n the $(n+1)$-dimensional and n-dimensional Lebesgue measures, respectively. Fix some k and let $A = [-k, k] \times \cdots \times [-k, k]$. Now, let $\varepsilon > 0$. By the uniform continuity of f on A, there exists some $\delta > 0$ such that $x, y \in A$ and $|x_i - y_i| < \delta$ for $1 \le i \le n$ imply $|f(x) - f(y)| < \varepsilon$. Fix a partition P of $[-k, k]$ with mesh $|P| < \delta$, and let $Q = P \times \cdots \times P$. Then, Q subdivides A into a finite number of distinct closed cells, say $A_1, \ldots, A_p$. (Note that the open cells corresponding to $A_1, \ldots, A_p$ are pairwise disjoint). For each $1 \le i \le p$ fix some $a_i \in A_i$, and let $I_i = [f(a_i) - \varepsilon, f(a_i) + \varepsilon]$. Then, $G_k \subseteq \bigcup_{i=1}^{p}(A_i \times I_i)$ holds, and so

$$\lambda_{n+1}(G_k) \le \sum_{i=1}^{p} \lambda_{n+1}(A_i \times I_i) = \sum_{i=1}^{p} \lambda_n(A_i) \cdot 2\varepsilon = (2k)^n \cdot 2\varepsilon$$

holds for all $\varepsilon > 0$. This shows that $\lambda_{n+1}(G_k) = 0$ for each k. To complete the proof, now apply Theorem 15.4 to $G_k \uparrow G$.

Problem 18.18. *Let X be a Hausdorff topological space, and let μ be a regular Borel measure on X. Show the following:*

a. *If A is an arbitrary subset of X, then*

$$\mu^*(A) = \inf\{\mu(\mathcal{O}): \mathcal{O} \text{ open and } A \subseteq \mathcal{O}\}.$$

b. *If A is a measurable subset of X with $\mu^*(A) < \infty$, then*

$$\mu^*(A) = \sup\{\mu(K): K \text{ compact and } K \subseteq A\}.$$

c. *If μ is σ-finite and A is a measurable subset of X, then*

$$\mu^*(A) = \sup\{\mu(K): K \text{ compact and } K \subseteq A\}.$$

Solution. (a) Since every σ-set is a Borel set, Problem 15.2 shows that

$$\mu^*(A) = \inf\{\mu(B): B \text{ is a Borel set satisfying } A \subseteq B\}.$$

Now, use property (2) of Definition 18.4.

(b) Let A be a measurable set with $\mu^*(A) < \infty$ and let $\varepsilon > 0$. Pick an open set V with $A \subseteq V$ and $\mu^*(V) < \mu^*(A) + \varepsilon$. Similarly, choose an open set W such that $V \setminus A \subseteq W \subseteq V$ and

$$\mu^*(W) < \mu^*(V \setminus A) + \varepsilon = \mu^*(V) - \mu^*(A) + \varepsilon < 2\varepsilon.$$

Next, pick a compact set C such that $C \subseteq V$ and $\mu^*(V) < \mu^*(C) + \varepsilon$. Set $K = C \cap W^c$, and note that K is a compact subset of A. Moreover,

$$0 \leq \mu^*(A) - \mu^*(K) = \mu^*(A \setminus K) \leq \mu^*(V \setminus K)$$
$$= \mu^*((V \setminus C) \cup W) \leq [\mu^*(V) - \mu^*(C)] + \mu^*(W) < 3\varepsilon$$

holds, and the desired conclusion follows.

(c) Straightforward using (b).

Problem 18.19. *If A is a (Lebesgue) measurable subset of $\mathbb{R}$ of positive measure and $0 < \delta < \lambda(A)$, then show that there exists a measurable subset B of A satisfying $\lambda(B) = \delta$.*

Solution. We shall present two solutions. The first one will employ the Axiom of Choice (via Zorn's Lemma); the second one will establish the validity of the conclusion without using the Axiom of Choice and without assuming that A is a measurable set.

(a) Consider a measurable subset A of $\mathbb{R}$ and some $\delta > 0$ satisfying $0 < \delta < \lambda(A)$. Since $\lambda(A \cap [-n, n]) \uparrow \lambda(A)$ holds, replacing A by some $A \cap [-n, n]$, we can assume that $\lambda(A) < \infty$ also holds.

Next, we shall denote by $\mathcal{A}$ the set of all collections $\mathcal{C}$ of pairwise disjoint measurable subsets of A such that:

a) $\lambda(C) > 0$ holds for each $C \in \mathcal{C}$ (and so $\mathcal{C}$ is at most countable); and

b) The Lebesgue measurable set $\bigcup_{C \in \mathcal{C}} C$ satisfies $\lambda(\bigcup_{C \in \mathcal{C}} C) \leq \delta$.

From Problem 15.18, it is easy to see that $\mathcal{A} \neq \emptyset$. Under the inclusion relation $\subseteq$ the set $\mathcal{A}$ is a partially ordered set. We claim that the partially ordered set $(\mathcal{A}, \subseteq)$ satisfies the hypothesis of Zorn's Lemma. To see this, let $\{\mathcal{C}_i : i \in I\}$ be a chain of $\mathcal{A}$ (i.e., for each pair $i, j \in I$ either $\mathcal{C}_i \subseteq \mathcal{C}_j$ or $\mathcal{C}_j \subseteq \mathcal{C}_i$ holds true). Our claim will be established, if we can show that $\mathcal{C} = \bigcup_{i \in I} \mathcal{C}_i \in \mathcal{A}$. Note first that if $B, C \in \mathcal{C}$, then $B, C \in \mathcal{C}_i$ must hold for at least one $i \in I$, and so $B \cap C = \emptyset$. In particular, it follows that $\mathcal{C}$ is at most countable. Now, if $B_1, \ldots, B_k \in \mathcal{C}$, then $B_1, \ldots, B_k \in \mathcal{C}_i$ also must hold for some i (why?), and so $\lambda(\bigcup_{r=1}^k B_r) \leq \lambda(\bigcup_{B \in \mathcal{C}_i} B) \leq \delta$. Since $\mathcal{C}$ is at most countable, it follows that $\lambda(\bigcup_{B \in \mathcal{C}} B) \leq \delta$. That is, $\mathcal{C} \in \mathcal{A}$.

Now, by Zorn's Lemma, the collection $\mathcal{A}$ has a maximal element, say $\mathcal{C}$. If $B = \bigcup_{C \in \mathcal{C}} C$, then we claim that the measurable set B satisfies $\lambda(B) = \delta$ (and this will complete the proof). To see the latter, assume by way of contradiction that $\lambda(B) < \delta$. Then, we have $0 < \eta = \delta - \lambda(B) \leq \lambda(A) - \lambda(B) = \lambda(A \setminus B)$ holds, and so by Problem 15.18 there exists a measurable subset D of $A \setminus B$

satisfying $0 < \lambda(D) < \eta$ (clearly, $D \notin C$). In view of $B \cap D = \emptyset$ and

$$\lambda(B \cup D) = \lambda(B) + \lambda(D) < \lambda(B) + \delta - \lambda(B) = \delta,$$

we see that $C_1 = C \cup \{D\} \in \mathcal{A}$. However, this contradicts the maximality property of C, and so $\lambda(B) = \delta$ must hold, as desired.

(b) For this solution the set A is an arbitrary subset of $\mathbf{R}$ satisfying $\lambda(A) > 0$. As in the preceding, we can assume that $A \subseteq [-k, k]$ holds for some k. Now, consider the function $f : [-k, k] \longrightarrow \mathbf{R}$ defined by

$$f(t) = \lambda(A \cap [-k, t]), \quad t \in [-k, k].$$

Clearly, $f(-k) = 0$ and $f(k) = \lambda(A)$. We claim that f is a continuous function. Indeed, if $-k \leq s < t \leq k$, then

$$f(t) = \lambda(A \cap [-k, t]) \leq \lambda(A \cap [-k, s]) + \lambda(A \cap (s, t]) \leq f(s) + t - s.$$

Therefore, $|f(s) - f(t)| \leq |t - s|$ holds for all $s, t \in [-k, k]$ and so f is a continuous function.

Finally, by the Intermediate Value Theorem, there exists some $-k \leq x \leq k$ such that the subset $B = A \cap [-k, x]$ of A (which is measurable if A is measurable) satisfies $f(x) = \lambda(B) = \delta$.

Problem 18.20. *Let E be a Lebesgue measurable subset of $\mathbf{R}$ of finite Lebesgue measure. Show that the function $f_E : \mathbf{R} \to \mathbf{R}$, defined by*

$$f_E(x) = \lambda(E \triangle (x + E)),$$

is uniformly continuous.

Solution.   The solution goes by steps.

(1) Assume first that $E = (a, b)$ is a bounded open subinterval of $\mathbf{R}$. In this case, an easy calculation shows that

$$f_E(x) = \begin{cases} 2|x|, & \text{if } |x| < b - a \\ 2(b - a), & \text{if } |x| \geq b - a. \end{cases}$$

This guarantees that f_E is uniformly continuous in this case.

(2) Assume that E and F are two Lebesgue measurable subsets of $\mathbf{R}$ of finite measure such that f_E and f_F are both uniformly continuous. Put $G = E \cup F$. We shall show that f_G is also uniformly continuous.

To see this, notice first that

$$\left|\chi_G(y) - \chi_{x+G}(y)\right| \le \left|\chi_E(y) - \chi_{x+E}(y)\right| + \left|\chi_F(y) - \chi_{x+F}(y)\right|$$

implies

$$\lambda\bigl(G\,\Delta(x + G)\bigr) \le \lambda\bigl(E\,\Delta(x + E)\bigr) + \lambda\bigl(F\,\Delta(x + F)\bigr).$$

Hence,

$$\begin{aligned}
\left|f_G(x) - f_G(y)\right| &= \left|\lambda(G\,\Delta(x + G)) - \lambda(G\,\Delta(y + G))\right| \\
&\le \lambda\bigl([G\,\Delta(x + G)]\,\Delta[G\,\Delta(y + G)]\bigr) \\
&= \lambda\bigl((x + G)\,\Delta(y + G)\bigr) = \lambda\bigl(G\,\Delta(y - x + G)\bigr) \\
&\le \lambda\bigl(E\,\Delta(y - x + E)\bigr) + \lambda\bigl(F\,\Delta(y - x + F)\bigr) \\
&= f_E(y - x) + f_F(y - x).
\end{aligned}$$

Since f_E and f_F are uniformly continuous, it follows that f_G is likewise uniformly continuous. (Actually, the continuity of f_E and f_F at zero is what is needed here.)

(3) By induction, we can show that if $E = \bigcup_{i=1}^{n} E_i$ with each E_i Lebesgue measurable having finite measure and $E_i \cap E_j = \emptyset$ if $i \ne j$, then f_E is uniformly continuous.

(4) Now, let $\epsilon > 0$. Pick a finite collection of pairwise disjoint bounded open intervals $I_1, \ldots, I_n$ such that the set $G = \bigcup_{i=1}^{n} I_i$ satisfies $\lambda(E\,\Delta G) < \epsilon$. Then, as previously, we have

$$\begin{aligned}
\left|f_E(x) - f_E(y)\right| &= \left|\lambda(E\,\Delta(x + E)) - \lambda(E\,\Delta(y + E))\right| \\
&\le \lambda\bigl(E\,\Delta(y - x + E)\bigr) \\
&\le \lambda(E\,\Delta G) + \lambda(G\,\Delta(y - x + G)) + \lambda((y - x + G)\,\Delta(y - x + E)) \\
&< 2\epsilon + \lambda(G\,\Delta(y - x + G)) \\
&= 2\epsilon + f_G(y - x).
\end{aligned}$$

This easily implies that f_E must be a uniformly continuous function.

19. CONVERGENCE IN MEASURE

Problem 19.1.   *Let $\{f_n\}$ be a sequence of measurable functions and let $f\colon X \to$ **R**. Assume that $\lim \mu^*(\{x \in X\colon |f_n(x) - f(x)| \ge \epsilon\}) = 0$ holds for every $\epsilon > 0$. Show that f is a measurable function.*

Solution. Pick a sequence $\{k_n\}$ of strictly increasing positive integers such that $\mu^*\left(\{x \in X : |f_k(x) - f(x)| \geq \frac{1}{n}\}\right) < 2^{-n}$ for all $k > k_n$. Set

$$E_n = \{x \in X : |f_{k_n}(x) - f(x)| \geq \tfrac{1}{n}\}$$

for each n and let $E = \bigcap_{m=1}^{\infty} \bigcup_{n=m}^{\infty} E_n$. Then,

$$\mu^*(E) \leq \mu^*\left(\bigcup_{n=m}^{\infty} E_n\right) \leq \sum_{n=m}^{\infty} \mu^*(E_n) \leq 2^{1-m}$$

holds for all m, so that $\mu^*(E) = 0$. Also, if $x \notin E$, then there exists some m such that $x \notin \bigcup_{n=m}^{\infty} E_n$, and so $|f_{k_n}(x) - f(x)| < \frac{1}{n}$ holds for each $n \geq m$. Therefore, $\lim f_{k_n}(x) = f(x)$ for each $x \notin E$, and so $f_{k_n} \longrightarrow f$ a.e. holds. The latter (by Theorem 16.6) easily implies that f is a measurable function.

Problem 19.2. *Assume that $\{f_n\} \subseteq \mathcal{M}$ satisfies $f_n \uparrow$ and $f_n \xrightarrow{\mu} f$. Show that $f_n \uparrow f$ a.e. holds.*

Solution. By Theorem 19.4, there exists a subsequence $\{f_{k_n}\}$ of the sequence $\{f_n\}$ with $f_{k_n} \longrightarrow f$ a.e. Since $f_n \uparrow$, it easily follows that $f_n \uparrow f$ a.e. holds.

Problem 19.3. *If $\{f_n\} \subseteq \mathcal{M}$ satisfies $f_n \xrightarrow{\mu} f$ and $f_n \geq 0$ a.e. for each n, then show that $f \geq 0$ a.e. holds.*

Solution. Since, by Theorem 19.4, some subsequence of $\{f_n\}$ converges almost everywhere to f, we must have $f \geq 0$ a.e.

Problem 19.4. *Let $\{f_n\} \subseteq \mathcal{M}$ and $\{g_n\} \subseteq \mathcal{M}$ satisfy $f_n \xrightarrow{\mu} f$, $g_n \xrightarrow{\mu} g$, and $f_n = g_n$ a.e. for each n. Show that $f = g$ a.e. holds.*

Solution. Since $f_n \xrightarrow{\mu} f$ implies $f_{k_n} \xrightarrow{\mu} f$ for each subsequence $\{f_{k_n}\}$ of $\{f_n\}$, by passing to two subsequences (if necessary), we can choose a strictly increasing sequence $\{k_n\}$ of positive integers such that $f_{k_n} \longrightarrow f$ a.e. and $g_{k_n} \longrightarrow g$ a.e. This easily implies $f = g$ a.e.

Problem 19.5. *Let $(X, \mathcal{S}, \mu)$ be a finite measure space. Assume that two sequences $\{f_n\}$ and $\{g_n\}$ of $\mathcal{M}$ satisfy $f_n \xrightarrow{\mu} f$ and $g_n \xrightarrow{\mu} g$. Show that $f_n g_n \xrightarrow{\mu} fg$. Is this statement true if $\mu^*(X) = \infty$?*

Solution. By Theorem 19.4, the only possible limit of $\{f_n g_n\}$ is fg. Consequently, if $f_n g_n \xrightarrow{\mu} fg$ does not hold, then there exist $\varepsilon > 0$ and $\delta > 0$ and some subsequence of $\{f_n g_n\}$ (which we shall denote by $\{f_n g_n\}$ again) such that

$$\mu^*\left(\{x \in X : |f_n(x)g_n(x) - f(x)g(x)| \geq \varepsilon\}\right) \geq \delta \qquad (\star)$$

holds for all n. In view of $f_n \xrightarrow{\mu} f$ and $g_n \xrightarrow{\mu} g$, Theorem 19.4 shows that for some subsequence $\{f_{k_n} g_{k_n}\}$ of $\{f_n g_n\}$ we must have $f_{k_n} g_{k_n} \longrightarrow fg$ a.e. Now, note that (by Theorem 19.5) $f_{k_n} g_{k_n} \xrightarrow{\mu} fg$ holds, contrary to $(\star)$. Thus, $f_n g_n \xrightarrow{\mu} fg$ holds.

If $\mu^*(X) = \infty$, then the conclusion is no longer true. An example: Take $X = (0, \infty)$ with the Lebesgue measure. Consider the functions $f_n(x) = \sqrt{x^4 + \frac{x}{n}}$ and $f(x) = x^2$. Then, $f_n \xrightarrow{\lambda} f$, while $f_n^2 \not\xrightarrow{\lambda} f^2$.

Problem 19.6. *Show that a sequence of measurable functions $\{f_n\}$ on a finite measure space converges to f in measure if and only if every subsequence of $\{f_n\}$ has in turn a subsequence which converges to f a.e.*

Solution. The conclusion follows immediately from Theorems 19.4 and 19.5.

Problem 19.7. *Define a sequence $\{f_n\}$ of $\mathcal{M}$ to be μ-**Cauchy** whenever for each $\epsilon > 0$ and $\delta > 0$ there exists some k (depending on ϵ and δ) such that $\mu^*(\{x \in X : |f_n(x) - f_m(x)| \geq \epsilon\}) < \delta$ holds for all $n, m \geq k$.*

Show that a sequence $\{f_n\}$ of $\mathcal{M}$ is a μ-Cauchy sequence if and only if there exists a measurable function f such that $f_n \xrightarrow{\mu} f$.

Solution. If $f_n \xrightarrow{\mu} f$, then the inclusion

$$\left\{x : |f_n(x) - f_m(x)| \geq 2\varepsilon\right\} \subseteq \left\{x : |f_n(x) - f(x)| \geq \varepsilon\right\} \cup \left\{x : |f_m(x) - f(x)| \geq \varepsilon\right\}$$

easily implies that $\{f_n\}$ is a μ-Cauchy sequence.

For the converse, assume that $\{f_n\}$ is a μ-Cauchy sequence. It suffices to show that $\{f_n\}$ has a subsequence that converges in measure (why?). To this end, start by selecting a subsequence $\{g_n\}$ of $\{f_n\}$ satisfying

$$\mu^*\left(\{x : |g_n(x) - g_m(x)| \geq 2^{-n}\}\right) < 2^{-n}$$

for all $m \geq n$. Let $E_n = \{x: |g_{n+1}(x) - g_n(x)| \geq 2^{-n}\}$. Also, let

$$F_n = \bigcup_{k=n}^{\infty} E_k = \{x: |g_{k+1}(x) - g_k(x)| \geq 2^{-k} \text{ holds for some } k \geq n\}.$$

Clearly, $\mu^*(F_n) \leq \sum_{k=n}^{\infty} \mu^*(E_k) \leq 2^{1-n}$ holds for all n, and hence the measurable set $F = \bigcap_{n=1}^{\infty} F_n$ satisfies $\mu^*(F) = 0$. Now, note for each fixed $x \notin F$ there exists some positive integer k_x such that $x \notin F_n$ holds for all $n \geq k_x$. Thus, for $n \geq k_x$, we have

$$\left|g_{n+p}(x) - g_n(x)\right| \leq \sum_{i=n}^{\infty} \left|g_{i+1}(x) - g_i(x)\right| \leq 2^{1-n}.$$

Therefore, $\{g_n(x)\}$ is a Cauchy sequence of real numbers for each $x \notin F$. Thus, there exists a function $g \in \mathcal{M}$ such that $g_n(x) \to g(x)$ holds for each $x \notin F$.

Now, if $n > k$ and $x \notin F_n$, then

$$\left|g_{n+1}(x) - g_{n+p}(x)\right| \leq \sum_{i=n+1}^{\infty} \left|g_{i+1} - g_i(x)\right| \leq 2^{-n}$$

implies that $|g_{n+1}(x) - g(x)| \leq 2^{-n} < 2^{-k}$ for all $n > k$. Thus,

$$\{x \in X: |g_{n+1}(x) - g(x)| \geq 2^{-k}\} \subseteq F_n$$

holds for all $n > k$. Finally, to see that $g_n \xrightarrow{\mu} g$ holds, note that for $n > k$, we have

$$\{x \in X: |g_n(x) - g(x)| \geq 2^{1-k}\}$$
$$\subseteq \{x \in X: |g_n(x) - g_{n+1}(x)| \geq 2^{-k}\} \cup \{x \in X: |g_{n+1}(x) - g(x)| \geq 2^{-k}\}$$
$$\subseteq E_n \cup F_n = F_n.$$

20. ABSTRACT MEASURABILITY

Problem 20.1. *Let $\mathcal{R}$ be a nonempty collection of subsets of a set X. Show that $\mathcal{R}$ is a ring if and only if $\mathcal{R}$ is closed under symmetric differences and finite intersections.*

Solution. Assume first that the nonempty collection $\mathcal{R}$ is a ring. That is, assume that $A, B \in \mathcal{R}$ imply $A \cup B \in \mathcal{R}$ and $A \setminus B \in \mathcal{R}$. Then the identities

$$A \triangle B = (A \setminus B) \cup (B \setminus A) \quad \text{and} \quad A \cap B = A \setminus (A \setminus B)$$

easily imply that $\mathcal{R}$ is closed under symmetric differences and finite intersections.

For the converse assume that $\mathcal{R}$ is closed under symmetric differences and finite intersections. Then, the identities

$$A \setminus B = A \triangle (A \cap B) \quad \text{and} \quad A \cup B = (A \triangle B) \triangle (A \cap B)$$

guarantee that $\mathcal{R}$ is a ring.

Problem 20.2. *If $\mathcal{R}$ is a ring of subsets of a set X, then show that the collection*

$$\mathcal{A} = \left\{ A \subseteq X \colon \text{Either } A \text{ or } A^c \text{ belongs to } \mathcal{R} \right\}$$

is an algebra of sets.

Solution. From the definition of $\mathcal{A}$, it easily follows that if $A \in \mathcal{A}$, then $A^c \in \mathcal{A}$, i.e., that $\mathcal{A}$ is closed under complementation.

Now, assume that $A, B \in \mathcal{A}$. If $A, B \in \mathcal{R}$, then since $\mathcal{R}$ (as being a ring) is closed under finite unions, we have $A \cup B \in \mathcal{R}$ and so $A \cup B \in \mathcal{A}$. If $A^c, B^c \in \mathcal{R}$, then $A^c \setminus (A^c \setminus B^c) \in \mathcal{R}$, and so

$$A \cup B = (A^c \cap B^c)^c = \left[A^c \setminus (A^c \setminus B^c) \right]^c \in \mathcal{A}.$$

Now, assume that $A \in \mathcal{R}$ and $B^c \in \mathcal{R}$. Then, $B^c \setminus A = B^c \cap A^c \in \mathcal{R}$, and consequently (from the definition of $\mathcal{A}$), $A \cup B = (A^c \cap B^c)^c \in \mathcal{A}$. The preceding show that $\mathcal{A}$ is an algebra.

Problem 20.3. *In the implication scheme of Figure 3.3 show that no other implication is true by verifying the following regarding an uncountable set X.*

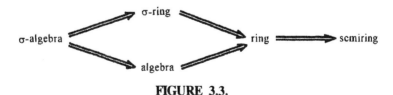

FIGURE 3.3.

a.   *The collection of all singleton subsets of X together with the empty set is a semiring but not a ring.*

b.   *The collection of all finite subsets of X is a ring but is neither an algebra nor a σ-ring.*

c.   *The collection of all subsets of X that are either finite or have finite complement is an algebra but is neither a σ-algebra nor a σ-ring.*

d.   *The collection of all at-most countable subsets of X is a σ-ring but not an algebra.*

e.   *The collection of all subsets of X that are either at-most countable or have at-most a countable complement is a σ-algebra (which is, in fact, the σ-algebra generated by the singletons).*

Solution.   (a) If A and B are singletons, then $A \cap B$ and $A \setminus B$ are either empty or singletons. This shows that the collection of all singletons together with the empty set is a semiring. However, it should be obvious that finite unions of singletons need not be a singleton, and so the collection of all singletons is not an algebra.

(b) Let $\mathcal{R}$ denote the collection of all finite subsets of (the infinite) set X. If $A, B \in \mathcal{R}$, then $A \cup B$ and $A \setminus B$ are finite sets and so $A \cup B$ and $A \setminus B$ belong to $\mathcal{R}$. This shows that $\mathcal{R}$ is a ring. Since the complement of a finite set is infinite, it follows that $\mathcal{R}$ is not closed under complementation, and so $\mathcal{R}$ is not an algebra.

To see that $\mathcal{R}$ is not a σ-ring, let $A = \{a_1, a_2, \ldots\}$ be a countable subset of X, and for each n let $A_n = \{a_n\} \in \mathcal{R}$. Then, $\bigcup_{n=1}^{\infty} A_n = A \notin \mathcal{R}$, and this shows that $\mathcal{R}$ is not a σ-ring.

(c) If $\mathcal{R}$ is the ring of all finite subsets, then by part (b) the collection

$$\mathcal{A} = \{A \subseteq X : \text{ Either } A \text{ or } A^c \text{ belongs to } \mathcal{R}\}$$

is an algebra of sets. To see that $\mathcal{A}$ is not a σ-ring (and hence neither a σ-algebra), let $A = \{a_1, a_2, \ldots\}$ be a countable subset of X such that $X \setminus A$ is an infinite set. Clearly, $A \notin \mathcal{A}$. On the other hand, if $A_n = \{a_n\}$, then $A_n \in \mathcal{A}$ and $\bigcup_{n=1}^{\infty} A_n = A \notin \mathcal{A}$. This shows that $\mathcal{A}$ is not a σ-ring.

(d) Let $\mathcal{C}$ denote the collection of all at-most countable subsets of X. Clearly, $A, B \in \mathcal{C}$ imply $A \setminus B \in \mathcal{C}$. Also, $\mathcal{C}$ is closed under countable unions (recall that an at-most countable union of sets each of which is at-most countable is at-most countable; see Theorem 2.6). Therefore, $\mathcal{C}$ is a σ-ring. However, when X is an uncountable set, $\mathcal{C}$ is not closed under complementation, and hence it cannot be an algebra.

(e) This is Problem 12.7.

Problem 20.4.   *Show that a Dynkin system is a σ-algebra if and only if it is closed under finite intersections.*

Solution. Let $\mathcal{D}$ be a Dynkin system that is closed under finite intersections. Since $\mathcal{D}$ is also closed under complementation, it is easy to see (by using the identity $A \cup B = (A^c \cap B^c)^c$) that $\mathcal{D}$ is in fact an algebra. So, if $A = \bigcup_{n=1}^{\infty} A_n$ with $\{A_n\} \subseteq \mathcal{D}$, then by letting $B_n = \bigcup_{k=1}^{n} A_k \in \mathcal{D}$, and noting that $B_n \uparrow A$, we see that $A \in \mathcal{D}$. In other words, $\mathcal{D}$ is a σ-algebra.

Problem 20.5. *Give an example of a Dynkin system which is not an algebra.*

Solution. Consider the set $X = \{1, 2, 3, 4\}$, and let

$$\mathcal{D} = \big\{\emptyset, \{1, 2\}, \{3, 4\}, \{1, 3\}, \{2, 4\}, X\big\}.$$

Then $\mathcal{D}$ is a Dynkin system (why?), which (since $\{1, 2\} \cup \{1, 3\} = \{1, 2, 3\}$ does not belong to $\mathcal{D}$) fails to be an algebra.

Problem 20.6. *A **monotone class** of sets is a family $\mathcal{M}$ of subsets of a set X such that if a sequence $\{A_n\}$ of $\mathcal{M}$ satisfies $A_n \uparrow A$ or $A_n \downarrow A$, then $A \in \mathcal{M}$. Establish the following properties regarding monotone classes:*

a. *We have the following implications:*

$$\sigma\text{-algebra} \implies \text{Dynkin system} \implies \text{monotone class}$$

Give examples to show that no other implication in the preceding scheme is true.

b. *An algebra is a monotone class if and only if it is a σ-algebra.*

c. *The σ-algebra $\sigma(\mathcal{A})$ generated by an algebra $\mathcal{A}$ is the smallest monotone class containing $\mathcal{A}$.*

Solution. (a) The implication scheme follows immediately from the definitions of the three classes of sets involved. An example of a Dynkin system which is not an algebra was exhibited in the preceding problem. Now, if $X = \{1, 2\}$, then the collection $\mathcal{M} = \big\{X, \{1\}\big\}$ is a monotone class but not a Dynkin system.

(b) Let $\mathcal{A}$ be an algebra of sets. If $\mathcal{A}$ is a σ-algebra, then it is clearly a monotone class. For the converse assume that the algebra $\mathcal{A}$ is a monotone class.

Assume $\{A_n\} \subseteq \mathcal{A}$ and put $A = \bigcup_{n=1}^{\infty} A_n$. Let $B_n = \bigcup_{k=1}^{n} A_k \in \mathcal{A}$ and note that $B_n \uparrow A$. Since $\mathcal{A}$ is a monotone class, it follows that $A \in \mathcal{A}$, and so $\mathcal{A}$ is a σ-algebra.

(c) Let $\mathcal{A}$ be an algebra of sets and let $\mathcal{M}$ be the smallest monotone class that contains $\mathcal{A}$, i.e., $\mathcal{M}$ is the intersection of the collection of all monotone classes that include $\mathcal{A}$. Clearly, $\mathcal{A} \subseteq \mathcal{M} \subseteq \sigma(\mathcal{A})$.

Let $\mathcal{C} = \big\{ B \in \mathcal{M}\colon\ B \setminus A \in \mathcal{M}$ for each $A \in \mathcal{A} \big\}$. An easy verification shows that $\mathcal{C}$ is a monotone class that includes $\mathcal{A}$, and so $\mathcal{M} = \mathcal{C}$. Now, let

$$\mathcal{D} = \big\{ B \in \mathcal{M}\colon\ M \setminus B \in \mathcal{M} \ \text{for each}\ M \in \mathcal{M} \big\}.$$

Again, $\mathcal{D}$ is a monotone class which (in view of $\mathcal{M} = \mathcal{C}$) satisfies $\mathcal{A} \subseteq \mathcal{D}$. Thus, $\mathcal{D} = \mathcal{M}$. This shows that $\mathcal{M}$ is, in fact, a Dynkin system. By Dynkin's Lemma 20.8, $\sigma(A) \subseteq \mathcal{M}$, and so $\mathcal{M} = \sigma(A)$.

Problem 20.7. *Show that if X and Y are two separable metric spaces, then $\mathcal{B}_{X \times Y} = \mathcal{B}_X \otimes \mathcal{B}_Y$.*

Solution. Assume that X and Y are two arbitrary topological spaces. For each subset A of X, let

$$\Sigma_A = \big\{ B \subseteq Y\colon\ A \times B \in \mathcal{B}_{X \times Y} \big\}.$$

From the identities $A \times (B \setminus C) = (A \times B) \setminus (A \times C)$, we see that if $B, C \in \Sigma_A$, then $B \setminus C \in \Sigma_A$. From $A \times \big(\bigcap_{n=1}^{\infty} B_n \big) = \bigcap_{n=1}^{\infty}(A \times B_n)$, it follows that Σ_A is closed under countable intersections. Observing that $\emptyset \in \Sigma_A$, we infer that Σ_A is a σ-ring. Clearly, Σ_A is a σ-algebra if and only if $Y \in \Sigma_A$.

Next, note that for any open subset $\mathcal{O}$ of X, $V \in \Sigma_{\mathcal{O}}$ for every open subset V of Y. Since Y is itself open, if $\mathcal{O}$ is open, then $\Sigma_{\mathcal{O}}$ is a σ-algebra of subsets of Y that includes the open subsets of Y. Thus, $\mathcal{B}_Y \subseteq \Sigma_{\mathcal{O}}$ for each open subset $\mathcal{O}$ of X. Now, let

$$\mathcal{A} = \big\{ A \subseteq X\colon\ \mathcal{B}_Y \subseteq \Sigma_A \big\}.$$

As we have just noticed, $V \in \mathcal{A}$ holds for each open subset V of X. Since (as easily checked) $\Sigma_A = \Sigma_{A^c}$ for each $A \in \mathcal{A}$, we see that $\mathcal{A}$ is closed under complementation. Moreover, if $\{A_n\} \subseteq \mathcal{A}$, then for any Borel subset B of Y, we have $A_n \times B \in \mathcal{B}_{X \times Y}$ for each n. Thus, in view of $\bigcap_{n=1}^{\infty}(A_n \times B) = \big(\bigcap_{n=1}^{\infty} A_n \big) \times B$, we obtain $B \in \Sigma_{\bigcap_{n=1}^{\infty} A_n}$. In other words, $\mathcal{A}$ is closed under countable intersections, and so $\mathcal{A}$ is a σ-algebra including the open subsets of X. This implies $\mathcal{B}_X \subseteq \mathcal{A}$.

We have just established the following: If A is a Borel subset of X and B is a Borel subset of Y, then $A \times B \in \mathcal{B}_{X \times Y}$. Therefore,

$$\mathcal{B}_X \otimes \mathcal{B}_Y \subseteq \mathcal{B}_{X \times Y}.$$

For the reverse inclusion, assume that X and Y are two separable metric spaces. Then every open subset of $X \times Y$ is an at-most countable union of sets of the form

$V \times U$, where V is an open subset of X and U an open subset of Y. Consequently, $\mathcal{B}_{X \times Y} \subseteq \mathcal{B}_X \otimes \mathcal{B}_Y$, from which it follows that $\mathcal{B}_X \otimes \mathcal{B}_Y = \mathcal{B}_{X \times Y}$.

Problem 20.8. *Show that the composition function of two measurable functions is measurable.*

Solution. Assume that $(X, \Sigma_1) \xrightarrow{f} (Y, \Sigma_2) \xrightarrow{g} (Z, \Sigma_3)$ are measurable functions. If $A \in \Sigma_3$, then $g^{-1}(A) \in \Sigma_2$, and so $(g \circ f)^{-1}(A) = f^{-1}\big(g^{-1}(A)\big) \in \Sigma_1$. This shows that $g \circ f$ is measurable.

Problem 20.9. *If (X, Σ) is a measurable space, then show that*

a. *the collection of all real-valued measurable functions defined on X is a function space and an algebra of functions, and*
b. *any real-valued function on X which is the pointwise limit of a sequence of $(\Sigma, \mathcal{B})$-measurable real-valued functions is itself $(\Sigma, \mathcal{B})$-measurable.*

Solution. Repeat the solution of Problem 17.17.

Problem 20.10. *Let (X, Σ) be a measurable space. A Σ-**simple function** is any measurable function $\phi: X \to \mathbf{R}$ having a finite range, i.e, if ϕ has finite range and its standard representation $\phi = \sum_{i=1}^{n} a_i \chi_{A_i}$ satisfies $A_i \in \Sigma$ for each i.*
Show that a function $f: X \to [0, \infty)$ is measurable if and only if there exists a sequence $\{\phi_n\}$ of Σ-simple functions such that $\phi_n(x) \uparrow f(x)$ holds for each $x \in X$.

Solution. The proof is identical to the proof of Theorem 17.7. Here it is.
For each n let $A_n^i = \big\{x \in X \colon (i-1)2^{-n} \le f(x) < i 2^{-n}\big\}$ for $i = 1, 2, \ldots, n2^n$, and note that $A_n^i \cap A_n^j = \emptyset$ if $i \ne j$. Since f is measurable, all the A_n^i belong to Σ.

Now, for each n define $\phi_n = \sum_{i=1}^{n2^n} 2^{-n}(i-1)\chi_{A_n^i}$, and note that $\{\phi_n\}$ is a sequence of Σ-simple functions. Also, an easy verification shows that $0 \le \phi_n(x) \le \phi_{n+1}(x) \le f(x)$ holds for all x and all n. Moreover, if x is fixed, then $0 \le f(x) - \phi_n(x) \le 2^{-n}$ holds for all sufficiently large n. This implies $\phi_n(x) \uparrow f(x)$ for all $x \in X$.

Problem 20.11. *Use Corollary 20.10 to show that if a measure μ is σ-finite, then μ^* is the one and only extension of μ to a measure on $\sigma(S)$.*

Solution. Let $\nu \colon \sigma(S) \to [0, \infty]$ be a measure satisfying $\nu(A) = \mu(A)$ for each $A \in S$. We shall establish that $\nu(A) = \mu^*(A)$ for each $A \in \sigma(S)$.

Fix $E \in S$ with $\mu(E) < \infty$ and let

$$S_E = \{E \cap A \colon A \in S\} = \{B \in S \colon B \subseteq E\}.$$

Clearly, S_E is a semiring of subsets of E and μ restricted to E is a measure. Moreover, we know (see Problem 15.7) that the outer measure generated by the measure space (E, S_E, μ) is simply the restriction of μ^* to $\mathcal{P}(E)$. In addition, we claim that if $\sigma(S_E)$ denotes the σ-algebra generated by S_E in $\mathcal{P}(E)$, then

$$\sigma(S_E) = \{A \cap E \colon A \in \sigma(S)\} = \{B \in \sigma(S) \colon B \subseteq E\}. \qquad (\star)$$

To see this, note first that since $\{B \in \sigma(S) \colon B \subseteq E\}$ is a σ-algebra containing S_E, we have

$$\sigma(S_E) \subseteq \{B \in \sigma(S) \colon B \subseteq E\}.$$

On the other hand, the collection

$$\mathcal{A} = \{A \subseteq X \colon A \cap E \in \sigma(S_E)\}$$

is a σ-algebra of subsets of X satisfying $S \subseteq \mathcal{A}$. Hence, $\sigma(S) \subseteq \mathcal{A}$. In particular, if $B \subseteq E$ satisfies $B \in \sigma(S)$, then $B \in \mathcal{A}$ and so $B = B \cap E \in \sigma(S_E)$. Consequently, $\{B \in \sigma(S) \colon B \subseteq E\} \subseteq \sigma(S_E)$, and the validity of $(\star)$ follows.

Next, note that since S_E is closed under finite intersections, $\mu^*(E) = \mu(E) = \nu(E) < \infty$, and $\nu(F) = \mu(F) = \mu^*(F)$ for all $F \in S_E$, it follows from Corollary 20.10 that $\nu(F) = \mu^*(F)$ for all $F \in \sigma(S)$ with $F \subseteq E$.

Now, let $\{E_n\}$ be a pairwise disjoint sequence of S satisfying $X = \bigcup_{n=1}^{\infty} E_n$ and $\mu(E_n) < \infty$ for each n. If $A \in \sigma(S)$, then by the preceding discussion we have $\nu(A \cap X_n) = \mu^*(A \cap X_n)$ for each n, and so

$$\nu(A) = \nu(A \cap X) = \nu\left(\bigcup_{n=1}^{\infty} A \cap X_n\right) = \sum_{n=1}^{\infty} \nu(A \cap X_n)$$

$$= \sum_{n=1}^{\infty} \mu^*(A \cap X_n) = \mu^*\left(\bigcup_{n=1}^{\infty} A \cap X_n\right)$$

$$= \mu^*(A \cap X) = \mu^*(A),$$

and we are finished. (For more about the extension of μ, see Problem 15.19.)

Problem 20.12. *Show that the uniform limit of a sequence of measurable functions from a measurable space into a metric space is measurable.*

Solution. Let $\{f_n\}$ be a sequence of measurable functions from a measurable space (X, Σ) into a metric space (Y, d). Suppose f is the uniform limit of $\{f_n\}$. That is, assume that for each $\epsilon > 0$ there exists some n_0 such that $d\big(f_n(x), f(x)\big) < \epsilon$ for all $x \in X$ and all $n \geq n_0$. By passing to a subsequence, we can assume that $d\big(f_n(x), f(x)\big) < \frac{1}{n}$ holds for each n and all $x \in X$.

Since the family of closed sets generates the Borel sets of Y, in order to establish the measurability of f, it suffices to prove that $f^{-1}(C) \in \Sigma$ for each closed set C. To this end, let C be a closed subset of Y.

Let $V_n = \big\{y \in Y \colon d(y, C) < \frac{1}{n}\big\}$. We claim that

$$f^{-1}(C) = \bigcap_{n=1}^{\infty} f^{-1}(V_n). \tag{$\star\star$}$$

To see this, assume $x \in f^{-1}(C)$, i.e., let $f(x) \in C$. From $d\big(f_n(x), C\big) \leq d\big(f_n(x), f(x)\big) < \frac{1}{n}$, we get $f_n(x) \in V_n$ or $x \in f_n^{-1}(V_n)$ for each n. Conversely, if $f_n(x) \in V_n$ for each n, then $d\big(f_n(x), C\big) < \frac{1}{n}$ for each n, and so if we pick some $c_n \in C$ with $d\big(f_n(x), c_n\big) < \frac{1}{n}$, then we have

$$d\big(f(x), C\big) \leq d\big(f(x), c_n\big) \leq d\big(f(x), f_n(x)\big) + d\big(f_n(x), c_n\big) < \tfrac{1}{n} + \tfrac{1}{n} = \tfrac{2}{n}$$

for each n. This implies $d\big(f(x), C\big) = 0$. Since C is a closed set, it follows that $f(x) \in C$, or $x \in f^{-1}(C)$.

Next, use the measurability of each f_n and the fact that each V_n is open to obtain that $f_n^{-1}(V_n) \in \Sigma$ for each n. Now, invoke $(\star\star)$ to conclude that $f^{-1}(C) \in \Sigma$.

Problem 20.13. *Let $f, g \colon X \to \mathbb{R}$ be two functions and let $\mathcal{B}$ denote the σ-algebra of all Borel sets of $\mathbb{R}$. Show that there exists a Borel measurable function $h \colon \mathbb{R} \to \mathbb{R}$ satisfying $f = h \circ g$ if and only if $f^{-1}(\mathcal{B}) \subseteq g^{-1}(\mathcal{B})$ holds.*

Solution. Assume $f = h \circ g$ holds for some Borel measurable function $h \colon \mathbb{R} \to \mathbb{R}$. Fix $D \in \mathcal{B}$ and note that $h^{-1}(D) \in \mathcal{B}$. Therefore, $f^{-1}(B) = g^{-1}\big(h^{-1}(B)\big) \subseteq g^{-1}(\mathcal{B})$, and so $f^{-1}(\mathcal{B}) \subseteq g^{-1}(\mathcal{B})$ holds.

For the converse, assume $f^{-1}(\mathcal{B}) \subseteq g^{-1}(\mathcal{B})$. The existence of the Borel measurable function h will be established by steps.
Step I: Assume $f = \chi_A$ for some $A \in f^{-1}(\mathcal{B})$.

Since $f^{-1}(\mathcal{B}) \subseteq g^{-1}(\mathcal{B})$ is true, there exists some $B \in \mathcal{B}$ such that $A = g^{-1}(B)$. Let $h = \chi_B$, and note that $h \circ g = f$.
Step II: Let $f = \sum_{i=1}^{n} a_i \chi_{A_i}$ with the A_i pairwise disjoint and $A_i \in f^{-1}(\mathcal{B})$ for each i. For each i choose some $B_i \in g^{-1}(\mathcal{B})$ such that $A_i = g^{-1}(B_i)$. If we consider the Borel step function $h = \sum_{i=1}^{n} a_i \chi_{B_i}$, then it is easy to see that $h \circ g = f$.

Step III: The general case.

The preceding problem applied with $\Sigma = f^{-1}(\mathcal{B})$ guarantees the existence of a sequence $\{\phi_n\}$ of $f^{-1}(\mathcal{B})$-simple functions satisfying $\phi_n(x) \uparrow f(x)$ for each $x \in X$. Now, by Step II, for each n there exists a Borel measurable function $h_n: \mathbb{R} \to \mathbb{R}$ such that $h_n \circ g = \phi_n$. Next, let

$$B = \left\{ x \in \mathbb{R}: \lim_{n \to \infty} h_n(x) = h(x) \text{ exists} \right\}.$$

It follows (as in Problem 16.7) that $B \in \mathcal{B}$ and $h_n(x)\chi_B(x) \to h(x)$ for each $x \in \mathbb{R}$. If we let $h(x) = 0$ for $x \notin B$, then $h: \mathbb{R} \to \mathbb{R}$ is Borel measurable and satisfies $h \circ g = f$.

Problem 20.14. *Let (X, Σ) be a measurable space, Y, Z_1, and Z_2 be separable metric spaces and Ψ a topological space. Now, assume also that the functions $f_i: X \times Y \to Z_i$, $(i = 1, 2)$, are Carathéodory functions and $g: Z_1 \times Z_2 \to \Psi$ is Borel measurable. Show that the function $h: X \times Y \to \Psi$, defined by*

$$h(x, y) = g\big(f_1(x, y), f_2(x, y)\big),$$

is jointly measurable.

Solution. By Theorem 20.15, each $f_i: X \times Y \to Z_i$ is jointly measurable. This implies that the function $F: X \times Y \to Z_1 \times Z_2$, defined by

$$F(x, y) = \big(f_1(x, y), f_2(x, y)\big)$$

is measurable (why?). Since $g: Z_1 \times Z_2 \to \Psi$ is $(\mathcal{B}_{Z_1 \times Z_2}, \mathcal{B}_\Psi)$-measurable and (by Problem 20.7) $\mathcal{B}_{Z_1} \otimes \mathcal{B}_{Z_2} = \mathcal{B}_{Z_1 \times Z_2}$, it follows that $h = g \circ F$ is likewise measurable.

Problem 20.15. *Let (X, Σ) be a measurable space and (Y, d) a separable metric space. Show that a function $f: X \to Y$ is measurable if and only if for each fixed $y \in Y$ the function $x \mapsto d(y, f(x))$, from X into $\mathbb{R}$, is measurable.*

Solution. Let $f:(X, \Sigma) \to Y$ be a function from a measurable space to a separable metric space. For each $y \in Y$ define the function $g_y: X \to \mathbb{R}$ by $g_y(x) = d\big(y, f(x)\big)$. Note that for each $r > 0$ and each $y \in Y$, we have

$$f^{-1}\big(B(y, r)\big) = \big\{ x \in X: f(x) \in B(y, r) \big\} = \big\{ x \in X: d\big(y, f(x)\big) < r \big\}$$
$$= \big\{ x \in X: g_y(x) < r \big\} = g_y^{-1}\big((-\infty, r)\big).$$

Assume that each g_y is measurable. Then, by the preceding identity, we have $f^{-1}(B(y, r)) = g_y^{-1}((-\infty, r)) \in \Sigma$ for each $y \in Y$ and all $r > 0$. Since Y is a separable metric space, every open set can be written as an at-most countable union of open balls, and so $f^{-1}(\mathcal{O}) \in \Sigma$ holds for each open set $\mathcal{O}$. By Theorem 20.6, f is a measurable function.

For the converse, suppose that f is a measurable function and let $y \in Y$. From $g_y^{-1}((-\infty, r)) = f^{-1}(B(y, r))$ if $r > 0$ and $g_y^{-1}((-\infty, r)) = \emptyset$ if $r \leq 0$, we easily infer that g_y is a measurable function for each $y \in Y$.

Problem 20.16. *Let $(X, \mathcal{S}, \mu)$ be a σ-finite measure space, where $\mathcal{S}$ is a σ-algebra. If $f: X \to \mathbf{R}$ is a Λ_μ-measurable function, then show that there exists a $\mathcal{S}$-measurable function $g: X \to \mathbf{R}$ such that $f = g$ a.e.*

Solution. We can assume $f(x) \geq 0$ for each $x \in X$ (otherwise, we apply the arguments below to f^+ and f^- separately). If $f = \chi_A$ for some $A \in \Lambda_\mu$, then an easy argument (using Theorem 15.11) shows that there exists a null set C such that $B = A \cup C \in \mathcal{S}$. So, if $g = \chi_B$, then g is $\mathcal{S}$-measurable and $f = g$ a.e. holds. It follows that if f is a Λ_μ-simple function, then there exists a $\mathcal{S}$-simple function g such that $f = g$ a.e.

Now, we consider the general case. By Problem 20.10 there exists a sequence $\{\phi_n\}$ of Λ_μ-simple functions satisfying $\phi_n(x) \uparrow f(x)$ for each $x \in X$. For each n fix a $\mathcal{S}$-simple function ψ_n such that $\psi_n = \phi_n$ a.e. By Theorem 15.11, for each n there exists a null set $A_n \in \mathcal{S}$ with $\psi_n(x) = \phi_n(x)$ for all $x \notin A_n$. Put $A = \bigcup_{n=1}^{\infty} A_n \in \mathcal{S}$, and note that A is a null set. Moreover, we have $\psi_n \chi_{A^c}(x) \uparrow f \chi_{A^c}(x)$ for each x. If $g = f \chi_{A^c}$, then (by Problem 20.9(b)) g is a $\mathcal{S}$-measurable function satisfying $f = g$ a.e.

CHAPTER 4

THE LEBESGUE INTEGRAL

21. UPPER FUNCTIONS

Problem 21.1. *Let L be the collection of all step functions ϕ such that there exist a finite number of members $A_1, \ldots, A_n$ of S all of finite measure and real numbers $a_1, \ldots, a_n$ such that $\phi = \sum_{i=1}^{n} a_i \chi_{A_i}$. Show that L is a function space. Is L an algebra of functions?*

Solution. Let $\phi = \sum_{i=1}^{n} a_i \chi_{A_i}$ and $\psi = \sum_{j=1}^{m} b_j \chi_{B_j}$, where the A_i and B_j belong to S and they all have finite measure. By Problem 12.14, there exist pairwise disjoint sets $C_1, \ldots, C_k$ of S such that each A_i and each B_j can be written as a union from the C_i. We can assume that $\bigcup_{r=1}^{k} C_r = \left[\bigcup_{i=1}^{n} A_i\right] \cup \left[\bigcup_{j=1}^{m} B_j\right]$. It is easy to see that ϕ and ψ can be written in the form $\phi = \sum_{r=1}^{k} c_r \chi_{C_r}$ and $\psi = \sum_{r=1}^{k} d_r \chi_{C_r}$. Now, everything follows from the equalities:

1. $\alpha\phi + \beta\psi = \sum_{r=1}^{k} (\alpha c_r + \beta d_r) \chi_{C_r}$;
2. $\phi \vee \psi = \sum_{r=1}^{k} (c_r \vee d_r) \chi_{C_r}$ and $\phi \wedge \psi = \sum_{r=1}^{k} (c_r \wedge d_r) \chi_{C_r}$; and
3. $\phi\psi = \sum_{r=1}^{k} c_r d_r \chi_{C_r}$.

Problem 21.2. *Consider the function $f: \mathbb{R} \to \mathbb{R}$ defined by $f(x) = 0$ if $x \notin (0, 1]$, and $f(x) = \sqrt{n}$ if $x \in (\frac{1}{n+1}, \frac{1}{n}]$ for some n. Show that f is an upper function and that $-f$ is not an upper function.*

Solution. Put $A_k = \left(\frac{1}{k+1}, \frac{1}{k}\right]$ and note that $\lambda(A_k) = \frac{1}{k(k+1)}$. Thus, if we let

$$\phi_n = \sum_{k=1}^{n} \sqrt{k} \chi_{A_k},$$

then $\{\phi_n\}$ is a sequence of step functions satisfying $\phi_n(x) \uparrow f(x)$ for each x. On the other hand, the relations

$$\int \phi_n \, d\lambda = \sum_{k=1}^{n} \sqrt{k} \cdot \frac{1}{k(k+1)} = \sum_{k=1}^{n} \frac{1}{\sqrt{k}(k+1)} \le \sum_{k=1}^{\infty} k^{-\frac{3}{2}} < \infty$$

guarantee that f is an upper function.

Since $-f$ is not bounded from below, there is no step function ϕ satisfying $\phi \le -f$. This implies that $-f$ cannot be an upper function.

Problem 21.3. *Compute $\int f \, d\lambda$ for the upper function f of the preceding exercise.*

Solution. We have $\int f \, d\lambda = \sum_{n=1}^{\infty} \frac{1}{(n+1)\sqrt{n}}$.

Problem 21.4. *Verify that every continuous function $f : [a, b] \to \mathbf{R}$ is an upper function—with respect to the Lebesgue measure on $[a, b]$.*

Solution. For each n let $P_n = \{x_0, x_1, \ldots, x_{2^n}\}$ be the partition that divides $[a, b]$ into 2^n subintervals all of the same length $(b - a)2^{-n}$; that is, $x_i = a + i(b - a)2^{-n}$. Let $m_i = \min\{f(x) \colon x \in [x_{i-1}, x_i]\}$, and then define

$$\phi_n = \sum_{i=1}^{2^n} m_i \chi_{[x_{i-1}, x_i)}.$$

Clearly, each ϕ_n is a step function. Using the uniform continuity of f, it is not difficult to see that $\phi_n(x) \uparrow f(x)$ holds for all $x \in [a, b)$. On the other hand, if $f(x) \le M$ holds for each x, then $\int \phi_n \, d\lambda \le M(b - a)$ holds for all n, implying that f is an upper function.

Problem 21.5. *Let A be a measurable set, and let f be an upper function. If $\chi_A \le f$ a.e., then show that $\mu^*(A) < \infty$.*

Solution. Choose a sequence $\{\phi_n\}$ of step functions with $\phi_n \uparrow f$ a.e. Then, $\phi_n \wedge \chi_A \uparrow f \wedge \chi_A = \chi_A$ a.e., and so, by Theorem 17.6,

$$\mu^*(A) = \lim_{n \to \infty} \int \phi_n \wedge \chi_A \, d\mu \le \lim_{n \to \infty} \int \phi_n \, d\mu = \int f \, d\mu < \infty.$$

Problem 21.6. *Let f be an upper function, and let A be a measurable set of finite measure such that $a \le f(x) \le b$ holds for each $x \in A$. Then, show that*

   a.   *$f \chi_A$ is an upper function, and*
   b.   *$a\mu^*(A) \le \int f \chi_A \, d\mu \le b\mu^*(A)$.*

Solution. (a) Pick a sequence $\{\phi_n\}$ of step functions with $\phi_n \uparrow f$ a.e. For each n define the step function $\psi_n = (\phi_n \chi_A) \wedge b\chi_A$. Then,

$$\int \psi_n \, d\mu \le \int b\chi_A \, d\mu = b\mu^*(A) < \infty.$$

Since $\psi_n \uparrow f\chi_A$ a.e. holds, $f\chi_A$ is an upper function.
(b) Apply the monotone property of the integral (Theorem 21.5) to the inequality $a\chi_A \le f\chi_A \le b\chi_A$.

Problem 21.7. *Let (X, S, μ) be a finite measure space, and let f be a positive measurable function. Show that f is an upper function if and only if there exists a real number M such that $\int \phi \, d\mu \le M$ holds for every step function ϕ with $\phi \le f$ a.e. Also, show that if this is the case, then*

$$\int f \, d\mu = \sup \left\{ \int \phi \, d\mu \colon \phi \text{ is a step function with } \phi \le f \text{ a.e.} \right\}.$$

Solution. If f is an upper function, then by Theorem 21.5 every step function ϕ with $\phi \le f$ a.e. satisfies $\int \phi \, d\mu \le \int f \, d\mu < \infty$.
   Conversely, by Theorem 17.7 there exists a sequence $\{\phi_n\}$ of simple functions with $\phi_n \uparrow f$ a.e. Since (X, S, μ) is a finite measure space, we know that each ϕ_n is a step function. If $\int \phi_n \, d\mu \le M$ holds for all n, then this readily implies that f is an upper function.
   The last formula is immediate from Theorem 21.5.

Problem 21.8. *Show that every monotone function $f\colon [a, b] \to \mathbf{R}$ is an upper function—with respect to the Lebesgue measure on $[a, b]$.*

Solution. We assume that $f\colon [a, b] \to \mathbf{R}$ is an increasing function. The "decreasing case" can be proven in a similar fashion and is left for the reader.
   By Problem 9.8, we know that the set D of all discontinuities of f is an at-most countable set. In particular, $\lambda(D) = 0$. Now, for each n, let P_n be the partition that subdivides $[a, b]$ into 2^n equal subintervals. That is, let $P_n = \{a_0^n, a_1^n, \ldots, a_{2^n}^n\}$,

where $a_i^n = a + \frac{b-a}{2^n} i$ for $i = 0, 1, \ldots, 2^n$. Next, for each $1 \le i \le 2^n$ let

$$m_i^n = \inf\{f(x)\colon x \in [a_{i-1}^n, a_i^n]\},$$

and put $\phi_n = \sum_{i=1}^{2^n} m_i^n \chi_{[a_{i-1}^n, a_i^n)}$. Clearly, each ϕ_n is a step function and, in view of the monotonicity of f, we have

$$\phi_n(x) \le \phi_{n+1}(x) \le f(x)$$

for all $x \in [a, b]$. Put $E = D \cup P_1 \cup P_2 \cup P_3 \cup \cdots$ and note that $\lambda(E) = 0$. We shall establish that $\phi_n(x) \uparrow f(x)$ for each $x \in [a, b] \setminus E$.

To this end, fix some $t \in [a, b] \setminus E$ and let $\epsilon > 0$. Since f is continuous at t, there exists some $\delta > 0$ such that

$$x \in [a, b] \text{ and } t - \delta < x < t + \delta \quad \text{imply} \quad f(t) - \epsilon < f(x) < f(t) + \epsilon. \quad (\star)$$

Next, pick some k such that $\frac{b-a}{2^k} < \delta$ for all $n \ge k$, and then choose the subinterval $[x_{i-1}, x_i]$ of P_k such that $t \in (x_{i-1}, x_i)$. From $(\star)$, it easily follows that

$$\phi_k(t) = \inf\{f(x)\colon x \in [x_{i-1}, x_i]\} \ge f(t) - \epsilon.$$

Therefore, $f(t) - \epsilon \le \phi_k(t) \le \phi_n(t) \le f(t)$ holds for all $n \ge k$, and this shows that $\phi_n(t) \uparrow f(t)$, as claimed.

Finally, note that the monotonicity of f guarantees that $m_i^n \le f(b)$ holds for all $1 \le i \le 2^n$. This implies

$$\int \phi_n \, d\lambda = \sum_{i=1}^{2^n} m_i^n (a_i^n - a_{i-1}^n) \le f(b) \sum_{i=1}^{2^n} (a_i^n - a_{i-1}^n) = f(b)(b - a) < \infty$$

for each n, and this establishes that f is an upper function.

22. INTEGRABLE FUNCTIONS

Problem 22.1. *Show by a counterexample that the integrable functions do not form an algebra.*

Solution. Consider the function $f\colon \mathbf{R} \to \mathbf{R}$ defined by $f(x) = 0$ if $x \notin (0, 1]$, and $f(x) = \sqrt{n}$ if $x \in (\frac{1}{n+1}, \frac{1}{n}]$ for some n. From Problem 21.2, we know that f is an integrable function. Now, note that f^2 is not an integrable function.

Problem 22.2.   *Let X be a nonempty set, and let δ be the Dirac measure on X with respect to the point a (see Example 13.4). Show that every function $f: X \to \mathbf{R}$ is integrable and that $\int f \, d\delta = f(a)$.*

Solution.   Note that $f = f(a)\chi_{\{a\}}$ a.e. holds. Consequently, the function f is integrable and $\int f \, d\delta = f(a)\delta(\{a\}) = f(a)$.

Problem 22.3.   *Let μ be the counting measure on $\mathbf{N}$ (see Example 13.3). Show that a function $f: \mathbf{N} \to \mathbf{R}$ is integrable if and only if $\sum_{n=1}^{\infty} |f(n)| < \infty$. Also, show that in this case $\int f \, d\mu = \sum_{n=1}^{\infty} f(n)$.*

Solution.   Let $f: \mathbf{N} \longrightarrow \mathbf{R}$. Since every function is measurable, f is integrable if and only if both f^+ and f^- are integrable. So, we can assume that $f(k) \geq 0$ holds for each k.

If $\phi_n = \sum_{k=1}^{n} f(k)\chi_{\{k\}}$, then $\{\phi_n\}$ is a sequence of step functions such that $\phi_n(k) \uparrow_n f(k)$ for each k, and

$$\int \phi_n \, d\mu = \sum_{k=1}^{n} f(k) \uparrow_n \sum_{k=1}^{\infty} f(k).$$

This shows that f is integrable if and only if $\sum_{k=1}^{\infty} f(k) < \infty$, and in this case $\int f \, d\mu = \sum_{k=1}^{\infty} f(k)$ holds.

Problem 22.4.   *Show that a measurable function f is integrable if and only if $|f|$ is integrable. Give an example of a nonintegrable function whose absolute value is integrable.*

Solution.   Apply Theorems 22.2 and 22.6.   For a counterexample: Let E be a non-Lebesgue measurable subset of $[0, 1]$ and consider the function $f = \chi_E - \chi_{[0,1] \setminus E}$.

Problem 22.5.   *Let f be an integrable function, and let $\{E_n\}$ be a sequence of disjoint measurable subsets of X. If $E = \bigcup_{n=1}^{\infty} E_n$, then show that*

$$\int_E f \, d\mu = \sum_{n=1}^{\infty} \int_{E_n} f \, d\mu.$$

Solution.   Let $F_n = \bigcup_{i=1}^{n} E_i$. Clearly, $|f\chi_{F_n}| \leq |f|$ for each n and $f\chi_{F_n} \longrightarrow f\chi_E$

a.e. Thus, by the Lebesgue Dominated Convergence Theorem, we have

$$\int_E f \, d\mu = \int f \chi_E \, d\mu = \lim_{n \to \infty} \int f \chi_{F_n} \, d\mu$$

$$= \lim_{n \to \infty} \left(\sum_{i=1}^{n} \int f \chi_{E_i} \, d\mu \right) = \sum_{n=1}^{\infty} \int_{E_n} f \, d\mu.$$

Problem 22.6. *Let f be an integrable function. Show that for each $\epsilon > 0$ there exists some $\delta > 0$ (depending on ϵ) such that $| \int_E f \, d\mu | < \epsilon$ holds for all measurable sets with $\mu^*(E) < \delta$.*

Solution. Consider an integrable function f and let $\varepsilon > 0$. From $0 \le |f| \wedge n \uparrow |f|$ and the Lebesgue Dominated Convergence Theorem we get $\int |f| \wedge n \, d\mu \uparrow \int |f| \, d\mu$. So, there exists some n_0 such that $\int (|f| - |f| \wedge n_0) \, d\mu < \frac{\varepsilon}{2}$ for all $n \ge n_0$. Now, put $\delta = \frac{\varepsilon}{2n_0}$ and note that if a measurable set E satisfies $\mu^*(E) < \delta$, then

$$\left| \int_E f \, d\mu \right| = \int_E |f| \, d\mu = \int_E (|f| - |f| \wedge n_0) \, d\mu + \int_E |f| \wedge n_0 \, d\mu$$

$$\le \int (|f| - |f| \wedge n_0) \, d\mu + \int_E n_0 \, d\mu$$

$$< \tfrac{\varepsilon}{2} + n_0 \mu^*(E) < \tfrac{\varepsilon}{2} + \tfrac{\varepsilon}{2} = \varepsilon,$$

as desired.

Problem 22.7. *Show that for every integrable function f the set*

$$\{x \in X : f(x) \ne 0\}$$

can be written as a countable union of measurable sets of finite measure—referred to as a σ-finite set.

Solution. Each $E_n = \left\{ x \in X : |f(x)| \ge \frac{1}{n} \right\}$ is a measurable set and, by Theorem 22.5, $\mu^*(E_n) < \infty$ holds. Now, observe that

$$\{x \in X : f(x) \ne 0\} = \bigcup_{n=1}^{\infty} E_n.$$

Problem 22.8. *Let $f : \mathbf{R} \to \mathbf{R}$ be integrable with respect to the Lebesgue measure. Show that the function $g : [0, \infty) \to \mathbf{R}$ defined by*

$$g(t) = \sup \left\{ \int |f(x + y) - f(x)| \, d\lambda(x) : |y| \le t \right\}$$

for $t \ge 0$ is continuous at $t = 0$.

Solution. Let $f: \mathbf{R} \longrightarrow \mathbf{R}$ be an upper function and let $\{\phi_n\}$ be a sequence of step functions with $\phi_n \uparrow f$ a.e. Fix some y, and note that $\phi_n(x + y) \uparrow f(x + y)$ holds for almost all x. Since $\chi_A(x + y) = \chi_{A-y}(x)$ and $\lambda(A) = \lambda(A - y)$, it follows that $f(x + y)$ as a function of x is integrable and $\int f(x + y) \, d\lambda(x) = \int f \, d\lambda$. Thus, if f is integrable, then $f(x + y)$ is integrable with respect to x for each fixed y and $\int f(x + y) \, d\lambda(x) = \int f \, d\lambda$ holds. In particular, for each fixed y we have $\int |f(x + y) - f(x)| \, d\lambda(x) \leq \int |f(x + y)| \, d\lambda(x) + \int |f(x)| \, d\lambda(x) = 2 \int |f| \, d\lambda < \infty$.

Now, for each integrable function $f: \mathbf{R} \longrightarrow \mathbf{R}$ and each $t \geq 0$, define

$$g_f(t) = \sup \left\{ \int |f(x + y) - f(x)| \, d\lambda(x) \colon |y| \leq t \right\} \geq 0.$$

Then, we have

$$g_{f+h}(t) \leq g_f(t) + g_h(t) \quad \text{and} \quad g_{\alpha f}(t) = |\alpha| g_f(t).$$

These relations show that the set

$$L = \left\{ f \colon f \text{ is integrable and } g_f \text{ is continuous at zero} \right\}$$

is a vector space. Moreover, L has the following approximation property:

- *If f is an integrable function such that for each $\varepsilon > 0$ there exists some $h \in L$ with $\int |f - h| \, d\lambda < \varepsilon$, then $f \in L$.*

Indeed, if f is such a function and $\varepsilon > 0$ is given, then choose $h \in L$ with $\int |f - h| \, d\lambda < \varepsilon$. Pick some $\delta > 0$ with $g_h(t) < \varepsilon$ whenever $0 < t < \delta$, and note that for $|y| \leq t$ we have

$$\int |f(x + y) - f(x)| \, d\lambda(x)$$

$$\leq \int |f(x + y) - h(x + y)| \, d\lambda(x) + \int |h(x + y) - h(x)| \, d\lambda(x)$$

$$+ \int |h - f| \, d\lambda < 3\varepsilon.$$

Thus, $g_f(t) \leq 3\varepsilon$ holds for all $0 < t < \delta$ so that $f \in L$.

Now, assume that $f = \chi_{[a,b)}$. If $0 < t < b - a$, then for $|y| \leq t$ we have

$$\int |f(x + y) - f(x)| \, d\lambda(x) = \int |\chi_{[a-y,b-y)}(x) - \chi_{[a,b)}(x)| \, d\lambda(x)$$

$$= \lambda([a - y, b - y) \Delta [a, b)) = 2|y| \leq 2t,$$

and so $g_f(t) \leq 2t$ holds for all $0 < t < b - a$, i.e., $f \in L$. By the approximation property, we have $\chi_A \in L$ for every σ-set A of finite measure, and hence, by the same property, $\chi_A \in L$ for every $A \in \Lambda$ with $\lambda(A) < \infty$ (see Problem 15.2). It follows that L contains the step functions. Since for every integrable function f and each $\varepsilon > 0$ there exists a step function ϕ with $\int |f - \phi| \, d\lambda < \varepsilon$, we infer that L consists of all the integrable functions.

Note. We basically verified here that the collection L satisfies properties (1), (2), and (3) of Theorem 22.12. This guarantees that L coincides with the vector space of all integrable functions.

Problem 22.9. *Let g be an integrable function and let $\{f_n\}$ be a sequence of integrable functions such that $|f_n| \leq g$ a.e. holds for all n. Show that if $f_n \xrightarrow{\mu} f$, then f is an integrable function and $\lim \int |f_n - f| \, d\mu = 0$ holds.*

Solution. By Theorem 19.4, the sequence $\{f_n\}$ has a subsequence that converges to f a.e., and so $|f| \leq g$ a.e. Thus, by Theorem 22.6, the function f is integrable.

Assume that for some $\varepsilon > 0$ there exists a subsequence $\{g_n\}$ of $\{f_n\}$ such that $\int |g_n - f| \, d\mu \geq \varepsilon$. By Theorem 19.4, there exists a subsequence $\{h_n\}$ of $\{g_n\}$ with $h_n \longrightarrow f$ a.e. Now, note that the Lebesgue Dominated Convergence Theorem implies $0 < \varepsilon \leq \int |h_n - f| \, d\mu \longrightarrow 0$, which is impossible. Hence, $\lim \int |f_n - f| \, d\mu = 0$.

Problem 22.10. *Establish the following generalization of Theorem 22.9: If $\{f_n\}$ is a sequence of integrable functions such that $\sum_{n=1}^{\infty} \int |f_n| \, d\mu < \infty$, then $\sum_{n=1}^{\infty} f_n$ defines an integrable function and*

$$\int \left(\sum_{n=1}^{\infty} f_n \right) d\mu = \sum_{n=1}^{\infty} \int f_n \, d\mu.$$

Solution. By Theorem 22.9, the series $g = \sum_{n=1}^{\infty} |f_n|$ defines an integrable function and $\left| \sum_{n=1}^{k} f_n \right| \leq g$ a.e. holds for each k. Since $\sum_{n=1}^{\infty} f_n$ is convergent for almost all points, it follows from the Lebesgue Dominated Convergence Theorem that $\sum_{n=1}^{\infty} f_n$ defines an integrable function and that

$$\int \left(\sum_{n=1}^{\infty} f_n \right) d\mu = \sum_{n=1}^{\infty} \int f_n \, d\mu.$$

Problem 22.11. *Let f be a positive (a.e.) measurable function, and let*

$$e_i = \mu^* \left(\{x \in X: 2^{i-1} < f(x) \leq 2^i\} \right)$$

for each integer i. Show that f is integrable if and only if $\sum_{i=-\infty}^{\infty} 2^i e_i < \infty$.

Solution. Let $E_i = \{x \in X: \ 2^{i-1} < f(x) \le 2^i\}$, $i = 0, \pm1, \pm2, \dots$. For each n let $\phi_n = \sum_{i=-n}^{n} 2^i \chi_{E_i}$. Then, there exists some function g with $\phi_n \uparrow g$ a.e. Clearly, g is a measurable function and $0 \le f \le g$ a.e. holds.

Assume that f is integrable. Then, each ϕ_n is a step function, and in view of $\phi_n \le 2f$ (why?), it follows that

$$\sum_{i=-\infty}^{\infty} 2^i e_i = \lim_{n\to\infty} \int \phi_n \, d\mu \le 2 \int f \, d\mu < \infty.$$

On the other hand, if $\sum_{i=-\infty}^{\infty} 2^i e_i < \infty$, then each ϕ_n is a step function, and so g is integrable. Since $0 \le f \le g$, Theorem 22.6 shows that f is also integrable.

Problem 22.12. *Let $\{f_n\}$ be a sequence of integrable functions satisfying $0 \le f_{n+1} \le f_n$ a.e. for each n. Then, show that $f_n \downarrow 0$ a.e. holds if and only if $\int f_n \, d\mu \downarrow 0$.*

Solution. Assume $\int f_n d\mu \downarrow 0$. Let $f_n \downarrow f$ a.e.; clearly, $f \ge 0$ a.e. It follows that $\int f \, d\mu = 0$, and thus (by Theorem 22.7) $f = 0$ a.e.

Problem 22.13. *Let f be an integrable function such that $f(x) > 0$ holds for almost all x. If A is a measurable set such that $\int_A f \, d\mu = 0$, then show that $\mu^*(A) = 0$.*

Solution. Let A be a measurable set satisfying $\int_A f \, d\mu = 0$. Next, consider the set $B = \{x \in A: \ f(x) \le 0\}$, and note that, by our hypothesis, $\mu^*(B) = 0$. Also, for each n put

$$A_n = \{x \in A: \ f(x) \ge \tfrac{1}{n}\}.$$

Then, $A = \left(\bigcup_{n=1}^{\infty} A_n\right) \cup B$, and $f\chi_{A_n} \le f\chi_A$ a.e. holds for each n. Thus,

$$0 \le \tfrac{1}{n}\mu^*(A_n) \le \int_{A_n} f \, d\mu \le \int_A f \, d\mu = 0,$$

which shows that $\mu^*(A_n) = 0$ for each n. This easily implies $\mu^*(A) = 0$.

Problem 22.14. *Let $(X, \mathcal{S}, \mu)$ be a finite measure space and let $f: X \to \mathbf{R}$ be an integrable function satisfying $f(x) > 0$ for almost all x. If $0 < \varepsilon \le \mu^*(X)$,* *then show that*

$$\inf\left\{\int_E f\, d\mu:\ E \in \Lambda_\mu\ \text{and}\ \mu^*(E) \geq \varepsilon\right\} > 0.$$

Solution. We can assume that $f(x) > 0$ holds for each $x \in X$. If for some $0 < \varepsilon \leq \mu^*(X)$ we have

$$\inf\left\{\int_E f\, d\mu:\ E \in \Lambda_\mu\ \text{and}\ \mu^*(E) \geq \varepsilon\right\} = 0,$$

then there exists a sequence $\{E_n\}$ of Λ_μ satisfying $\mu^*(E_n) \geq \epsilon$ and $\int_{E_n} f\, d\mu < \frac{1}{2^n}$ for each n. Put $F_n = \bigcup_{k=n}^{\infty} E_k$ and note that:

- *each F_n is measurable;*
- *$F_{n+1} \subseteq F_n$ holds for each n; and*
- *$\mu^*(F_n) \geq \mu^*(E_n) \geq \varepsilon$ holds for each n.*

If $F = \bigcap_{n=1}^{\infty} F_n$, then F is a measurable set and (by Theorem 15.4(2)) we have

$$\mu^*(F) = \lim_{n\to\infty} \mu^*(F_n) \geq \varepsilon > 0. \qquad (\star)$$

From $f\chi_{F_n} \leq \sum_{k=n}^{\infty} f\chi_{E_k}$, we infer that

$$\int f\chi_{F_n}d\mu \leq \sum_{k=n}^{\infty}\int f\chi_{E_k}d\mu \leq \sum_{k=n}^{\infty} \frac{1}{2^k} = \frac{1}{2^{n-1}},$$

and so $\int f\chi_{F_n} \downarrow 0$. The latter implies $f\chi_{F_n} \downarrow 0$ a.e. (see Problem 22.12), and since $f\chi_{F_n} \downarrow f\chi_F$, we infer that $f\chi_F = 0$ a.e. In view of $f(x) > 0$ for each x, the latter (in view of Problem 22.13) implies $\mu^*(F) = 0$, contrary to $(\star)$, and the desired conclusion follows.

Problem 22.15. *Let f be a positive integrable function. Define $\nu\colon \Lambda \to [0, \infty)$ by $\nu(A) = \int_A f\, d\mu$ for each $A \in \Lambda$. Show that*

a. *(X, Λ, ν) is a measure space.*

b. *If Λ_ν denotes the σ-algebra of all ν-measurable subsets of X, then show that $\Lambda \subseteq \Lambda_\nu$. Give an example for which $\Lambda \neq \Lambda_\nu$.*

c. *If $\mu^*\big(\{x \in X\colon f(x) = 0\}\big) = 0$, then show that $\Lambda = \Lambda_\nu$.*

d. *If g is an integrable function with respect to the measure space (X, Λ, ν),* *then show that fg is integrable with respect to the measure space (X, S, μ), and that*

$$\int g \, d\nu = \int gf \, d\mu.$$

Solution. (a) This part follows immediately from Problem 22.5.

(b) The measure ν has initial domain Λ. Hence, by Theorem 15.3, $\Lambda \subseteq \Lambda_\nu$ holds.

Consider $\mathbb{R}$ with the Lebesgue measure, and let $f = \chi_{(1,2)}$. Since, in this case, $\nu([0, 1]) = 0$, it follows that every subset of $[0, 1]$ is a ν-null set (and hence ν-measurable). On the other hand, not every subset of $[0, 1]$ is Lebesgue measurable. Thus, $\Lambda \neq \Lambda_\nu$ holds in this case.

(c) First, observe that ν is a finite measure. Now, let $A \in \Lambda_\nu$ with $\nu^*(A) = 0$. By Theorem 15.11, there exists some $B \in \Lambda$ such that $A \subseteq B$ and $\nu(B) = 0$. The relation $\int_B f \, d\mu = \nu(B) = 0$ combined with Problem 22.13, shows that $\mu^*(A) = 0$. Thus, $A \in \Lambda$. Now, if $V \in \Lambda_\nu$, then pick some $W \in \Lambda$ with $V \subseteq W$ and $\nu(W) = \nu^*(V)$ (Theorem 15.11). Note that $\nu^*(W \setminus V) = 0$, and so, by the preceding discussion, $W \setminus V \in \Lambda$. Finally, $V = W \setminus (W \setminus V) \in \Lambda$ holds, which shows that $\Lambda = \Lambda_\nu$.

(d) We shall assume $g(x) \geq 0$ and $f(x) \geq 0$ for all x. Pick a sequence $\{\phi_n\}$ of ν-step functions such that $0 \leq \phi_n \uparrow g$ ν-a.e. Let

$$G = \{x \in X \colon f(x) > 0\}.$$

Clearly, $G \in \Lambda$. Since $G^c = \{x \in X \colon f(x) = 0\}$, it follows that $\nu(G^c) = \int_{G^c} f \, d\mu = 0$. Since f is strictly positive on G, the arguments of part (c) show that whenever $A \subseteq G$, we have:

1) If $A \in \Lambda_\nu$, then $A \in \Lambda$, and

2) By Problem 22.13, $\nu^*(A) = 0$ if and only if $\mu^*(A) = 0$ (and in this case $A \in \Lambda$).

In particular, it follows that $\phi_n f \uparrow fg$ μ-a.e. holds. Now, if $A \in \Lambda_\nu$ satisfies $\nu^*(A) < \infty$, then

$$\int \chi_A \, d\nu = \nu(A \cap G) = \int_{A \cap G} f \, d\mu = \int \chi_A f \, d\mu.$$

This implies that if ϕ is a ν-step function, then ϕf is μ-integrable, and that $\int \phi \, d\nu = \int \phi f \, d\mu$ holds. Now, note that $0 \leq \phi_n f \uparrow fg$ μ-a.e. and $\int \phi_n d\nu = \int \phi_n f \, d\mu$, show that fg is μ-integrable and that $\int g \, d\nu = \int gf \, d\mu$ holds.

Problem 22.16.   *Let I be an interval of $\mathbf{R}$, and let $f: I \to \mathbf{R}$ be an integrable function with respect to the Lebesgue measure. For a pair of real numbers a and b with $a \neq 0$, let $J = \{(x - b)/a: x \in I\}$. Show that the function $g: J \to \mathbf{R}$ defined by $g(x) = f(ax+b)$ for $x \in J$ is integrable and that $\int_I f\, d\lambda = |a| \int_J g\, d\lambda$ holds.*

Solution.   Assume first that $f = \chi_A$ for some measurable set $A \subseteq I$. Clearly, $\frac{1}{a}(A - b) \subseteq J$. Thus, in view of the identity, $\chi_A(ax + b) = \chi_{\frac{1}{a}(A-b)}(x)$, it follows from Problem 15.5 that

$$\int_J g\, d\lambda = \tfrac{1}{|a|}\lambda(A) = \tfrac{1}{|a|}\int_I f\, d\lambda.$$

Thus, the formula is true for the characteristic function of a measurable set. It follows that it is also true for step functions.

Now, let f be an upper function. Choose a sequence $\{\phi_n\}$ of step functions with $\phi_n \uparrow f$ a.e. on I. If $\psi_n(x) = \phi_n(ax + b)$ for $x \in J$, then ψ_n is a step function on J and $\psi_n \uparrow g$ a.e. holds on J. (Note that if $B \subseteq I$ satisfies $\lambda(B) = 0$, then by Problem 15.5, we have $\lambda\big(\frac{1}{a}(B - b)\big) = \frac{1}{|a|}\lambda(B) = 0$.) Therefore,

$$|a| \int_J g\, d\lambda = |a| \lim_{n \to \infty} \int_J \psi_n\, d\lambda = \lim_{n \to \infty} \int_I \phi_n\, d\lambda = \int_I f\, d\lambda.$$

Thus, the formula holds true for every integrable function f on I.

Problem 22.17.   *Let $(X, \mathcal{S}, \mu)$ be a finite measure space. For every pair of measurable functions f and g let*

$$d(f, g) = \int \frac{|f - g|}{1 + |f - g|}\, d\mu.$$

a.   *Show that $(\mathcal{M}, d)$ is a metric space.*
b.   *Show that a sequence $\{f_n\}$ of measurable functions (i.e., $\{f_n\} \subseteq \mathcal{M}$) satisfies $f_n \overset{\mu}{\longrightarrow} f$ if and only if $\lim d(f_n, f) = 0$.*
c.   *Show that $(\mathcal{M}, d)$ is a complete metric space. That is, show that if a sequence $\{f_n\}$ of measurable functions satisfies $d(f_n, f_m) \to 0$ as n, $m \to \infty$, then there exists a measurable function f such that $\lim d(f_n, f) = 0$.*

Solution.   (a) We assume that functions equal μ-a.e. are considered identical. Only the triangle inequality needs verification. To this end, let $f, g, h \in \mathcal{M}$. The triangle inequality follows immediately from the inequality

$$\frac{|f(x) - g(x)|}{1 + |f(x) - g(x)|} \leq \frac{|f(x) - h(x)|}{1 + |f(x) - h(x)|} + \frac{|h(x) - g(x)|}{1 + |h(x) - g(x)|} \,,\,.$$

For details see the solution of Problem 9.11.

(b) Start by observing that for $x \geq 0$ and $\varepsilon > 0$ we have:

$$x \geq \varepsilon \quad \Longleftrightarrow \quad \frac{x}{1+x} \geq \frac{\varepsilon}{1+\varepsilon}.$$

Now, assume that $\lim d(f_n, f) = 0$ holds. Then, the inequality

$$\mu^*\big(\{x \in X \colon |f_n(x) - f(x)| \geq \varepsilon\}\big) = \mu^*\big(\{x \in X \colon \tfrac{|f_n(x) - f(x)|}{1 + |f_n(x) - f(x)|} \geq \tfrac{\varepsilon}{1+\varepsilon}\}\big)$$
$$\leq \tfrac{1+\varepsilon}{\varepsilon} \cdot d(f_n, f)$$

easily implies that $f_n \xrightarrow{\mu} f$.

For the converse, assume $f_n \xrightarrow{\mu} f$. If $\lim d(f_n, f) \neq 0$, then there exists some $\varepsilon > 0$ and some subsequence $\{g_n\}$ of $\{f_n\}$ with $d(g_n, f) \geq \varepsilon$ for all n. By passing to a subsequence, we can assume that $g_n \longrightarrow f$ a.e. (Theorem 19.4). In view of $\frac{|g_n - f|}{1 + |g_n - f|} \leq 1$ and the finiteness of the measure space, the Lebesgue Dominated Convergence Theorem yields $0 < \varepsilon \leq \lim d(g_n, f) = 0$, which is absurd. Hence, $\lim d(f_n, f) = 0$ holds.

(c) Assume $d(f_n, f_m) \longrightarrow 0$ as $n, m \longrightarrow \infty$. The inequality

$$\mu^*\big(\{x \in X \colon |f_n(x) - f_m(x)| \geq \varepsilon\}\big) \leq \tfrac{1+\varepsilon}{\varepsilon} \cdot d(f_n, f_m)$$

shows that $\{f_n\}$ is a μ-Cauchy sequence. Thus, by Problem 19.7, $f_n \xrightarrow{\mu} f$ holds for some f, and by (b) above, $\lim d(f_n, f) = 0$ also holds.

Conversely, if $\lim d(f_n, f) = 0$ holds, then by part (b) above $f_n \xrightarrow{\mu} f$, which implies that $\{f_n\}$ is a μ-Cauchy sequence.

Problem 22.18. *Let $f \colon \mathbf{R} \to \mathbf{R}$ be a Lebesgue integrable function. For each finite interval I let $f_I = \frac{1}{\lambda(I)} \int_I f \, d\lambda$ and $E_I = \{x \in I \colon f(x) > f_I\}$. Show that*

$$\int_I |f - f_I| \, d\lambda = 2 \int_{E_I} (f - f_I) \, d\lambda.$$

Solution.   We follow the notation of the problem. Start by observing that

$$\int_{E_I} (f - f_I)\, d\lambda + \int_{I \setminus E_I} (f - f_I)\, d\lambda = \int_I (f - f_I)\, d\lambda$$

$$= \int_I f\, d\lambda - \int_I f_I\, d\lambda$$

$$= \int_I f\, d\lambda - \int_I f\, d\lambda = 0.$$

Consequently,

$$\int_{E_I} (f - f_I)\, d\lambda = \int_{I \setminus E_I} (f_I - f)\, d\lambda.$$

Now, note that

$$\int_I |f - f_I|\, d\lambda = \int_{E_I} |f - f_I|\, d\lambda + \int_{I \setminus E_I} |f - f_I|\, d\lambda$$

$$= \int_{E_I} (f - f_I)\, d\lambda + \int_{I \setminus E_I} (f_I - f)\, d\lambda$$

$$= \int_{E_I} (f - f_I)\, d\lambda + \int_{E_I} (f - f_I)\, d\lambda = 2 \int_{E_I} (f - f_I)\, d\lambda.$$

Problem 22.19.   *Let* $f : [0, \infty) \to \mathbb{R}$ *be a Lebesgue integrable function such that* $\int_0^t f(x)\, d\lambda(x) = 0$ *for each* $t \geq 0$. *Show that* $f(x) = 0$ *holds for almost all* x.

Solution.   Start by observing that

$$\int_{[a,b)} f\, d\lambda = \int_{[0,b)} f\, d\lambda - \int_{[0,a)} f\, d\lambda = 0$$

holds for each interval $[a, b)$. By Problem 22.5, we see that $\int_A f\, d\lambda = 0$ holds for each σ-set A. From Problem 15.2 (and the Lebesgue Dominated Convergence Theorem), we see that $\int_A f\, d\lambda = 0$ holds for each Lebesgue measurable subset A of $\mathbb{R}$.

Now, let $X = \{x \in \mathbb{R}: f(x) > 0\}$ and $Y = \{x \in X: f(x) < 0\}$. Clearly, X and Y are both Lebesgue measurable sets, and

$$\int_X f \, d\lambda = \int_Y f \, d\lambda = 0.$$

Now, invoke Problem 22.13 to obtain $\lambda(X) = \lambda(Y) = 0$. Therefore, $f(x) = 0$ holds for almost all x.

Problem 22.20. *Let $(X, \mathcal{S}, \mu)$ be a measure space and let $f, f_1, f_2, \ldots$ be non-negative integrable functions such that $f_n \to f$ a.e. and $\lim \int f_n \, d\mu = \int f \, d\mu$. If E is a measurable set, then show that*

$$\lim_{n \to \infty} \int_E f_n \, d\mu = \int_E f \, d\mu.$$

Solution. Assume that the integrable functions $f, f_1, f_2, \ldots$ are non-negative satisfying the hypotheses of the problem and let E be a measurable set. Then the functions $f \chi_E, f_1 \chi_E, f_2 \chi_E, \ldots$ are non-negative and integrable (because $0 \leq f \chi_E \leq f$ and $0 \leq f_n \chi_E \leq f_n$) and $f_n \chi_E \longrightarrow f \chi_E$ holds. Using Fatou's Lemma, we see that

$$\int_E f \, d\mu = \int \liminf f_n \chi_E \, d\mu \leq \liminf \int f_n \chi_E \, d\mu = \liminf \int_E f_n \, d\mu. \qquad (\star)$$

Similarly, we have

$$\int_{E^c} f \, d\mu \leq \liminf \int_{E^c} f_n \, d\mu. \qquad (\star\star)$$

Therefore,

$$\begin{aligned}
\int f \, d\mu = \int_E f \, d\mu + \int_{E^c} f \, d\mu &\leq \liminf \int_E f_n \, d\mu + \liminf \int_{E^c} f_n \, d\mu \\
&\leq \liminf \left(\int_E f_n \, d\mu + \int_{E^c} f_n \, d\mu \right) \\
&= \liminf \int f_n \, d\mu \\
&= \int f \, d\mu,
\end{aligned}$$

where the second inequality holds by virtue of Problem 4.7(b). It follows that

$$\int_E f\,d\mu + \int_{E^c} f\,d\mu = \liminf \int_E f_n\,d\mu + \liminf \int_{E^c} f_n\,d\mu,$$

and from $(\star)$ and $(\star\star)$, we see that

$$\liminf \int_E f_n\,d\mu = \int_E f\,d\mu.$$

Now, let $\{g_n\}$ be a subsequence of $\{f_n\}$. Then,

$$g_n \longrightarrow f \quad \text{a.e.} \quad \text{and} \quad \lim_{n\to\infty} \int g_n\,d\mu = \int f\,d\mu.$$

By the preceding conclusion, we infer that

$$\liminf \int_E g_n\,d\mu = \int_E f\,d\mu,$$

and so there exists a subsequence $\{g_{k_n}\}$ of the sequence $\{g_n\}$ such that $\lim \int_E g_{k_n}\,d\mu = \int_E f\,d\mu$.

Thus, we have demonstrated that every subsequence of the bounded sequence of real numbers $\{\int_E f_n\,d\mu\}$ has a convergent subsequence to $\int_E f\,d\mu$. This means that

$$\lim_{n\to\infty} \int_E f_n\,d\mu = \int_E f\,d\mu$$

holds; see Problem 4.2.

Problem 22.21. *If a Lebesgue integrable function $f:[0,1] \to \mathbf{R}$ satisfies $\int_0^1 x^{2n} f(x)\,d\lambda(x) = 0$ for each $n = 0, 1, 2, \ldots$, then show that $f = 0$ a.e.*

Solution. Let an integrable function $f:[0,1] \longrightarrow \mathbf{R}$ satisfy

$$\int_0^1 x^{2n} f(x)\,d\lambda(x) = 0 \quad \text{for } n = 0, 1, 2, \ldots.$$

Since the algebra of functions generated by $\{1, x^2\}$ is uniformly dense in $C[0,1]$ (see Problem 11.5), it follows that $\int_0^1 g(x)f(x)\,d\lambda(x) = 0$ holds for all g in

$C[0, 1]$. Consider the two measurable sets

$$E = \{x \in [0, 1]: \ f(x) > 0\} \quad \text{and} \quad F = \{x \in [0, 1]: \ f(x) < 0\}.$$

We have to show that $\lambda(E) = \lambda(F) = 0$. We shall establish that $\lambda(E) = 0$ holds and leave the identical arguments for F to the reader.

Pick a sequence $\{K_n\}$ of compact sets and a sequence $\{O_n\}$ of open sets of $[0, 1]$ satisfying $K_n \subseteq E \subseteq O_n$ for each n, $K_n \uparrow$, $O_n \downarrow$, and $\lambda(E) = \lim \lambda(K_n) = \lim \lambda(O_n)$. (Here we use the regularity of the Lebesgue measure.) For each n there exists (by Theorem 10.8) a continuous function $g_n: [0, 1] \longrightarrow [0, 1]$ satisfying $g_n(x) = 1$ for each $x \in K_n$ and $f(x) = 0$ for each $x \notin O_n$. Clearly, $|g_n f| \leq |f|$ and $g_n f \longrightarrow f \chi_E$ a.e. By the Lebesgue Dominated Convergence Theorem, we get

$$\lim_{n \to \infty} \int_0^1 g_n(x) f(x) \, d\lambda(x) = \int_0^1 f(x) \chi_E(x) \, d\lambda(x) = \int_E f \, d\lambda.$$

Taking into account that $\int_0^1 g_n(x) f(x) \, d\lambda(x) = 0$ holds for all n, we infer that $\int_E f \, d\lambda = 0$. Now, invoke Problem 22.13 to infer that $\lambda(E) = 0$, as claimed.

Problem 22.22. *For each n consider the partition*

$$\left\{0, 2^{-n}, 2 \cdot 2^{-n}, 3 \cdot 2^{-n}, \ldots, (2^n - 1) \cdot 2^{-n}, 1\right\}$$

of the interval $[0, 1]$ and define the function $r_n: [0, 1] \to \mathbb{R}$ by $r_n(1) = -1$ and

$$r_n(x) = (-1)^{k-1} \quad \text{for} \quad (k - 1)2^{-n} \leq x < k2^{-n} \quad (k = 1, 2, \ldots, 2^n).$$

   a. *Draw the graphs of r_1 and r_2.*
   b. *Show that if $f: [0, 1] \to \mathbb{R}$ is a Lebesgue integrable function, then*

$$\lim_{n \to \infty} \int_0^1 r_n(x) f(x) \, d\lambda(x) = 0.$$

Solution. (a) The graphs of r_1 and r_2 are shown in Figure 4.1.
(b) By Theorem 22.12, it suffices (how?) to establish the claim for the case $f = \chi_{[a,b)}$, where $[a, b)$ is a subinterval of $[0, 1]$. Clearly, $\int_0^1 r_n(x) \chi_{[a,b)} \, d\lambda(x) = \int_a^b r_n(x) \, dx$. Therefore, it suffices to show that $\lim \int_a^b r_n(x) \, dx = 0$ holds for each $0 \leq a < b \leq 1$.

To this end, fix $0 \leq a < b \leq 1$ and let $\varepsilon > 0$. Fix n_0 such that $2^{-n_0} < \min\{\varepsilon, \frac{b-a}{4}\}$. Pick $n \geq n_0$ and consider the partition $\{0, \frac{1}{2^n}, \frac{2}{2^n}, \ldots, \frac{2^n-1}{2^n}, 1\}$; for

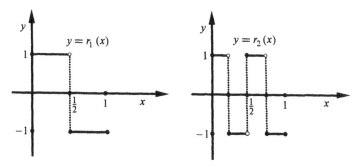

FIGURE 4.1.   The Graphs of r_1 and r_2

simplicity, let $x_i = \frac{i}{2^n}$ and note that the points a and b are related to the x_i as shown in Figure 4.2.

Since for any three consecutive points x_{i-1}, x_i, x_{i+1} we have $\int_{x_{i-1}}^{x_i} r_n(x)\,dx = 0$, we see that $\int_a^b r_n(x)\,dx = \int_a^{x_k} r_n(x)\,dx + \int_c^b r_n(x)\,dx$, where $c = x_{m-1}$ or $c = x_m$; see Figure 4.2. Hence,

$$\left| \int_a^b r_n(x)\,dx \right| \le \int_a^{x_k} |r_n(x)|\,dx + \int_c^b |r_n(x)|\,dx$$
$$= (x_k - a) + (b - c) < \varepsilon + 2\varepsilon = 3\varepsilon$$

for each $n \ge n_0$. This means that $\lim \int_a^b r_n(x)\,dx = 0$, as desired.

Problem 22.23.   *Let $\{\epsilon_n\}$ be a sequence of real numbers such that $0 < \epsilon_n < 1$ for each n. Also, let us say that a sequence $\{A_n\}$ of Lebesgue measurable subsets of $[0, 1]$ is consistent with the sequence $\{\epsilon_n\}$ if $\lambda(A_n) = \epsilon_n$ for each n. Establish the following properties of $\{\epsilon_n\}$:*

  a.   *The sequence $\{\epsilon_n\}$ converges to zero if and only if there exists a consistent sequence $\{A_n\}$ of measurable subsets of $[0, 1]$ such that $\sum_{n=1}^\infty \chi_{A_n}(x) < \infty$ for almost all x.*

  b.   *The series $\sum_{n=1}^\infty \epsilon_n$ converges in $\mathbf{R}$ if and only if for each consistent sequence $\{A_n\}$ of measurable subsets of $[0, 1]$ we have $\sum_{n=1}^\infty \chi_{A_n}(x) < \infty$ for almost all x.*

Solution.   (a) If $\varepsilon_n \longrightarrow 0$, then let $A_n = (0, \varepsilon_n)$ $(n = 1, 2, \ldots)$, and note that $\lambda(A_n) = \varepsilon_n$ holds for each n and that $\sum_{n=1}^\infty \chi_{A_n}(x) < \infty$ for each $x \in [0, 1]$.

FIGURE 4.2.

For the converse, assume that there exists a consistent sequence of measurable subsets $\{A_n\}$ of $[0, 1]$ satisfying $\sum_{n=1}^{\infty} \chi_{A_n}(x) < \infty$ for almost all x. For each n let $B_n = \bigcup_{k=n}^{\infty} A_k$ and note that $B_n \downarrow B = \bigcap_{k=1}^{\infty} B_k$. If $\lambda(B) > 0$ holds, then note that $\sum_{n=1}^{\infty} \chi_{A_n}(x) = \infty$ for each $x \in B$ (why?), which contradicts our hypothesis. Thus, $\lambda(B) = 0$. From the continuity of the measure (Theorem 15.4), we see that $\lambda(B_n) \downarrow 0$. In view of $A_n \subseteq B_n$, we have $0 < \varepsilon_n = \lambda(A_n) \leq \lambda(B_n)$ for each n, and so $\lim \varepsilon_n = 0$.

(b) Assume $\sum_{n=1}^{\infty} \varepsilon_n < \infty$ and that $\{A_n\}$ is a consistent sequence of measurable subsets of $[0, 1]$. Then,

$$\sum_{n=1}^{\infty} \int_{[0,1]} \chi_{A_n} d\lambda = \sum_{n=1}^{\infty} \lambda(A_n) = \sum_{n=1}^{\infty} \varepsilon_n < \infty,$$

and so, by the series version of Levi's Theorem 22.9, we have $\sum_{n=1}^{\infty} \chi_{A_n}(x) < \infty$ for almost all x.

For the converse, assume that for every consistent sequence $\{A_n\}$ of measurable subsets of $[0, 1]$, we have $\sum_{n=1}^{\infty} \chi_{A_n}(x) < \infty$ for almost all x. Suppose by way of contradiction that $\sum_{n=1}^{\infty} \varepsilon_n = \infty$. Using an inductive argument (how?), we see that there is a sequence $\{k_n\}$ of strictly increasing natural numbers such that $\sum_{i=k_n+1}^{k_{n+1}} \varepsilon_i > 1$ holds for each n. Next, for each n we can choose (how?) subintervals $A_{k_n+1}, A_{k_n+2}, \ldots, A_{k_{n+1}}$ of $[0, 1]$ such that $\lambda(A_i) = \varepsilon_i$ for $k_n + 1 \leq i \leq k_{n+1}$ and $\bigcup_{i=k_n+1}^{k_{n+1}} A_i = [0, 1]$. Now, note that the sequence of measurable sets $\{A_n\}$ is consistent with $\{\varepsilon_n\}$ and $\sum_{n=1}^{\infty} \chi_{A_n}(x) = \infty$ holds for each $x \in [0, 1]$, contrary to our hypothesis. So, $\sum_{n=1}^{\infty} \varepsilon_n < \infty$ must hold.

Problem 22.24. *Let $(X, \mathcal{S}, \mu)$ be a finite measure space and let $f: X \to \mathbf{R}$ be a measurable function.*

   a. *Show that if f^n is integrable for each n and that $\lim \int f^n d\mu$ exists in $\mathbf{R}$, then $|f(x)| \leq 1$ holds for almost all x.*

   b. *If f^n is integrable for each n, then show that $\int f^n d\mu = c$ (a constant) for $n = 1, 2, \ldots$ if and only if $f = \chi_A$ for some measurable subset A of X.*

Solution. Keep in mind that f^n denotes the function $f^n: X \longrightarrow \mathbf{R}$ defined by $f^n(x) = [f(x)]^n$ for each $x \in X$.

(a) Assume that f^n is Lebesgue integrable for each n and that $\lim \int f^n d\mu$ exists in $\mathbf{R}$. Assume by way of contradiction that the measurable set

$$E = \{x \in X : |f(x)| > 1\}$$

satisfies $\mu^*(E) > 0$. From the identity $E = \bigcup_{k=1}^{\infty} E_k$, where

$$E_k = \left\{x \in X : |f(x)| \geq 1 + \tfrac{1}{k}\right\},$$

we see that there exists some $\delta > 1$ such that the measurable set

$F = \{x \in X : |f(x)| > \delta\}$ satisfies $\mu^*(F) > 0$. Now, note $f^{2n} \geq \delta^{2n} \chi_F$ holds for each n, and so from

$$\delta^{2n} \mu^*(F) = \int \delta^{2n} \chi_F \, d\mu \leq \int f^{2n} \, d\mu,$$

we infer that $\lim \int f^{2n} \, d\mu = \infty$, contradicting the existence in $\mathbf{R}$ of the limit $\lim \int f^n \, d\mu$. Hence, $|f(x)| \leq 1$ must hold for almost all x.

(b) Assume $\int f^n \, d\mu = c$ holds for each $n = 1, 2, \ldots$. By part (a), we know that $|f(x)| \leq 1$ holds for almost all $x \in X$. Now, define the sets $A = \{x \in X : f(x) = 1\}$, $B = \{x \in X : f(x) = -1\}$, and $C = \{x \in X : |f(x)| < 1\}$. Then, for each n we have

$$\int f^n \, d\mu = \int_A f^n \, d\mu + \int_B f^n \, d\mu + \int_C f^n \, d\mu$$

$$= \int_A 1 \, d\mu + \int_B (-1)^n \, d\mu + \int_C f^n \, d\mu$$

$$= \mu^*(A) + (-1)^n \mu^*(B) + \int_C f^n \, d\mu = c.$$

Since $f^n(x) \longrightarrow 0$ holds for each $x \in C$, it follows from the Lebesgue Dominated Convergence Theorem that $\lim \int_C f^n \, d\mu = 0$. Hence,

$$\lim_{n \to \infty} \left[\mu^*(A) + (-1)^n \mu^*(B) \right] = c.$$

Since $\lim(-1)^n$ does not exist, we infer that $\mu^*(B) = 0$, and therefore, $c = \mu^*(A) = \mu^*(A) + \int_C f^n \, d\mu$ for each n. In particular, we have $\int_C f^2 \, d\mu = 0$, and so $f(x) = 0$ must hold for almost all $x \in C$. The latter implies that $f = \chi_A$ a.e. holds.

23.   THE RIEMANN INTEGRAL AS A LEBESGUE INTEGRAL

Problem 23.1.   *Let $f : [a, b] \to \mathbf{R}$ be Riemann integrable. Show that f is Riemann integrable on every closed subinterval of $[a, b]$. Also, show that*

$$\int_c^d f(x) \, dx = \int_c^e f(x) \, dx + \int_e^d f(x) \, dx$$

holds for every three points c, d, and e of $[a, b]$.

Solution.   Let $f:[a,b] \longrightarrow \mathbb{R}$ be a Riemann integrable function and let $[u, v]$ be a closed subinterval of $[a, b]$. If $f:[u, v] \longrightarrow \mathbb{R}$ is discontinuous at some point $x \in [u, v]$, then $f:[a, b] \longrightarrow \mathbb{R}$ is also discontinuous at the point x—note that, in this case, there exists a sequence $\{x_n\}$ of $[u, v]$ (and hence of $[a, b]$) such that $\{f(x_n)\}$ does not converge to $f(x)$. Thus, the set D of all points of discontinuity of $f:[u, v] \longrightarrow \mathbb{R}$ is a subset of the set of all points of discontinuity of $f:[a, b] \longrightarrow \mathbb{R}$. Since $f:[a, b] \longrightarrow \mathbb{R}$ is Riemann integrable, we know that $\lambda(D) = 0$, and so (by Theorem 23.7) the function $f:[u, v] \longrightarrow \mathbb{R}$ is Riemann integrable.

Now, assume that $a \le c < e < d \le b$. Since $f:[c, e] \longrightarrow \mathbb{R}$ is Riemann integrable (and hence Lebesgue integrable), there exists a sequence of step functions $\{\phi_n\}$ over $[c, e]$ (i.e., $\phi_n(x) = 0$ holds for $x \notin [c, e]$) with $\phi_n(x) \uparrow f(x)$ for almost all $x \in [c, e]$. Similarly, there exists a sequence of step functions $\{\psi_n\}$ over $[e, d]$ such that $\psi_n(x) \uparrow f(x)$ holds for almost all $x \in [e, d]$. Then, $\{\phi_n + \psi_n\}$ is a sequence of step functions over $[c, d]$ satisfying $\phi_n(x) + \psi_n(x) \uparrow f(x)$ for almost all $x \in [c, d]$. Therefore,

$$
\begin{aligned}
\int_c^d f(x)\,dx &= \int_{[c,d]} f\,d\lambda = \lim_{n \to \infty} \int_{[c,d]} (\phi_n + \psi_n)\,d\lambda \\
&= \lim_{n \to \infty} \int_{[c,d]} \phi_n\,d\lambda + \lim_{n \to \infty} \int_{[c,d]} \psi_n\,d\lambda \\
&= \int_{[c,e]} f\,d\lambda + \int_{[e,d]} f\,d\lambda \\
&= \int_c^e f(x)\,dx + \int_e^d f(x)\,dx.
\end{aligned}
$$

Now, the equality $\int_c^d f(x)\,dx = \int_c^e f(x)\,dx + \int_e^d f(x)\,dx$ for arbitrary elements c, d, and e of $[a, b]$ can be obtained by considering all possible cases. We prove it for one such case and leave the rest for the reader. Assume that $a \le e < c < d \le b$. Then, by the preceding case, we have

$$
\int_e^d f(x)\,dx = \int_e^c f(x)\,dx + \int_c^d f(x)\,dx = -\int_c^e f(x)\,dx + \int_c^d f(x)\,dx,
$$

from which it follows that $\int_c^e f(x)\,dx + \int_e^d f(x)\,dx = \int_c^d f(x)\,dx$.

Problem 23.2.   *Let $f:[a, b] \to \mathbb{R}$ be Riemann integrable. Then, show that*

$$
\int_a^b f(x)\,dx = \lim_{n \to \infty} \frac{b-a}{n} \sum_{i=1}^n f\left(a + \frac{i(b-a)}{n}\right).
$$

Solution. The conclusion follows from Theorem 23.5 by observing that

$$\frac{b-a}{n} \sum_{i=1}^{n} f\left(a + i\frac{b-a}{n}\right) = S_f(P_n, T_n),$$

where the partition $P_n = \{x_0, x_1, \ldots, x_n\}$ and $T = \{t_1, \ldots, t_n\}$ satisfy $x_i = a + i\frac{b-a}{n}$ $(0 \le i \le n)$ and $t_i = x_i$ $(1 \le i \le n)$.

Problem 23.3. *Let $\{f_n\}$ be a sequence of Riemann integrable functions on $[a, b]$ such that $\{f_n\}$ converges uniformly to a function f. Show that f is Riemann integrable and that*

$$\lim_{n \to \infty} \int_a^b f_n(x)\,dx = \int_a^b f(x)\,dx.$$

Solution. Choose some k such that $|f_k(x) - f_n(x)| < 1$ holds for all $n > k$ and all $x \in [a, b]$. Thus, $|f_k(x) - f(x)| \le 1$ holds for all $x \in [a, b]$. Since f_k is bounded, it is easy to see that there exists some $M > 0$ such that $|f(x)| \le M$ holds for all $x \in [a, b]$.

If $D_n \subseteq [a, b]$ denotes the set of discontinuities of f_n, then (by Theorem 23.7) $D = \bigcup_{n=1}^{\infty} D_n$ satisfies $\lambda(D) = 0$. Since each f_n is continuous on $[a, b] \setminus D$, it follows from Theorem 9.2 that f is continuous on $[a, b] \setminus D$, i.e., f is continuous almost everywhere. By Theorem 23.7, f is Riemann integrable.

For the last part, let $\epsilon > 0$. Pick some k such that $|f_n(x) - f(x)| < \epsilon$ for all $n \ge k$ and all $x \in [a, b]$. So, for $n \ge k$ we have

$$\left| \int_a^b f_n(x)\,dx - \int_a^b f(x)\,dx \right| \le \int_a^b |f_n(x) - f(x)|\,dx < \epsilon(b - a),$$

and this shows that $\lim_{n \to \infty} \int_a^b f_n(x)\,dx = \int_a^b f(x)\,dx$.

Problem 23.4. *For each n, let $f_n(x): [0, 1] \to \mathbb{R}$ be defined by $f_n(x) = \frac{nx^{n-1}}{1+x}$ for all $x \in [0, 1]$. Then, show that $\lim \int_0^1 f_n(x)\,dx = \frac{1}{2}$.*

Solution. Integrating by parts, we get

$$\int_0^1 f_n(x)\,dx = \frac{x^n}{1+x}\Big|_0^1 + \int_0^1 \frac{x^n}{(1+x)^2}\,dx = \frac{1}{2} + \int_0^1 \frac{x^n}{(1+x)^2}\,dx. \qquad (\star)$$

Since $0 \le \frac{x^n}{(1+x)^2} \le 1$ holds for all $x \in [0, 1]$ and $\lim \frac{x^n}{(1+x)^2} = 0$ for each x in

$[0, 1)$, the Lebesgue Dominated Convergence Theorem yields $\lim \int_0^1 \frac{x^n}{(1+x)^2} \, dx = 0$.
Thus, from $(\star)$, we see that $\lim \int_0^1 f_n(x) \, dx = \frac{1}{2}$.

Problem 23.5. *Let* $f : [a, b] \rightarrow \mathbf{R}$ *be an increasing function. Show that* f *is Riemann integrable.*

Solution. Let $P_n = \{x_0, x_1, \ldots, x_n\}$ be the partition that subdivides $[a, b]$ into n subintervals of equal length $\frac{b-a}{n}$. Since f is increasing, note that $m_i = f(x_{i-1})$ and $M_i = f(x_i)$ hold for each $1 \le i \le n$. Next, observe that

$$0 \le I^*(f) - I_*(f) \le S^*(f, P_n) - S_*(f, P_n)$$
$$= \sum_{i=1}^n [f(x_i) - f(x_{i-1})](x_i - x_{i-1})$$
$$= [f(b) - f(a)] \cdot \frac{b-a}{n},$$

holds for each n. Thus, $I^*(f) - I_*(f) = 0$ and so f is Riemann integrable.

An alternate proof goes as follows: According to Problem 9.8 the set of discontinuities of f is at-most countable—and hence, it has Lebesgue measure zero. Now, Theorem 23.7 guarantees that f is Riemann integrable. (See also Problem 21.8.)

Problem 23.6 (The Fundamental Theorem of Calculus). *If* $f : [a, b] \rightarrow \mathbf{R}$ *is a Riemann integrable function, then define its* **area function** $A : [a, b] \rightarrow \mathbf{R}$ *by* $A(x) = \int_a^x f(t) \, dt$ *for each* $x \in [a, b]$. *Show that*
  a. A *is a uniformly continuous function.*
  b. *If* f *is continuous at some point* c *of* $[a, b]$, *then* A *is differentiable at* c *and* $A'(c) = f(c)$ *holds.*
  c. *Give an example of a Riemann integrable function* f *whose area function* A *is differentiable and satisfies* $A' \ne f$.

Solution. (a) Choose some $M > 0$ with $|f(x)| \le M$ for each x in $[a, b]$. The uniform continuity of A follows from the inequalities

$$\left| A(x) - A(y) \right| = \left| \int_x^y f(t) \, dt \right| \le \left| \int_x^y |f(t)| \, dt \right| \le M |x - y|.$$

(b) Let f be continuous at some point $c \in [a, b]$ and let $\varepsilon > 0$. Choose some $\delta > 0$ such that $|f(x) - f(c)| < \varepsilon$ holds whenever $x \in [a, b]$ satisfies $|x - c| < \delta$. Then, for $x \in [a, b]$ with $0 < |x - c| < \delta$ we have $\left| f(t) - f(c) \right| < \epsilon$ for all t in the interval with endpoints x and c, and so

$$
\begin{aligned}
\left| \frac{A(x) - A(c)}{x - c} - f(c) \right| &= \frac{1}{|x - c|} \left| \int_x^c f(t)\,dt - f(c)(x - c) \right| \\
&= \frac{1}{|x-c|} \left| \int_x^c [f(t) - f(c)]\,dt \right| \\
&\leq \frac{1}{|x-c|} \cdot \varepsilon |x - c| = \varepsilon.
\end{aligned}
$$

This shows that $A'(c)$ exists and that $A'(c) = f(c)$ holds.

(c) We consider the function $f: \mathbb{R} \to \mathbb{R}$ of Problem 9.7 defined by $f(x) = 0$ if x is irrational and $f(x) = \frac{1}{n}$ if $x = \frac{m}{n}$ with $n > 0$ and with the integers m and n without having any common factors other than ± 1. It was proven in Problem 9.7 that f is continuous at every irrational and discontinuous at every rational. This implies that f restricted on an arbitrary closed interval $[c, d]$ is continuous almost everywhere and $f = 0$ a.e. From Theorems 23.6 and 23.7, we infer that f is Riemann integrable over $[c, d]$ and $\int_c^d f(x)\,dx = \int f\,d\lambda = 0$.

In particular, if $[a, b]$ is any closed interval, then $A(x) = \int_a^x f(t)\,dt = 0$ for each $x \in [a, b]$. Thus, $A'(x) = 0$ for each $x \in [a, b]$, and consequently $A'(x) \neq f(x)$ at each rational number x in $[a, b]$.

Problem 23.7 (Arzelà). *Let $\{f_n\}$ be a sequence of Riemann integrable functions on $[a, b]$ such that $\lim f_n(x) = f(x)$ holds for each $x \in [a, b]$ and f is Riemann integrable. Also, assume that there exists a constant M such that $|f_n(x)| \leq M$ holds for all $x \in [a, b]$ and all n. Show that*

$$
\lim_{n \to \infty} \int_a^b f_n(x)\,dx = \int_a^b f(x)\,dx.
$$

Solution. Using Theorem 23.6 and the Lebesgue Dominated Convergence Theorem, we see that

$$
\int_a^b f(x)\,dx = \int_{[a,b]} f\,d\lambda = \lim_{n \to \infty} \int_{[a,b]} f_n\,d\lambda = \lim_{n \to \infty} \int_a^b f_n(x)\,dx.
$$

Problem 23.8. *Determine the lower and upper Riemann integrals for the function $f: [0, 1] \longrightarrow \mathbb{R}$ defined by $f(x) = 0$ if x is a rational number and $f(x) = 1$ if x is an irrational number.*

Solution. Let P be a partition of $[0, 1]$. Since each interval contains rational and irrational numbers, we have $m_i = 0$ and $M_i = 1$ for all i. Thus, $S^*(f, P) = 1$ and $S_*(f, P) = 0$ for all partitions P. Therefore, $I^*(f) = 1$ and $I_*(f) = 0$.

Problem 23.9. *Let C be the Cantor set (see Example 6.15). Show that χ_C is Riemann integrable over $[0, 1]$, and that $\int_0^1 \chi_C\, dx = 0$.*

Solution. Note that χ_C is continuous at every point of $[0, 1] \setminus C$ and discontinuous at every point of C. Since $\lambda(C) = 0$, it follows from Theorem 23.7 that χ_C is Riemann integrable over $[0, 1]$. Since $\chi_C = 0$ a.e. holds, we see that

$$\int_0^1 \chi_C(x)\, dx = \int_{[0,1]} \chi_C\, d\lambda = 0.$$

Problem 23.10. *Let $0 < \epsilon < 1$, and consider the ϵ-Cantor set C_ϵ of $[0, 1]$. Show that χ_{C_ϵ} is not Riemann integrable over $[0, 1]$. Also, determine $I_*(\chi_{C_\epsilon})$ and $I^*(\chi_{C_\epsilon})$.*

Solution. Consider the ε-Cantor set for some $0 < \varepsilon < 1$. Since C_ε is nowhere dense in $[0, 1]$, it is easy to see that χ_{C_ε} is discontinuous at every point of C_ε and continuous at every point of $[0, 1] \setminus C_\varepsilon$. Since $\lambda(C_\varepsilon) = \varepsilon > 0$, it follows from Theorem 23.7 that χ_{C_ε} is not Riemann integrable.

Now, let $P = \{x_0, x_1, \ldots, x_n\}$ be a partition of $[0, 1]$. Since C_ε is nowhere dense, it follows that $m_i = 0$ for each $1 \leq i \leq n$. Thus, $S_*(\chi_{C_\varepsilon}, P) = 0$ for each partition P, and so $I_*(\chi_{C_\varepsilon}) = 0$. Clearly, $M_i = 1$ if $[x_{i-1}, x_i] \cap C_\varepsilon \neq \emptyset$, and $M_i = 0$ if $[x_{i-1}, x_i] \cap C_\varepsilon = \emptyset$. Since $C_\varepsilon = \bigcup_{i=1}^n [x_{i-1}, x_i] \cap C_\varepsilon$, it follows that

$$\varepsilon = \lambda(C_\varepsilon) \leq \sum_{i=1}^n \lambda\big([x_{i-1}, x_i] \cap C_\varepsilon\big) \leq \sum_{i=1}^n M_i(x_i - x_{i-1}) = S^*(\chi_{C_\varepsilon}, P),$$

and so $I^*(\chi_{C_\varepsilon}) \geq \varepsilon$.

On the other hand, if $0 < \delta < 1 - \varepsilon$, then there exist pairwise disjoint open subintervals $(a_1, b_1), \ldots, (a_n, b_n)$ such that

$$[a_i, b_i] \subseteq [0, 1] \setminus C_\varepsilon \ (1 \leq i \leq n) \quad \text{and} \quad -\sum_{i=1}^n (b_i - a_i) > 1 - \varepsilon - \delta.$$

The endpoints of all these subintervals together with 0 and 1 form a partition P

of $[0, 1]$ such that

$$\varepsilon \le I^*(\chi_{C_\varepsilon}) \le S^*(\chi_{C_\varepsilon}, P) \le 1 - (1 - \varepsilon - \delta) = \varepsilon + \delta.$$

Since $0 < \delta < 1 - \varepsilon$ is arbitrary, it easily follows that $I^*(\chi_{C_\varepsilon}) = \varepsilon$.

Problem 23.11.   *Give a proof of the Riemann integrability of a continuous function based upon its uniform continuity (Theorem 7.7).*

Solution.   Let $\varepsilon > 0$. Since (by Theorem 7.7) f is uniformly continuous, there is some $\delta > 0$ such that x, $y \in [a, b]$ and $|x - y| < \delta$ imply $|f(x) - f(y)| < \varepsilon$. Let P be a partition of $[a, b]$ with mesh $|P| < \delta$. Then, $M_i - m_i < \varepsilon$ holds for each $1 \le i \le n$ (why?), and so

$$0 \le I^*(f) - I_*(f) \le \sum_{i=1}^n (M_i - m_i)(x_i - x_{i-1}) < \varepsilon(b - a).$$

Since $\varepsilon > 0$ is arbitrary, we see that $I^*(f) = I_*(f)$ holds, and therefore f is Riemann integrable.

Problem 23.12.   *Establish the following change of variable formula for the Riemann integral of continuous functions: If $[a, b] \xrightarrow{g} [c, d] \xrightarrow{f} \mathbf{R}$ are continuous functions with g continuously differentiable (i.e., g has a continuous derivative), then*

$$\int_a^b f(g(x))g'(x)\, dx = \int_{g(a)}^{g(b)} f(u)\, du.$$

Solution.   We shall apply the Fundamental Theorem of Calculus in connection with the Chain Rule. We consider the two functions $F, G: [a, b] \to \mathbf{R}$ defined by

$$F(x) = \int_{g(a)}^{g(x)} f(u)\, du \quad \text{and} \quad G(x) = \int_a^x f(g(x))g'(x)\, dx$$

for all $x \in [a, b]$. Next, we shall compute the derivatives of F and G separately. For the derivative of F we use the Fundamental Theorem of Calculus and the Chain Rule to get $F'(x) = f(g(x))g'(x)$ for each $x \in [a, b]$. For the derivative of G, the Fundamental Theorem of Calculus yields $G'(x) = f(g(x))g'(x)$ for each $x \in [a, b]$. So, $F'(x) = G'(x)$ for all $x \in [a, b]$.

   The latter implies that there exists a constant c such that $F(x) = G(x) + c$ for all $x \in [a, b]$. Letting $x = a$ and taking into account that $F(a) = G(a) = 0$, we get $c = 0$. Thus, $F(x) = G(x)$ for all $x \in [a, b]$. Finally, letting $x = b$, we obtain

$$\int_a^b f\big(g(x)\big)g'(x)\,dx = \int_{g(a)}^{g(b)} f(u)\,du,$$

as desired.

Problem 23.13. *Let* $f : [0, \infty) \to \mathbb{R}$ *be a continuous function such that* $\lim_{x\to\infty} f(x) = \delta$. *Show that* $\lim_{n\to\infty} \int_0^a f(nx)\,dx = a\delta$ *for each* $a > 0$.

Solution. Fix $a > 0$ and then define the sequence of continuous functions $\{f_n\}$ by $f_n(x) = f(nx)$. Clearly, $\lim_{n\to\infty} f_n(x) = \delta$ holds for all $x \in (0, a]$. We claim that the sequence of functions $\{f_n\}$ is uniformly bounded on the interval $[0, a]$. Indeed, since $\lim_{x\to\infty} f(x) = \delta$ holds, there exists a number $M > 0$ such that $|f(x)| < |\delta| + 1$ whenever $x > M$. Also, since f is a continuous function, it is bounded on the interval $[0, M]$. Thus, there exists a constant C such that $|f(x)| \le C$ holds for all x, and hence $|f_n(x)| = |f(nx)| \le C$ holds for all x. Now, an application of the Lebesgue Dominated Convergence yields

$$\lim_{n\to\infty} \int_0^a f(nx)\,dx = \lim_{n\to\infty} \int_0^a f_n(x)\,dx = \int_0^a \delta\,dx = a\delta.$$

Problem 23.14. *Let* $f : [0, \infty) \to \mathbb{R}$ *be a real-valued continuous function such that* $f(x + 1) = f(x)$ *for all* $x \ge 0$. *If* $g : [0, 1] \to \mathbb{R}$ *is an arbitrary continuous function, then show that*

$$\lim_{n\to\infty} \int_0^1 g(x)f(nx)\,dx = \left(\int_0^1 g(x)\,dx\right) \cdot \left(\int_0^1 f(x)\,dx\right).$$

Solution. Let the functions f and g satisfy the hypotheses of the problem. Observe first that an easy inductive argument establishes that $f(x + k) = f(x)$ holds for all $x \ge 0$ and all non-negative integers k.

The change of variable $u = nx$ yields

$$\int_0^1 g(x)f(nx)\,dx = \frac{1}{n} \int_0^n g(\tfrac{u}{n})f(u)\,du$$

$$= \frac{1}{n} \sum_{i=1}^n \int_{i-1}^i g(\tfrac{u}{n})f(u)\,du.$$

Letting $t = u - i + 1$, we get

$$\int_{i-1}^{i} g(\tfrac{u}{n}) f(u)\, du = \int_0^1 g(\tfrac{t+i-1}{n}) f(t+i-1)\, dt = \int_0^1 g(\tfrac{t+i-1}{n}) f(t)\, dt.$$

Consequently,

$$\int_0^1 g(x) f(nx)\, dx = \int_0^1 \Big[\sum_{i=1}^n \tfrac{1}{n} g(\tfrac{t+i-1}{n})\Big] f(t)\, dt = \int_0^1 h_n(t)\, dt, \qquad (\star)$$

where $h_n(t) = \big[\sum_{i=1}^n \tfrac{1}{n} g(\tfrac{t+i-1}{n})\big] f(t)$. Clearly, h_n is a continuous function defined on $[0, 1]$. In addition, note that if $|g(x)| \le K$ and $|f(x)| \le K$ hold for each $x \in [0, 1]$, then

$$|h_n(t)| \le K^2 \quad \text{for all} \quad t \in [0, 1],$$

i.e., the sequence $\{h_n\}$ is uniformly bounded on $[0, 1]$. Next, note that $0 \le t \le 1$ implies $\frac{i-1}{n} \le \frac{t+i-1}{n} \le \frac{i}{n}$. Thus, if m_i^n and M_i^n denote the minimum and maximum values of g, respectively, on the closed interval $[\frac{i-1}{n}, \frac{i}{n}]$, then

$$m_i^n \le g(\tfrac{t+i-1}{n}) \le M_i^n$$

holds for each $0 \le t \le 1$. Next, put

$$R_n = \sum_{i=1}^n \tfrac{1}{n} m_i^n \quad \text{and} \quad S_n = \sum_{i=1}^n \tfrac{1}{n} M_i^n,$$

and note that R_n and S_n are two Riemann sums—the smallest and largest ones, respectively—for the function g corresponding to the partition $\{0, \frac{1}{n}, \frac{2}{n}, \dots, \frac{n-1}{n}, 1\}$. Hence, $\lim_{n\to\infty} R_n = \lim_{n\to\infty} S_n = \int_0^1 g(x)\, dx$. From

$$\big|h_n(t) - R_n \cdot f(t)\big| = \Big|\Big[\sum_{i=1}^n \tfrac{1}{n} g(\tfrac{t+i-1}{n})\Big] f(t) - R_n \cdot f(t)\Big|$$

$$= \Big(\Big[\sum_{i=1}^n \tfrac{1}{n} g(\tfrac{t+i-1}{n})\Big] - R_n\Big) \cdot |f(t)|$$

$$\le (S_n - R_n)|f(t)|,$$

we see that $\lim_{n\to\infty} h_n(t) = f(t)\int_0^1 g(x)\,dx$—in fact, the sequence $\{h_n\}$ converges uniformly (why?).

Now, use $(\star)$ and the Lebesgue Dominated Convergence Theorem to obtain

$$
\begin{aligned}
\lim_{n\to\infty}\int_0^1 g(x)f(nx)\,dx &= \lim_{n\to\infty}\int_0^1 h_n(t)\,dt \\
&= \int_0^1 \Big[\lim_{n\to\infty} h_n(t)\Big]\,dt \\
&= \int_0^1 \Big[f(t)\int_0^1 g(x)\,dx\Big]\,dt \\
&= \Big(\int_0^1 f(t)\,dt\Big)\cdot\Big(\int_0^1 g(x)\,dx\Big).
\end{aligned}
$$

Problem 23.15. *Let $f:[0,1] \to [0,\infty)$ be Riemann integrable on every closed subinterval of $(0,1]$. Show that f is Lebesgue integrable over $[0,1]$ if and only if $\lim_{\epsilon\downarrow 0}\int_\epsilon^1 f(x)\,dx$ exists in $\mathbf{R}$. Also, show that if this is the case, then we have $\int f\,d\lambda = \lim_{\epsilon\downarrow 0}\int_\epsilon^1 f(x)\,dx$.*

Solution. Assume that f is Lebesgue integrable. Let $\{\varepsilon_n\}$ be an arbitrary sequence of $(0,1]$ with $\varepsilon_n \downarrow 0$. For each n, we consider the upper function $g_n = f\chi_{[\varepsilon_n,1]}$. Then, $g_n \uparrow f$ a.e. holds and so, by Theorem 21.6, we have

$$
\int f\,d\lambda = \lim_{n\to\infty}\int g_n\,d\lambda = \lim_{n\to\infty}\int_{\varepsilon_n}^1 f(x)\,dx.
$$

This shows that $\lim_{\varepsilon\downarrow 0}\int_\varepsilon^1 f(x)\,dx$ exists and that

$$
\lim_{\varepsilon\downarrow 0}\int_\varepsilon^1 f(x)\,dx = \int f\,d\lambda.
$$

Conversely, assume that the limit exists. Let $\varepsilon_n = \frac{1}{n}$ and consider the sequence of upper functions $\{g_n\}$ as previously (i.e., let $g_n = f\chi_{[\varepsilon_n,1]}$). Then, $g_n \uparrow f$ a.e. and

$$
\lim_{n\to\infty}\int g_n\,d\lambda = \lim_{n\to\infty}\int_{\varepsilon_n}^1 f(x)\,dx = \lim_{\varepsilon\downarrow 0}\int_\varepsilon^1 f(x)\,dx < \infty.
$$

By Theorem 21.6, f is an upper function, and hence, Lebesgue integrable.

Problem 23.16.   *As an application of the preceding problem, show that the function $f: [0, 1] \to \mathbb{R}$ defined by $f(x) = x^p$ if $x \in (0, 1]$ and $f(0) = 0$ is Lebesgue integrable if and only if $p > -1$. Also, show that if f is Lebesgue integrable, then*

$$\int f \, d\lambda = \frac{1}{1+p}.$$

Solution.   If $0 < \varepsilon < 1$, then note that $\int_\varepsilon^1 x^p \, dx = \frac{1 - \varepsilon^{p+1}}{p+1}$ for $p \neq -1$ and $\int_\varepsilon^1 x^{-1} \, dx = -\ln \varepsilon$. Thus, $\lim_{\varepsilon \downarrow 0} \int_\varepsilon^1 x^p \, dx$ exists if and only if $p > -1$, and, in this case, the limit is $\frac{1}{p+1}$. The conclusion now follows immediately from the preceding problem.

Problem 23.17.   *Let $f: [0, 1] \to \mathbb{R}$ be a function and define $g: [0, 1] \to \mathbb{R}$ by $g(x) = e^{f(x)}$.*

   a.   *Show that if f is measurable (or Borel measurable), then so is g.*
   b.   *If f is Lebesgue integrable, is then g necessarily Lebesgue integrable?*
   c.   *Give an example of an essentially unbounded function f which is continuous on $(0, 1]$ such that f^n is Lebesgue integrable for each $n = 1, 2, \ldots$. (A function f is "essentially unbounded," if for each positive real number $M > 0$ the set $\{x \in [0, 1] : |f(x)| > M\}$ has positive measure.)*

Solution.   (a) Let $h(x) = e^x$ and note that $g = h \circ f$. The conclusion follows from the identity $(h \circ f)^{-1}(V) = f^{-1}\big(g^{-1}(V)\big)$ and the fact that h is a continuous function.

   (b) The measurable function g need not be necessarily Lebesgue integrable. Here is an example. Consider the function $f: [0, 1] \to \mathbb{R}$ defined by $f(x) = \frac{1}{\sqrt{x}}$; at $x = 0$ we let $f(0) = 0$. If $0 < \varepsilon < 1$, then the change of variable $t = \sqrt{x}$ yields

$$\int_\varepsilon^1 f(x) \, dx = \int_\varepsilon^1 \frac{dx}{\sqrt{x}} = 2 \int_{\sqrt{\varepsilon}}^1 dt = 2(1 - \sqrt{\varepsilon}).$$

Therefore, from Problem 23.16, we see that f is Lebesgue integrable and $\int f \, d\lambda = 2$. On the other hand, for each $0 < \epsilon < 1$ the change of variable $u = \frac{1}{\sqrt{x}}$ yields

$$\int_\varepsilon^1 g(x) \, d\lambda(x) = \int_\varepsilon^1 e^{\frac{1}{\sqrt{x}}} \, dx = 2 \int_1^{\frac{1}{\sqrt{\varepsilon}}} \frac{e^u}{u^3} \, du \geq 2 \int_1^{\frac{1}{\sqrt{\varepsilon}}} e^u \, du = 2\big(e^{\frac{1}{\sqrt{\varepsilon}}} - 1\big).$$

This implies $\lim_{\varepsilon \downarrow 0} \int_\varepsilon^1 e^{\frac{1}{\sqrt{x}}} \, dx = \infty$, and so by Problem 23.15 the function g is not Lebesgue integrable over $[0, 1]$.

(c) The function $f: [0, 1] \to \mathbf{R}$ defined by $f(x) = \ln x$ for $0 < x \leq 1$ and $f(0) = 0$ satisfies $\int_0^1 f^n(x) \, d\lambda(x) = (-1)^n n!$ for each $n = 1, 2, \ldots$.

Problem 23.18.   *Let $f: [0, 1] \to \mathbf{R}$ be Lebesgue integrable (with respect to the Lebesgue measure). Assume that f is differentiable at $x = 0$ and $f(0) = 0$. Show that the function $g: [0, 1] \to \mathbf{R}$ defined by $g(x) = x^{-\frac{3}{2}} f(x)$ for $x \in (0, 1]$ and $g(0) = 0$ is Lebesgue integrable.*

Solution.   Start by observing that (by Problem 23.16) the function $h(x) = x^{-\frac{1}{2}}$ for $x \in (0, 1]$ is Lebesgue integrable over $[0, 1]$. Since $f(0) = 0$ and $f'(0)$ exists, there exist $0 < \delta < 1$ and $M > 0$ such that $|f(x)| \leq Mx$ for all $0 \leq x \leq \delta$. Since for $\delta \leq x \leq 1$ we have $x^{-\frac{3}{2}} \leq \delta^{-\frac{3}{2}}$, we can assume that $M > \delta^{-\frac{3}{2}}$. Now, note that for $0 < x \leq 1$, we have

$$|g(x)| = \left| x^{-\frac{3}{2}} f(x) \right| \leq M \begin{cases} x^{-\frac{1}{2}} & \text{if } 0 < x \leq \delta \\ |f(x)| & \text{if } \delta < x \leq 1 \end{cases}$$

$$\leq M \big(h + |f| \big)(x).$$

Since $h + |f|$ is integrable and (obviously) g is measurable, Theorem 22.6 guarantees that g is also Lebesgue integrable.

Problem 23.19.   *Let $f: [a, b] \times [c, d] \to \mathbf{R}$ be a continuous function. Show that the Riemann integral of f can be computed with two iterated integrations. That is, show that*

$$\int_a^b \int_c^d f(x, y) \, dx \, dy = \int_a^b \left[\int_c^d f(x, y) \, dy \right] dx = \int_c^d \left[\int_a^b f(x, y) \, dx \right] dy.$$

Generalize this to a continuous function of n variables.

Solution.   Note first that the functions

$$x \longmapsto \int_c^d f(x, y) \, dy \quad \text{and} \quad y \longmapsto \int_a^b f(x, y) \, dx$$

are both continuous—and so both iterated integrals are well defined. Indeed, since the function f is uniformly continuous, given $\varepsilon > 0$ there exists some $\delta > 0$ such that $|x_1 - x_2| < \delta$ and $|y_1 - y_2| < \delta$ imply $|f(x_1, y_1) - f(x_2, y_2)| < \varepsilon$.

Thus, if $|x_1 - x_2| < \delta$ and $|y_1 - y_2| < \delta$ both hold, then

$$\left| \int_c^d f(x_1, y)\, dy - \int_c^d f(x_2, y)\, dy \right| < \varepsilon(c - d)$$

and

$$\left| \int_a^b f(x, y_1)\, dx - \int_a^b f(x, y_2)\, dx \right| < \varepsilon(b - a).$$

Let $P = \{x_0, x_1, \ldots, x_n\}$ be a partition of $[a, b]$ and $Q = \{y_0, y_1, \ldots, y_k\}$ be a partition of $[c, d]$. Put $R_{ij} = [x_{i-1}, x_i] \times [y_{j-1}, y_j]$, and then define

$$m_{ij} = \inf\{f(x, y): (x, y) \in R_{ij}\} \text{ and } M_{ij} = \sup\{f(x, y): (x, y) \in R_{ij}\}.$$

From the inequality $m_{ij} \leq f(x, y) \leq M_{ij}$ for $(x, y) \in R_{ij}$, it follows that

$$m_{ij}(y_j - y_{j-1}) \leq \int_{y_{j-1}}^{y_j} f(x, y)\, dy \leq M_{ij}(y_j - y_{j-1})$$

for all $x_{i-1} \leq x \leq x_i$, and so

$$m_{ij}(x_i - x_{i-1})(y_j - y_{j-1}) \leq \int_{x_{i-1}}^{x_i} \left(\int_{y_{j-1}}^{y_j} f(x, y)\, dy \right) dx$$

$$\leq M_{ij}(x_i - x_{i-1})(y_j - y_{j-1}).$$

Consequently, we have

$$S_*(f, P \times Q) = \sum_{i=1}^n \sum_{j=1}^k m_{ij}(x_i - x_{i-1})(y_j - y_{j-1})$$

$$\leq \sum_{i=1}^n \sum_{j=1}^k \int_{x_{i-1}}^{x_i} \left(\int_{y_{j-1}}^{y_j} f(x, y)\, dy \right) dx$$

$$= \int_a^b \left(\int_c^d f(x, y)\, dy \right) dx$$

$$\leq \sum_{i=1}^n \sum_{j=1}^k M_{ij}(x_i - x_{i-1})(y_j - y_{j-1})$$

$$= S^*(f, P \times Q).$$

Since P and Q are arbitrary and f is Riemann integrable, it follows that

$$\int_a^b \int_c^d f(x, y)\, dx\, dy = \int_a^b \left(\int_c^d f(x, y)\, dy \right) dx.$$

The other equality can be proven in a similar manner.

Problem 23.20. *Assume that $f: [a, b] \to \mathbf{R}$ and $g: [a, b] \to \mathbf{R}$ are two continuous functions such that $f(x) \le g(x)$ for each $x \in [a, b]$. Let*

$$A = \{(x, y) \in \mathbf{R}^2 \colon x \in [a, b] \text{ and } f(x) \le y \le g(x)\}.$$

a. *Show that A is a closed set—and hence, a measurable subset of $\mathbf{R}^2$.*
b. *If $h: A \to \mathbf{R}$ is a continuous function, then show that h is Lebesgue integrable over A and that*

$$\int_A h\, d\lambda = \int_a^b \left[\int_{f(x)}^{g(x)} h(x, y)\, dy \right] dx.$$

Solution.   (a) Let $\{(x_n, y_n)\}$ be a sequence of A such that $x_n \to x$ and $y_n \to y$. From the inequality $f(x_n) \le y_n \le g(x_n)$ and the continuity of f and g, it follows that $f(x) \le y \le g(x)$, i.e., $(x, y) \in A$. Thus, A is a closed set.
(b) Let $c < \inf\{f(x) \colon x \in [a, b]\}$ and $d > \sup\{g(x) \colon x \in [a, b]\}$. Consequently, $A \subseteq [a, b] \times [c, d] = E$. Extend h to E by $h(x, y) = 0$ if $(x, y) \notin A$, and note that the set of all discontinuities of h is a subset of

$$D = \{(x, y) \in \mathbf{R}^2 \colon a \le x \le b \text{ and } y = f(x) \text{ or } y = g(x)\}.$$

By Problem 18.17, $\lambda(D) = 0$, and so h is Riemann integrable on E (and hence, Lebesgue integrable). Now, by modifying the arguments of Problem 23.19, we easily see that

$$\int_A h\, d\lambda = \int_a^b \int_c^d h(x, y)\, dx\, dy = \int_a^b \left(\int_c^d h(x, y)\, dy \right) dx$$

$$= \int_a^b \left(\int_{f(x)}^{g(x)} h(x, y)\, dy \right) dx.$$

Problem 23.21.   *Let $f: [a, b] \to \mathbf{R}$ be a differentiable function—with one-sided derivatives at the end points. If the derivative f' is uniformly bounded on $[a, b]$,* *then show that f' is Lebesgue integrable and that*

$$\int_{[a,b]} f' \, d\lambda = f(b) - f(a).$$

Solution.   Let $f: \mathbf{R} \longrightarrow \mathbf{R}$ be a differentiable function such that for some $M > 0$ we have $|f'(x)| \le M$ for all $x \in [a, b]$. By letting $f(x) = f(a) + f'(a)(x - a)$ for $x < a$ and $f(x) = f(b) + f'(b)(x - b)$ for $x > b$, we can assume that f is defined (and is differentiable) on $\mathbf{R}$.

Next, consider the sequence of differentiable functions $\{f_n\}$ defined by

$$f_n(x) = n\left[f(x + \tfrac{1}{n}) - f(x)\right] = \frac{f(x + \tfrac{1}{n}) - f(x)}{\tfrac{1}{n}}, \quad x \in \mathbf{R},$$

and note that $f_n(x) \longrightarrow f'(x)$ holds for each $x \in \mathbf{R}$. Also, by the Mean Value Theorem, it is easy to see that $|f_n(x)| \le M$ holds for each x. Consequently, by the Lebesgue Dominated Convergence Theorem, f' is Lebesgue integrable over $[a, b]$ and

$$\int_{[a,b]} f' \, d\lambda = \lim_{n \to \infty} \int_a^b f_n(x) \, dx. \qquad (\star)$$

Now, using the change of variable $u = x + \tfrac{1}{n}$, we see that

$$\int_a^b f_n(x) \, dx = n\left[\int_a^b f(x + \tfrac{1}{n}) \, dx - \int_a^b f(x) \, dx\right]$$

$$= n\left[\int_{a+\frac{1}{n}}^{b+\frac{1}{n}} f(u) \, du - \int_a^b f(x) \, dx\right]$$

$$= n\left[\int_b^{b+\frac{1}{n}} f(x) \, dx - \int_a^{a+\frac{1}{n}} f(x) \, dx\right]$$

$$= \frac{\int_b^{b+\frac{1}{n}} f(x) \, dx}{\tfrac{1}{n}} - \frac{\int_a^{a+\frac{1}{n}} f(x) \, dx}{\tfrac{1}{n}} \longrightarrow f(b) - f(a),$$

where the last limit is justified by virtue of the Fundamental Theorem of Calculus. A glance at $(\star)$ guarantees that $\int_{[a,b]} f' \, d\lambda = f(b) - f(a)$, and we are finished.

Problem 23.22. *Let* $f, g: [a, b] \to \mathbb{R}$ *be two Lebesgue integrable functions satisfying*

$$\int_a^x f(t)\,d\lambda(t) \leq \int_a^x g(t)\,d\lambda(t)$$

for all $x \in [a, b]$. *If* $\phi: [a, b] \to \mathbb{R}$ *is a non-negative decreasing function, then show that the functions* ϕf *and* ϕg *are both Lebesgue integrable over* $[a, b]$ *and that they satisfy*

$$\int_a^x \phi(t)f(t)\,d\lambda(t) \leq \int_a^x \phi(t)g(t)\,d\lambda(t)$$

for all $x \in [a, b]$.

Solution. Since ϕ is decreasing there exists some $M > 0$ satisfying $|\phi(t)| \leq M$ for each $t \in [a, b]$. Since f and g are Lebesgue integrable, it follows from the inequalities $|\phi(t)f(t)| \leq M|f(t)|$ and $|\phi(t)g(t)| \leq M|g(t)|$ for each $t \in [a, b]$ that ϕf and ϕg are both Lebesgue integrable functions over $[a, b]$.

To obtain the desired inequality, fix $x \in [a, b]$. Assume first that ϕ is a non-negative decreasing function of the form $\phi = \sum_{i=1}^k c_i \chi_{[a_{i-1}, a_i)}$, where $\{a = a_0 < a_1 < \cdots < a_k = b\}$ is a partition of $[a, b]$. Since f is decreasing, we know that $c_1 \geq c_2 \geq \cdots \geq c_k \geq 0$. Clearly,

$$\phi = (c_1 - c_2)\chi_{[a,a_1)} + (c_2 - c_3)\chi_{[a,a_2)} + \cdots + (c_{k-1} - c_k)\chi_{[a,a_{k-1})} + c_k\chi_{[a,b)}$$

$$= \sum_{i=1}^k \gamma_i \chi_{[a,a_i)},$$

with $\gamma_i \geq 0$ for each i. Pick $1 \leq m \leq k$ such that $a_{m-1} \leq x < a_m$, and note that

$$\int_a^x \phi(t)f(t)\,d\lambda(t) = \sum_{i=1}^{m-1} \gamma_i \int_a^{a_i} f(t)\,d\lambda(t) + \gamma_m \int_a^x f(t)\,d\lambda(t)$$

$$\leq \sum_{i=1}^{m-1} \gamma_i \int_a^{a_i} g(t)\,d\lambda(t) + \gamma_m \int_a^x g(t)\,d\lambda(t)$$

$$= \int_a^x \phi(t)g(t)\,d\lambda(t).$$

Now, we consider the general case. Fix $x \in [a, b)$. As in the solution of Problem 21.8, we see that there exists a sequence $\{\phi_n\}$ of non-negative decreasing step functions (as above) satisfying $\phi_n(t) \uparrow \phi(t)$ for almost all $t \in [a, b]$. Since $|\phi_n(t)f(t)| \le M|f(t)|$, $|\phi_n(t)g(t)| \le M|g(t)|$, $\phi_n(t)f(t) \to \phi(t)f(t)$, and since $\phi_n(t)g(t) \to \phi(t)g(t)$ for almost all $t \in [a, b]$, it follows from the inequality

$$\int_a^x \phi_n(t)f(t)\,d\lambda(t) \le \int_a^x \phi_n(t)g(t)\,d\lambda(t)$$

and the Lebesgue Dominated Convergence Theorem that

$$\int_a^x \phi(t)f(t)\,d\lambda(t) = \lim_{n\to\infty} \int_a^x \phi_n(t)f(t)\,d\lambda(t)$$

$$\le \lim_{n\to\infty} \int_a^x \phi_n(t)g(t)\,d\lambda(t) = \int_a^x \phi(t)g(t)\,d\lambda(t).$$

24. APPLICATIONS OF THE LEBESGUE INTEGRAL

Problem 24.1. *Show that*

$$\int_0^\infty x^{2n} e^{-x^2}\,dx = \frac{(2n)!}{2^{2n}n!} \cdot \frac{\sqrt{\pi}}{2}$$

holds for $n = 0, 1, 2, \ldots$.

Solution. We shall establish the formula by induction on n. For $n = 0$ the formula is true by virtue of Theorem 24.6. If the formula is true for some $n \ge 0$, then an integration by parts yields

$$\int_0^r x^{2(n+1)} e^{-x^2}\,dx = -\tfrac{1}{2}\int_0^r x^{2n+1}\,d\!\left(e^{-x^2}\right)$$

$$= -\tfrac{1}{2} r^{2n+1} e^{-r^2} + \tfrac{2n+1}{2}\int_0^r x^{2n} e^{-x^2}\,dx$$

for each $r > 0$. This implies

$$\int_0^\infty x^{2(n+1)} e^{-x^2}\,dx = \lim_{r\to\infty}\int_0^r x^{2(n+1)} e^{-x^2}\,dx = \tfrac{2n+1}{2}\int_0^\infty x^{2n} e^{-x^2}\,dx$$

$$= \frac{(2n+1)(2n+2)}{2^2(n+1)} \cdot \frac{(2n)!}{2^{2n}n!} \cdot \frac{\sqrt{\pi}}{2}$$

$$= \frac{[2(n+1)]!}{2^{2(n+1)}(n+1)!} \cdot \frac{\sqrt{\pi}}{2}.$$

Problem 24.2. *Show that $\int_0^\infty e^{-tx^2}\,dx = \frac{1}{2}\sqrt{\frac{\pi}{t}}$ for each $t > 0$.*

Solution. Let $u = x\sqrt{t}$. Then, $\int_0^r e^{-tx^2}\,dx = \frac{1}{\sqrt{t}}\int_0^{r\sqrt{t}} e^{-u^2}\,du$ holds for each $r > 0$. Therefore,

$$\int_0^\infty e^{-tx^2}\,dx = \lim_{r\to\infty}\int_0^r e^{-tx^2}\,dx = \frac{1}{\sqrt{t}}\lim_{r\to\infty}\int_0^{r\sqrt{t}} e^{-u^2}\,du$$
$$= \frac{1}{\sqrt{t}}\cdot\frac{\sqrt{\pi}}{2} = \frac{1}{2}\sqrt{\frac{\pi}{t}}.$$

Problem 24.3. *Show that $f(x) = \frac{\ln x}{x^2}$ is Lebesgue integrable over $[1,\infty)$ and that $\int f\,d\lambda = 1$.*

Solution. Since $\frac{\ln x}{x^2} \geq 0$ holds for each $x \geq 1$, it suffices (in view of Theorem 24.3) to show that $\int_1^\infty \frac{\ln x}{x^2}\,dx$ exists.

If $r > 1$, then an integration by parts yields

$$\int_1^r \frac{\ln x}{x^2}\,dx = -\int_1^r \ln x\,d\left(\frac{1}{x}\right) = -\frac{\ln x}{x}\Big|_1^r + \int_1^r \frac{1}{x^2}\,dx = 1 - \frac{1}{r} - \frac{\ln r}{r}.$$

Therefore,

$$\int f\,d\lambda = \int_1^\infty \frac{\ln x}{x^2}\,dx = \lim_{r\to\infty}\int_1^r \frac{\ln x}{x^2}\,dx = \lim_{r\to\infty}\left(1 - \frac{1}{r} - \frac{\ln r}{r}\right) = 1.$$

Problem 24.4. *Show that*

$$\lim_{n\to\omega}\int_0^n \left(1 + \frac{x}{n}\right)^n e^{-2x}\,dx = 1.$$

Solution. Note that

$$\int_0^\infty e^{-x}\,dx = \lim_{r\to\infty}\int_0^r e^{-x}\,dx = \lim_{r\to\infty}(1 - e^{-r}) = 1.$$

Therefore, the function e^{-x} is Lebesgue integrable over $[0,\infty)$. Now, let $g_n(x) = \left(1 + \frac{x}{n}\right)^n e^{-2x}\chi_{[0,n]}(x)$, and note that each g_n is Lebesgue integrable over $[0,\infty)$.

From elementary calculus, we know that $\left(1 + \frac{x}{n}\right)^n \uparrow e^x$ for each $x \geq 0$, and so $g_n(x) \uparrow e^{-x}$ holds for each $x \geq 0$. Thus,

$$\lim_{n \to \infty} \int_0^n \left(1 + \tfrac{x}{n}\right)^n e^{-x}\, dx = \lim_{n \to \infty} \int g_n\, d\lambda = \int_0^\infty e^{-x}\, dx = 1.$$

Problem 24.5. *Let* $f : [0, \infty) \to (0, \infty)$ *be a continuous, decreasing, and Lebesgue integrable function. Show that*

$$\lim_{x \to \infty} \frac{1}{f(x)} \int_x^\infty f(s)\, ds = 0 \quad \text{if and only if} \quad \lim_{x \to \infty} \frac{f(x + t)}{f(x)} = 0 \ \text{for each } t > 0.$$

Solution. Assume that $\lim_{x \to \infty} \frac{1}{f(x)} \int_x^\infty f(s)\, ds = 0$ and let $t > 0$ be fixed. Since f is decreasing, we see that $f(x + t) \leq f(s)$ for all $x \leq s \leq x + t$, and so

$$t f(x + t) = \int_x^{x+t} f(x + t)\, ds \leq \int_x^{x+t} f(s)\, ds.$$

Consequently, we have

$$0 < \frac{f(x+t)}{f(x)} \leq \frac{1}{t} \cdot \frac{\int_x^{x+t} f(s)\, ds}{f(x)} \leq \frac{1}{t} \cdot \frac{\int_x^\infty f(s)\, ds}{f(x)},$$

from which it follows that $\lim_{x \to \infty} \frac{f(x+t)}{f(x)} = 0$. For the converse, assume that $\lim_{x \to \infty} \frac{f(x+t)}{f(x)} = 0$ holds for each fixed t, and, for simplicity, let us write $F(x) = \frac{1}{f(x)} \int_x^\infty f(s)\, ds$ for each $x \in [0, \infty)$. Fix $\varepsilon > 0$ and then choose some $0 < \delta < 1$ such that $\frac{\delta}{1-\delta} < \varepsilon$. (Since $\lim_{\delta \to 0^+} \frac{\delta}{1-\delta} = 0$ such a δ always exists.) From $\lim_{x \to \infty} \frac{f(x+\delta)}{f(x)} = 0$, we infer that there exists some $M > 0$ such that $\frac{f(x+\delta)}{f(x)} < \delta$ holds for all $x > M$. That is, $f(x + \delta) \leq \delta f(x)$ holds for each $x > M$. Now, note that for $x > M$, we have

$$F(x) = \tfrac{1}{f(x)} \int_x^{x+\delta} f(s)\, ds + \tfrac{1}{f(x)} \int_{x+\delta}^\infty f(s)\, ds$$

$$= \tfrac{1}{f(x)} \int_x^{x+\delta} f(s)\, ds + \tfrac{1}{f(x)} \int_x^\infty f(u + \delta)\, du$$

$$\leq \tfrac{1}{f(x)} \int_x^{x+\delta} f(s)\, ds + \tfrac{1}{f(x)} \int_x^\infty \delta f(u)\, du$$

$$= \tfrac{1}{f(x)} \int_x^{x+\delta} f(s)\, ds + \delta F(x).$$

Consequently, if $x > M$, then

$$(1 - \delta)F(x) \le \tfrac{1}{f(x)} \int_x^{x+\delta} f(s)\,ds \le \tfrac{1}{f(x)} \int_x^{x+\delta} f(x)\,ds = \delta,$$

and so $0 < F(x) \le \frac{\delta}{1-\delta} < \varepsilon$ holds for all $x > M$. Thus, $\lim_{x \to \infty} F(x) = 0$.

Problem 24.6. *Show that the improper Riemann integrals*

$$\int_0^\infty \cos(x^2)\,dx \quad and \quad \int_0^\infty \sin(x^2)\,dx$$

*(which are known as the **Fresnel integrals**) both exist. Also, show that* $\cos(x^2)$ *and* $\sin(x^2)$ *are not Lebesgue integrable over* $[0, \infty)$.

Solution. We shall work with $\int_0^\infty \sin(x^2)\,dx$. Similar arguments will establish the corresponding result for $\int_0^\infty \cos(x^2)\,dx$.

Let $0 < s < t$. The substitution $u = x^2$ followed by an integration by parts gives

$$\left| \int_s^t \sin(x^2)\,dx \right| = \tfrac{1}{2} \left| \int_{s^2}^{t^2} \tfrac{\sin u}{\sqrt{u}}\,du \right| = \tfrac{1}{2} \left| \left[\tfrac{\cos u}{\sqrt{u}} \Big|_{s^2}^{t^2} - \int_{s^2}^{t^2} \cos u\,d\big(u^{-\frac{1}{2}}\big) \right] \right|$$

$$\le \tfrac{1}{2} \left[\tfrac{1}{s} + \tfrac{1}{t} + \int_{s^2}^{t^2} d\big(u^{-\frac{1}{2}}\big) \right] = \tfrac{1}{t}.$$

This inequality, combined with Theorem 24.1, guarantees the existence of the improper Riemann integral $\int_0^\infty \sin(x^2)\,dx$. The inequality

$$\int_{\sqrt{k\pi - \pi}}^{\sqrt{k\pi}} \left| \sin(x^2) \right|\,dx = \tfrac{1}{2} \int_{k\pi-\pi}^{k\pi} \tfrac{|\sin u|}{\sqrt{u}}\,du \ge \tfrac{1}{2\sqrt{\pi k}} \int_{k\pi-\pi}^{k\pi} |\sin x|\,dx = \tfrac{1}{\sqrt{\pi k}}$$

implies

$$\int_0^{\sqrt{n\pi}} \left| \sin(x^2) \right|\,dx = \sum_{k=1}^n \int_{\sqrt{k\pi-\pi}}^{\sqrt{k\pi}} \left| \sin(x^2) \right|\,dx \ge \tfrac{1}{\sqrt{\pi}} \sum_{k=1}^n \tfrac{1}{\sqrt{k}},$$

which shows that $\int_0^\infty \left| \sin(x^2) \right|\,dx$ does not exist in $\mathbb{R}$—and hence, that $\sin(x^2)$ is not Lebesgue integrable over $[0, \infty)$.

Problem 24.7. *Show that* $\int_0^\infty \frac{\sin^2 x}{x^2}\,dx = \frac{\pi}{2}$.

Solution.   Consider the function

$$f(x) = \begin{cases} 1 & \text{if } x = 0 \\ \frac{\sin^2 x}{x^2} & \text{if } 0 < x \leq 1 \\ \frac{1}{x^2} & \text{if } x > 1, \end{cases}$$

and note that f is Lebesgue integrable over $[0, \infty)$. In view of the inequality $0 \leq \frac{\sin^2 x}{x^2} \leq f(x)$, we see that the function $\frac{\sin^2 x}{x^2}$ is Lebesgue integrable over $[0, \infty)$.

Now, note that for each r, $\varepsilon > 0$, we have

$$\int_\varepsilon^r \tfrac{\sin^2 x}{x^2}\, dx = -\int_\varepsilon^r \sin^2 x\, d\left(\tfrac{1}{x}\right) = -\tfrac{\sin^2 x}{x}\Big|_\varepsilon^r + \int_\varepsilon^r \tfrac{2\sin x \cos x}{x}\, dx$$

$$= \tfrac{\sin^2 \varepsilon}{\varepsilon} - \tfrac{\sin^2 r}{r} + \int_{2\varepsilon}^{2r} \tfrac{\sin x}{x}\, dx.$$

Thus, by Theorem 24.8, we see that

$$\int_0^\infty \tfrac{\sin^2 x}{x^2}\, dx = \lim_{\substack{r \to \infty \\ \varepsilon \to 0^+}} \int_\varepsilon^r \tfrac{\sin^2 x}{x^2}\, dx = \int_0^\infty \tfrac{\sin x}{x}\, dx = \tfrac{\pi}{2}.$$

Problem 24.8.   *Let (X, S, μ) be an arbitrary measure space, T a metric space, and $f: X \times T \to \mathbf{R}$ a function. Assume that $f(\cdot, t)$ is a measurable function for each $t \in T$ and $f(x, \cdot)$ is a continuous function for each $x \in X$. Assume also that there exists an integrable function g such that for each $t \in T$ we have $|f(x, t)| \leq g(x)$ for almost all $x \in X$. Show that the function $F: T \to \mathbf{R}$, defined by*

$$F(t) = \int_X f(x, t)\, d\mu(x),$$

is a continuous function.

Solution.   Let $t_n \longrightarrow t$ in T. Define the function $g_n: X \longrightarrow \mathbf{R}$ by the formula $g_n(x) = f(x, t_n)$. By our assumptions each g_n is integrable, $|g_n| \leq g$ a.e., and $g_n(x) \longrightarrow f(x, t)$ holds for each $x \in X$. Thus, by the Lebesgue Dominated Convergence Theorem, we have

$$F(t_n) = \int f(x, t_n)\, d\mu(x) = \int g_n\, d\mu \longrightarrow \int f(x, t)\, d\mu(x) = F(t).$$

This shows that F is a continuous function.

Problem 24.9. *Show that*

$$\int_0^\infty \frac{e^{-x} - e^{-xt}}{x}\, dx = \ln t$$

holds for each t > 0.

Solution. Consider the function $f(x, t) = \frac{e^{-x} - e^{-xt}}{x}$ for $x > 0$ and $t > 0$. Observe that the value $f(0, t) = t - 1$ extends f continuously to the point $(0, t)$, $t > 0$. Next, note that the function $g(x, t)$, defined by

$$g(x, t) = \begin{cases} |f(x, t)| & \text{if } 0 \le x \le 1 \text{ and } t > 0 \\ e^{-x} + e^{-xt} & \text{if } x > 1 \text{ and } t > 0, \end{cases}$$

is Lebesgue integrable for each $t > 0$. Moreover, $|f(x, t)| \le g(x, t)$ holds. This implies that

$$F(t) = \int_0^\infty f(x, t)\, dx = \int_0^\infty \frac{e^{-x} - e^{-xt}}{x}\, dx$$

exists both as an improper Riemann integral and as a Lebesgue integral; see also Theorem 24.3.

Next, note that $\frac{\partial f}{\partial t}(x, t) = e^{-xt}$ holds for all $x > 0$ and all $t > 0$. The inequality $0 \le e^{-xt} \le e^{-xa}$ for all $t > a > 0$ and all $x \ge 0$, coupled with Theorem 24.5, shows that

$$F'(t) = \int_0^\infty \frac{\partial f}{\partial t}(x, t)\, dx = \int_0^\infty e^{-xt}\, dx = \frac{1}{t}$$

holds for all $t > 0$. Thus, $F(t) = \ln t + C$. Since $F(1) = 0$, it follows that $C = 0$ and so

$$F(t) = \int_0^\infty \frac{e^{-x} - e^{-xt}}{x}\, dx = \ln t.$$

Problem 24.10. *For each $t > 0$, let $F(t) = \int_0^\infty \frac{e^{-xt}}{1+x^2}\, dx$.*

    a. *Show that the integral exists as an improper Riemann integral and as a Lebesgue integral.*

    b. *Show that F has a second-order derivative and that $F''(t) + F(t) = \frac{1}{t}$ holds for each $t > 0$.*

Solution.   (a) The integrability of F follows from Theorem 24.3 and the inequality

$$\left|\frac{e^{-xt}}{1+x^2}\right| \le \frac{1}{1+x^2}.$$

(b) If $f(x,t) = \frac{e^{-xt}}{1+x^2}$, then

$$\frac{\partial f}{\partial t}(x,t) = \frac{-xe^{-xt}}{1+x^2} \quad \text{and} \quad \frac{\partial^2 f}{\partial t^2}(x,t) = \frac{x^2 e^{-xt}}{1+x^2}.$$

Since $\left|\frac{\partial f}{\partial t}(x,t)\right| \le e^{-xt}$ and $\left|\frac{\partial^2 f}{\partial t^2}(x,t)\right| \le e^{-xt}$ both hold, by applying Theorem 24.5 twice we get

$$F''(t) = \int_0^\infty \frac{\partial^2 f}{\partial t^2}(x,t)\,dx = \int_0^\infty \frac{x^2 e^{-xt}}{1+x^2}\,dx.$$

Consequently,

$$F''(t) + F(t) = \int_0^\infty \frac{x^2 e^{-xt}}{1+x^2}\,dx + \int_0^\infty \frac{e^{-xt}}{1+x^2}\,dx = \int_0^\infty e^{-xt}\,dx = \frac{1}{t}.$$

Problem 24.11.   *Show that the improper Riemann integral $\int_0^{\frac{\pi}{2}} \ln(t\cos x)\,dx$ exists for each $t > 0$ and that it is also a Lebesgue integral. Also, show that*

$$\int_0^{\frac{\pi}{2}} \ln(t\cos x)\,dx = \frac{\pi}{2}\ln\!\left(\frac{t}{2}\right)$$

holds for all $t > 0$.

Solution.   Let $f(x,t) = \ln(t\cos x)$ for $0 \le x < \frac{\pi}{2}$, $t > 0$, and let $g(x) = \left(\frac{\pi}{2} - x\right)^{-\frac{1}{2}}$ for $0 \le x < \frac{\pi}{2}$. An easy argument shows that the improper Riemann integral (and hence, the Lebesgue integral) of g exists over $[0, \frac{\pi}{2})$. Also, L'Hôpital's Rule shows that $\lim_{x\uparrow\frac{\pi}{2}}\left[\frac{f(x,t)}{g(x)}\right] = 0$. Thus, for each $t > 0$ there exists some $0 < x_0 < \frac{\pi}{2}$ such that $|f(x,t)| \le g(x)$ holds for all $x_0 < x < \frac{\pi}{2}$. Since $f(x,t)$ is continuous for $0 \le x < \frac{\pi}{2}$, an easy application of Theorem 22.6 guarantees that

$$F(t) = \int_0^{\frac{\pi}{2}} \ln(t\cos x)\,dx$$

exists both as a Lebesgue and as an improper Riemann integral. Next, note that $\frac{\partial f}{\partial t}(x, t) = \frac{1}{t}$, and that for $0 < a < t$ we have $\left|\frac{\partial f}{\partial t}(x, t)\right| \leq \frac{1}{a}$. Thus, by Theorem 24.5, we have

$$F'(t) = \int_0^{\frac{\pi}{2}} \frac{\partial f}{\partial t}(x, t)\, dx = \frac{\pi}{2t}$$

for each $t > 0$, and therefore $F(t) = \frac{\pi}{2} \ln t + C$.

Since $\int_0^{\frac{\pi}{2}} \ln(\cos x)\, dx = \int_0^{\frac{\pi}{2}} \ln(\sin x)\, dx$ (why?), it follows that

$$
\begin{aligned}
2C = 2F(1) &= 2\int_0^{\frac{\pi}{2}} \ln(\cos x)\, dx \\
&= \int_0^{\frac{\pi}{2}} \ln(\cos x)\, dx + \int_0^{\frac{\pi}{2}} \ln(\sin x)\, dx \\
&= \int_0^{\frac{\pi}{2}} \ln\left(\frac{\sin 2x}{2}\right) dx \\
&= \int_0^{\frac{\pi}{2}} \ln(\sin 2x)\, dx - \frac{\pi}{2} \ln 2 \\
&= \frac{1}{2} \int_0^{\pi} \ln(\sin x)\, dx - \frac{\pi}{2} \ln 2 \\
&= C - \frac{\pi}{2} \ln 2
\end{aligned}
$$

Thus, $C = -\frac{\pi}{2} \ln 2$, and so

$$\int_0^{\frac{\pi}{2}} \ln(t \cos x)\, dx = \frac{\pi}{2} \ln t - \frac{\pi}{2} \ln 2 = \frac{\pi}{2} \ln\left(\frac{t}{2}\right)$$

holds for each $t > 0$.

Problem 24.12. *Show that for each $t \geq 0$ the improper Riemann integral $\int_0^{\infty} \frac{\sin xt}{x(1+x^2)}\, dx$ exists as a Lebesgue integral and that*

$$\int_0^{\infty} \frac{\sin xt}{x(x^2 + 1)}\, dx = \frac{\pi}{2}(1 - e^{-t}).$$

Solution.   For each $t \in \mathbb{R}$ let $F(t) = \int_0^{\infty} \frac{\sin xt}{x(1+x^2)}\, dx$. From

$$\left|\frac{\sin xt}{x(1 + x^2)}\right| \leq \frac{|xt|}{|x(1 + x^2)|} = \frac{|t|}{1 + x^2},$$

we see that F is indeed a well defined real-valued function on $\mathbf{R}$ and that the integral defining F exists both as a Lebesgue integral and as an improper Riemann integral. Moreover, the relations

$$\frac{\partial}{\partial t}\left[\frac{\sin xt}{x(1+x^2)}\right] = \frac{\cos xt}{1+x^2} \quad \text{and} \quad \left|\frac{\cos xt}{1+x^2}\right| \le \frac{1}{1+x^2}$$

in connection with Theorem 24.5 guarantee that F is a differentiable function and

$$F'(t) = \int_0^\infty \frac{\partial}{\partial t}\left[\frac{\sin xt}{x(1+x^2)}\right] dx = \int_0^\infty \frac{\cos xt}{1+x^2}\, dx$$

holds for each $t \in \mathbf{R}$.

Since $\frac{\partial}{\partial t}\left[\frac{\cos xt}{1+x^2}\right] = -\frac{x\sin xt}{1+x^2}$ and the natural dominating function in the inequality $\left|-\frac{x\sin xt}{1+x^2}\right| \le \frac{|x|}{1+x^2}$ is not Lebesgue integrable over $[0, \infty)$, we cannot use Theorem 24.5 to conclude that

$$F''(t) = -\int_0^\infty \frac{x\sin xt}{1+x^2}\, dx. \qquad (\star)$$

As a matter of fact, the identity

$$\frac{x\sin xt}{1+x^2} = \frac{x^2\sin xt}{x(1+x^2)} = \frac{\sin xt}{x} - \frac{\sin xt}{x(1+x^2)} \qquad (\star\star)$$

shows that, on one hand, the function $x \mapsto \frac{x\sin xt}{1+x^2}$ is not Lebesgue integrable over $[0, \infty)$ for each $t > 0$ and, on the other hand, that

$$\int_0^\infty \frac{x\sin xt}{1+x^2}\, dx = \int_0^\infty \frac{\sin xt}{x}\, dx = -\int_0^\infty \frac{\sin xt}{x(1+x^2)}\, dx = \frac{\pi}{2} - F(t) \quad (\dagger)$$

for each $t > 0$.

We shall establish the validity of $(\star)$ for each $t > 0$ using another method. For each n, let

$$G_n(t) = \int_0^n \frac{\cos xt}{1+x^2}\, dx.$$

Clearly, $G_n(t) \to \int_0^\infty \frac{\cos xt}{1+x^2}\, dx = F'(t)$ for each $t \in \mathbf{R}$. Now, from Theorem 24.5

and (⋆⋆), we see that

$$G'_n(t) = -\int_0^n \frac{x \sin xt}{1 + x^2} \, dx = -\int_0^n \frac{\sin xt}{x} \, dx + \int_0^n \frac{\sin xt}{x(1 + x^2)} \, dx$$

$$= -\int_0^{nt} \frac{\sin x}{x} \, dx + \int_0^n \frac{\sin xt}{x(1 + x^2)} \, dx,$$

and consequently for each $t > 0$, we have

$$\lim_{n \to \infty} G'_n(t) = -\int_0^\infty \frac{\sin x}{x} \, dx + \int_0^\infty \frac{\sin xt}{x(1 + x^2)} \, dx = -\tfrac{\pi}{2} + F(t) = g(t).$$

We claim that for each $a > 0$ the sequence of derivatives $\{G'_n\}$ converges uniformly to the function $g(t) = -\tfrac{\pi}{2} + F(t)$ on the open interval (a, ∞). To see this, fix $a > 0$ and let $\epsilon > 0$. Choose some $x_0 > 1$ such that

$$\left| \int_s^\infty \frac{\sin x}{x} \, dx \right| < \epsilon \quad \text{and} \quad \left| \int_s^\infty \frac{\sin xt}{x(1 + x^2)} \, dx \right| \le \int_s^\infty \frac{dx}{1 + x^2} < \epsilon$$

for all $s > x_0$. Now, if we fix some natural number k satisfying $k \ge x_0$ and $ka > x_0$, then for each $n \ge k$ and all $t > a$, we have

$$\left| G'_n(t) - g(t) \right| = \left| \int_{nt}^\infty \frac{\sin x}{x} \, dx - \int_n^\infty \frac{\sin xt}{x(1 + x^2)} \, dx \right| < 2\epsilon.$$

This shows that $\{G'_n\}$ converges uniformly to the function $g(t) = -\tfrac{\pi}{2} + F(t)$.

Finally, using Problem 9.29, we get $F''(t) = \left[\lim_{n \to \infty} G_n(t) \right]' = -\tfrac{\pi}{2} + F(t)$, or $F''(t) - F(t) = -\tfrac{\pi}{2}$ for each $t > 0$. Solving the differential equation, we obtain

$$F(t) = \tfrac{\pi}{2} + c_1 e^{-t} + c_2 e^t \quad \text{for each } t > 0.$$

Since F and F' are continuous at zero (why?), it follows from $F(0) = 0$ and $F'(0) = \int_0^\infty \frac{dx}{1+x^2} = \tfrac{\pi}{2}$ and the preceding formula of $F(t)$ that $c_1 = -\tfrac{\pi}{2}$ and $c_2 = 0$. Hence, $F(t) = \tfrac{\pi}{2}\left(1 - e^{-t}\right)$ for each $t \ge 0$.

Problem 24.13. *The **Gamma function** for $t > 0$ is defined by an integral as follows:*

$$\Gamma(t) = \int_0^\infty x^{t-1} e^{-x} \, dx.$$

a. *Show that the integral*

$$\int_0^\infty x^{t-1}e^{-x}\,dx = \lim_{\substack{r\to\infty \\ \epsilon\to 0^+}} \int_\epsilon^r x^{t-1}e^{-x}\,dx$$

exists as an improper Riemann integral (and hence, as a Lebesgue integral).

b. *Show that $\Gamma(\frac{1}{2}) = \sqrt{\pi}$.*

c. *Show that $\Gamma(t+1) = t\Gamma(t)$ holds for all $t > 0$, and use this conclusion to establish $\Gamma(n+1) = n!$ for $n = 1, 2, \ldots$.*

d. *Show that Γ is differentiable at every $t > 0$ and that*

$$\Gamma'(t) = \int_0^\infty x^{t-1}e^{-x}\ln x\,dx$$

holds.

e. *Show that Γ has derivatives of all order and that*

$$\Gamma^{(n)}(t) = \int_0^\infty x^{t-1}e^{-x}(\ln x)^n\,dx$$

holds for $n = 1, 2, \ldots$ and all $t > 0$.

Solution. (a) Since $x^{t-1}e^{-x} \le x^{t-1}$ holds for $0 < x \le 1$, it follows from Problem 23.16 that $\int_0^1 x^{t-1}e^{-x}\,dx$ exists both as an improper Riemann integral and as a Lebesgue integral.

Now, for each fixed $t > 0$ we have $\lim_{x\to\infty} x^{t-1}e^{-\frac{x}{2}} = 0$. Thus, there exists some $M > 0$ (depending upon t) such that $0 \le x^{t-1}e^{-\frac{x}{2}} \le M$ holds for all $x \ge 1$. Hence, $x^{t-1}e^{-x} \le Me^{-\frac{x}{2}}$ holds for each $x \ge 1$. This shows that $\int_1^\infty x^{t-1}e^{-x}\,dx$ exists both as an improper Riemann integral and as a Lebesgue integral for each $t > 0$.

The preceding show that $\int_0^\infty x^{t-1}e^{-x}\,dx$ exists both as an improper Riemann integral and as a Lebesgue integral.

(b) Substitute $u = x^{\frac{1}{2}}$ to get

$$\Gamma(\tfrac{1}{2}) = \int_0^\infty x^{-\frac{1}{2}}e^{-x}\,dx = 2\int_0^\infty e^{-u^2}\,du = \sqrt{\pi}.$$

(c) Integrating by parts, we get

$$\begin{aligned}
\Gamma(t+1) &= \int_0^\infty x^t e^{-x}\,dx = -\int_0^\infty x^t\,d\!\left(e^{-x}\right) \\
&= -x^t e^{-x}\Big|_0^\infty + \int_0^\infty tx^{t-1}e^{-x}\,dx = t\int_0^\infty x^{t-1}e^{-x}\,dx \\
&= t\Gamma(t).
\end{aligned}$$

Consequently, we see that

$$\Gamma(n+1) = n!\Gamma(1) = n! \int_0^\infty e^{-x}\,dx = n!.$$

(d) and (e). Note that $\frac{\partial^n}{\partial t^n}\left(x^{t-1}e^{-x}\right) = x^{t-1}e^{-x}(\ln x)^n$ holds for all $t > 0$ and all $x > 0$.

Now, let $a < t < b$ be fixed and consider the continuous function $h(x,t) = \frac{\partial^n}{\partial t^n}(x^{t-1}e^{-x}) = x^{t-1}e^{-x}(\ln x)^n$, $a < t < b$, $x > 0$. We claim that there exists a Lebesgue integrable function $g:(0,\infty) \longrightarrow (0,\infty)$ such that $|h(x,t)| \le g(x)$ holds for all $x > 0$ and all $a < t < b$. If this is the case, then Theorem 24.5 allows us to "*differentiate under the integral sign*," and since $0 < a < b$ are arbitrary this shows that Γ must have derivatives of all orders and that the desired formulas hold. So, we must construct a positive Lebesgue integrable function g over $(0,\infty)$ such that $|h(x,t)| \le g(x)$ holds for each $a < t < b$ and each $x > 0$.

Note that for $x \ge 1$, we have $0 \le x^{t-1} \le x^b$. Using L'Hôpital's rule, we see that

$$\lim_{x\to\infty} x^b e^{-\frac{x}{2}}\left(\ln x\right)^n = \lim_{x\to\infty}\frac{x^b}{e^{\frac{x}{4}}} \cdot \lim_{x\to\infty}\frac{(\ln x)^n}{e^{\frac{x}{4}}} = 0 \cdot 0 = 0,$$

and so there exists some $M > 0$ such that $x^b e^{-\frac{x}{2}}(\ln x)^n \le M$ for all $x \ge 1$. Therefore,

$$\left|h(x,t)\right| \le \left|x^{t-1}e^{-x}\left(\ln x\right)^n\right| \le Me^{-\frac{x}{2}}$$

holds for all $x \ge 1$ and all $a < t < b$.

For the rest of our discussion, we shall need two facts from calculus.

$$\lim_{x\to 0^+} x^a\left(\ln x\right)^n = 0 \quad \text{and} \quad \int_{0^+}^1 x^{a-1}\left(\ln x\right)^n dx = \frac{(-1)^n n!}{a^{n+1}}.$$

Both can be proven by induction. For this limit use induction and L'Hôpital's rule by observing that

$$\lim_{x\to 0^+} x^a \ln x = \lim_{x\to 0^+}\frac{(\ln x)'}{(x^{-a})'} = \lim_{x\to 0^+} -\frac{x^a}{a} = 0$$

and

$$\lim_{x\to 0^+} x^a\left(\ln x\right)^{n+1} = \lim_{x\to 0^+}\frac{[(\ln x)^{n+1}]'}{(x^{-a})'} = \frac{n+1}{a}\lim_{x\to 0^+} x^a\left(\ln x\right)^n.$$

For the integral, use induction and take into account that

$$\int_{0+}^{1} x^{a-1} \ln x \, dx = \tfrac{1}{a} \int_{0+}^{1} \ln x \, d(x^a) = \tfrac{1}{a} x^a \ln x \Big|_{0+}^{1} - \tfrac{1}{a} \int_{0+}^{1} x^{a-1} \, dx = -\tfrac{1}{a^2}$$

and

$$\int_{0+}^{1} x^{a-1} \left(\ln x\right)^{n+1} dx = \tfrac{1}{a} \int_{0+}^{1} \left(\ln x\right)^{n+1} d(x^a)$$

$$= \tfrac{1}{a} x^a \left(\ln x\right)^{n+1} \Big|_{0+}^{1} - \tfrac{n+1}{a} \int_{0+}^{1} x^{a-1} \left(\ln x\right)^{n} dx$$

$$= -\tfrac{n+1}{a} \int_{0+}^{1} x^{a-1} \left(\ln x\right)^{n} dx.$$

Since either $(\ln x)^n \geq 0$ holds for all $x \in (0, 1]$ or $(\ln x)^n \leq 0$ holds for all $x \in (0, 1]$, it follows that the function $\phi(x) = x^{a-1}(\ln x)^n$, $x \in (0, 1]$, is Lebesgue integrable over $(0, 1]$. Now, let

$$g(x) = \begin{cases} x^{a-1} |\ln x|^n & \text{if } 0 < x \leq 1 \\ Me^{-\frac{x}{2}} & \text{if } x \geq 1 \end{cases},$$

and note that g is Lebesgue integrable over $(0, \infty)$. To finish the proof, notice that

$$\left| h(x, t) \right| \leq g(x)$$

holds for all $x > 0$ and all $a < t < b$.

Problem 24.14. *Let $f: [0, 1] \to \mathbf{R}$ be a Lebesgue integrable function and define the function $F: [0, 1] \to \mathbf{R}$ by $F(t) = \int_0^1 f(x) \sin(xt) \, d\lambda(x)$.*

a. *Show that the integral defining F exists and that F is a uniformly continuous function.*

b. *Show that F has derivatives of all orders and that*

$$F^{(2n)}(t) = (-1)^n \int_0^1 x^{2n} f(x) \sin(xt) \, d\lambda(x)$$

and

$$F^{(2n-1)}(t) = (-1)^n \int_0^1 x^{2n-1} f(x) \cos(xt) \, d\lambda(x)$$

for $n = 1, 2, \ldots$ and each $t \in [0, 1]$.

  c.   *Show that $F = 0$ (i.e., $F(t) = 0$ for all $t \in [0, 1]$) if and only if $f = 0$*
      a.e.

Solution.   (a) Note that for each fixed $t \in [0, 1]$ the function $x \mapsto \sin(xt)$
is continuous and hence, measurable. The inequality $|f(x)\sin(xt)| \le |f(x)|$
guarantees that $x \mapsto f(x)\sin(xt)$ is integrable for each $t \in [0, 1]$. So, F is a
well-defined function.
  For the uniform continuity of F note that

$$
\begin{aligned}
|F(t) - F(s)| &= \left| \int_0^1 f(x)\sin(xt)\, d\lambda(x) - \int_0^1 f(x)\sin(xs)\, d\lambda(x) \right| \\
&= \left| \int_0^1 f(x)\big[\sin(xt) - \sin(xs)\big]\, d\lambda(x) \right| \\
&\le \int_0^1 |f(x)|\,|\sin(xt) - \sin(xs)|\, d\lambda(x) \\
&\le \int_0^1 |f(x)|\,|xt - xs|\, d\lambda(x) \\
&= \left[\int_0^1 |f(x)|\, d\lambda(x) \right]|t - s|
\end{aligned}
$$

holds for all $s, t \in [0, 1]$.
  (b) Consider the function of two variables $h(x, t) = f(x)\sin(xt)$. Then an easy
inductive argument shows that for each $n = 1, 2, \ldots$ we have

$$
\frac{\partial^{2n} h(x,t)}{\partial t^{2n}} = (-1)^n x^{2n} f(x)\sin(xt) \quad \text{and} \quad \frac{\partial^{2n-1} h(x,t)}{\partial t^{2n-1}} = (-1)^n x^{2n-1} f(x)\cos(xt)
$$

for each $t \in [0, 1]$ and almost all x. This implies $\left| \frac{\partial^n h(x,t)}{\partial t^n} \right| \le |x^n f(x)| = g_n(x)$ for
all $t \in [0, 1]$ and almost all x. Since g_n is Lebesgue integrable for each n, it easily
follows from Theorem 24.5 that we can "differentiate under the integral sign" and
get the desired formulas.
  (c) Assume $F(t) = 0$ for each $t \in [0, 1]$. Then $F^{(2n)}(t) = 0$ for all n, and so
from (b) we get $\int_0^1 x^{2n} f(x)\sin(xt)\, d\lambda(x) = 0$ for each n and all $t \in [0, 1]$. Letting
$t = 1$, we get

$$
\int_0^1 x^{2n}\big[f(x)\sin x\big]\, d\lambda(x) = 0
$$

for each n. Now, invoke Problem 22.21 to conclude that $f(x)\sin x = 0$ for almost
all x. Since $\sin x > 0$ for each $0 < x \le 1$, we easily infer that $f(x) = 0$ for
almost all x.

25. APPROXIMATING INTEGRABLE FUNCTIONS

Problem 25.1.   *Let $f: \mathbb{R} \to \mathbb{R}$ be a Lebesgue integrable function. Show that*

$$\lim_{t \to \infty} \int f(x) \cos(xt) \, d\lambda(x) = \lim_{t \to \infty} \int f(x) \sin(xt) \, d\lambda(x) = 0.$$

Solution.   By Theorem 25.2, it suffices to establish the result for the special case $f = \chi_{[a,b)}$. So, let $f = \chi_{[a,b)}$, where $-\infty < a < b < \infty$. In this case, for each $t > 0$ we have

$$\left| \int f(x) \cos(xt) \, d\lambda(x) \right| = \left| \int_a^b \cos(xt) \, dx \right|$$
$$= \left| \tfrac{\sin(xt)}{t} \big|_{x=a}^{x=b} \right| = \left| \tfrac{\sin(bt) - \sin(at)}{t} \right| \le \tfrac{2}{t},$$

and so $\lim_{t \to \infty} \int f(x) \cos(xt) \, d\lambda(x) = 0$ holds. In a similar fashion, we can show that $\lim_{t \to \infty} \int f(x) \sin(xt) \, d\lambda(x) = 0$.

Problem 25.2.   *A function $f: \mathcal{O} \to \mathbb{R}$ (where $\mathcal{O}$ is a nonempty open subset of $\mathbb{R}^n$) is said to be a C^∞-**function** if f has continuous partial derivatives of all orders.*

   a. *Consider the function $\rho: \mathbb{R} \to \mathbb{R}$ defined by $\rho(x) = \exp[\frac{1}{x^2-1}]$ if $|x| < 1$ and $\rho(x) = 0$ if $|x| \ge 1$. Then show that ρ is a C^∞-function such that $\text{Supp} \, \rho = [-1, 1]$.*
   b. *For $\epsilon > 0$ and $a \in \mathbb{R}$ show that the function $f(x) = \rho(\frac{x-a}{\epsilon})$ is also a C^∞-function with $\text{Supp} \, f = [a - \epsilon, a + \epsilon]$.*

Solution.   (a) We shall establish that $\rho^{(n)}(1)$ exists for each n.
   Start by observing that, by L'Hôpital's Rule, $\lim_{t \to \infty} t^k e^{-\frac{1}{2}t} = 0$ holds for $k = 0, 1, 2, \dots$. Notice that if for $0 < x < 1$ we let $t = \frac{1}{1-x}$, then we have the inequality

$$\left| \tfrac{x^m e^{\frac{1}{x^2-1}}}{(x^2-1)^k(x-1)} \right| \le \left| \tfrac{e^{-\frac{1}{2(1-x)}}}{(1-x)^{k+1}} \right| = t^{k+1} e^{-\frac{1}{2}t},$$

from which it follows that

$$\lim_{x \uparrow 1} \tfrac{x^m e^{\frac{1}{x^2-1}}}{(x^2-1)^k(x-1)} = \lim_{t \uparrow \infty} t^{k+1} e^{-\frac{1}{2}t} = 0 \quad \text{for } k, m = 0, 1, 2, \dots . \qquad (\star)$$

   Now, by a simple induction argument, we see that for $-1 < x < 1$ the derivative $\rho^{(n)}(x)$ is a finite sum of terms of the form $\frac{x^m e^{\frac{1}{x^2-1}}}{(x^2-1)^k}$. Using $(\star)$ and another simple inductive argument, we can also see that $\rho^{(n)}(1) = 0$ holds for $n = 1, 2, \ldots$.

(b) Note that: $f(x) \neq 0$ if and only if $-1 < \frac{x-a}{\varepsilon} < 1$ if and only if $a - \varepsilon < x < a + \varepsilon$. Therefore, Supp $f = [a - \varepsilon, a + \varepsilon]$.

Problem 25.3. *Let $[a, b]$ be an interval, $\epsilon > 0$ such that $a + \epsilon < b - \epsilon$, and ρ as in the previous exercise. Define $h: \mathbf{R} \to \mathbf{R}$ by $h(x) = \int_a^b \rho(\frac{t-x}{\varepsilon}) dt$ for all $x \in \mathbf{R}$. Then show that*

   a.   Supp $h \subseteq [a - \epsilon, b + \epsilon]$,
   b. $h(x) = c$ *(a constant function) for all $x \in [a + \epsilon, b - \epsilon]$,*
   c. h *is a C^∞-function and $h^{(n)}(x) = \int_a^b \frac{\partial^n}{\partial x^n} \rho(\frac{t-x}{\varepsilon}) dt$ holds for all $x \in \mathbf{R}$, and*
   d.   *the C^∞-function $f = h/c$ satisfies $0 \leq f(x) \leq 1$ for all $x \in \mathbf{R}$, $f(x) = 1$ for all $x \in [a + \epsilon, b - \epsilon]$, and $\int |\chi_{[a,b)} - f| d\lambda < 4\epsilon$.*

Solution. For simplicity, let $g_x: \mathbf{R} \longrightarrow \mathbf{R}$ be the function defined by $g_x(t) = \rho(\frac{t-x}{\varepsilon})$, and so $h(x) = \int_a^b g_x(t) dt$.

(a) By part (b) of the preceding problem, we know that Supp $g_x = [x - \varepsilon, x + \varepsilon]$. Thus, if $a < t < b$ and $x \notin [a - \varepsilon, b + \varepsilon]$, then $g_x(t) = \rho(\frac{t-x}{\varepsilon}) = 0$ (since $\left|\frac{t-x}{\varepsilon}\right| \geq 1$). This implies that $h(x) = 0$ holds for all $x \notin [a - \varepsilon, b + \varepsilon]$, so that Supp $h \subseteq [a - \varepsilon, b + \varepsilon]$.

(b) If $a + \varepsilon < x < b - \varepsilon$, then Supp $g_x = [x - \varepsilon, x + \varepsilon]$ and so

$$h(x) = \int_a^b g_x(t) dt = \int_{x-\varepsilon}^{x+\varepsilon} \rho\left(\tfrac{t-x}{\varepsilon}\right) dt = \varepsilon \int_{-1}^1 \rho(u) du = c > 0.$$

(c) Since every partial derivative $\frac{\partial^n}{\partial x^n} \rho(\frac{t-x}{\varepsilon})$ is continuous, it must be bounded on $[a, b]$ (and hence, on $\mathbf{R}$). Now, the desired conclusion follows from Theorem 24.5.

(d) Since Supp $g_x = [x - \varepsilon, x + \varepsilon]$ and g_x is a positive function, it follows that

$$0 \leq h(x) = \int_a^b g_x(t) dt = \int_a^b \rho\left(\tfrac{t-x}{\varepsilon}\right) dt \leq \int_{x-\varepsilon}^{x+\varepsilon} \rho\left(\tfrac{t-x}{\varepsilon}\right) dt = c$$

holds for all x. Thus, $f = h/c$ satisfies $0 \leq f(x) \leq 1$ for all x.

Finally, observe that

$$\left|\chi_{(a,b)} - f\right| \leq \chi_{(a-\varepsilon,a+\varepsilon)} + \chi_{(b-\varepsilon,b+\varepsilon)}$$

holds, and so $\int |\chi_{[a,b)} - f| d\lambda = \int |\chi_{(a,b)} - f| d\lambda < 4\varepsilon$.

The graph of f is shown in Figure 4.3.

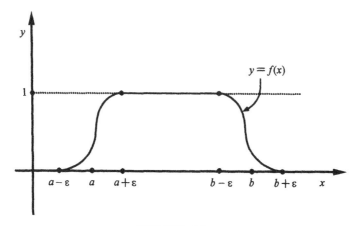

FIGURE 4.3.

Problem 25.4. *Let* $f: \mathbf{R} \to \mathbf{R}$ *be an integrable function with respect to the Lebesgue measure, and let* $\epsilon > 0$. *Show that there exists a* C^∞-*function g such that* $\int |f - g| \, d\lambda < \epsilon$.

Solution. Let $f: \mathbf{R} \longrightarrow \mathbf{R}$ be a Lebesgue integrable function and let $\varepsilon > 0$. By Theorem 25.2, there is a step function $\phi = \sum_{i=1}^n c_i \chi_{[a_i, b_i)}$ (with $c_i \neq 0$ for each i) such that $\int |f - \phi| \, d\lambda < \varepsilon$. Now, by the preceding problem, for each i there exists a C^∞-function g_i with compact support such that $\int |\chi_{[a_i, b_i)} - g_i| \, d\lambda < \frac{\varepsilon}{n|c_i|}$. Now, note that the C^∞-function $g = \sum_{i=1}^n c_i g_i$ has compact support and satisfies

$$\int |f - g| \, d\lambda \leq \int |f - \phi| \, d\lambda + \int |\phi - g| \, d\lambda$$

$$< \varepsilon + \int \left| \sum_{i=1}^n c_i \chi_{[a_i, b_i)} - \sum_{i=1}^n c_i g_i \right| \, d\lambda$$

$$\leq \varepsilon + \sum_{i=1}^n |c_i| \int |\chi_{[a_i, b_i)} - g_i| \, d\lambda$$

$$< \varepsilon + \sum_{i=1}^n |c_i| \frac{\varepsilon}{n|c_i|}$$

$$= \varepsilon + \varepsilon = 2\varepsilon.$$

Problem 25.5. *The purpose of this problem is to establish the following general result. If* $f: \mathbf{R}^n \to \mathbf{R}$ *is an integrable function (with respect to the*

Lebesgue measure) and $\epsilon > 0$, then there exists a C^∞-function g such that $\int |f - g| \, d\lambda < \epsilon$.

a. *Let $a_i < b_i$ for $i = 1, \ldots, n$, and let $I = \prod_{i=1}^{n}(a_i, b_i)$. Choose $\epsilon > 0$ such that $a_i + \epsilon < b_i - \epsilon$ for each i. Use Problem 25.3 to select for each i a C^∞-function $f_i: \mathbb{R} \to \mathbb{R}$ such that $0 \le f_i(x) \le 1$ for all x, $f_i(x) = 1$ if $x \in [a+\epsilon, b_i -\epsilon]$, and Supp $f_i \subseteq [a_i -\epsilon, b_i +\epsilon]$. Now, define $h: \mathbb{R}^n \to \mathbb{R}$ by $h(x_1, \ldots, x_n) = \prod_{i=1}^{n} f_i(x_i)$. Then show that h is a C^∞-function on $\mathbb{R}^n$ and that*

$$\int |\chi_I - h| \, d\lambda \le 2\left[\prod_{i=1}^{n}(b_i - a_i + 2\epsilon) - \prod_{i=1}^{n}(b_i - a_i) \right].$$

b. *Let $f: \mathbb{R}^n \to \mathbb{R}$ be Lebesgue integrable, and let $\epsilon > 0$. Then use part (a) to show that there exists a C^∞-function g with compact support such that*

$$\int |f - g| \, d\lambda < \epsilon.$$

Solution. (a) Clearly, h is a C^∞-function. Let $A = \prod_{i=1}^{n}(a_i - \varepsilon, b_i + \varepsilon)$, $B = \prod_{i=1}^{n}(a_i + \varepsilon, b_i + \varepsilon)$, and $C = \prod_{i=1}^{n}(a_i - \varepsilon, b_i - \varepsilon)$. Now, the desired conclusion follows from the inequality

$$|\chi_I - h| \le (\chi_A - \chi_B) + (\chi_A - \chi_C).$$

(b) Let f be an integrable function and let $\varepsilon > 0$. Pick a step function of the form $\phi = \sum_{i=1}^{k} c_i \chi_{I_i}$ (where each I_i is a finite open interval of $\mathbb{R}^n$) such that $\int |f - \phi| \, d\lambda < \varepsilon$. From part (a) it follows that there exists a C^∞-function g with compact support such that $\int |\phi - g| \, d\lambda < \varepsilon$. Consequently, $\int |f - g| \, d\lambda < 2\varepsilon$.

Problem 25.6. *Let μ be a regular Borel measure on $\mathbb{R}^n$, f a μ-integrable function, and $\epsilon > 0$. Show that there exists a C^∞-function g such that $\int |f-g| \, d\mu < \epsilon$.*

Solution. Let $I = \prod_{i=1}^{n}[a_i, b_i]$ be a finite closed interval. Given $\delta > 0$, pick $\varepsilon > 0$ such that the closed interval $J = \prod_{i=1}^{n}[a_i - 2\varepsilon, b_i + 2\varepsilon]$ satisfies $\mu(J \setminus I) = \mu(J) - \mu(I) < \delta$. (This is always possible since $\prod_{i=1}^{n}[a_i - \frac{1}{k}, b_i + \frac{1}{k}] \downarrow_k I$.) As in Problem 25.5, there exists a C^∞-function $h: \mathbb{R}^n \to \mathbb{R}$ such that $0 \le h(x) \le 1$ for all $x \in \mathbb{R}^n$, $h(x) = 1$ for $x \in I$, and Supp $h \subseteq J$. Therefore, if $f = \chi_I$, then

$$\int |\chi_I - h| \, d\mu = \int (h - \chi_I) \, d\mu \le \int (\chi_J - \chi_I) \, d\mu = \mu(J) - \mu(I) < \delta.$$

Thus, the desired result holds true for the characteristic functions of the finite closed intervals.

Now, let $I = \prod_{i=1}^{n}[a_i, b_i)$ be finite. Since $\prod_{i=1}^{n}[a_i, b_i - \frac{1}{k}] \uparrow_k I$, it follows that the approximation result is also true for the characteristic functions of sets of the form $\prod_{i=1}^{n}[a_i, b_i)$. Since these sets form a semiring and every open set is a σ-set (for this semiring), it is not difficult to see that the result is true for the characteristic functions of open sets of finite measure. The regularity of μ guarantees the validity of the approximation result for characteristic functions of μ-measurable sets of finite μ-measure. This in turn implies that the result holds true for μ-step functions. Finally, since for each μ-integrable function f and each $\varepsilon > 0$, there exists some μ-step function ϕ with $\int |f - \phi| \, d\mu < \varepsilon$, it follows that the C^{∞}-functions with compact support satisfy the desired approximation property.

Problem 25.7. *Let $f: [a, b] \to \mathbb{R}$ be a Lebesgue integrable function, and let $\epsilon > 0$. Show that there exists a polynomial p such that $\int |f - p| \, d\lambda < \epsilon$, where the integral is considered, of course, over $[a, b]$.*

Solution. Let $f: [a, b] \longrightarrow \mathbb{R}$ be integrable (over $[a, b]$), and let $\varepsilon > 0$. By Theorem 25.3 there exists a continuous function $g: [a, b] \longrightarrow \mathbb{R}$ such that $\int |f - g| \, d\lambda < \varepsilon$. Now, by Corollary 11.6, there exists a polynomial p such that $|g(x) - p(x)| < \varepsilon$ holds for all $x \in [a, b]$. Thus,

$$\int |f - p| \, d\lambda \le \int |f - g| \, d\lambda + \int |g - p| \, d\lambda < \varepsilon + \varepsilon(b - a) = \varepsilon(1 + b - a),$$

and our conclusion follows.

26. PRODUCT MEASURES AND ITERATED INTEGRALS

Problem 26.1. *Let $(X, \mathcal{S}, \mu)$ and (Y, Σ, ν) be two measure spaces, and assume that $A \times B \in \Lambda_\mu \otimes \Lambda_\nu$.*

   a. *Show that $\mu^*(A) \cdot \nu^*(B) \le (\mu \times \nu)^*(A \times B)$.*
   b. *Show that if $\mu^*(A) \cdot \nu^*(B) \ne 0$, then $(\mu \times \nu)^*(A \times B) = \mu^*(A) \cdot \nu^*(B)$.*
   c. *Give an example for which $(\mu \times \nu)^*(A \times B) \ne \mu^*(A) \cdot \nu^*(B)$.*

Solution.   (a) We have $\mathcal{S} \otimes \Sigma \subseteq \Lambda_\mu \otimes \Lambda_\nu$. Let $A \times B \in \Lambda_\mu \otimes \Lambda_\nu$. Also, let $\{A_n \times B_n\}$ be a sequence of $\mathcal{S} \otimes \Sigma$ such that $A \times B \subseteq \bigcup_{n=1}^{\infty} A_n \times B_n$. Since (by Theorem 26.1) $\mu^* \times \nu^*$ is a measure on the semiring $\Lambda_\mu \otimes \Lambda_\nu$, it follows from

Theorem 13.8 that

$$\mu^* \times \nu^*(A \times B) \leq \sum_{n=1}^{\infty} \mu^* \times \nu^*(A_n \times B_n) = \sum_{n=1}^{\infty} \mu \times \nu(A_n \times B_n).$$

Consequently, we see that

$$\mu^*(A) \cdot \nu^*(B) = \mu^* \times \nu^*(A \times B)$$

$$\leq \inf\left\{ \sum_{n=1}^{\infty} \mu \times \nu(A_n \times B_n) \colon \{A_n \times B_n\} \subseteq \mathcal{S} \otimes \Sigma \text{ and } A \times B \subseteq \bigcup_{n=1}^{\infty} A_n \times B_n \right\}$$

$$= (\mu \times \nu)^*(A \times B).$$

(b) If $0 < \mu^*(A) < \infty$ and $0 < \nu^*(B) < \infty$, then

$$(\mu \times \nu)^*(A \times B) = \mu^*(A) \cdot \nu^*(B)$$

holds true by virtue of Theorem 26.2. On the other hand, if either $\mu^*(A) = \infty$ or $\nu^*(B) = \infty$, then—by (a)—the equality holds with both sides equal to ∞.
(c) Let $X = Y = \{0\}$, $\mathcal{S} = \{\emptyset\}$, and $\Sigma = \mathcal{P}(Y)$. Also, let $\mu = 0$ on $\mathcal{S}$ (the only choice!) and $\nu = 0$ on Σ. Now, note that $\mu^*(X) \cdot \nu^*(Y) = \infty \cdot 0 = 0$, while $(\mu \times \nu)^*(X \times Y) = \infty$.

Problem 26.2. *Let $(X, \mathcal{S}, \mu)$ and (Y, Σ, ν) be two σ-finite measure spaces. Then show that $(\mu \times \nu)^*(A \times B) = \mu^*(A) \cdot \nu^*(B)$ holds for each $A \times B$ in $\Lambda_\mu \otimes \Lambda_\nu$.*

Solution. Let $\{X_n\} \subseteq \Lambda_\mu$ and $\{Y_n\} \subseteq \Lambda_\nu$ satisfy $X_n \uparrow X$, $Y_n \uparrow Y$, $\mu^*(X_n) < \infty$, and $\nu^*(Y_n) < \infty$ for each n. Using Theorems 15.4 and 26.2, we see that

$$(\mu \times \nu)^*(A \times B) = \lim_{n\to\infty} (\mu \times \nu)^*\big((A \cap X_n) \times (B \cap Y_n)\big)$$

$$= \lim_{n\to\infty} \left[\mu^*(A \cap X_n) \cdot \nu^*(B \cap Y_n) \right]$$

$$= \mu^*(A) \cdot \nu^*(B)$$

for each $A \times B \in \Lambda_\mu \otimes \Lambda_\nu$.

Problem 26.3. *Let $(X, \mathcal{S}, \mu)$ and (Y, Σ, ν) be two measure spaces. Assume that A and B are subsets of X and Y, respectively, such that $0 < \mu^*(A) < \infty$, and $0 < \nu^*(B) < \infty$. Then show that $A \times B$ is $\mu \times \nu$-measurable if and only if both A*

and B are measurable in their corresponding spaces. Is the preceding conclusion
true if either A or B has measure zero?

Solution. If $A \in \Lambda_\mu$ and $B \in \Lambda_\nu$, then by Theorem 26.3, $A \times B \in \Lambda_{\mu \times \nu}$. For
the converse, assume that $A \times B \in \Lambda_{\mu \times \nu}$. We claim first that $(\mu \times \nu)^*(A \times B) < \infty$.
To see this, pick two sequences $\{A_n\} \subseteq S$ and $\{B_m\} \subseteq \Sigma$ with $A \subseteq \bigcup_{n=1}^\infty A_n$,
$\sum_{n=1}^\infty \mu^*(A_n) < \mu^*(A) + 1$, $B \subseteq \bigcup_{m=1}^\infty B_m$, and $\sum_{m=1}^\infty \nu^*(B_m) < \nu^*(B) + 1$.
Now, from $A \times B \subseteq \bigcup_{n=1}^\infty \bigcup_{m=1}^\infty A_n \times B_m$, we see that

$$(\mu \times \nu)^*(A \times B) \le \sum_{n=1}^\infty \sum_{m=1}^\infty (\mu \times \nu)^*(A_n \times B_m)$$

$$= \sum_{n=1}^\infty \sum_{m=1}^\infty \mu(A_n) \cdot \nu(B_m) = \left[\sum_{n=1}^\infty \mu(A_n) \right] \cdot \left[\sum_{m=1}^\infty \nu(B_m) \right]$$

$$< \left[\mu^*(A) + 1 \right] \cdot \left[\nu^*(B) + 1 \right] < \infty.$$

Therefore, by Theorem 26.4, $(A \times B)^y$ is μ-measurable for ν-almost all y. Since
$(A \times B)^y = A$ holds for all $y \in B$ and $\nu^*(B) > 0$, it follows that A is μ-
measurable. Similarly, B is ν-measurable.

Finally, if $\mu^*(A) = 0$, $A \ne \emptyset$, and $A \times B \in \Lambda_{\mu \times \nu}$, then B need not be
necessarily ν-measurable. An example: Take $X = Y = \mathbb{R}$ with $\mu = \nu = \lambda$. If
$E \subseteq [0, 1]$ is nonmeasurable, then $\{0\} \times E$ is a $\mu \times \nu$-null set (since $\{0\} \times E \subseteq$
$\{0\} \times [0, 1]$), and so $\{0\} \times E$ is a $\mu \times \nu$-measurable set.

Problem 26.4. *Let (X, S, μ) and (Y, Σ, ν) be two σ-finite measure spaces, and
let $f: X \times Y \to \mathbb{R}$ be a $\mu \times \nu$-measurable function. Show that for μ-almost all x
the function f_x is a ν-measurable function. Similarly, show that for ν-almost all
y the function f^y is μ-measurable.*

Solution. We can assume $f(x, y) \ge 0$ for each $(x, y) \in X \times Y$. Since (in
this case) the product measure is σ-finite, there exists a sequence $\{A_n\}$ of $\mu \times \nu$-
measurable sets with $A_n \uparrow X \times Y$ and $(\mu \times \nu)^*(A_n) < \infty$ for each n.

Now, by Fubini's Theorem, the function $(f \wedge \chi_{A_n})_x$ is ν-integrable for μ-almost
all x. Since $(f \wedge \chi_{A_n})_x \uparrow f_x$, it follows that f_x is ν-measurable for μ-almost all
x.

Problem 26.5. *Show that if $f(x, y) = (x^2 - y^2)/(x^2 + y^2)^2$, with $f(0, 0) = 0$,
then*

$$\int_0^1 \left[\int_0^1 f(x, y) \, dx \right] dy = -\frac{\pi}{4} \quad and \quad \int_0^1 \left[\int_0^1 f(x, y) \, dy \right] dx = \frac{\pi}{4}.$$

Solution. Note that

$$\int_0^1 \left[\int_0^1 f(x, y)\, dx \right] dy = \int_0^1 \left[-\frac{x}{x^2+y^2} \Big|_{x=0}^{x=1} \right] dy = -\int_0^1 \frac{1}{1+y^2}\, dy = -\frac{\pi}{4}$$

and

$$\int_0^1 \left[\int_0^1 f(x, y)\, dy \right] dx = \int_0^1 \left[\frac{y}{x^2+y^2} \Big|_{y=0}^{y=1} \right] dx = \int_0^1 \frac{1}{1+x^2}\, dx = \frac{\pi}{4}.$$

Problem 26.6. *Let $X = Y = \mathbb{N}$, $S = \Sigma = $ the collection of all subsets of $\mathbb{N}$, and $\mu = \nu = $ the counting measure. Give an interpretation of Fubini's Theorem in this case.*

Solution. Let $f: \mathbb{N} \times \mathbb{N} \longrightarrow \mathbb{R}$ be a non-negative $\mu \times \nu$-integrable function. Then, by Problem 22.3 and Fubini's Theorem, we see that

$$\int f\, d(\mu \times \nu) = \int \int f\, d\mu d\nu = \int \left[\sum_{m=1}^{\infty} f(n, m) \right] d\nu(n) = \sum_{n=1}^{\infty} \sum_{m=1}^{\infty} f(n, m).$$

On the other hand, if $f: \mathbb{N} \times \mathbb{N} \longrightarrow \mathbb{R}$ is a non-negative function such that

$$\sum_{n=1}^{\infty} \sum_{m=1}^{\infty} f(n, m) < \infty,$$

then it follows from Tonelli's Theorem that f is $\mu \times \nu$-integrable. *Conclusion*: A function $f: \mathbb{N} \times \mathbb{N} \longrightarrow \mathbb{R}$ is $\mu \times \nu$-integrable if and only if $\sum_{n=1}^{\infty} \sum_{m=1}^{\infty} |f(n, m)| < \infty$, and in this case

$$\int f\, d(\mu \times \nu) = \sum_{n=1}^{\infty} \sum_{m=1}^{\infty} f(n, m) = \sum_{m=1}^{\infty} \sum_{n=1}^{\infty} f(n, m).$$

Problem 26.7. *Establish the following result, known as* **Cavalieri's Principle**. *Let (X, S, μ) and (Y, Σ, ν) be two measure spaces, and let E and F be two $\mu \times \nu$-measurable subsets of $X \times Y$ of finite measure. If $\nu^*(E_x) = \mu^*(F_x)$ holds for μ-almost all x, then*

$$(\mu \times \nu)^*(E) = (\mu \times \nu)^*(F).$$

Solution. By Theorem 26.4, we have

$$(\mu \times \nu)^*(E) = \int_X \nu^*(E_x) \, d\mu(x) = \int_X \nu^*(F_x) \, d\mu(x) = (\mu \times \nu)^*(F).$$

Problem 26.8. *For this problem* λ *denotes the Lebesgue measure on* **R**. *Let* $(X, \mathcal{S}, \mu)$ *be a* σ-*finite measure space, and let* $f: X \to \mathbf{R}$ *be a measurable function such that* $f(x) \geq 0$ *holds for all* $x \in X$. *Then show that*

    a. *The set* $A = \{(x, y) \in X \times \mathbf{R}: 0 \leq y \leq f(x)\}$, *called the* **ordinate set** *of* f, *is a* $\mu \times \lambda$-*measurable subset of* $X \times \mathbf{R}$.

    b. *The set* $B = \{(x, y) \in X \times \mathbf{R}: 0 \leq y < f(x)\}$ *is a* $\mu \times \lambda$-*measurable subset of* $X \times \mathbf{R}$ *and* $(\mu \times \lambda)^*(A) = (\mu \times \lambda)^*(B)$ *holds.*

    c. *The graph of* f, *i.e., the set* $G = \{(x, f(x)): x \in X\}$, *is a* $\mu \times \lambda$-*measurable subset of* $X \times \mathbf{R}$.

    d. *If* f *is* μ-*integrable, then* $(\mu \times \lambda)^*(A) = \int f \, d\mu$ *holds.*

    e. *If* f *is* μ-*integrable, then* $(\mu \times \lambda)^*(G) = 0$ *holds.*

Solution. If $g: X \longrightarrow \mathbf{R}$ is an arbitrary positive measurable function, then we shall write $A_g = \{(x, y) \in X \times \mathbf{R}: 0 \leq y \leq g(x)\}$ and

$$B_g = \{(x, y) \in X \times \mathbf{R}: 0 \leq y < g(x)\}.$$

Observe that if $f_n(x) \uparrow f(x)$ and $h_n(x) \downarrow f(x)$ hold for each $x \in X$, then $B_{f_n} \uparrow B_f$ and $A_{h_n} \downarrow A_f$.

Assume first that g is a positive simple function. Let $g = \sum_{i=1}^n a_i \chi_{C_i}$ be the standard representation of g, where $a_i > 0$ for each $1 \leq i \leq n$. Then, it is easy to see that

$$A_g = (X \times \{0\}) \cup (C_1 \times [0, a_1]) \cup (C_2 \times [0, a_2]) \cup \cdots \cup (C_n \times [0, a_n]) \quad (\star)$$

and

$$B_g = (C_1 \times [0, a_1]) \cup (C_2 \times [0, a_2]) \cup \cdots \cup (C_n \times [0, a_n]). \quad (\star\star)$$

By Theorem 26.3, both A_g and B_g are $\mu \times \lambda$-measurable subsets of $X \times \mathbf{R}$.

(a) First assume that f is a bounded measurable function. That is, assume that there exists some $M > 0$ such that $0 \leq f(x) \leq M$ holds for all $x \in X$. By Theorem 17.7 there exists a sequence $\{\psi_n\}$ of simple functions with $\psi_n(x) \uparrow M - f(x)$ for each $x \in X$. Thus, the sequence $\{\phi_n\}$ of simple functions, defined by $\phi_n(x) = M - \psi_n(x)$, satisfies $\phi_n(x) \downarrow f(x)$ for each $x \in X$. This implies $A_{\phi_n} \downarrow A_f$. Since (by the preceding discussion) each A_{ϕ_n} is $\mu \times \lambda$-measurable, we see that in this case A_f is likewise a $\mu \times \lambda$-measurable set.

Now, let f be an arbitrary positive measurable function. For each n, let $f_n = f \wedge n\mathbf{1}$, and note that (by the preceding case) each A_{f_n} is $\mu \times \lambda$-measurable. To infer that A_f is a $\mu \times \lambda$-measurable set, observe that $A_{f_n} \uparrow A_f$ holds.

(b) By Theorem 17.7 there exists a sequence $\{s_n\}$ of simple functions such that $0 \leq s_n(x) \uparrow f(x)$ holds for all $x \in X$. Clearly, $B_{s_n} \uparrow B_f$ holds. Since each B_{s_n} is $\mu \times \lambda$-measurable, it follows that B_f is likewise $\mu \times \lambda$-measurable.

Next, we shall establish the equality $(\mu \times \lambda)^*(A_f) = (\mu \times \lambda)^*(B_f)$ by cases.

CASE 1. Assume $\mu^*(X) < \infty$.

Clearly, $(\mu \times \lambda)^*(X \times \{0\}) = \mu^*(X) \cdot \lambda(\{0\}) = 0$. Also, assume that $0 \leq f(x) \leq M < \infty$ holds for all x. Then, there exist two sequences $\{\phi_n\}$ and $\{\psi_n\}$ of step functions with $0 \leq \phi_n(x) \uparrow f(x)$ and $\psi_n(x) \downarrow f(x)$ for all $x \in X$. Clearly, $B_{\phi_n} \uparrow B_f$ and $A_{\psi_n} \downarrow A_f$. Now, use $(\star)$ and $(\star\star)$ in connection with Theorem 26.3 and the Lebesgue Dominated Convergence Theorem to see that

$$\int \phi_n \, d\mu = (\mu \times \lambda)^*\left(B_{\phi_n}\right) \uparrow (\mu \times \lambda)^*(B_f) = \int f \, d\mu$$

and

$$\int \psi_n \, d\mu = (\mu \times \lambda)^*\left(A_{\psi_n}\right) \downarrow (\mu \times \lambda)^*(A_f) = \int f \, d\mu.$$

Thus, in this case $(\mu \times \lambda)^*(A_f) = (\mu \times \lambda)^*(B_f) = \int f \, d\mu$ holds.

CASE 2. Assume $\mu^*(X) < \infty$ and that f is a positive μ-measurable function.

For each n let $f_n = f \wedge n\mathbf{1}$. Note that $B_{f_n} \uparrow B_f$ and $A_{f_n} \uparrow A_f$. By the preceding case, we have $(\mu \times \lambda)^*\left(A_{f_n}\right) = (\mu \times \lambda)^*\left(B_{f_n}\right)$ for each n. Thus, from Theorem 15.4, it follows that

$$(\mu \times \lambda)^*(A_f) = (\mu \times \lambda)^*(B_f) = \lim_{n \to \infty} \int f_n \, d\mu.$$

CASE 3. The general case. Here we shall use the hypothesis that μ is σ-finite.

Choose a sequence $\{E_n\}$ of measurable subsets of X with $E_n \uparrow X$ and $\mu^*(E_n) < \infty$ for each n. Let $g_n = f\chi_{E_n}$, and observe that $B_{g_n} \uparrow B_f$ and $A_{f_n} \uparrow A_f$. Using the preceding case and Theorem 15.4, we see that

$$(\mu \times \lambda)^*(B_f) = \lim_{n \to \infty} (\mu \times \lambda)^*\left(B_{g_n}\right) = \lim_{n \to \infty} (\mu \times \lambda)^*\left(A_{f_n}\right) = (\mu \times \lambda)^*(A_f).$$

Also, it should be noted here that if f is integrable, then

$$(\mu \times \lambda)^*(B_f) = (\mu \times \lambda)^*(A_f) = \int f \, d\mu.$$

(c) From the identity $G = A_f \setminus B_f$, it follows that the graph G of f is $\mu \times \lambda$-measurable.

(d) The equality follows from the discussion in part (b).

(e) From $G = A_f \setminus B_f$ and part (d), we see that

$$(\mu \times \lambda)^*(G) = (\mu \times \lambda)^*(A_f) - (\mu \times \lambda)^*(B_f) = 0.$$

Problem 26.9. *Let $g: X \to \mathbb{R}$ be a μ-integrable function, and let $h: Y \to \mathbb{R}$ be a ν-integrable function. Define $f: X \times Y \to \mathbb{R}$ by $f(x, y) = g(x)h(y)$ for each x and y. Show that f is $\mu \times \nu$-integrable and that*

$$\int f \, d(\mu \times \nu) = \left(\int_X g \, d\mu \right) \cdot \left(\int_Y h \, d\nu \right).$$

Solution. We can assume $g \geq 0$ and $h \geq 0$. Choose a sequence $\{\phi_n\}$ of μ-step functions and a sequence $\{\psi_n\}$ of ν-step functions with $\phi_n \uparrow g$ and $\psi_n \uparrow h$. Then, $\{\phi_n \psi_n\}$ is a sequence of $\mu \times \nu$-step functions such that $\phi_n \psi_n \uparrow gh$. The conclusion now follows from the relation

$$\int \phi_n \psi_n d(\mu \times \nu) = \left(\int \phi_n \, d\mu \right) \cdot \left(\int \psi_n \, d\nu \right) \uparrow \left(\int g \, d\mu \right) \cdot \left(\int h \, d\nu \right).$$

Problem 26.10. *Use Tonelli's Theorem to verify that*

$$\int_\epsilon^r \frac{\sin x}{x} \, dx = \int_0^\infty \left(\int_\epsilon^r e^{-xy} \sin x \, dx \right) dy$$

holds for each $0 < \epsilon < r$. By letting $\epsilon \to 0^+$ and $r \to \infty$ (and justifying your steps) give another proof of the formula

$$\int_0^\infty \frac{\sin x}{x} \, dx = \frac{\pi}{2}.$$

Solution.   Fix $0 < \varepsilon < r$ and consider the function

$$g_{\varepsilon,r}(x, y) = \begin{cases} e^{-xy} & \text{if } (x, y) \in [\varepsilon, r] \times [0, 1] \\ e^{-\varepsilon y} & \text{if } (x, y) \in [\varepsilon, r] \times [1, \infty). \end{cases}$$

Clearly, the continuous (and hence, measurable) function $f(x, y) = e^{-xy} \sin x$ satisfies $\left| f(x, y) \right| \leq g_{\varepsilon,r}(x, y)$ for all $(x, y) \in [\varepsilon, r] \times [0, \infty)$. From

$$\int_{\varepsilon}^{r} \left[\int_{1}^{\infty} g_{\varepsilon,r}(x, y)\, dy \right] dx \leq \int_{\varepsilon}^{r} \frac{1}{\varepsilon}\, dx = \frac{r - \varepsilon}{\varepsilon} < \infty$$

and Tonelli's Theorem, we see that the function $g_{\varepsilon,r}$ is Lebesgue integrable over $[\varepsilon, r] \times [0, \infty)$. So, the function $f(x, y)$ is integrable over $[\varepsilon, r] \times [0, \infty)$. Now Fubini's Theorem guarantees that

$$\int_{\varepsilon}^{r} \left(\int_{0}^{\infty} e^{-xy} \sin x\, dy \right) dx = \int_{0}^{\infty} \int_{\varepsilon}^{r} \left(e^{-xy} \sin x\, dx \right) dy. \qquad (\star)$$

Using the elementary integral

$$\int e^{-\alpha t} \sin t\, dt = -\frac{\alpha \sin t + \cos t}{1 + \alpha^2}\, e^{-\alpha t}$$

and performing the innermost integrations in $(\star)$, we get

$$\int_{\varepsilon}^{r} \frac{\sin x}{x}\, dx = \int_{0}^{\infty} \left[-\frac{y \sin x + \cos x}{1 + y^2}\, e^{-xy} \Big|_{x=\varepsilon}^{x=r} \right] dy$$

$$= \int_{0}^{\infty} \frac{y \sin \varepsilon + \cos \varepsilon}{1 + y^2}\, e^{-\varepsilon y}\, dy - \int_{0}^{\infty} \frac{y \sin r + \cos r}{1 + y^2}\, e^{-ry}\, dy,$$

and consequently,

$$\int_{\varepsilon}^{r} \frac{\sin x}{x}\, dx = \sin \varepsilon \int_{0}^{\infty} \frac{y e^{-\varepsilon y}}{1 + y^2}\, dy + \cos \varepsilon \int_{0}^{\infty} \frac{e^{-\varepsilon y}}{1 + y^2}\, dy - \int_{0}^{\infty} \frac{y \sin r + \cos r}{1 + y^2}\, e^{-ry}\, dy. \qquad (\star\star)$$

We shall compute the limits of the three terms in the right-hand side of $(\star\star)$ as $r \to \infty$ and $\varepsilon \to 0^+$.

We start by computing $\lim_{\varepsilon \to 0^+} \sin \varepsilon \int_{0}^{\infty} \frac{y e^{-\varepsilon y}}{1 + y^2}\, dy$. To this end, let $\eta > 0$. Since $\lim_{y \to \infty} \frac{y}{1 + y^2} = 0$, there exists some $y_0 > 0$ such that $0 < \frac{y}{1 + y^2} < \eta$ holds for all $y \geq y_0$. Now, from $\lim_{\varepsilon \to 0^+} y_0 \sin \varepsilon = 0$ and $\lim_{\varepsilon \to 0^+} \frac{\sin \varepsilon}{\varepsilon} = 1$, we see that there exists some $0 < \delta < 1$ such that

$$0 < \epsilon < \delta \quad \text{implies} \quad 0 < y_0 \sin \epsilon < \eta \quad \text{and} \quad \tfrac{\sin \epsilon}{\epsilon} < 2.$$

Now, if $0 < \varepsilon < \delta$, then (by taking into account that $0 \le \tfrac{ye^{-\varepsilon y}}{1+y^2} \le 1$ for $y \ge 0$) we infer that

$$\left| \sin \varepsilon \int_0^\infty \tfrac{ye^{-\varepsilon y}}{1+y^2}\, dy \right| \le \left| \sin \varepsilon \int_0^{y_0} \tfrac{ye^{-\varepsilon y}}{1+y^2}\, dy \right| + \left| \sin \varepsilon \int_{y_0}^\infty \tfrac{ye^{-\varepsilon y}}{1+y^2}\, dy \right|$$

$$\le y_0 \sin \varepsilon + \eta \sin \varepsilon \int_{y_0}^\infty e^{-\varepsilon y}\, dy$$

$$\le y_0 \sin \varepsilon + \eta \sin \varepsilon \int_0^\infty e^{-\varepsilon y}\, dy$$

$$= y_0 \sin \varepsilon + \eta \tfrac{\sin \varepsilon}{\varepsilon} < \eta + 2\eta = 3\eta.$$

That is, $\lim_{\varepsilon \to 0^+} \sin \varepsilon \int_0^\infty \tfrac{ye^{-\varepsilon y}}{1+y^2}\, dy = 0$.

For the second limit, note that $\left| \tfrac{e^{-\varepsilon y}}{1+y^2} \right| \le \tfrac{1}{1+y^2}$ holds for each $y \in [0, \infty)$. Thus, in view of the Lebesgue integrability of the function $h(y) = \tfrac{1}{1+y^2}$ over $[0, \infty)$, Theorem 24.4 yields

$$\lim_{\varepsilon \to 0^+}\left[\cos \varepsilon \int_0^\infty \tfrac{e^{-\varepsilon y}}{1+y^2}\, dy \right] = \lim_{\varepsilon \to 0^+}\left[\cos \varepsilon \right] \cdot \lim_{\varepsilon \to 0^+}\left[\int_0^\infty \tfrac{e^{-\varepsilon y}}{1+y^2}\, dy \right]$$

$$= 1 \cdot \int_0^\infty \lim_{\varepsilon \to 0^+}\left[\tfrac{e^{-\varepsilon y}}{1+y^2} \right] dy = \int_0^\infty \tfrac{dy}{1+y^2} = \tfrac{\pi}{2}.$$

For the third limit, note that for each $r \ge 1$ and each $y \ge 0$, we have

$$\left| \tfrac{y \sin r + \cos r}{1+y^2}\, e^{-ry} \right| \le \tfrac{1+y}{1+y^2} e^{-y} \le 2e^{-y},$$

and so by the Lebesgue integrability of $g(y) = 2e^{-y}$ over $[0, \infty)$, it follows from Theorem 24.4 that

$$\lim_{r \to \infty} \int_0^\infty \tfrac{y \sin r + \cos r}{1+y^2} e^{-ry}\, dy = \int_0^\infty \lim_{r \to \infty}\left[\tfrac{y \sin r + \cos r}{1+y^2} e^{-ry} \right] dy$$

$$= \int_0^\infty 0\, dy = 0.$$

Finally, from (**), we see that

$$\int_0^\infty \frac{\sin x}{x}\, dx = \lim_{\substack{\varepsilon \to 0^+ \\ r \to \infty}} \int_\varepsilon^r \frac{\sin x}{x}\, dx = \lim_{\varepsilon \to 0^+}\left[\sin\varepsilon \int_0^\infty \frac{y}{1+y^2} e^{-\varepsilon y}\, dy\right] +$$

$$+ \lim_{\varepsilon \to 0^+}\left[\cos\varepsilon \int_0^\infty \frac{e^{-\varepsilon y}}{1+y^2}\, dy\right] - \lim_{r \to \infty}\int_0^\infty \frac{y\sin r + \cos r}{1+y^2} e^{-ry}\, dy$$

$$= 0 + \tfrac{\pi}{2} + 0 = \tfrac{\pi}{2}.$$

Problem 26.11.   *Show that if* $f(x, y) = ye^{-(1+x^2)y^2}$ *for each x and y, then*

$$\int_0^\infty \left[\int_0^\infty f(x, y)\, dx\right] dy = \int_0^\infty \left[\int_0^\infty f(x, y)\, dy\right] dx.$$

Use the preceding equality to give an alternate proof of the formula

$$\int_0^\infty e^{-x^2}\, dx = \frac{\sqrt{\pi}}{2}.$$

Solution.   Note that

$$\int_0^\infty \left[\int_0^\infty ye^{-(1+x^2)y^2}\, dy\right] dx = \tfrac{1}{2}\int_0^\infty \frac{1}{1+x^2}\, dx = \tfrac{\pi}{4}$$

and

$$\int_0^\infty \left[\int_0^\infty ye^{-(1+x^2)y^2}\, dx\right] dy = \left(\int_0^\infty e^{-x^2}\, dx\right)\cdot\left(\int_0^\infty e^{-y^2}\, dy\right)$$

$$= \left(\int_0^\infty e^{-x^2}\, dx\right)^2.$$

Since $ye^{-(1+x^2)y^2} \geq 0$ holds for all $x \geq 0$ and $y > 0$, Tonelli's Theorem shows that

$$\left(\int_0^\infty e^{-x^2}\, dx\right)^2 = \tfrac{\pi}{4}.$$

Problem 26.12.   *Show that*

$$\int_0^\infty \left(\int_0^r e^{-xy^2}\sin x\, dx\right) dy = \int_0^r \left(\int_0^\infty e^{-xy^2}\sin x\, dy\right) dx$$

holds for all $r > 0$. By letting $r \to \infty$ show that

$$\int_0^\infty \frac{\sin x}{\sqrt{x}}\, dx = \frac{\sqrt{2\pi}}{2}.$$

In a similar manner show that $\int_0^\infty \frac{\cos x}{\sqrt{x}}\, dx = \sqrt{2\pi}/2$.

Solution. Since

$$\int_0^\infty \left(\int_0^r e^{-xy^2}\, dx\right) dy = \int_0^1 \left(\int_0^r e^{-xy^2}\, dx\right) dy + \int_1^\infty \left(\tfrac{1}{y^2}\int_0^{ry^2} e^{-t}\, dt\right) dy$$

$$\leq \int_0^1 \left(\int_0^r 1\, dx\right) dy + \int_1^\infty \left(\tfrac{1}{y^2}\int_0^\infty e^{-t}\, dt\right) dy$$

$$\leq r + \int_1^\infty \tfrac{1}{y^2}\, dy = r + 1,$$

it follows from Tonelli's Theorem that e^{-xy^2} is integrable over $[0, r] \times [0, \infty)$. In view of $\left|e^{-xy^2} \sin x\right| \leq e^{-xy^2}$, we see that $e^{-xy^2} \sin x$ is also integrable over $[0, r] \times [0, \infty)$, and the stated identity follows from Fubini's Theorem.

Performing the innermost integrations and using the elementary integral

$$\int e^{-\alpha t} \sin t\, dt = -\tfrac{\alpha \sin t + \cos t}{1+\alpha^2}\, e^{-\alpha t},$$

we get

$$\int_0^\infty \tfrac{dy}{1+y^4} - \int_0^\infty \tfrac{y^2 \sin r + \cos r}{1+y^4}\, e^{-ry^2}\, dy = \tfrac{\sqrt{\pi}}{2}\int_0^r \tfrac{\sin x}{\sqrt{x}}\, dx. \qquad (\star)$$

Since $\left|\tfrac{y^2 \sin r + \cos r}{1+y^4}e^{-ry^2}\right| \leq \tfrac{1+y^2}{1+y^4} = f(y)$ holds, and f is Lebesgue integrable over $[0, \infty)$, it follows from Theorem 24.4 that

$$\lim_{r\to\infty}\left[\int_0^r \tfrac{dy}{1+y^4} - \int_0^\infty \tfrac{y^2 \sin r + \cos r}{1+y^4}\, e^{-ry^2}\, dy\right]$$

$$= \int_0^\infty \tfrac{dy}{1+y^4} - \lim_{r\to\infty}\left[\int_0^\infty \tfrac{y^2 \sin r + \cos r}{1+y^4}\, e^{-ry^2}\, dy\right]$$

$$= \int_0^\infty \tfrac{dy}{1+y^4} - \int_0^\infty \lim_{r\to\infty}\left[\tfrac{y^2 \sin r + \cos r}{1+y^4}\, e^{-ry^2}\right] dy$$

$$= \int_0^\infty \tfrac{dy}{1+y^4} - \int_0^\infty 0\, dy = \int_0^\infty \tfrac{dy}{1+y^4}.$$

Now, using the elementary integral

$$\int \frac{dy}{1+y^4} = \frac{1}{4\sqrt{2}}\left[\ln\left(\frac{1+y\sqrt{2}+y^2}{1-y\sqrt{2}+y^2}\right) + 2\arctan\frac{y\sqrt{2}}{1-y^2}\right]$$

and an easy computation, we see that

$$\int_0^\infty \frac{dy}{1+y^4} = \int_0^{1^-} \frac{dy}{1+y^4} + \int_{1^+}^\infty \frac{dy}{1+y^4} = \frac{\pi\sqrt{2}}{4}.$$

Thus, from $(\star)$, we see that

$$\int_0^\infty \frac{\sin x}{\sqrt{x}} \, dx = \lim_{r\to\infty} \int_0^r \frac{\sin x}{\sqrt{x}} \, dx = \frac{2}{\sqrt{\pi}} \cdot \frac{\pi\sqrt{2}}{4} = \frac{\sqrt{2\pi}}{2}.$$

Problem 26.13. *Using the conclusions of the preceding problem (and an appropriate change of variable), show that the values of the* **Fresnel integrals** *(see Problem 24.6) are*

$$\int_0^\infty \sin(x^2) \, dx = \int_0^\infty \cos(x^2) \, dx = \frac{\sqrt{2\pi}}{4}.$$

Solution. Using the change of variable $x = \sqrt{u}$, we get

$$\int_0^r \sin(x^2) \, dx = \frac{1}{2}\int_0^{\sqrt{r}} \frac{\sin u}{\sqrt{u}} \, du \quad \text{and} \quad \int_0^r \cos(x^2) \, dx = \frac{1}{2}\int_0^{\sqrt{r}} \frac{\cos u}{\sqrt{u}} \, du.$$

Now, let $r \to \infty$ and use the preceding problem.

Problem 26.14. *Let $X = Y = [0, 1]$, $\mu = $ the Lebesgue measure on $[0, 1]$, and $\nu = $ the counting measure on $[0, 1]$. Consider the "diagonal" $\Delta = \{(x, x): x \in X\}$ of $X \times Y$. Then show that*

   a. *Δ is a $\mu \times \nu$-measurable subset of $X \times Y$, and hence, χ_Δ is a non-negative $\mu \times \nu$-measurable function.*

   b. *Both iterated integrals $\int\int \chi_\Delta \, d\mu d\nu$ and $\int\int \chi_\Delta \, d\nu d\mu$ exist.*

   c. *The function χ_Δ is not $\mu \times \nu$-integrable. Why doesn't this contradict Tonelli's Theorem?*

Solution. (a) Consider the two sets

$$A = \{(x, y) \in X \times Y: x > y\} \quad \text{and} \quad B = \{(x, y) \in X \times Y: x < y\}.$$

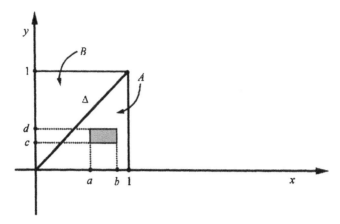

FIGURE 4.4.

Note that $A = \bigcup [a, b] \times [c, d]$, where the union extends over all rectangles $[a, b] \times [c, d]$ with rational end points and $a > d$; see Figure 4.4.

Clearly, the collection of all such rectangles is countable. Since each rectangle is $\mu \times \nu$-measurable, it follows that A is $\mu \times \nu$-measurable. Similarly, the set B is $\mu \times \nu$-measurable. Hence, $\Delta = X \times Y \setminus A \cup B$ is a $\mu \times \nu$-measurable set.

(b) Note that

$$\int \int \chi_\Delta \, d\mu \, d\nu = \int_0^1 \left[\int_0^1 \chi_\Delta(x, y) \, d\mu(x) \right] d\nu(y)$$

$$= \int_0^1 \left[\int_0^1 \chi_{\{y\}} d\mu(x) \right] d\nu(y) = \int_0^1 0 \cdot d\nu(y) = 0,$$

and

$$\int \int \chi_\Delta \, d\nu \, d\mu = \int_0^1 \left[\int_0^1 \chi_\Delta(x, y) \, d\nu(y) \right] d\mu(x)$$

$$= \int_0^1 \left[\int_0^1 \chi_{\{x\}} \, d\nu(y) \right] d\mu(x) = \int_0^1 1 \cdot d\mu(x) = 1.$$

(c) Fubini's Theorem combined with part (b) shows that χ_Δ is not integrable over $X \times Y$ (i.e., $(\mu \times \nu)^*(\Delta) = \infty$ must hold). This does not contradict Tonelli's Theorem because ν is not a σ-finite measure.

Problem 26.15. *Let $f: \mathbf{R} \to \mathbf{R}$ be Borel measurable. Then show that the functions $f(x + y)$ and $f(x - y)$ are both $\lambda \times \lambda$-measurable.*

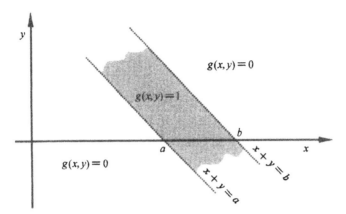

$g(x, y) = 0$

$g(x, y) = 1$

$g(x, y) = 0$

b

a

$x + y = b$

$x + y = a$

y

x

FIGURE 4.5.

Solution. Since f is the limit of a sequence of step functions, it suffices to establish the result for characteristic functions of measurable sets of finite measure. The regularity of the Lebesgue measure allows us to reduce it to the characteristic functions of open sets of finite measure. Finally, this can be reduced to the case when $f = \chi_{(a,b)}$ for some finite open interval (a, b).

The $\lambda \times \lambda$-measurability of the function $g(x, y) = \chi_{(a,b)}(x + y)$ follows easily from the graph shown in Figure 4.5.

The $\lambda \times \lambda$-measurability of $f(x - y)$ can be proven in a similar manner.

CHAPTER 5

NORMED SPACES AND
L_p-SPACES

27. NORMED SPACES AND BANACH SPACES

Problem 27.1. *Let X be a normed space. Then show that X is a Banach space if and only if its unit sphere $\{x \in X: \|x\| = 1\}$ is a complete metric space (under the induced metric $d(x, y) = \|x - y\|$).*

Solution. Let $S = \{x \in X: \|x\| = 1\}$, and note that S is a closed set. Clearly, if X is a Banach space, then S is a complete metric space.

Conversely, assume that S is complete. Let $\{x_n\}$ be a Cauchy sequence of X. In view of the inequality

$$\big| \|x_n\| - \|x_m\| \big| \le \|x_n - x_m\|,$$

we see that $\{\|x_n\|\}$ is a Cauchy sequence of real numbers. If $\lim \|x_n\| = 0$, then $\lim x_n = 0$. So, we can assume that $\delta = \lim \|x_n\| > 0$. In this case, we can also assume that there exists some $M > 0$ such that $\frac{1}{\|x_n\|} \le M$ and $\|x_n\| \le M$ both hold for each n. The inequalities

$$\begin{aligned}
\left\| \frac{x_n}{\|x_n\|} - \frac{x_m}{\|x_m\|} \right\| &= \frac{\|(\|x_m\|\|x_n\| - \|x_n\|\|x_m\|)\|}{\|x_n\| \cdot \|x_m\|} \\
&\le M^2 \big\| \|x_m\|(x_n - x_m) - (\|x_n\| - \|x_m\|)x_m \big\| \\
&\le 2M^3 \|x_n - x_m\|,
\end{aligned}$$

show that the sequence $\left\{ \frac{x_n}{\|x_n\|} \right\}$ is a Cauchy sequence of S. If x is its limit in S, then $x_n = \|x_n\| \cdot \frac{x_n}{\|x_n\|} \longrightarrow \delta x$ holds in X, and so X is a Banach space.

Problem 27.2. *Let X be a normed vector space. Fix $a \in X$ and a nonzero scalar α.*

a. *Show that the mappings $x \mapsto a + x$ and $x \mapsto \alpha x$ are both homeomorphisms.*

b. *If A and B are two sets with either A or B open and α and β are nonzero scalars, then show that $\alpha A + \beta B$ is an open set.*

Solution. (a) Observe that $\|(a + x) - (a + y)\| = \|x - y\|$ holds for all x and y. This shows that $x \mapsto a + x$ is, in fact, an isometry.

Also notice that for all $x, y \in X$ we have $\|\alpha x - \alpha y\| = |\alpha| \cdot \|x - y\|$. This easily implies that $x \mapsto \alpha x$ is a homeomorphism.

(b) Assume first that B is an open set. Since the mapping $x \mapsto a + x$ is a homeomorphism, we know that $a + B$ is an open set for each $a \in X$. This implies that the set $A + B = \bigcup_{a \in A}(a + B)$ is an open set for each subset A of X.

Now, assume that B is an open set and that α and β are nonzero scalars. Since the mapping $x \mapsto \beta x$ is a homeomorphism, the set βB is an open set. So, by the preceding discussion, $\alpha A + \beta B$ must be an open set.

Problem 27.3. *Let X be a normed vector space, and let $B = \{x \in X: \|x\| < 1\}$ be its open unit ball. Show that $\overline{B} = \{x \in X: \|x\| \le 1\}$.*

Solution. Repeat the solution of Problem 6.2.

Problem 27.4. *Let X be a normed space, and let $\{x_n\}$ be a sequence of X such that $\lim x_n = x$ holds. If $y_n = n^{-1}(x_1 + \cdots + x_n)$ for each n, then show that $\lim y_n = x$.*

Solution. Let $\varepsilon > 0$. Choose some k with $\|x_n - x\| < \varepsilon$ for all $n > k$. Fix some $m > k$ so that $\|\frac{1}{n}[(x_1 - x) + \cdots + (x_k - x)]\| < \varepsilon$ holds for all $n > m$. Thus, if $n > m$, then

$$
\begin{aligned}
\|y_n - x\| &= \left\|\frac{1}{n}[(x_1 - x) + \cdots + (x_k - x) + (x_{k+1} - x) + \cdots + (x_n - x)]\right\| \\
&\le \left\|\frac{1}{n}[(x_1 - x) + \cdots + (x_k - x)]\right\| + \frac{1}{n}[\|x_{k+1} - x\| + \cdots + \|x_n - x\|] \\
&< \varepsilon + \varepsilon = 2\varepsilon.
\end{aligned}
$$

That is, $\lim y_n = x$ holds. (See also Problem 4.11.)

Problem 27.5. *Assume that two vectors x and y in a normed space satisfy $\|x + y\| = \|x\| + \|y\|$. Then show that*

$$\|\alpha x + \beta y\| = \alpha \|x\| + \beta \|y\|$$

holds for all scalars $\alpha \ge 0$ and $\beta \ge 0$.

Solution. Assume $\|x + y\| = \|x\| + \|y\|$ holds for two vectors x and y in a normed space, and let $\alpha \ge 0$ and $\beta \ge 0$. Without loss of generality, we can suppose that $\alpha \geq \beta \geq 0$. From the triangle inequality, it follows that

$$\|\alpha x + \beta y\| \leq \alpha \|x\| + \beta \|y\|.$$

Next, notice that

$$\begin{aligned}
\|\alpha x + \beta y\| &= \|\alpha(x+y) + (\beta - \alpha)y\| \\
&\geq \|\alpha(x+y)\| - \|(\beta - \alpha)y\| \\
&= \alpha \|x+y\| - (\alpha - \beta)\|y\| = \alpha(\|x\| + \|y\|) - (\alpha - \beta)\|y\| \\
&= \alpha \|x\| + \beta \|y\|.
\end{aligned}$$

Hence, $\|\alpha x + \beta y\| = \alpha \|x\| + \beta \|y\|$, as desired.

Problem 27.6. *Let X be the vector space of all real-valued functions defined on* $[0, 1]$ *having continuous first-order derivatives. Show that* $\|f\| = |f(0)| + \|f'\|_\infty$ *is a norm on X that is equivalent to the norm* $\|f\|_\infty + \|f'\|_\infty$.

Solution. The verification of the norm properties of $\|\cdot\|$ are straightforward. From

$$f(x) = f(0) + \int_0^x f'(t)\,dt,$$

we see that $|f(x)| \leq |f(0)| + \|f'\|_\infty$ holds for each $x \in [0, 1]$, and consequently, $\|f\|_\infty \leq |f(0)| + \|f'\|_\infty$ holds.

The equivalence of the two norms follows from the inequalities

$$\begin{aligned}
\|f\|_\infty + \|f'\|_\infty &\leq |f(0)| + 2\|f'\|_\infty \leq 2(|f(0)| + \|f'\|_\infty) \\
&= 2\|f\| \leq 2(\|f\|_\infty + \|f'\|_\infty).
\end{aligned}$$

Problem 27.7. *A series $\sum_{n=1}^\infty x_n$ in a normed space is said to* **converge** *to x if* $\lim \|x - \sum_{i=1}^n x_i\| = 0$. *As usual, we write* $x = \sum_{n=1}^\infty x_n$. *A series $\sum_{n=1}^\infty x_n$ is said to be* **absolutely summable** *if* $\sum_{n=1}^\infty \|x_n\| < \infty$ *holds.*

Show that a normed space X is a Banach space if and only if every absolutely summable series is convergent.

Solution. Let X be a Banach space, and let $\sum_{n=1}^\infty x_n$ be an absolutely summable series. For each n let $s_n = \sum_{i=1}^n x_i$. The inequality

$$\left\| s_{n+p} - s_n \right\| = \left\| \sum_{i=n+1}^{n+p} x_i \right\| \leq \sum_{i=n+1}^{n+p} \|x_i\|$$

implies that $\{s_n\}$ is a Cauchy sequence, and hence, convergent in X.

For the converse, let $\{x_n\}$ be a Cauchy sequence in a normed space X whose absolutely summable series are convergent. By passing to a subsequence (if necessary), we can assume that $\|x_{n+1} - x_n\| < 2^{-n}$ holds for each n. Put $x_0 = 0$, and note that $\sum_{n=0}^{\infty} \|x_{n+1} - x_n\| < \infty$. Thus, $\lim_{n\to\infty} \sum_{i=0}^{n-1}(x_{i+1} - x_i) = \lim_{n\to\infty} x_n$ exists in X so that X is a Banach space.

Problem 27.8. *Show that a closed proper vector subspace of a normed vector space is nowhere dense.*

Solution. Let E be a proper closed subspace of a normed space X. Assume that E has an interior point a. Then, there exists some $r > 0$ such that $B(a, r) \subseteq E$. Now, if y is an arbitrary nonzero element of X, then $a + \frac{r}{2\|y\|}y \in B(a, r) \subseteq E$, and so $y = \frac{2\|y\|}{r}\left[(a + \frac{r}{2\|y\|}y) - a\right] \in E$. That is, $E = X$ holds, a contradiction. Thus, $E^\circ = \emptyset$.

Problem 27.9. *Assume that $f:[0, 1] \to \mathbb{R}$ is a continuous function which is not a polynomial. By Corollary 11.6 we know that there exists a sequence of polynomials $\{p_n\}$ that converges uniformly to f. Show that the set of natural numbers*

$$\{k \in \mathbb{N}: k = \text{degree of } p_n \text{ for some } n\}$$

is countable.

Solution. Let $f:[0, 1] \to \mathbb{R}$ be a continuous function which is not a polynomial, and let $\{p_n\}$ be a sequence of polynomials that converges uniformly to f on $[0, 1]$. Assume by way of contradiction that the set of natural numbers

$$K = \{k \in \mathbb{N}: k = \text{degree of } p_n \text{ for some } n\}$$

is bounded. This means that there exists some $m \in \mathbb{N}$ such that every p_n has degree at-most m. So, if V is the finite dimensional vector subspace generated in $C[0, 1]$ by the functions $\{1, x, x^2, \ldots, x^m\}$, then $\{p_n\} \subseteq V$ holds. Now, a glance at Theorem 27.7 guarantees that V is a closed subspace of $C[0, 1]$, and thus the (sup) norm limit f of $\{p_n\}$ must lie in V. That is, f must be a polynomial of degree at-most m, contrary to our hypothesis. Hence, K is not bounded, and therefore it must be a countable set; see Theorem 2.4.

Problem 27.10. *This problem describes some important classes of subsets of a vector space. A nonempty subset A of a vector space X is said to be:*
  a. **symmetric,** *if $x \in A$ implies $-x \in A$, i.e., if $A = -A$;*

　　b.　**convex**, *if* $x, y \in A$ *implies* $\lambda x + (1 - \lambda)y \in A$ *for all* $0 \le \lambda \le 1$, *i.e.,*
　　　　for every two vectors $x, y \in A$ *the line segment joining* x *and* y *lies in*
　　　　A; *and*
　　c.　**circled** (*or* **balanced**) *if* $x \in A$ *implies* $\lambda x \in A$ *for each* $|\lambda| \le 1$.

Establish the following:

　　i.　*A circled set is symmetric.*
　　ii.　*A convex and symmetric set containing zero is circled.*
　　iii.　*A nonempty subset* A *of a vector space is convex if and only if* $aA + bA =$
　　　　$(a + b)A$ *holds for all scalars* $a \ge 0$ *and* $b \ge 0$.
　　iv.　*If* A *is a convex subset of a normed space, then the closure* $\overline{A}$ *and the*
　　　　interior A° *of* A *are also convex sets.*

Solution.　(i) Let A be a circled set. Since $|-1| = 1 \le 1$, it follows that
$-x = (-1)x \in A$ for each $x \in A$. Thus, A is a symmetric set.

　　(ii) Let A be a convex symmetric set containing zero. Fix $x \in A$ and $|\lambda| \le 1$.
If $0 \le \lambda \le 1$, then $\lambda x = \lambda x + (1 - \lambda)0 \in A$ and if $-1 \le \lambda < 0$, then
$\lambda x = (-\lambda)(-x) + (1 + \lambda)0 \in A$. So, A is a circled set.

　　(iii) Let A be a subset of a vector space. Assume first that A is a convex set, and
let $a > 0$ and $b > 0$. If $x \in (a + b)A = \{(a + b)u : u \in A\}$, then for some $u \in A$,
we have $x = (a + b)u = au + bu \in aA + bA$, and so $(a + b)A \subseteq aA + bA$
is always true. Now, let $x \in aA + bA$. Then, there exist $u, v \in A$ such that
$x = au + bv$. Since A is convex, we have $z = \frac{a}{a+b}u + \frac{b}{a+b}v \in A$, and so
$x = (a+b)\left[\frac{a}{a+b}u + \frac{b}{a+b}v\right] = (a+b)z \in (a+b)A$. Therefore, $aA+bA \subseteq (a+b)A$
is also true, and consequently $aA + bA = (a + b)A$.

　　Next, suppose that $aA + bA = (a + b)A$ holds true for all $a \ge 0$ and $b \ge 0$.
Let $x, y \in A$ and $0 \le \lambda \le 1$. Letting $a = \lambda$ and $b = 1 - \lambda$, we see that

$$\lambda x + (1 - \lambda)y = ax + by \in (a + b)A = A.$$

This shows that A is a convex set.

　　(iv) Let A be a convex subset of a normed space. We show first that $\overline{A}$ is a convex
set. To this end, let $x, y \in \overline{A}$ and fix $0 \le \lambda \le 1$. Pick two sequences $\{x_n\}$ and $\{y_n\}$
of A such that $x_n \to x$ and $y_n \to y$. Put $z_n = \lambda x_n + (1 - \lambda)y_n$ and note that $\{z_n\}$ is
a sequence of A. Since the function $f : X \to X$, defined by $f(u) = \lambda u + (1 - \lambda)u$,
is continuous (see Problem 27.2), it follows that $z_n \to \lambda x + (1 - \lambda)y$. This implies
$\lambda x + (1 - \lambda)y \in \overline{A}$, so that $\overline{A}$ is a convex set.

　　Next, we shall show that A° is a convex set. Fix $0 \le \lambda \le 1$. Since A° is an
open set, it follows (from Problem 27.2) that the set $\lambda A^\circ + (1 - \lambda)A^\circ$ is also an
open set, which (since A is convex) is contained in A. Since A° is the largest open
set contained in A, we infer that $\lambda A^\circ + (1 - \lambda)A^\circ \subseteq A^\circ$. This shows that A° is
a convex set.

Problem 27.11. *This problem describes all norms on a vector space X that are equivalent to a given norm. So, let $(X, \| \cdot \|)$ be a normed vector space. Let A be a norm bounded convex symmetric subset of X having zero as an interior point (relative to the topology generated by the norm $\| \cdot \|$). Define the function $p_A: X \to \mathbb{R}$ by*

$$p_A(x) = \inf\{\lambda > 0: x \in \lambda A\}.$$

Establish the following:

    a. *The function p_A is a well-defined norm on X.*

    b. *The norm p_A is equivalent to $\| \cdot \|$, i.e., there exist two constants $C > 0$ and $K > 0$ such that $C\|x\| \leq p_A(x) \leq K\|x\|$ holds for all $x \in X$.*

    c. *The closed unit ball of p_A is the closure of A, i.e., $\{x \in X: p_A(x) \leq 1\} = \overline{A}$.*

    d. *Let $\||\cdot\||$ be a norm on X which is equivalent to $\| \cdot \|$, and consider the norm bounded nonempty symmetric convex set $B = \{x \in X: \||x\|| \leq 1\}$. Then zero is an interior point of B and $\||x\|| = p_B(x)$ holds for each $x \in X$.*

Solution. Assume that A is a norm bounded convex symmetric subset of a normed space $(X, \| \cdot \|)$ such that zero is an interior point of A.

(a) Pick some $r > 0$ such that $B(0, 2r) = \{x \in X: \|x\| < 2r\} \subseteq A$. If $x \in X$ is a nonzero vector, then $\frac{r}{\|x\|}x \in B(0, 2r) \subseteq A$, and so $x \in \frac{\|x\|}{r} A$. This shows that the set $\{\lambda > 0: x \in \lambda A\}$ is nonempty and so the formula $p_A(x) = \inf\{\lambda > 0: x \in \lambda A\}$ is well defined and satisfies

$$p_A(x) \leq r\|x\| \qquad\qquad (\star)$$

for all $x \in X$. Next, we shall show that p_A is a norm on X.

Clearly, $p_A(x) \geq 0$ and $p_A(0) = 0$. Now, if $p_A(x) = 0$, then there exist a sequence $\{a_n\} \subseteq A$ and a sequence $\{\lambda_n\}$ of positive real numbers satisfying $\lambda_n \to 0$ and $x = \lambda_n a_n$ for each n. Since A is a norm bounded set, it easily follows that $x = \lim \lambda_n a_n = 0$. Thus, $p_A(x) = 0$ if and only if $x = 0$.

Next, we shall show that $p_A(\alpha x) = |\alpha| p_A(x)$ holds for all $\alpha \in \mathbb{R}$ and all $x \in X$. Since A is symmetric, we have

$$\{\lambda > 0: \lambda x \in A\} = \{\lambda > 0: \lambda(-x) \in A\},$$

and so for proving $p_A(\alpha x) = |\alpha| p_A(x)$, we can suppose without loss of generality that $\alpha > 0$. Now, note that

$$\begin{aligned} p_A(\alpha x) &= \inf\{\lambda > 0: \alpha x \in \lambda A\} = \inf\{\lambda > 0: x \in \tfrac{\lambda}{\alpha} A\} \\ &= \alpha \inf\{\tfrac{\lambda}{\alpha}: x \in \tfrac{\lambda}{\alpha} A\} = \alpha \inf\{\mu > 0: x \in \mu A\} \\ &= \alpha p_A(x). \end{aligned}$$

For the triangle inequality, let x, $y \in X$ and fix $\epsilon > 0$. Choose $\lambda > 0$ and $x \in \lambda A$ such that $\lambda < p_A(x) + \epsilon$. Likewise, pick some $\mu > 0$ such that $y \in \mu A$ and $\mu < p_A(y) + \epsilon$. From Problem 27.10 we know that $x + y \in \lambda A + \mu A = (\lambda + \mu)A$, and so

$$p_A(x + y) \leq \lambda + \mu < \big[p_A(x) + \epsilon\big] + \big[p_A(y) + \epsilon\big] = p_A(x) + p_A(y) + 2\epsilon.$$

Since $\epsilon > 0$ is arbitrary, we infer that $p_A(x + y) \leq p_A(y) + p_A(y)$.

(b) Let $x \in X$ and fix some $M > 0$ such that $\|a\| \leq M$ holds for all $a \in A$. Now, if $\lambda > 0$ satisfies $x \in \lambda A$, then there exists some $y \in A$ such that $x = \lambda y$. Hence, $\|x\| = \lambda \|y\| \leq \lambda M$, or $\lambda \geq \|x\|/M$. This implies $p_A(x) \geq \frac{1}{M}\|x\|$. Now, combine this inequality with ($\star$) to establish that p_A is a norm equivalent to $\| \cdot \|$.

(c) Assume first that $p_A(x) \leq 1$ and $x \neq 0$. Then for each n there exist $0 < \lambda_n \leq 1 + \frac{1}{n}$ and $a_n \in A$ such that $x = \lambda_n a_n$. By passing to a subsequence, we can assume $\lambda_n \to \lambda$. Since $x \neq 0$ and A is a norm bounded set, it easily follows that $0 < \lambda \leq 1$. Now, note that the sequence $\{a_n\} \subseteq A$ satisfies $a_n = \frac{1}{\lambda_n}x \to \frac{1}{\lambda}x$, and so $\frac{1}{\lambda}x \in \overline{A}$. Since $\overline{A}$ is also a convex set (see Problem 27.10), we see that $x = \lambda\big(\frac{1}{\lambda}x\big) + (1 - \lambda)0 \in \overline{A}$.

Now, let $x \in \overline{A}$. Then, there exists a sequence $\{x_n\} \subseteq A$ such that $\|x_n - x\| \to 0$. Since $\| \cdot \|$ is equivalent to p_A, we also have $p_A(x_n - x) \to 0$. In particular, $p_A(x_n) \to p_A(x)$. Now, notice that since $x_n \in A$, we have $p_A(x_n) \leq 1$ for each n. This implies $p_A(x) \leq 1$. Therefore, the closed unit ball of p_A is $\overline{A}$.

(d) Let $\|| \cdot \||$ be a norm on X which is equivalent to $\| \cdot \|$. It is easy to check that the closed unit ball B of $\|| \cdot \||$ is a bounded convex and symmetric set containing zero as an interior point. We shall show next that $\||x\|| = p_B(x)$ holds for each $x \in X$.

To see this, let $x \in X$ be a nonzero vector. Since $x/\||x\|| \in B$, we see that $p_B(x)/\||x\|| = p_B\big(x/\||x\||\big) \leq 1$, and so $p_B(x) \leq \||x\||$. On the other hand, there exist a sequence $\{\lambda_n\}$ of positive real numbers and a sequence $\{b_n\}$ of B such that $\lambda_n \to p_B(x)$, $b_n \in B$ and $x = \lambda_n b_n$ for each n. Since $\||x\|| = \lambda_n \||b_n\|| \leq \lambda_n$, we easily infer that $\||x\|| \leq p_B(x)$. Hence, $p_B(x) = \||x\||$ for each $x \in X$.

28. OPERATORS BETWEEN BANACH SPACES

Problem 28.1. *Let X and Y be two Banach spaces and let $T: X \to Y$ be a bounded linear operator. Show that either T is onto or else $T(X)$ is a meager set.*

Solution. Assume that $T(X)$ is not a meager set. Then, we have to show that $T(X) = Y$ holds.

Let $V = \{x \in X: \|x\| \leq 1\}$. Since (by assumption) $T(X)$ is not a meager set, some $\overline{nT(V)} = n\overline{T(V)}$ has an interior point. This implies that $\overline{T(V)}$ has an interior point. So, there exists some $y_0 \in \overline{T(V)}$ and some $r > 0$ such that

$B(y_0, 2r) \subseteq \overline{T(V)} = -\overline{T(V)}$. Note that if $y \in Y$ satisfies $\|y\| < 2r$, then $y - y_0 = -(y_0 - y) \in \overline{T(V)}$ and so $y = (y - y_0) + y_0 \in \overline{T(V)} + \overline{T(V)} \subseteq 2\overline{T(V)}$. (The last inclusion follows, of course, from the identity $V + V = 2V$.) Consequently, we have established that $\{y \in Y: \|y\| < r\} \subseteq \overline{T(V)}$. From the linearity of T, we infer that

$$\left\{ y \in Y: \ \|y\| < 2^{-n}r \right\} \subseteq 2^{-n}\overline{T(V)} = \overline{T(2^{-n}V)} \qquad (\star\star)$$

holds for each n.

Next, let $y \in Y$ be fixed such that $\|y\| < 2^{-1}r = \frac{r}{2}$. From $(\star\star)$, we know that $y \in \overline{T(2^{-1}V)}$. So, for some vector $x_1 \in 2^{-1}V$ we have $\|y - T(x_1)\| < 2^{-2}r$. Now, proceed inductively. Assume that $x_n \in 2^{-n}V$ has been selected such that $\left\|y - \sum_{i=1}^{n} T(x_i)\right\| < 2^{-n-1}r$. From $(\star\star)$ it follows that $y - \sum_{i=1}^{n} T(x_i) \in \overline{T(2^{-n-1}V)}$, and so there exists some $x_{n+1} \in 2^{-n-1}V$ such that $\left\|y - \sum_{i=1}^{n+1} T(x_i)\right\| < 2^{-n-2}r$. Thus, there exists a sequence $\{x_n\}$ of X such that $\|x_n\| \leq 2^{-n}$ and

$$\left\| y - \sum_{i=1}^{n} T(x_i) \right\| = \left\| y - T\left(\sum_{i=1}^{n} x_i\right) \right\| < 2^{-n-1}r$$

hold for each n. Now, for each n let $s_n = x_1 + \cdots + x_n$ and note that the relation

$$\left\| s_{n+p} - s_n \right\| = \left\| \sum_{i=n+1}^{n+p} x_i \right\| \leq \sum_{i=n+1}^{n+p} \|x_i\| \leq \sum_{i=n+1}^{\infty} 2^{-i} = 2^{-n}$$

shows that $\{s_n\}$ is a Cauchy sequence of X. Since X is a Banach space, the sequence $\{s_n\}$ converges; let $s = \lim s_n$. Clearly, $\|s\| \leq \sum_{n=1}^{\infty} \|x_n\| \leq 1$ (i.e., $s \in V$), and by the continuity and linearity of T, we see that

$$T(s) = \lim_{n \to \infty} T(s_n) = \lim_{n \to \infty} \sum_{i=1}^{n} T(x_i) = y.$$

That is, $y \in T(V)$, and so $\{y \in Y: \|y\| < \frac{r}{2}\} \subseteq T(V) \subseteq T(X)$. Since $T(X)$ is a vector subspace of Y, the latter inclusion implies that $T(X) = Y$ must hold.

Problem 28.2. *Let X be a Banach space, $T: X \to X$ a bounded operator, and I the identity operator on X. If $\|T\| < 1$, then show that $I - T$ is invertible.*

Solution. If $A, B \in L(X, X)$, then the inequalities

$$\|ABx\| \leq \|A\| \cdot \|Bx\| \leq \|A\| \cdot \|B\| \cdot \|x\|$$

easily imply that $\|AB\| \leq \|A\| \cdot \|B\|$. In particular, if a sequence $\{A_n\}$ of operators of $L(X, X)$ satisfies $\lim A_n = A$ in $L(X, X)$ and $B \in L(X, X)$, then the inequality

$$\|BA_n - BA\| = \|B(A_n - A)\| \leq \|B\| \cdot \|A_n - A\|$$

shows that $\lim BA_n = BA$. Similarly, $\lim A_n B = AB$.

Now, assume $T \in L(X, X)$ satisfies $\|T\| < 1$. In view of the inequality $\|T^n\| \leq \|T\|^n$, it follows that

$$\sum_{n=0}^{\infty} \|T^n\| \leq \sum_{n=0}^{\infty} \|T\|^n = \frac{1}{1-\|T\|} < \infty.$$

Thus, $\sum_{n=0}^{\infty} T^n$ is an absolutely summable series. Since $L(X, X)$ is a Banach space, $S = \sum_{n=0}^{\infty} T^n$ converges in $L(X, X)$; see Problem 27.7. Moreover,

$$(I - T)S = \lim_{n \to \infty} (I - T)\left(\sum_{i=0}^{n} T^i\right) = \lim_{n \to \infty} \left(I - T^{n+1}\right) = I,$$

and similarly $S(I - T) = I$. Therefore, $S = (I - T)^{-1}$.

Problem 28.3. *On $C[0, 1]$ consider the two norms*

$$\|f\|_{\infty} = \sup\{|f(x)|\colon x \in [0, 1]\} \quad and \quad \|f\|_1 = \int_0^1 |f(x)|\, dx.$$

Then show that the identity operator $I\colon (C[0, 1], \|\cdot\|_{\infty}) \to (C[0, 1], \|\cdot\|_1)$ is continuous, onto, but not open. Why doesn't this contradict the Open Mapping Theorem?

Solution. Clearly, I is onto, and in view of the inequality $\|f\|_1 \leq \|f\|_{\infty}$, we see that I is also continuous.

For the rest of the proof, we need to show that $\left(C[0, 1], \|\cdot\|_1\right)$ is not a Banach space. To establish this, consider the sequence $\{f_n\}$ of continuous functions whose graphs are shown in Figure 5.1.

The inequality $\|f_{n+p} - f_n\|_1 \leq \frac{1}{n}$ shows that $\{f_n\}$ is a Cauchy sequence for the norm $\|\cdot\|_1$. Assume by way of contradiction that $\lim \|f_n - f\|_1 = 0$ holds for some $f \in C[0, 1]$.

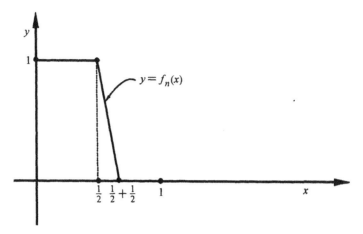

FIGURE 5.1.

Let $a \in (0, \frac{1}{2})$. If $f(a) \neq 1$, then there exist some $\varepsilon > 0$ and some $0 < \delta < \min\{a, \frac{1}{2} - a\}$ such that $|f(x) - 1| \geq \varepsilon$ holds whenever $|x - a| < \delta$. Now, note that $2\delta\varepsilon \leq \int_0^1 |f_n(x) - f(x)| \, dx = \|f_n - f\|_1$ for all sufficiently large n, contrary to $\lim \|f_n - f\|_1 = 0$. Thus, $f(a) = 1$ holds for all $a \in (0, \frac{1}{2})$. Similarly, $f(a) = 0$ for all $a \in (\frac{1}{2}, 1)$. Now, it is readily seen that f cannot be a continuous function, contrary to $f \in C[0, 1]$. Thus, $\{f_n\}$ does not converge in $C[0, 1]$ with respect to the $\| \cdot \|_1$ norm.

Finally, $I : (C[0, 1], \| \cdot \|_\infty) \longrightarrow (C[0, 1], \| \cdot \|_1)$ cannot be an open mapping. Since otherwise, $\| \cdot \|_1$ and $\| \cdot \|_\infty$ would be equivalent norms, and therefore $(C[0, 1], \| \cdot \|_1)$ would be a Banach space.

Problem 28.4. *Let X be the vector space of all real-valued functions on* $[0, 1]$ *that have continuous derivatives with the sup norm. Also, let* $Y = C[0, 1]$ *with the sup norm. Define* $D : X \to Y$ *by* $D(f) = f'$.

    a. *Show that D is an unbounded linear operator.*
    b. *Show that D has a closed graph.*
    c. *Why doesn't the conclusion in (b) contradict the Closed Graph Theorem?*

Solution. (a) The standard properties of differentiation guarantee that D is a linear operator. Now, for each n let $f_n(x) = x^n$. Then, $f_n \in X$ and $\|f_n\|_\infty = \sup\{|f_n(x)| : 0 \leq x \leq 1\} = 1$ for each n. Now, notice that $D(f_n)(x) = nx^{n-1}$ holds for each n, and from this it follows that

$$\|D\| \geq \|D(f_n)\|_\infty = \sup\{nx^{n-1} : 0 \leq x \leq 1\} = n,$$

Therefore, $\|D\| = \infty$, and so D is an unbounded operator.

(b) To see that D has a closed graph, assume $f_n \to 0$ in X and $Df_n = f'_n \to g$ in Y. That is, $\{f_n\}$ converges uniformly to zero, and $\{f'_n\}$ converges uniformly to g. We have to show that $g = 0$.

From $\int_0^x f'_n(t)\,dt = f_n(x) - f_n(0)$ (and Problem 9.16), it follows that

$$\int_0^x g(t)\,dt = \lim_{n\to\infty} \int_0^x f'_n(t)\,dt = \lim_{n\to\infty} \big[f_n(x) - f_n(0)\big] = 0$$

holds for all $x \in [0, 1]$. Differentiating, we get $g(x) = 0$ for each $x \in [0, 1]$, as required. (See also Problem 9.29.)

(c) The conclusion in (b) does not contradict the Closed Graph Theorem since X is not a Banach space. For instance, we know (from Corollary 11.6) that every function $f \in C[0, 1]$ is the uniform limit of a sequence of polynomials. So, if $f \in C[0, 1]$ is a nondifferentiable continuous function and $\{p_n\}$ is a sequence of polynomials that converges uniformly to f, then $\{p_n\}$ is a Cauchy sequence of X which cannot converge in X.

Problem 28.5. *Consider the mapping $T: C[0, 1] \to C[0, 1]$ defined by $Tf(x) = x^2 f(x)$ for all $f \in C[0, 1]$ and each $x \in [0, 1]$.*

  a. *Show that T is a bounded linear operator.*
  b. *If $I: C[0, 1] \to C[0, 1]$ denotes the identity operator (i.e., $I(f) = f$ for each $f \in C[0, 1]$), then show that $\|I + T\| = 1 + \|T\|$.*

Solution.   (a) From the identities

$$T(f + g)(x) = x^2(f + g)(x) = x^2 f(x) + x^2 g(x) = \big(Tf + Tg\big)(x)$$

and

$$T(\alpha f)(x) = \alpha x^2 f(x) = \big(\alpha Tf\big)(x),$$

we easily infer that T is a linear operator. For the norm of T, note that for each $f \in C[0, 1]$ we have

$$\|Tf\|_\infty = \sup_{x\in[0,1]} |Tf(x)| = \sup_{x\in[0,1]} x^2|f(x)| \le \sup_{x\in[0,1]} |f(x)| = \|f\|_\infty,$$

and so $\|T\| \le 1$. On the other hand, for the constant function $\mathbf{1}$, we have

$$\|T\| \ge \|T\mathbf{1}\|_\infty = \sup_{x\in[0,1]} x^2 = 1.$$

Thus, $\|T\| = 1$, and so T is a bounded operator.

(b) Clearly, $\|I + T\| \le \|I\| + \|T\| = 1 + 1 = 2$. Moreover, we have

$$\|I + T\| \ge \|(I + T)\mathbf{1}\|_\infty = \sup_{x \in [0,1]} (1 + x^2) = 2,$$

and so $\|I + T\| = 1 + \|T\| = 2$ holds true.

Problem 28.6. *Let X be a vector space which is complete in each of the two norms $\|\cdot\|_1$ and $\|\cdot\|_2$. If there exists a real number $M > 0$ such that $\|x\|_1 \le M\|x\|_2$ holds for all $x \in X$, then show that the two norms must be equivalent.*

Solution. The identity operator $I : (X, \|\cdot\|_2) \longrightarrow (X, \|\cdot\|_1)$ is one-to-one, continuous, and onto. By the Open Mapping Theorem it is a homeomorphism, and the conclusion follows.

Problem 28.7. *Let X, Y, and Z be three Banach spaces. Assume that $T: X \to Y$ is a linear operator and $S: Y \to Z$ is a bounded, one-to-one linear operator. Show that T is a bounded operator if and only if the composite linear operator $S \circ T$ (from X into Z) is bounded.*

Solution. Assume that $S \circ T$ is a bounded operator. Let $x_n \to 0$ in X and $T(x_n) \to y$ in Y. Using that $S \circ T$ and S are both continuous, we get

$$S(y) = \lim_{n \to \infty} S\big(T(x_n)\big) = \lim_{n \to \infty} S \circ T(x_n) = 0.$$

Since S is one-to-one, we infer that $y = 0$, and hence—by the Closed Graph Theorem—the operator T is continuous.

Problem 28.8. *An operator $P: V \to V$ on a vector space is said to be a **projection** if $P^2 = P$ holds. Also, a closed vector subspace Y of a Banach space is said to be **complemented** if there exists another closed subspace Z of X such that $Y \oplus Z = X$.*

Show that a closed subspace of a Banach space is complemented if and only if it is the range of a continuous projection.

Solution. Let Y be a closed subspace of a Banach space X. Assume first that there exists a continuous projection $P: X \to X$ whose range is Y, i.e., $P(X) = Y$. From $P^2 = P$, it follows that $Y = \{y \in X : y = Py\}$.

If $I: X \to X$ denotes the identity operator, let $Z = (I - P)(X)$, the range of the continuous operator $I - P$. Clearly, Z is a vector subspace of X and in view of $x = Px + (I - P)(x)$, we see that $Y + Z = X$. Now, if $u \in Y \cap Z$, then

$u = z - Pz$ for some $z \in Z$, and so $u = P(u) = P(z - Pz) = P(z) - P^2(z) = 0$. This means that $Y \oplus Z = X$. Finally, to see that Z is also closed, assume that a sequence $\{z_n\}$ of X satisfies $(I - P)(z_n) \to z$. Then, the continuity of P implies $0 = P(I - P)(z_n) \to Pz$, and so $Pz = 0$. Hence, $z = (I - P)(z) \in Z$, proving that Z is also closed. Thus, Y is a complemented closed subspace.

For the converse, assume that Z is a closed subspace such that $Y \oplus Z = X$. So, for each $x \in X$ there exist $y \in Y$ and $z \in Z$ (both uniquely determined) such that $x = y + z$. Define an operator $P: X \to X$ via the formula $P(x) = y$, where y satisfies $x - y \in Z$. We claim that P is a continuous projection whose range is Y.

Notice first that P is a linear operator. Also, $P^2(x) = P(y) = y = P(x)$ holds for each $x \in X$, so that P is a projection. Clearly, the range of P is Y. To finish the proof, we must show that P is also continuous. For this, it suffices to show (in view of the Closed Graph Theorem) that P has a closed graph.

To this end, assume that a sequence $\{x_n\}$ of X satisfies $x_n \to x$ and $P(x_n) \to y$ in X. For each n let $x_n = y_n + z_n$, where $y_n \in Y$ and $z_n \in Z$. Clearly, $y_n = P(x_n)$ for each n. Since Y is a closed subspace, it follows that $y \in Y$. Now, from $z_n = x_n - y_n \to x - y$ and the closedness of Z, we infer that $z = x - y \in Z$. Thus, $x = y + z$, and so $y = P(x)$. This shows that P has a closed graph, and we are done.

29. LINEAR FUNCTIONALS

Problem 29.1. *Let $f: X \to \mathbb{R}$ be a linear functional defined on a vector space X. The **kernel** of f is the vector subspace*

$$\text{Ker } f = f^{-1}(\{0\}) = \{x \in X: \ f(x) = 0\}.$$

If X is a normed space and $f: X \to \mathbb{R}$ is nonzero linear functional, establish the following:

    a.   *f is continuous if and only if its kernel is a closed subspace of X.*

    b.   *f is discontinuous if and only if its kernel is dense in X.*

Solution. (a) Clearly, if f is continuous, then its kernel $f^{-1}(\{0\})$ is a closed set. For the converse, assume that $f \neq 0$ and that $f^{-1}(\{0\})$ is a closed set. Pick some $e \in X$ with $f(e) = 1$.

Suppose by way of contradiction that $\|f\| = \infty$. Then, there exists a sequence $\{x_n\}$ of X with $\|x_n\| = 1$ and $|f(x_n)| \geq n$ for each n. Note that the sequence $\{y_n\}$, defined by $y_n = e - \frac{x_n}{f(x_n)}$, satisfies $y_n \in f^{-1}(\{0\})$ for each n and $y_n \longrightarrow e$. Since the set $f^{-1}(\{0\})$ is closed, it follows that $e \in f^{-1}(\{0\})$, and so $f(e) = 0$, which is a contradiction. Thus, f is a continuous linear functional.

(b) If Ker f is dense in X, then Ker f is not closed and hence, by part (a), f is not continuous. For the converse, assume that f is a discontinuous linear functional, i.e., $\|f\| = \infty$. This implies (as in the previous part) that there exists a sequence $\{x_n\}$ of X satisfying $\|x_n\| = 1$ and $|f(x_n)| \geq n$ for each n. Now, if $x \in X$ and $y_n = x - \frac{f(x)}{f(x_n)} x_n$, then $\{y_n\}$ is a sequence in Ker f and satisfies $y_n \to x$. This shows that Ker f is dense in X.

Problem 29.2. *Show that a linear functional f on a normed space X is discontinuous if and only if for each $a \in X$ and each $r > 0$, we have*

$$f(B(a, r)) = \{f(x): \|a - x\| < r\} = \mathbf{R}.$$

Solution. Let f be a linear functional on a normed space X and let $B = B(0, 1) = \{x \in X: \|x\| < 1\}$. Assume that f is discontinuous. Fix $a \in X$ and $r > 0$. From the relation $B(a, r) = a + r B(0, 1) = a + r B$ and the linearity of f, it follows that $f\big(B(a, r)\big) = \mathbf{R}$ holds if and only if $f(B) = \Re$.

We claim first that $f(B)$ is unbounded from above in $\mathbf{R}$. To see this, assume by way of contradiction that there exists some $M > 0$ such that $f(x) \leq M$ holds for each $x \in B$. Note that if $x \in X$ satisfies $\|x\| \leq 1$, then $\pm \frac{1}{2} x \in B$, and so from

$$\pm \tfrac{1}{2} f(x) = f\big(\pm \tfrac{1}{2} x\big) \leq \tfrac{M}{2},$$

we see that $|f(x)| \leq M$ holds for all $x \in X$ with $\|x\| \leq 1$. That is, $\|f\| = \sup\{|f(x)|: \|x\| \leq 1\} \leq M < \infty$, and so f is a continuous linear functional, a contradiction. Thus, $f(B)$ is unbounded from above in $\mathbf{R}$. Now, let $\alpha > 0$ be an arbitrary positive real number. By the above, there exists some $x \in B$ satisfying $f(x) > \alpha$. Now, note that the element $y = \frac{\alpha}{f(x)} x \in B$ satisfies $f(y) = \alpha$ (and, of course, $-y \in B$ satisfies $f(-y) = -\alpha$). Consequently, $f(B) = \mathbf{R}$.

For the converse, assume that $f\big(B(a, r)\big) = \mathbf{R}$ holds for each $a \in X$ and each $r > 0$. In particular, from

$$\infty = \sup\{|f(x)|: \|x\| < \tfrac{1}{2}\} \leq \sup\{|f(x)|: \|x\| \leq 1\} = \|f\|,$$

we see that $\|f\| = \infty$. Thus, f is unbounded and so (by Theorem 28.6) f is a discontinuous linear functional.

Problem 29.3. *Let $f, f_1, f_2, \ldots, f_n$ be linear functionals defined on a common vector space X. Show that there exist constants $\lambda_1, \ldots, \lambda_n$ satisfying $f = \sum_{i=1}^{n} \lambda_i f_i$ (i.e., f lies in the linear span of $f_1, \ldots, f_n$) if and only if $\bigcap_{i=1}^{n}$ Ker $f_i \subseteq$ Ker f.*

Solution. If $f = \sum_{i=1}^{n} \lambda_i f_i$ holds, then clearly $\bigcap_{i=1}^{n} \operatorname{Ker} f_i \subseteq \operatorname{Ker} f$. For the converse, assume $\bigcap_{i=1}^{n} \operatorname{Ker} f_i \subseteq \operatorname{Ker} f$. Let

$$V = \left\{ y \in \mathbb{R}^n \colon \exists\ x \in X \text{ such that } y = \big(f_1(x),\ f_2(x),\ \ldots,\ f_n(x) \big) \right\}.$$

It is easy to verify that V is a vector subspace of $\mathbb{R}^n$. Now, define the linear functional $g \colon V \to \mathbb{R}$ via the formula

$$g\big(f_1(x),\ f_2(x),\ \ldots,\ f_n(x) \big) = f(x).$$

Notice that g is well defined. To see this, assume

$$\big(f_1(x),\ f_2(x),\ \ldots,\ f_n(x) \big) = \big(f_1(y),\ f_2(y),\ \ldots,\ f_n(y) \big).$$

Then, $f_i(x - y) = 0$ for each i, and so $x - y \in \bigcap_{i=1}^{n} \operatorname{Ker} f_i$. From our hypothesis, it follows that $x - y \in \operatorname{Ker} f$, which means that $f(x) = f(y)$. Now, it is a routine matter to verify that g is linear.

Denote by g again a linear extension of g to all of $\mathbb{R}^n$. This implies that there exist scalars $\lambda_1, \ldots, \lambda_n$ such that $g(z_1, \ldots, z_n) = \sum_{i=1}^{n} \lambda_i z_i$ holds for all $(z_1, \ldots, z_n) \in \mathbb{R}^n$. In particular, we have

$$f(x) = g\big(f_1(x),\ f_2(x),\ \ldots,\ f_n(x) \big) = \sum_{i=1}^{n} \lambda_i f_i(x)$$

for all $x \in X$, as desired.

Problem 29.4. *Prove the converse of Theorem 28.7. That is, show that if X and Y are (nontrivial) normed spaces and $L(X, Y)$ is a Banach space, then Y is a Banach space.*

Solution. Let $\{y_n\}$ be a Cauchy sequence of Y. Pick some f in X^* with $f \neq 0$, and then consider the sequence of operators $\{T_n\}$ of $L(X, Y)$ defined by $T_n(x) = f(x)y_n$. The inequality

$$\big\| T_n(x) - T_m(x) \big\| = \big\| f(x)(y_n - y_m) \big\| \leq \|f\| \cdot \|y_n - y_m\| \cdot \|x\|,$$

shows that $\|T_n - T_m\| \leq \|f\| \cdot \|y_n - y_m\|$, and so $\{T_n\}$ is a Cauchy sequence of $L(X, Y)$. By the completeness of $L(X, Y)$, there exists some $T \in L(X, Y)$ with $\lim T_n = T$. Now, pick some $e \in X$ with $f(e) = 1$, and note that

$$\lim_{n \to \infty} T_n(e) = \lim_{n \to \infty} y_n = T(e).$$

Problem 29.5. *The Banach space $B(\mathbb{N})$ is denoted by ℓ_∞. That is, ℓ_∞ is the Banach space consisting of all bounded sequences with the sup norm. Consider the collections of vectors*

$$c_0 = \{x = (x_1, x_2, x_3, \ldots) \in \ell_\infty \colon\ x_n \to 0\},\ \ and$$

$$c = \{x = (x_1, x_2, x_3, \ldots) \in \ell_\infty \colon\ \lim x_n\ exists\ in\ \mathbb{R}\}.$$

Show that c_0 and c are both closed vector subspaces of ℓ_∞.

Solution. It should be obvious that c_0 and c are vector subspaces of ℓ_∞ (and that c_0 is a vector subspace of c). What needs verification is their closedness. To see that c_0 is closed, assume that a sequence $\{x^n\}$ of c_0, where $x^n = (x_1^n, x_2^n, \ldots)$, satisfies $\|x^n - x\|_\infty \to 0$. If $x = (x_1, x_2, \ldots)$, we must show that $\lim x_n = 0$.

To this end, let $\epsilon > 0$. Fix some k such that $\|x^n - x\| < \epsilon$ for all $n \geq k$; clearly $|x_i^n - x_i| < \epsilon$ also holds for all $n \geq k$ and all i. Since $\lim_{i \to \infty} x_i^k = 0$, there exists some $m \geq k$ such that $|x_i^k| < \epsilon$ holds for all $i \geq m$. Now, notice that if $i \geq m$, then

$$|x_i| \leq |x_i - x_i^k| + |x_i^k| < \epsilon + \epsilon = 2\epsilon.$$

This shows that $\lim x_n = 0$, as desired.

Next, we shall establish that c is closed. For simplicity, for a sequence $x = (x_1, x_2, \ldots) \in c$ we shall write $x_\infty = \lim x_n$. Now, assume that a sequence $\{x^n\}$ in c satisfies $x^n \to x = (x_1, x_2, \ldots) \in \ell_\infty$. We must show that $\lim x_n$ exists in $\mathbb{R}$.

Start by fixing some $\epsilon > 0$. Then, there exists some k such that

$$\|x^n - x\|_\infty < \epsilon \quad \text{holds for all}\ \ n \geq k. \tag{$\star$}$$

This implies $\|x^n - x^m\|_\infty < 2\epsilon$ for all $n, m \geq k$, and so $|x_i^n - x_i^m| < 2\epsilon$ for all $n, m \geq k$ and each i. Consequently, $|x_\infty^n - x_\infty^m| \leq 2\epsilon$ for all $n, m \geq k$. This shows that $\{x_\infty^n\}$ is a Cauchy sequence of real numbers. Let $x_\infty = \lim x_\infty^n$ and note that $|x_\infty^n - x_\infty| \leq 2\epsilon$ holds for all $n \geq k$.

We claim that $x_n \to x_\infty$. To see this, let again $\epsilon > 0$ and choose k so that $(\star)$ is true. Next, fix some $r \geq k$ such that $|x_n^k - x_\infty^k| < \epsilon$ holds for all $n \geq r$. Now, note that if $n \geq r$, then

$$|x_n - x_\infty| \leq |x_n - x_n^k| + |x_n^k - x_\infty^k| + |x_\infty^k - x_\infty| < \epsilon + \epsilon + 2\epsilon = 4\epsilon.$$

This shows that $x_n \to x_\infty$, and so $x \in c$. Therefore, c is a closed subspace of ℓ_∞.

Problem 29.6. *Let c denote the vector subspace of ℓ_∞ consisting of all convergent sequences (see Problem 29.5). Define the **limit functional** $L \colon c \to \mathbb{R}$*

by

$$L(x) = L(x_1, x_2, \ldots) = \lim_{n \to \infty} x_n,$$

and $p: \ell_\infty \to \mathbf{R}$ *by* $p(x) = p(x_1, x_2, \ldots) = \limsup x_n.$

   a. *Show that L is a continuous linear functional, where c is assumed equipped with the sup norm.*

   b. *Show that p is sublinear and that* $L(x) = p(x)$ *holds for each* $x \in c$.

   c. *By the Hahn–Banach Theorem 29.2 there exists a linear extension of L to all of* ℓ_∞ *(which we shall denote by L again) satisfying* $L(x) \le p(x)$ *for all* $x \in \ell_\infty$. *Establish the following properties of the extension L:*

      i. *For each* $x \in \ell_\infty$, *we have*

$$\liminf_{n \to \infty} x_n \le L(x) \le \limsup_{n \to \infty} x_n.$$

     ii. *L is a positive linear functional, i.e.,* $x \ge 0$ *implies* $L(x) \ge 0$.

    iii. *L is a continuous linear functional (and in fact* $\|L\| = 1$).

Solution.   (a) Clearly, L is a linear functional. Moreover, if $x = (x_1, x_2, \ldots) \in c$, then $|x_n| \le \|x\|_\infty = \sup_m |x_m|$ for each n, and so $|L(x)| = \lim |x_n| \le \|x\|_\infty$. This shows that L is a continuous linear functional. (Since $L(1, 1, 1, \ldots) = 1$, it is easy to see that $\|L\| = 1$.)

(b) The sublinearity of p follows immediately from Problem 4.7. The equality $p(x) = L(x) = \lim x_n$ for each $x \in c$ should be also obvious.

(c) We shall establish the stated properties.

   (i) If $x \in \ell_\infty$, then notice that

$$-L(x) = L(-x) \le \limsup_{n \to \infty}(-x_n) = -\liminf_{n \to \infty} x_n,$$

and so $\liminf x_n \le L(x) \le \limsup x_n$ holds true.

   (ii) If $x = (x_1, x_2, \ldots) \ge 0$ (i.e., if $x_n \ge 0$ for each n), then it follows from Problem 4.8 and the preceding conclusion that $L(x) \ge \liminf x_n \ge 0$. That is, L is a positive linear functional.

   (iii) If $\|x\|_\infty \le 1$ (i.e., if $|x_n| \le 1$ for each n), then it follows from part (i) and Problem 4.8 that

$$-1 \le \liminf_{n \to \infty} x_n \le L(x) \le \limsup_{n \to \infty} x_n \le 1,$$

and so $|L(x)| \le 1$. This implies $\|L\| = \sup\{|L(x)|: \|x\|_\infty \le 1\} \le 1$. Since $L(1, 1, 1, \ldots) = 1$, we easily infer that $\|L\| = 1$.

Problem 29.7. *Generalize Problem 29.6 as follows. Show that there exists a linear functional* $\mathcal{L}im: \ell_\infty \to \mathbf{R}$ *(called a* **Banach–Mazur limit***) with the following properties.*

a. $\mathcal{L}im$ *is a positive linear functional of norm one.*
b. *For each* $x = (x_1, x_2, \ldots) \in \ell_\infty$, *we have*

$$\liminf_{n \to \infty} \frac{x_1 + x_2 + \cdots + x_n}{n} \leq \mathcal{L}im(x) \leq \limsup_{n \to \infty} \frac{x_1 + x_2 + \cdots + x_n}{n}.$$

In particular, $\mathcal{L}im$ *is an extension of the limit functional L.*
c. *For each* $x = (x_1, x_2, \ldots) \in \ell_\infty$, *we have*

$$\mathcal{L}im(x_1, x_2, x_3, \ldots) = \mathcal{L}im(x_2, x_3, x_4, \ldots).$$

Solution. For each $x = (x_1, x_2, \ldots) \in \ell_\infty$, let

$$A(x) = \left(x_1, \frac{x_1 + x_2}{2}, \ldots, \frac{x_1 + x_2 + \cdots + x_n}{n}, \ldots \right)$$

be the sequence of averages of x. If we define $p: \ell_\infty \to \mathbf{R}$ via the formula

$$p(x) = \limsup A(x) = \limsup_{n \to \infty} \frac{x_1 + x_2 + \cdots + x_n}{n},$$

then a glance at Problem 4.7 guarantees that p is a sublinear functional. Moreover, it is easy to see that if $x \in c$, then

$$L(x) = \lim_{n \to \infty} x_n = p(x).$$

Now, by the Hahn–Banach Theorem 29.2, L has an extension $\mathcal{L}im: \ell_\infty \to \mathbf{R}$ satisfying $\mathcal{L}im(x) \leq p(x)$ for each $x \in \ell_\infty$. Properties (a) and (b) can be established exactly as in the solution of Problem 29.6.

To verify (c), let $x = (x_1, x_2, \ldots) \in \ell_\infty$ and put $y = (x_2, x_3, \ldots)$. Then, an easy computation shows that

$$A(x - y) = \left(x_1 - x_2, \frac{x_1 - x_3}{2}, \frac{x_1 - x_4}{3}, \ldots, \frac{x_1 - x_{n+1}}{n}, \ldots \right).$$

Since $x = (x_1, x_2, \ldots)$ is a bounded sequence, the latter implies

$$p(x - y) = \limsup \frac{x_1 - x_{n+1}}{n} = 0.$$

Hence, $\mathcal{L}im(x - y) \leq p(x - y) = 0$. Similarly, $L(y - x) \leq 0$, and so $\mathcal{L}im(x) - \mathcal{L}im(y) = \mathcal{L}im(x - y) = 0$. Thus, $\mathcal{L}im(x) = \mathcal{L}im(y)$, as desired.

Problem 29.8. *Let X be a normed vector space. Show that if X^* is separable (in the sense that it contains a countable dense subset), then X is also separable.*

Solution. Let $\{f_1, f_2, \ldots\}$ be a countable dense subset of X^*. For each n choose some $x_n \in X$ with $\|x_n\| = 1$ and $|f_n(x_n)| \geq \frac{1}{2}\|f_n\|$, and let Y be the closed subspace generated by $\{x_1, x_2, \ldots\}$. We claim that $Y = X$.

To see this, assume by way of contradiction that $Y \neq X$. Fix some $a \notin Y$ with $\|a\| = 1$. By Theorem 29.5, there exists some $f \in X^*$ with $f(y) = 0$ for all $y \in Y$ and $f(a) \neq 0$. Given $\varepsilon > 0$ choose some n with $\|f - f_n\| < \varepsilon$, and note that

$$\left|f_n(a)\right| \leq \|f_n\| \leq 2\left|f_n(x_n)\right| = 2\left|(f_n - f)(x_n)\right| \leq 2\|f_n - f\| < 2\varepsilon.$$

Thus, $|f(a)| \leq |f(a) - f_n(a)| + |f_n(a)| < 3\varepsilon$ holds for all $\varepsilon > 0$, and so $f(a) = 0$, a contradiction. Therefore, $Y = X$ holds.

Now, note that the collection of all finite linear combinations of the countable set $\{x_1, x_2, \ldots\}$ with rational coefficients is a countable dense subset of X.

Problem 29.9. *Show that a Banach space X is reflexive if and only if X^* is reflexive.*

Solution. Assume that X^* is reflexive. If $X \neq X^{**}$, then by Theorem 29.5 there exists some nonzero $F \in X^{***}$ with $F(x) = 0$ for each $x \in X$. Since X^* is reflexive, there exists a nonzero $x^* \in X^*$ so that $F(f) = f(x^*)$ holds for all $f \in X^{**}$. In particular,

$$x^*(x) = \hat{x}(x^*) = F(x) = 0$$

holds for all $x \in X$, and so $x^* = 0$, a contradiction. Therefore, X must be a reflexive Banach space.

Problem 29.10. *This problem describes the adjoint of a bounded operator. If $T: X \to Y$ is a bounded operator between two normed spaces, then the **adjoint** of T is the operator $T^*: Y^* \to X^*$ defined by $(T^*f)(x) = f(Tx)$ for all $f \in Y^*$ and all $x \in X$. (Writing $h(x) = \langle x, h \rangle$, the definition of the adjoint operator is written in "duality" notation as*

$$\langle Tx, f \rangle = \langle x, T^*f \rangle$$

for all $f \in Y^*$ and all $x \in X$.)

a. *Show that $T^*: Y^* \to X^*$ is a well-defined bounded linear operator whose norm coincides with that of T, i.e., $\|T^*\| = \|T\|$.*

b. *Fix some $g \in X^*$ and some $u \in Y$ and define $S: X \to Y$ by $S(x) = g(x)u$. Show that S is a bounded linear operator satisfying $\|S\| = \|g\| \cdot \|u\|$. (Any such operator S is called a **rank-one** operator.)*

c. *Describe the adjoint of the operator S defined in part (b).*

d. *Let $A = [a_{ij}]$ be an $m \times n$ matrix with real entries. As usual, we consider the adjoint A^* as a (bounded) linear operator from $\mathbf{R}^n$ to $\mathbf{R}^m$. Describe A^*.*

Solution.   As usual, we shall denote from simplicity $T(x)$ by Tx.

(a) Fix $f \in Y^*$. Then, for $x, y \in X$ and $\alpha, \beta \in \mathbf{R}$, we have

$$(T^*f)(\alpha x + \beta y) = f(T(\alpha x + \beta y)) = f(\alpha Tx + \beta Ty)$$
$$= \alpha f(Tx) + \beta f(Ty) = \alpha(T^*f)(x) + \beta(T^*f)(y),$$

so that T^*f is a linear functional on X. To see that T^*f is also continuous, notice that

$$\left|(T^*f)(x)\right| = \left|f(Tx)\right| \leq \|f\| \cdot \|T(x)\| \leq \|f\| \cdot \|T\| \cdot \|x\|$$

holds for all $x \in X$. This shows that T^*f is a bounded (and hence, continuous) linear functional and that $\|T^*f\| \leq \|T\| \cdot \|f\|$ holds true for each $f \in Y^*$.

The last inequality also shows that $T^*: Y^* \to X^*$ is a bounded operator and that $\|T^*\| \leq \|T\|$. For the reverse inequality, let $x \in X$ satisfy $\|x\| \leq 1$. By Theorem 29.4 there exists some $h \in Y^*$ satisfying $\|h\| = 1$ and $h(Tx) = \|Tx\|$. So,

$$\|T^*\| \geq \|T^*h\| \geq \|T^*h(x)\| = \left\|h(Tx)\right\| = \|Tx\|,$$

for each $x \in X$ with $\|x\| \leq 1$. This implies $\|T\| = \sup\{\|Tx\|: \|x\| \leq 1\} \leq \|T^*\|$. Hence, $\|T^*\| = \|T\|$.

(b) It is a routine matter to verify that S is linear. From

$$\|S(x)\| = \|g(x)u\| = \left|g(x)\right| \cdot \|u\|$$
$$\leq \|g\| \cdot \|x\| \cdot \|u\| = \big(\|g\| \cdot \|u\|\big)\|x\|,$$

we see that S is a bounded operator and that $\|S\| \leq \|g\| \cdot \|u\|$. Now, if $x \in X$

satisfies $\|x\| \leq 1$, then we have

$$\|S\| \geq \|S(x)\| = \|g(x)u\| \geq |g(x)| \cdot \|u\|,$$

and so $\|S\| \geq \sup\{|g(x)| \cdot \|u\|: x \in X$ and $\|x\| \leq 1\} = \|g\| \cdot \|u\|$. The preceding show that $\|S\| = \|g\| \cdot \|u\|$.

(c) Note that for each $f \in Y^*$ and each $x \in X$ we have

$$(S^*f)(x) = f(Sx) = f(g(x)u) = f(u)g(x) = \big[f(u)g\big](x).$$

So, $S^*f = f(u)g$ holds for all $f \in Y^*$.

(d) Let $A = [a_{ij}]$ be an $m \times n$ real matrix. Note that the norm dual of $\mathbf{R}^n$ is again $\mathbf{R}^n$, where every $y = (y_1, \ldots, y_n) \in \mathbf{R}^n$ defines a linear functional on $\mathbf{R}^n$ via the formula

$$y(x) = \langle x, y \rangle = x \cdot y = \sum_{i=1}^{n} x_i y_i.$$

This easily implies that the adjoint A^* of A is an $n \times m$ matrix $B = [b_{ij}]$ that satisfies the duality identity $\langle Ax, y \rangle = \langle x, A^*y \rangle$, or $Ax \cdot y = x \cdot By$. That is, the elements of B satisfy the equation

$$\sum_{j=1}^{m} \sum_{i=1}^{n} a_{ij} x_i y_j = \sum_{i=1}^{n} \sum_{j=1}^{m} b_{ji} y_j x_i$$

for all $x \in \mathbf{R}^n$ and all $y \in \mathbf{R}^m$. This easily implies $b_{ij} = a_{ji}$ for all i and j. Therefore, A^* is the transpose of A, i.e. $A^* = A^t$.

30. BANACH LATTICES

Problem 30.1. *Let X be a vector lattice, and let $f: X^+ \rightarrow [0, \infty)$ be an additive function (that is, $f(x + y) = f(x) + f(y)$ holds for all $x, y \in X^+$). Then show that there exists a unique linear functional g on X such that $g(x) = f(x)$ holds for all $x \in X^+$.*

Solution. Note first that if $x \geq y \geq 0$ holds, then

$$f(x) = f\big(y + (x - y)\big) = f(y) + f(x - y) \geq f(y).$$

Also, the arguments of the proof of Lemma 18.7 show that $f(rx) = rf(x)$ holds for all $x \in X^+$ and all rational numbers $r \geq 0$.

Now, let $\alpha > 0$ and $x \geq 0$. Pick two sequences $\{r_n\}$ and $\{t_n\}$ of rational numbers with $0 \leq r_n \uparrow \alpha$ and $t_n \downarrow \alpha$. Then, the inequality $r_n x \leq \alpha x \leq t_n x$ implies

$$r_n f(x) = f(r_n x) \leq f(\alpha x) \leq f(t_n x) = t_n f(x),$$

from which it follows that $\alpha f(x) = f(\alpha x)$ holds.

Now, define $g: X \longrightarrow \mathbb{R}$ by

$$g(x) = f(x^+) - f(x^-).$$

Note that if $x = y - z$ holds with $y, z \in X^+$, then the relation $x^+ + z = y + x^-$, coupled with the additivity of f on X^+, shows that $f(x^+) + f(z) = f(y) + f(x^-)$. That is,

$$g(x) = f(x^+) - f(x^-) = f(y) - f(z).$$

In particular, for $x, y \in X$ we have

$$
\begin{aligned}
g(x + y) = g\big(x^+ + y^+ - (x^- + y^-)\big) &= f(x^+ + y^+) - f(x^- + y^-) \\
&= f(x^+) + f(y^+) - f(x^-) - f(y^-) \\
&= \big[f(x^+) - f(x^-)\big] + \big[f(y^+) - f(y^-)\big] \\
&= g(x) + g(y).
\end{aligned}
$$

Moreover, for $\alpha > 0$ we have

$$g(\alpha x) = f(\alpha x^+) - f(\alpha x^-) = \alpha\big[f(x^+) - f(x^-)\big] = \alpha g(x),$$

and if $\alpha < 0$, then

$$g(\alpha x) = -\alpha g(-x) = -\alpha\big[g(x^- - x^+)\big] = -\alpha\big[f(x^-) - f(x^+)\big] = \alpha g(x).$$

Thus, g is a linear functional on X, which is clearly a unique extension of f.

Problem 30.2. *A vector lattice is called* **order complete** *if every nonempty subset that is bounded from above has a least upper bound (also called the supremum of the set).*

Show that if X is a vector lattice, then its order dual $X^\sim$ is an order complete vector lattice.

Solution. Let A be a nonempty subset of $X^\sim$ that is bounded from above by some $g \in X^\sim$. By replacing A with the set $\{g - f: f \in A\}$, we can assume that $A \subseteq X_+^\sim$. Let B denote the collection of all finite suprema of A, i.e., $f \in B$ if and only if there exist $f_1, \ldots, f_n \in A$ with $f = \bigvee_{i=1}^n f_i$. Clearly, $f \leq g$ also holds for all $f \in B$. Next, define $h: X^+ \longrightarrow \mathbf{R}^+$ by

$$h(x) = \sup\{f(x): f \in B\}$$

for each $x \in X^+$. Clearly, $0 \leq h(x) \leq g(x)$ holds.

Let $x, y \in X^+$. Since $f(x + y) = f(x) + f(y) \leq h(x) + h(y)$ holds for all $f \in B$, we see that

$$h(x + y) \leq h(x) + h(y).$$

On the other hand, given $\varepsilon > 0$ choose $f_1, f_2 \in B$ such that $h(x) - \varepsilon < f_1(x)$ and $h(y) - \varepsilon < f_2(y)$. Taking into account that $f_1 \vee f_2 \in B$, we see that

$$h(x) + h(y) - 2\varepsilon \leq f_1(x) + f_2(y) \leq f_1 \vee f_2(x) + f_1 \vee f_2(y)$$
$$= f_1 \vee f_2(x + y) \leq h(x + y)$$

holds, for all $\varepsilon > 0$. Thus,

$$h(x) + h(y) \leq h(x + y)$$

also holds, and so $h(x + y) = h(x) + h(y)$.

By the preceding problem, h extends uniquely to a positive linear functional. Clearly, $f \leq h$ holds for all $f \in A$. On the other hand, if $f \leq \phi$ holds for all $f \in A$, then $f \leq \phi$ also holds for all $f \in B$. This easily implies $h \leq \phi$. That is, $h = \sup A$ holds in $X^\sim$.

Problem 30.3. *Show that the collection of all bounded functions on $[0, 1]$ is an ideal of $\mathbf{R}^{[0,1]}$. Also, show that $C[0, 1]$ is a vector sublattice of $\mathbf{R}^{[0,1]}$ but not an ideal.*

Solution. Let $f: [0, 1] \longrightarrow \mathbf{R}$ be a bounded function. If $|f(x)| \leq M$ holds for all $x \in [0, 1]$ and $g \in \mathbf{R}^{[0,1]}$ satisfies $|g| \leq |f|$, then $|g(x)| \leq M$ also holds for all $x \in [0, 1]$. This implies that the space of all bounded functions is an ideal of $\mathbf{R}^{[0,1]}$.

The function $\chi_{(0,\frac{1}{2})}$ is not a continuous function and satisfies $0 \leq \chi_{(0,\frac{1}{2})} \leq \mathbf{1}$, where $\mathbf{1}$ denotes the constant function one on $[0, 1]$. Hence, $C[0, 1]$ is not an ideal of $\mathbf{R}^{[0,1]}$.

Problem 30.4. *Let X be a vector lattice. Show that a norm $\| \cdot \|$ on X is a lattice norm if and only if it satisfies the following two properties:*

a. *If $0 \le x \le y$, then $\|x\| \le \|y\|$, and*
b. *$\|x\| = \| |x| \|$ holds for all $x \in X$.*

Solution. Assume that $\| \cdot \|$ is a lattice norm. Clearly, $0 \le x \le y$ implies $\|x\| \le \|y\|$. Also, $|x| = \big||x|\big|$ holds, and so $\|x\| \le \big\||x|\big\| \le \|x\|$.

Conversely, assume (a) and (b) to be true. If $|x| \le |y|$, then

$$\|x\| = \big\||x|\big\| \le \big\||y|\big\| = \|y\|$$

so that $\| \cdot \|$ is a lattice norm.

Problem 30.5. *Show that in a normed vector lattice X its positive cone X^+ is a closed set.*

Solution. From Theorem 30.1(3) we see that

$$\big|x^- - y^-\big| = \big|(-x)^+ - (-y)^+\big| \le \big|x - y\big|.$$

This implies that the function $x \longmapsto x^-$ from X into X is (uniformly) continuous. Thus, $X^+ = \{x \in X \colon x^- = 0\}$ is a closed set.

Problem 30.6. *Let X be a normed vector lattice. Assume that $\{x_n\}$ is a sequence of X such that $x_n \le x_{n+1}$ holds for all n. Show that if $\lim x_n = x$ holds in X, then the vector x is the least upper bound of the sequence $\{x_n\}$ in X. In symbols, $x_n \uparrow x$ holds.*

Solution. Assume that $\{x_n\}$ satisfies $x_n \le x_{n+1}$ for each n and $\lim x_n = x$. Then, $x_{n+p} - x_n \ge 0$ holds for all n and all p and $\lim_{p \to \infty}(x_{n+p} - x_n) = x - x_n$. Since (by Problem 30.5) the positive cone X^+ is closed, we see that $x - x_n \ge 0$, or $x \ge x_n$ for each n. This shows that x is an upper bound for the sequence $\{x_n\}$.

To see that x is the least upper bound for the sequence $\{x_n\}$, assume that $y \ge x_n$ holds for each n. So, $y - x_n \ge 0$ holds for all n and $\lim(y - x_n) = y - x$. Using once more that X^+ is closed, we get $y - x \in X^+$. That is, $y - x \ge 0$, or $y \ge x$. Therefore, $x = \sup\{x_n\}$, or $x_n \uparrow x$ holds true, as desired.

Problem 30.7. *Assume that $x_n \to x$ holds in a Banach lattice and let $\{\epsilon_n\}$ be a sequence of strictly positive real numbers. Show that there exists a subsequence $\{x_{k_n}\}$ of $\{x_n\}$ and some positive vector u such that $|x_{k_n} - x| \le \epsilon_n u$ holds for each n.*

Solution. An easy inductive argument guarantees the existence of a subsequence $\{x_{k_n}\}$ of $\{x_n\}$ satisfying $\|x_{k_n} - x\| \le \epsilon_n 2^{-n}$ for each n. Now, notice that the series of positive vectors $\sum_{n=1}^{\infty}(\epsilon_n)^{-1}|x_{k_n}-x|$ is absolutely summable. Since X is a Banach space, $u=\sum_{n=1}^{\infty}(\epsilon_n)^{-1}|x_{k_n}-x|$ exists in X. Now, a glance at Problem 30.6 shows that $(\epsilon_n)^{-1}|x_{k_n}-x| \le u$ for each n. Thus, $|x_{k_n}-x| \le \epsilon_n u$ holds for each n, as desired.

Problem 30.8. *Let $T: X \to Y$ be a positive operator between two normed vector lattices, i.e, $x \ge 0$ in X implies $Tx \ge 0$ in Y. If X is a Banach lattice, then show that T is continuous.*

Solution. Let $T: X \longrightarrow Y$ be a positive operator, where X is a Banach lattice and Y is a normed vector lattice. Assume by way of contradiction that T is not continuous. Then, there exist a sequence $\{x_n\}$ of X and some $\epsilon > 0$ satisfying $x_n \to 0$ and $\|Tx_n\| \ge \epsilon$ for each n. By Problem 30.7 there exists a subsequence $\{y_n\}$ of $\{x_n\}$ and some $u \in X^+$ satisfying $|y_n| \le \frac{1}{n}u$ for each n. Now, notice that the positivity of T implies $|Ty_n| \le T|y_n| \le \frac{1}{n}Tu$ for each n, and so $\|Ty_n\| \le \frac{1}{n}\|Tu\|$ for each n. Since $\frac{1}{n}\|Tu\| \to 0$, it follows that $\|Ty_n\| \to 0$ contrary to $\|Ty_n\| > \epsilon$ for each n. Thus, T is a continuous operator.

Problem 30.9. *Show that any two complete lattice norms on a vector lattice must be equivalent.*

Solution. If $\|\cdot\|_1$ and $\|\cdot\|_2$ are two complete lattice norms on a vector lattice X, then, by Problem 30.8, the identity operator $I: (X, \|\cdot\|_1) \longrightarrow (X, \|\cdot\|_2)$ is a homeomorphism. That is, $\|\cdot\|_1$ and $\|\cdot\|_2$ are two equivalent norms.

Problem 30.10. *The **averaging operator** $A: \ell_\infty \to \ell_\infty$ is defined by*

$$A(x) = \left(x_1, \frac{x_1+x_2}{2}, \frac{x_1+x_2+x_3}{3}, \dots, \frac{x_1+x_2+\cdots+x_n}{n}, \dots\right)$$

for each $x = (x_1, x_2, \dots) \in \ell_\infty$. Establish the following:

a. *A is a positive operator.*
b. *A is a continuous operator.*
c. *The vector space*

$$V = \left\{x = (x_1, x_2, \dots) \in \ell_\infty: \left\{\tfrac{x_1+x_2+\cdots+x_n}{n}\right\} \text{ converges in } \mathbf{R}\right\}$$

is a closed subspace of ℓ_∞. Is $V = \ell_\infty$?

Solution. (a) If $x = (x_1, x_2, \dots) \ge 0$, then $x_i \ge 0$ for each i and so $\frac{x_1+x_2+\cdots+x_n}{n} \ge 0$ for each n. This implies $A(x) \ge 0$, and so A is a positive operator.

(b) By Problem 30.8 every positive operator on a Banach lattice is continuous. Therefore, A (as a positive operator) is continuous.

(c) We know from Problem 29.5 that the vector space of all convergent sequences

$$c = \{x = (x_1, x_2, \ldots) \in \ell_\infty\colon \lim_{n \to \infty} x_n \text{ exists in } \mathbb{R}\}$$

is a closed subspace of ℓ_∞. Clearly, $V = A^{-1}(c)$. Since A is continuous, the latter guarantees that V is a closed subspace of ℓ_∞.

There are bounded sequences having divergent sequences of averages. Here is an example:

$$\big(1, -1, 1, 1, -1, -1, 1, 1, 1, 1, 1, 1, 1, -1, -1, -1, -1, -1, -1, \ldots\big).$$

Hence, V is a proper closed subspace of ℓ_∞.

Problem 30.11. *This problem shows that for a normed vector lattice X its norm dual X^* may be a proper ideal of its order dual $X^\sim$. Let X be the collection of all sequences $\{x_n\}$ such that $x_n = 0$ for all but a finite number of terms (depending on the sequence). Show that:*

a. *X is a function space.*
b. *X equipped with the sup norm is a normed vector lattice, but not a Banach lattice.*
c. *If $f\colon X \to \mathbb{R}$ is defined by $f(x) = \sum_{n=1}^\infty n x_n$ for each $x = \{x_n\} \in X$, then f is a positive linear functional on X that is not continuous.*

Solution. (a) Routine.

(b) If $\mathbf{x}_n = (1, \frac{1}{2}, \frac{1}{3}, \ldots, \frac{1}{n}, \ldots)$, then $\{\mathbf{x}_n\}$ is Cauchy sequence of X that does not converge in X.

(c) Clearly, f is a positive linear functional. If $\mathbf{e}_n$ denotes the sequence whose n^{th} component is one and every other zero, then $\|\mathbf{e}_n\|_\infty = 1$ and $n = f(\mathbf{e}_n) \le \|f\|$. That is, $\|f\| = \infty$.

Problem 30.12. *Determine the norm completion of the normed vector lattice of the preceding problem.*

Solution. Let

$$c_0 = \big\{x = (x_1, x_2, \ldots) \in \ell_\infty\colon \lim_{n \to \infty} x_n = 0\big\}.$$

Clearly, c_0 is a vector sublattice of ℓ_∞. Also, it is not difficult to see that c_0 is a closed subspace, and so c_0 is a Banach lattice (with the sup norm). We claim that c_0 is the norm completion of the normed vector lattice X of the preceding problem.

To see this, note first that X is a vector sublattice of c_0. Now, let $x = (x_1, x_2, \ldots) \in c_0$ and let $\varepsilon > 0$. Choose some n with $|x_k| < \varepsilon$ for all $k \geq n$, and note that the element $y = (x_1, \ldots, x_n, 0, 0, \ldots) \in X$ satisfies $\|x - y\|_\infty \leq \varepsilon$. Thus, X is dense in c_0, and our claim follows.

Problem 30.13. *Let $C_c(X)$ be the normed vector lattice—with the sup norm—of all continuous real-valued functions on a Hausdorff locally compact topological space X. Determine the norm completion of $C_c(X)$.*

Solution. Consider the vector space of functions

$$c_0(X) = \left\{ f \in C(X) \colon \forall \, \varepsilon > 0 \, \exists \, K \text{ compact with } |f(x)| < \varepsilon \text{ for } x \notin K \right\}.$$

Clearly, $c_0(X)$ is a vector sublattice of $B(X)$. We claim that $c_0(X)$ is a closed subspace. To see this, let $\{f_n\} \subseteq c_0(X)$ satisfy $f_n \longrightarrow f$ in $B(X)$, and let $\varepsilon > 0$. By Theorem 9.2, $f \in C(X)$. Pick some n with $\|f - f_n\|_\infty < \varepsilon$, and then select a compact set K with $|f_n(x)| < \varepsilon$ for $x \notin K$. Thus,

$$\left| f(x) \right| \leq \left| f(x) - f_n(x) \right| + \left| f_n(x) \right| < 2\varepsilon$$

holds for each $x \notin K$, and so $f \in c_0(X)$. Therefore, $c_0(X)$ (with the sup norm) is a Banach lattice.

Clearly, $C_c(X)$ is a vector sublattice of $c_0(X)$, and we claim that $C_c(X)$ is dense in $c_0(X)$. To see this, let $f \in c_0(X)$ and let $\varepsilon > 0$. Choose some compact set K with $|f(x)| < \varepsilon$ for all $x \notin K$, and then use Theorem 10.8 to pick some $g \in C_c(X)$ with $g(x) = 1$ for all $x \in K$ and $0 \leq g(x) \leq 1$ for $x \notin K$. Then, $fg \in C_c(X)$ and $\|fg - f\|_\infty \leq \varepsilon$ holds, proving that $\overline{C_c(X)} = c_0(X)$. Thus, $c_0(X)$ is the norm completion of $C_c(X)$.

Problem 30.14. *Let X and Y be two vector lattices, and let $T \colon X \to Y$ be a linear operator. Show that the following statements are equivalent:*

   a. $T(x \vee y) = T(x) \vee T(y)$ *holds for all $x, y \in X$*
   b. $T(x \wedge y) = T(x) \wedge T(y)$ *holds for all $x, y \in X$.*
   c. $T(x) \wedge T(y) = 0$ *holds in Y whenever $x \wedge y = 0$ holds in X.*
   d. $|T(x)| = T(|x|)$ *holds for all $x \in X$.*

*(A linear operator T that satisfies the preceding equivalent statements is referred to as a **lattice homomorphism**.)*

Solution. $(1) \Longrightarrow (2)$ From the identity (a) of Problem 9.1, we get

$$T(x \wedge y) = T(x + y - x \vee y) = T(x) + T(y) - T(x \vee y)$$
$$= T(x) + T(y) - T(x) \vee T(y) = T(x) \wedge T(y).$$

$(2) \Longrightarrow (3)$ If $x \wedge y = 0$, then

$$T(x) \wedge T(y) = T(x \wedge y) = T(0) = 0.$$

$(3) \Longrightarrow (4)$ Using the identity (e) of Problem 9.1, we see that

$$
\begin{aligned}
|T(x)| &= |T(x^+) - T(x^-)| = T(x^+) \vee T(x^-) - T(x^+) \wedge T(x^-) \\
&= T(x^+) \vee T(x^-) = T(x^+) + T(x^-) - T(x^+) \wedge T(x^-) \\
&= T(x^+) + T(x^-) = T(x^+ + x^-) = T(|x|).
\end{aligned}
$$

$(4) \Longrightarrow (1)$ From the identity (f) of Problem 9.1, we get

$$
\begin{aligned}
T(x \vee y) &= T\left(\tfrac{1}{2}[x + y + |x - y|]\right) = \tfrac{1}{2}\left[T(x) + T(y) + T(|x - y|)\right] \\
&= \tfrac{1}{2}\left[T(x) + T(y) + |T(x) - T(y)|\right] = T(x) \vee T(y).
\end{aligned}
$$

Problem 30.15. *Let ℓ_∞ be the Banach lattice of all bounded real sequences; that is, $\ell_\infty = B(\mathbb{N})$, and let $\{r_1, r_2, \ldots\}$ be an enumeration of the rational numbers of $[0, 1]$. Show that the mapping $T: C[0, 1] \to \ell_\infty$ defined by $T(f) = (f(r_1), f(r_2), \ldots)$ is a lattice isometry that is not onto.*

Solution. Clearly, T is a linear operator. Let $f \in C[0, 1]$. Since f is a continuous function and the set of all rational numbers of $[0, 1]$ is a dense set, it easily follows that

$$
\begin{aligned}
\|T(f)\|_\infty &= \sup\{|f(r_n)|: n = 1, 2, \ldots\} \\
&= \sup\{|f(x)|: x \in [0, 1]\} = \|f\|_\infty.
\end{aligned}
$$

In addition, note that

$$|T(f)| = (|f(r_1)|, |f(r_2)|, \ldots) = (|f|(r_1), |f|(r_2), \ldots) = T(|f|),$$

which shows that T is a lattice isometry.

To see that T is not onto, note that

$$T(f) \neq (0, 1, 0, 1, \ldots)$$

holds for each $f \in C[0, 1]$.

Problem 30.16. *Let X be a normed vector lattice. Then show that an element $x \in X$ satisfies $x \geq 0$ if and only if $f(x) \geq 0$ holds for each continuous positive linear functional f on X.*

Solution. If $x \geq 0$ holds, then clearly $f(x) \geq 0$ also holds for each $0 \leq f \in X^\sim$.

For the converse, assume that x is fixed and satisfies $f(x) \geq 0$ for each $f \in X^\sim_+$. Let $0 \leq f \in X^\sim$ be fixed. Since $-g(x) \leq 0$ holds for all $0 \leq g \leq f$, it follows from Theorem 30.3 that

$$0 \leq f(x^-) = \sup\{-g(x): g \in X^\sim \text{ and } 0 \leq g \leq f\} \leq 0.$$

That is, $f(x^-) = 0$ holds for all $0 \leq f \in X^\sim$, and consequently $f(x^-) = 0$ for all $f \in X^*$. From Theorem 29.4, we see that $x^- = 0$. Thus, $x = x^+ - x^- = x^+ \geq 0$, as required.

Problem 30.17. *Let X be a Banach lattice. If $0 \leq x \in X$, then show that*

$$\|x\| = \sup\{f(x): 0 \leq f \in X^* \text{ and } \|f\| = 1\}.$$

Solution. Let $x \geq 0$. In view of the inequality $|f(x)| \leq |f|(x) \leq \|f\| \cdot \|x\|$, we have

$$\begin{aligned}
\|x\| &= \sup\{|f(x)|: f \in X^* \text{ and } \|f\| = 1\} \\
&\leq \sup\{|f|(x): f \in X^* \text{ and } \|f\| = 1\} \\
&= \sup\{f(x): 0 \leq f \in X^* \text{ and } \|f\| = 1\} \leq \|x\|,
\end{aligned}$$

and the conclusion follows.

Problem 30.18. *Assume that $\varphi: [0, 1] \to \mathbb{R}$ is a strictly monotone continuous function and that $T: C[0, 1] \to C[0, 1]$ is a continuous linear operator. If $T(\varphi f) = \varphi T(f)$ holds for each $f \in C[0, 1]$ (where φf denotes the pointwise product of φ and f). Show that there exists a unique function $h \in C[0, 1]$ satisfying $T(f) = hf$ for all $f \in C[0, 1]$.*

Solution. Taking $f = 1$, the constant one function, and letting $h = T\mathbf{1}$, we obtain $T(\varphi) = h\varphi$, and by induction $T(\varphi^n) = h\varphi^n$ for each $n \geq 0$. Hence, by the linearity of T, we see that

$$T(P(\varphi)) = hP(\varphi) \tag{$\star$}$$

for each polynomial P of one variable. Since the function φ is strictly increasing, the algebra $\mathcal{A} = \{P(\varphi)\colon\ P\ \text{polynomial}\}$ separates the points and contains the constant function $\mathbf{1}$. Consequently, by the Stone–Weierstrass Theorem 11.5, $\mathcal{A}$ is dense in $C[0, 1]$. From $(\star)$, it easily follows that $T(f) = hf$ for each $f \in C[0, 1]$.

Problem 30.19. *If $f \in C[0, 1]$, then the polynomials*

$$B_n(x) = \sum_{k=0}^{n} \binom{n}{k} f(\tfrac{k}{n}) x^k (1 - x)^{n-k},$$

where $\binom{n}{k}$ is the binomial coefficient defined by $\binom{n}{k} = \frac{n!}{k!(n-k)!}$, are known as the **Bernstein polynomials** *of f.*

Show that if $f \in C[0, 1]$, then the sequence $\{B_n\}$ of Bernstein polynomials of f converges uniformly to f.

Solution. Let $\{T_n\}$ be the sequence of positive operators from $C[0, 1]$ into $C[0, 1]$ defined by

$$T_n f(t) = \sum_{k=0}^{n} \binom{n}{k} f(\tfrac{k}{n}) t^k (1 - t)^{n-k}$$

for all $f \in C[0, 1]$ and each $t \in [0, 1]$. We must show that

$$\lim \|T_n f - f\|_\infty = 0$$

holds for each $f \in C[0, 1]$. By Korovkin's Theorem 30.13, it suffices to establish that $\lim \|T_n f - f\|_\infty = 0$ holds for $f = 1$, x, and x^2.

To do this, we need some elementary identities. First note that by the binomial theorem

$$\sum_{k=0}^{n} \binom{n}{k} t^k (1 - t)^{n-k} = \left[t + (1 - t)\right]^n = 1 \qquad (\star)$$

holds for all t. Differentiating $(\star)$, we get

$$\sum_{k=0}^{n} \binom{n}{k}\left[kt^{k-1}(1 - t)^{n-k} - (n - k)t^k(1 - t)^{n-k-1}\right]$$

$$= \sum_{k=0}^{n} \binom{n}{k} t^{k-1}(1 - t)^{n-k-1}(k - nt) = 0.$$

Multiplication by $t(1-t)$ yields

$$\sum_{k=0}^{n} \binom{n}{k} t^k (1-t)^{n-k} (k-nt) = 0,$$

and by using $(\star)$, we see that

$$\sum_{k=0}^{n} \binom{n}{k} \frac{k}{n} t^k (1-t)^{n-k} = t. \qquad (\star\star)$$

Differentiating $(\star\star)$ yields

$$\sum_{k=0}^{n} \binom{n}{k} \frac{k}{n} t^{k-1} (1-t)^{n-k-1} (k-nt) = 1,$$

and multiplying by $t(1-t)$, we get

$$\sum_{k=0}^{n} \binom{n}{k} \frac{k}{n} t^k (1-t)^{k-n} (k-nt) = t(1-t).$$

That is,

$$\sum_{k=0}^{n} \binom{n}{k} \left(\frac{k}{n}\right)^2 t^k (1-t)^{k-n} - t \sum_{k=0}^{n} \binom{n}{k} \frac{k}{n} t^k (1-t)^{k-n} = \frac{t(1-t)}{n},$$

and by taking into account $(\star\star)$, we see that

$$\sum_{k=0}^{n} \binom{n}{k} \left(\frac{k}{n}\right)^2 t^k (1-t)^{n-k} - t^2 = \frac{t(t-1)}{n}. \qquad (\star\star\star)$$

The identities $(\star)$, $(\star\star)$, and $(\star\star\star)$ can be rewritten as follows:

$$T_n \mathbf{1} = \mathbf{1}, \quad T_n x = x, \quad \text{and} \quad \left[T_n x^2 - x^2\right](t) = \frac{t(1-t)}{n}.$$

Now, note that these identities readily imply that $\lim \|T_n f - f\|_\infty = 0$ holds for $f = \mathbf{1}$, x, and x^2.

Problem 30.20. *Let $T: C[0,1] \to C[0,1]$ be a positive operator. Show that if $Tf = f$ holds true when f equals $\mathbf{1}$, x, and x^2, then T is the identity operator (that is, $Tf = f$ holds for each $f \in C[0,1]$).*

Solution. For each n, let $T_n = T$. Clearly, $\lim T_n f = f$ holds in $C[0, 1]$ when $f = 1$, x, and x^2. By Korovkin's Theorem 30.13, we have $Tf = \lim T_n f = f$ for each $f \in C[0, 1]$.

Problem 30.21 (Korovkin). *Let $\{T_n\}$ be a sequence of positive operators from $C[0, 1]$ into $C[0, 1]$ satisfying $T_n 1 = 1$. If there exists some $c \in [0, 1]$ such that $\lim T_n g = 0$ holds for the function $g(t) = (t - c)^2$, then show that $\lim T_n f = f(c) \cdot 1$ holds for all $f \in C[0, 1]$.*

Solution. Let $f \in C[0, 1]$ and let $\varepsilon > 0$. It suffices to show that there exist constants C_1 and C_2 such that

$$\left\| T_n f - f(c) \cdot 1 \right\|_\infty \le \varepsilon + C_1 \left\| T_n 1 - 1 \right\|_\infty + C_2 \left\| T_n g \right\|_\infty$$

holds for all n.

Set $M = \| f \|_\infty$. By the continuity of f at the point c there exists some $\delta > 0$ such that $-\varepsilon < f(t) - f(c) < \varepsilon$ holds whenever $t \in [0, 1]$ satisfies $|t - c| < \delta$. Next, observe that

$$-\varepsilon - \tfrac{2M}{\delta^2}(t - c)^2 \le f(t) - f(c) \le \varepsilon + \tfrac{2M}{\delta^2}(t - c)^2 \qquad \textbf{(a)}$$

holds for all $t \in [0, 1]$. (To see this, repeat the arguments in the proof of Theorem 30.13.) Since each T_n is positive and linear, it follows from **(a)** that

$$-\varepsilon T_n 1 - \tfrac{2M}{\delta^2} T_n g \le T_n f - f(c) \cdot T_n 1 \le \varepsilon T_n 1 + \tfrac{2M}{\delta^2} T_n g.$$

Put $C = \tfrac{2M}{\delta^2}$, and note that

$$\left| T_n f - f(c) \cdot T_n 1 \right| \le \varepsilon T_n 1 + C T_n g = \varepsilon 1 + \varepsilon \left| T_n 1 - 1 \right| + C T_n g.$$

Consequently,

$$\begin{aligned} \left| T_n f - f(c) \cdot 1 \right| &\le \left| T_n f - f(c) \cdot T_n 1 \right| + \left| f(c) \right| \cdot \left| T_n 1 - 1 \right| \\ &\le \varepsilon 1 + (\varepsilon + |f(c)|) \left| T_n 1 - 1 \right| + C T_n g, \end{aligned}$$

and so

$$\left\| T_n f - f(c) \cdot 1 \right\|_\infty \le \varepsilon + (\varepsilon + |f(c)|) \left\| T_n 1 - 1 \right\|_\infty + C \left\| T_n g \right\|_\infty,$$

31. L_p-SPACES

Problem 31.1.   *Let $f \in L_p(\mu)$, and let $\epsilon > 0$. Show that*

$$\mu^*(\{x \in X: |f(x)| \geq \epsilon\}) \leq \epsilon^{-p} \int |f|^p \, d\mu.$$

Solution.   Consider the measurable set $E = \{x \in X: |f(x)| \geq \varepsilon\}$, and note that $E = \{x \in X: |f(x)|^p \geq \varepsilon^p\}$. Thus,

$$\int |f|^p \, d\mu \geq \int \chi_E |f|^p \, d\mu \geq \int \varepsilon^p \chi_E \, d\mu = \varepsilon^p \mu^*(E).$$

Problem 31.2.   *Let $\{f_n\}$ be a sequence of some $L_p(\mu)$-space with $1 \leq p < \infty$. Show that if $\lim \|f_n - f\|_p = 0$ holds in $L_p(\mu)$, then $\{f_n\}$ converges in measure to f.*

Solution.   From the preceding problem, we see that

$$\mu^*\big(\{x \in X: |f_n(x) - f(x)| \geq \varepsilon\}\big) \leq \varepsilon^{-p} \int |f_n - f|^p \, d\mu$$

holds. Clearly, this inequality shows that $f_n \xrightarrow{\mu} f$ holds whenever $\lim \|f_n - f\|_p = 0$.

Problem 31.3.   *Let $(X, \mathcal{S}, \mu)$ be a measure space and consider the set*

$$E = \big\{\chi_A: A \in \Lambda_\mu \text{ with } \mu^*(A) < \infty\big\}.$$

Show that E is a closed subset of $L_1(\mu)$ (and hence, a complete metric space in its own right). Use this conclusion and the identity

$$\mu(A \triangle B) = \int \big|\chi_A - \chi_B\big| \, d\mu = \|\chi_A - \chi_B\|_1$$

to provide an alternate solution to Problem 14.12(c).

Solution.   Assume that $\{\chi_{A_n}\}$ is a sequence of E such that $\int |\chi_{A_n} - f| \, d\mu \to 0$ holds for some $f \in L_1(\mu)$. By Lemma 31.6 there exists a subsequence $\{\chi_{A_{k_n}}\}$ of $\{\chi_{A_n}\}$ such that $\chi_{A_{k_n}} \to f$ a.e. This implies (how?) that $f = \chi_A$ a.e. for some $A \in \Lambda_\mu$ with $\mu^*(A) < \infty$. Thus, $f \in E$ and so E is a closed subset of $L_1(\mu)$.

Problem 31.4.   *Show that equality holds in the inequality*

$$a^t b^{1-t} \le ta + (1-t)b, \ 0 < t < 1; \ a \ge 0; \ b \ge 0$$

if and only if $a = b$. Use this to show that if $f \in L_p(\mu)$ and $g \in L_q(\mu)$, where $1 < p, q < \infty$ and $\frac{1}{p} + \frac{1}{q} = 1$, then $\int |fg| \, d\mu = \|f\|_p \cdot \|g\|_q$ holds if and only if there exist two constants C_1 and C_2 (not both zero) such that $C_1|f|^p = C_2|g|^q$ holds.

Solution.   Clearly, if $a = b \ge 0$, then $a^t b^{t-1} = ta + (1-t)b = a$ holds. For the converse, let $a^t b^{1-t} = ta + (1-t)b$ hold for some $a, b > 0$. Put $y = \frac{a}{b}$, and rewrite the given equality as $1 - t + ty - y^t = 0$. Since the function $f(x) = 1 - t + tx - x^t$ for $x \ge 0$ (and some fixed $0 < t < 1$) attains its minimum when $x = 1$ (see the proof of Lemma 31.2), it follows that $y = \frac{a}{b} = 1$, and so $a = b$. Thus, $a^t b^{1-t} = ta + (1-t)b$ holds if and only if $a = b$.

For the second part, assume first that there exist two constants C_1 and C_2 (which are not both zero) such that $C_1|f|^p = C_2|g|^q$. We can assume $C_1 > 0$ and $C_2 \ge 0$. Then, we have

$$\int |fg| \, d\mu = \int \left(\tfrac{C_2}{C_1}\right)^{\frac{1}{p}} |g|^{\frac{q}{p}} |g| \, d\mu = \left(\tfrac{C_2}{C_1}\right)^{\frac{1}{p}} \int |g|^q \, d\mu$$

$$= \left[\int \left(\tfrac{C_2}{C_1}\right) |g|^q \, d\mu\right]^{\frac{1}{p}} \cdot \left[\int |g|^q \, d\mu\right]^{\frac{1}{q}}$$

$$= \left(\int |f|^p \, d\mu\right)^{\frac{1}{p}} \cdot \left(\int |g|^q \, d\mu\right)^{\frac{1}{q}} = \|f\|_p \cdot \|g\|_q.$$

For the converse, assume $\int |fg| \, d\mu = \|f\|_p \cdot \|g\|_q$. If either f or g is zero, then the conclusion is trivial. (If $f = 0$, then put $C_1 = 1$ and $C_2 = 0$.) So, we can assume $f \ne 0$ and $g \ne 0$. Taking $t = \frac{1}{p}$, $a = \left(\frac{|f(x)|}{\|f\|_p}\right)^p$, and $b = \left(\frac{|g(x)|}{\|g\|_q}\right)^q$, the inequality $a^t b^{1-t} \le ta + (1-t)b$ gives

$$0 \le \tfrac{1}{p}\left(\tfrac{|f(x)|}{\|f\|_p}\right)^p + \tfrac{1}{q}\left(\tfrac{|g(x)|}{\|g\|_q}\right)^q - \tfrac{|f(x)g(x)|}{\|f\|_p \cdot \|g\|_q}.$$

Integrating (and using our hypothesis), we get

$$0 \le \int \left[\tfrac{1}{p}\left(\tfrac{|f(x)|}{\|f\|_p}\right)^p + \tfrac{1}{q}\left(\tfrac{|g(x)|}{\|g\|_q}\right)^q - \tfrac{|f(x)g(x)|}{\|f\|_p \cdot \|g\|_q}\right] d\mu(x)$$

$$= \tfrac{1}{p} + \tfrac{1}{q} - \tfrac{\int |f(x)g(x)| \, d\mu(x)}{\|f\|_p \cdot \|g\|_q} = 1 - 1 = 0.$$

Consequently,

$$\frac{|f(x)g(x)|}{\|f\|_p \cdot \|g\|_q} = \frac{1}{p}\left(\frac{|f(x)|}{\|f\|_p}\right)^p + \frac{1}{q}\left(\frac{|g(x)|}{\|g\|_q}\right)^q$$

holds for almost all x, and by the first part of the problem, we see that $\left(\frac{|f(x)|}{\|f\|_p}\right)^p = \left(\frac{|g(x)|}{\|g\|_q}\right)^q$ holds, so that

$$\left(\|g\|_q\right)^q |f(x)|^p = \left(\|f\|_p\right)^p |g(x)|^q$$

holds for almost all x, as required.

Problem 31.5. *Assume that $\mu^*(X) = 1$ and $0 < p < q \leq \infty$. If f is in $L_q(\mu)$, then show that $\|f\|_p \leq \|f\|_q$ holds.*

Solution. Assume $\mu^*(X) = 1$ and $0 < p < q < \infty$. Let $f \in L_q(\mu)$. From Theorem 31.14, we know that $L_q(\mu) \subseteq L_p(\mu)$, and so $f \in L_p(\mu)$.

Put $r = \frac{q}{p} > 1$, and then choose $s > 1$ so that $\frac{1}{r} + \frac{1}{s} = 1$ holds. Since $|f|^p \in L_r(\mu)$ and $\mathbf{1} \in L_s(\mu)$, it follows from Hölder's inequality that

$$\left(\|f\|_p\right)^p = \int |f|^p \, d\mu = \int |f|^p \cdot 1 \, d\mu \leq \left(\int |f|^{pr} \, d\mu\right)^{\frac{1}{r}} \cdot \left(\int \mathbf{1}^s \, d\mu\right)^{\frac{1}{s}}$$

$$= \left(\int |f|^{pr} \, d\mu\right)^{\frac{1}{r}} = \left(\int |f|^q \, d\mu\right)^{\frac{p}{q}} = \left(\|f\|_q\right)^p.$$

Consequently, $\|f\|_p \leq \|f\|_q$ holds.

If $q = \infty$, then

$$\|f\|_p = \left(\int |f|^p \, d\mu\right)^{\frac{1}{p}} \leq \left(\int \left(\|f\|_\infty\right)^p d\mu\right)^{\frac{1}{p}} = \|f\|_\infty = \|f\|_q$$

also holds in this case.

Problem 31.6. *Let $f \in L_1(\mu) \cap L_\infty(\mu)$. Then show that*
   a.   *$f \in L_p(\mu)$ for each $1 < p < \infty$.*
   b.   *If $\mu^*(X) < \infty$, then $\lim_{p \to \infty} \|f\|_p = \|f\|_\infty$ holds.*

Solution. (a) If $M = \|f\|_\infty$, then the inequality

$$|f|^p = |f|^{p-1} \cdot |f| \leq M^{p-1} \cdot |f|$$

shows that $f \in L_p(\mu)$ for each $1 < p < \infty$.

(b) Let $\{p_n\}$ be a sequence of positive real numbers satisfying $p_n > 1$ for each n and $\lim p_n = \infty$. From the inequality

$$\|f\|_{p_n} = \left(\int |f|^{p_n}\, d\mu\right)^{\frac{1}{p_n}} \leq \|f\|_\infty [\mu^*(X)]^{\frac{1}{p_n}},$$

it follows that

$$\limsup \|f\|_{p_n} \leq \|f\|_\infty.$$

Let $0 < \varepsilon < M$. Then, the measurable set

$$E = \left\{x \in X\colon |f(x)| \geq \|f\|_\infty - \varepsilon\right\}$$

satisfies $\mu^*(E) > 0$. From $\left(\|f\|_\infty - \varepsilon\right)^{p_n} \chi_E \leq |f|^{p_n}$, we see that $\left(\|f\|_\infty - \varepsilon\right)$ $[\mu^*(E)]^{\frac{1}{p_n}} \leq \|f\|_{p_n}$, and so $\|f\|_\infty - \varepsilon \leq \liminf \|f\|_{p_n}$ holds for all $0 < \varepsilon < M$. That is,

$$\|f\|_\infty \leq \liminf \|f\|_{p_n}.$$

Thus, $\limsup \|f\|_{p_n} \leq \|f\|_\infty \leq \liminf \|f\|_{p_n}$ holds. This shows that $\lim \|f\|_{p_n} = \|f\|_\infty$, and from this it follows that $\lim_{p \to \infty} \|f\|_p = \|f\|_\infty$.

Problem 31.7. *Let $f \in L_2[0, 1]$ satisfy $\|f\|_2 = 1$ and $\int_0^1 f(x)\, d\lambda(x) \geq \alpha > 0$. Also, for each $\beta \in \mathbb{R}$ let $E_\beta = \{x \in [0, 1]\colon f(x) \geq \beta\}$. If $0 < \beta < \alpha$, show that*

$$\lambda(E_\beta) \geq (\beta - \alpha)^2.$$

(This inequality is known in the literature as the **Paley–Zygmund Lemma***.)*

Solution. Assume $f \in L_2[0, 1]$ satisfies the stated properties and let $0 < \beta < \alpha$. Then, note that

$$f - \beta \leq (f - \beta)\chi_{E_\beta} \leq f\chi_{E_\beta},$$

and so, from Hölder's inequality, it follows that

$$0 < \alpha - \beta \leq \int_0^1 f(x)\, d\lambda(x) - \beta = \int_0^1 \left[f(x) - \beta\right] d\lambda(x) \leq \int_0^1 f(x)\chi_{E_\beta}(x)\, d\lambda(x)$$

$$\leq \|f\|_2 \cdot \left[\lambda(E_\beta)\right]^{\frac{1}{2}} = \left[\lambda(E_\beta)\right]^{\frac{1}{2}}.$$

This implies $\lambda(E_\beta) \geq (\alpha - \beta)^2$.

Problem 31.8.   *Show that for $1 \le p < \infty$ each ℓ_p is a separable Banach lattice.*

Solution.   Let e_n denote the sequence whose n^{th} component is one and every other is zero. Also, denote by E the set of all finite linear combinations of $\{e_1, e_2, \ldots\}$ with rational coefficients. Clearly, E is a countable set, which we claim is also dense in ℓ_p whenever $1 \le p < \infty$.

To see this, let $x = (x_1, x_2, \ldots) \in \ell_p$ $(1 \le p < \infty)$, and let $\varepsilon > 0$. Fix some natural number n with $\sum_{i=n+1}^{\infty} |x_i|^p < \frac{\varepsilon^p}{2}$. Then, pick rational numbers $r_1, r_2, \ldots, r_n$ with $\sum_{i=1}^{n} |x_i - r_i|^p < \frac{\varepsilon^p}{2}$, and note that the element $a = (r_1, r_2, \ldots, r_n, 0, 0, \ldots) = r_1 e_1 + r_2 e_2 + \cdots + r_n e_n \in E$ satisfies

$$\|x - a\|_p = \left(\sum_{i=1}^{n} |x_i - r_i|^p + \sum_{i=n+1}^{\infty} |x_i|^p \right)^{\frac{1}{p}} < \left(\frac{\varepsilon^p}{2} + \frac{\varepsilon^p}{2} \right)^{\frac{1}{p}} = \varepsilon.$$

That is, the countable set E is dense in ℓ_p, and so each ℓ_p $(1 \le p < \infty)$ is a separable Banach lattice.

Problem 31.9.   *Show that ℓ_∞ is not separable.*

Solution.   Let $E = \{x_1, x_2, \ldots\}$ be a countable subset of ℓ_∞. Write $x_n = (x_1^n, x_2^n, \ldots)$ for each n. Now, define

$$y_n = \begin{cases} 0 & \text{if } |x_n^n| \ge 1 \\ 2 & \text{if } |x_n^n| < 1 . \end{cases}$$

Clearly, $y = (y_1, y_2, \ldots) \in \ell_\infty$ and

$$\|y - x_n\|_\infty \ge |y_n - x_n^n| \ge |y_n - |x_n^n|| \ge 1$$

holds for each n. Thus, $B(y, 1) \cap E = \emptyset$, and this shows that no countable subset of ℓ_∞ can be dense.

An alternate way of proving that ℓ_∞ is not separable is as follows: Consider the set F of all sequences whose coordinates are zero or one. By Problem 2.8, the set F is uncountable, and it is not difficult to see that $\|x - y\|_\infty = 1$ holds for each pair $x, y \in F$ with $x \ne y$. It follows that $\{B(x, 1) \colon x \in F\}$ is an uncountable collection of pairwise disjoint open balls. This easily implies that every dense subset of ℓ_∞ must be uncountable.

Problem 31.10.   *Show that $L_\infty([0, 1])$ (with the Lebesgue measure) is not separable.*

Solution. Write $f_x = \chi_{[0,x]}$, $0 < x < 1$. Since $\|f_x - f_y\|_\infty = 1$ holds whenever $x \neq y$, it follows that $\{B(f_x, 1): x \in (0, 1)\}$ is an uncountable collection of pairwise disjoint open balls of $L_\infty([0, 1])$. This easily implies that every dense subset of $L_\infty([0, 1])$ must be uncountable, and so $L_\infty([0, 1])$ is not a separable Banach lattice.

Problem 31.11. *Let X be a Hausdorff locally compact topological space, and fix a point $a \in X$. Let μ be the measure on X defined on all subsets of X by $\mu(A) = 1$ if $a \in A$ and $\mu(A) = 0$ if $a \notin A$. In other words, μ is the Dirac measure (see Example 13.4). Show that μ is a regular Borel measure and that $\mathrm{Supp}\,\mu = \{a\}$.*

Solution. The regularity of μ will be established first.
1) Clearly, $\mu(A) \leq 1$ holds for each $A \subseteq X$.
2) Let $B \subseteq X$. If $a \in B$, then

$$1 = \mu(B) \leq \inf\{\mu(\mathcal{O}): \mathcal{O} \text{ open and } B \subseteq \mathcal{O}\} \leq \mu(X) = 1.$$

On the other hand, if $a \notin B$, then use the open set $X \setminus \{a\}$ to see that

$$0 = \mu(B) \leq \inf\{\mu(\mathcal{O}): \mathcal{O} \text{ open and } B \subseteq \mathcal{O}\} \leq \mu(X \setminus \{a\}) = 0.$$

3) Let $B \subseteq X$. If $a \notin B$, then each subset C of B satisfies $\mu(C) = 0$, and so

$$0 = \sup\{\mu(K): K \text{ compact and } K \subseteq B\} \leq \mu(B) = 0.$$

Now, if $a \in B$, then using that $\{a\}$ is a compact subset of B we see that

$$1 = \mu(\{a\}) \leq \sup\{\mu(K): K \text{ compact and } K \subseteq B\} \leq \mu(B) = 1.$$

Thus, μ is a regular Borel measure. Since $\mu(X \setminus \{a\}) = 0$, it is easy to see that $\mathrm{Supp}\,\mu = \{a\}$ holds.

Problem 31.12. *If $g \in C^1[a, b]$ and $f \in L_1[a, b]$, then*

    a. *show that the function $F: [a, b] \to \mathbb{R}$ defined by $F(x) = \int_a^x f(t)\, d\lambda(t)$ is uniformly continuous, and*

    b. *establish the following "Integration by Parts" formula:*

$$\int_a^b g(x)f(x)\, d\lambda(x) = g(x)F(x)\Big|_a^b - \int_a^b g'(x)F(x)\, dx.$$

Solution. (a) The uniform continuity of F follows immediately from Problem 22.6.

(b) Start by choosing some constant $C > 0$ such that $|g(x)| \leq C$ and $|g'(x)| \leq C$ hold for each $x \in [a, b]$. Now, by Theorem 25.3 there exists a sequence of continuous functions $\{f_n\}$ satisfying $\lim \int_a^b |f - f_n| \, d\lambda = 0$. From Lemma 31.6, we can suppose (by passing to a subsequence if necessary) that there exists some function $0 \leq h \in L_1[a, b]$ satisfying $|f_n| \leq h$ a.e. for each n and $f_n \to f$ a.e. Let $F_n(x) = \int_a^x f_n(t) \, dt$, and note that by the "standard" Integration by Parts Formula we have

$$\int_a^b g(x) f_n(x) \, d\lambda(x) = \int_a^b g(x) f_n(x) \, dx = g(x) F_n(x) \Big|_a^b - \int_a^b g'(x) F_n(x) \, dx.$$

$$(\star)$$

From $|g f_n| \leq Ch \in L_1[a, b]$, $g f_n \to gf$ a.e., and the Lebesgue Dominated Convergence Theorem, it follows that

$$\lim_{n \to \infty} \int_a^b g(x) f_n(x) \, dx = \int_a^b g(x) f(x) \, d\lambda(x).$$

Likewise, the Lebesgue Dominated Convergence Theorem implies

$$F_n(x) = \int_a^x f_n(t) \, dt \longrightarrow \int_a^x f(t) \, d\lambda(t) = F(x).$$

for each $x \in [a, b]$. Observing that $|g' F_n| \leq C \int_a^b h \, d\lambda$ and $g' F_n \to g' F$, the Lebesgue Dominated Convergence Theorem once more yields

$$\lim_{n \to \infty} \int_a^b g'(x) F_n(x) \, dx = \int_a^b g'(x) F(x) \, dx.$$

Finally, letting $n \to \infty$ in $(\star)$, we obtain

$$\int_a^b g(x) f(x) \, d\lambda(x) = g(x) F(x) \Big|_a^b - \int_a^b g'(x) F(x) \, dx,$$

as desired.

Problem 31.13. *Let μ be a regular Borel measure on $\mathbf{R}^n$. Then show that the collection of all real-valued functions on $\mathbf{R}^n$ that are infinitely many times differentiable is norm dense in $L_p(\mu)$ for each $1 \leq p < \infty$.*

Solution. Let S be the semiring consisting of the sets of the form $\prod_{i=1}^{n}[a_i, b_i)$. By Theorem 15.10, the outer measure generated by $(\mathbf{R}^n, S, \mu)$ agrees with μ on the σ-algebra B of all Borel sets of $\mathbf{R}^n$. Thus, what needs to be shown is that given $I = \prod_{i=1}^{n}[a_i, b_i)$ and $\varepsilon > 0$, there exists some C^∞-function f with compact support such that $\|\chi_I - f\|_p < \varepsilon$.

To this end, let $I = \prod_{i=1}^{n}[a_i, b_i)$ and let $\varepsilon > 0$. The arguments of the first part of the solution of Problem 25.6 show that there is a C^∞-function $f : \mathbf{R}^n \longrightarrow [0, 1]$ satisfying $\int |\chi_I - f| \, d\mu < 2^{-p}\varepsilon^p$. Since $|\chi_I - f| \le 2$ holds, it follows that

$$\|\chi_I - f\|_p = \left(\int |\chi_I - f|^p \, d\mu \right)^{\frac{1}{p}} = \left(\int |\chi_I - f|^{p-1} \cdot |\chi_I - f| \, d\mu \right)^{\frac{1}{p}}$$

$$\le 2 \left(\int |\chi_I - f| \, d\mu \right)^{\frac{1}{p}} < 2 \cdot 2^{-1}\varepsilon = \varepsilon,$$

Problem 31.14. *Let (X, S, μ) be a measure space with $\mu^*(X) = 1$. Assume that a function $f \in L_1(\mu)$ satisfies $f(x) \ge M > 0$ for almost all x. Then show that $\ln(f) \in L_1(\mu)$ and that $\int \ln(f) \, d\mu \le \ln(\int f \, d\mu)$ holds.*

Solution. The function $g(t) = t - 1 - \ln t$, $t > 0$, attains its minimum value at $t = 1$. Thus, $0 = g(1) \le g(t) = t - 1 - \ln t$ holds for all $t > 0$, and so $\ln t \le t - 1$. Replacing t by $\frac{1}{t}$, the last inequality yields $1 - \frac{1}{t} \le \ln t$. Therefore,

$$1 - \tfrac{1}{t} \le \ln t \le t - 1 \qquad\qquad (\star)$$

holds for each $t > 0$.

Since the function $\ln x$ is continuous on $(0, \infty)$ and f is a measurable function, it follows that $\ln(f)$ is a measurable function. (See the solution of Problem 16.8.) Replacing t by $\frac{f(x)}{\|f\|_1}$ in $(\star)$, we see that

$$1 - \tfrac{\|f\|_1}{f(x)} \le \ln(f(x)) - \ln(\|f\|_1) \le \tfrac{f(x)}{\|f\|_1} - 1 \qquad\qquad (\star\star)$$

holds for almost all x. From our assumptions, it is easy to see that both functions $1 - \frac{\|f\|_1}{f(x)}$ and $\frac{f(x)}{\|f\|_1}$ are integrable. Thus, from $(\star\star)$ and Theorem 22.6, it follows that $\ln(f) \in L_1(\mu)$.

Finally, integrating the right inequality of $(\star\star)$ (and taking into account that $\mu^*(X) = 1$), we see that

$$\int \ln(f) \, d\mu - \ln(\|f\|_1) \le \int \tfrac{f}{\|f\|_1} \, d\mu - 1 = 0.$$

That is,

$$\int \ln(f)\, d\mu \leq \ln\big(\|f\|_1\big) = \ln\big(\int f\, d\mu\big)$$

holds, as required.

Problem 31.15. Theorem 31.7 states that: *If* $1 \leq p < \infty$, f *in* $L_p(\mu)$, $\{f_n\} \subseteq$ $L_p(\mu)$, $f_n \longrightarrow f$ *a.e., and* $\lim \|f_n\|_p = \|f\|_p$, *then* $\lim \|f_n - f\|_p = 0$.
Show with an example that this theorem is false when $p = \infty$.

Solution. Consider the sequence $\{f_n\}$ of $L_\infty([0, 1])$ defined by $f_n = \chi_{(\frac{1}{n}, 1]}$. Then $f_n \longrightarrow \mathbf{1}$ a.e., and $\|f_n\|_\infty = 1 \to 1 = \|\mathbf{1}\|_\infty$. However, $\|f_n - \mathbf{1}\|_\infty = 1$ holds for each n.

Problem 31.16. *This exercise presents a necessary and sufficient condition for the mapping* $g \longmapsto F_g$ *from* $L_\infty(\mu)$ *into* $L_1^*(\mu)$ *(defined by* $F_g(f) = \int fg\, d\mu$*) to be an isometry.*

a. *Show that for each* $g \in L_\infty(\mu)$ *the linear functional* $F_g(f) = \int fg\, d\mu$, *for* $f \in L_1(\mu)$, *is a bounded linear functional on* $L_1(\mu)$ *such that* $\|F_g\| \leq \|g\|_\infty$ *holds.*

b. *Consider a nonempty set* X *and* μ *the measure defined on every subset of* X *by* $\mu(\emptyset) = 0$ *and* $\mu(A) = \infty$ *if* $A \neq \emptyset$. *Then show that* $L_1(\mu) = \{0\}$ *and* $L_\infty(\mu) = B(X)$ *[the bounded functions on* X*] and conclude from this that* $g \in L_\infty(\mu)$ *satisfies* $\|F_g\| = \|g\|_\infty$ *if and only if* $g = 0$.

c. *Let us say that a measure space* $(X, \mathcal{S}, \mu)$ *has the* **finite subset property** *whenever every measurable set of infinite measure has a measurable subset of finite positive measure.*
   Show that the linear mapping $g \longmapsto F_g$ *from* $L_\infty(\mu)$ *into* $L_1^*(\mu)$ *is a lattice isometry if and only if* $(X, \mathcal{S}, \mu)$ *has the finite subset property.*

Solution. (a) Let $g \in L_\infty(\mu)$. Then, for each $f \in L_1(\mu)$, we have $|fg| \leq \|g\|_\infty \cdot |f|$, and so

$$|F_g(f)| = \left| \int fg\, d\mu \right| \leq \int |fg|\, d\mu \leq \|g\|_\infty \int |f|\, d\mu = \|g\|_\infty \cdot \|f\|_1.$$

That is, F_g is a bounded linear functional on $L_1(\mu)$, and $\|F_g\| \leq \|g\|_\infty$ holds.
(b) Since every nonempty set has infinite measure, it is easy to see that there is only one step function. Namely, the constant function zero. That is, $L_1(\mu) = \{0\}$ holds. On the other hand, since every one-point set has infinite measure, each equivalence class of $L_\infty(\mu)$ consists precisely of one function. This implies that $L_\infty(\mu) = B(X)$.

Finally, note that in view of $L_1^*(\mu) = \{0\}$, we must have $F_g = 0$ for each $g \in L_\infty(\mu)$. Thus, $\|F_g\| = \|g\|_\infty$ holds if and only if $\|g\|_\infty = 0$ (i.e., if and only if $g = 0$).

(c) Assume that a measure space $(X, \mathcal{S}, \mu)$ has the finite subset property. Let $0 < g \in L_\infty(\mu)$ and let $0 < \varepsilon < \|g\|_\infty$. The set

$$E = \left\{ x \in X \colon |g(x)| \geq \|g\|_\infty - \varepsilon \right\}$$

is measurable and $\mu^*(E) > 0$ holds. By the finite subset property, there exists a measurable set F with $F \subseteq E$ and $0 < \mu^*(F) < \infty$. Put $f = \frac{\mathrm{Sgn}\, g \cdot \chi_F}{\mu^*(F)} \in L_1(\mu)$, and note that $\|f\|_1 = 1$. Therefore,

$$\|F_g\| \geq |F_g(f)| = \left| \int fg\, d\mu \right| = \int_F \left[\frac{|g|}{\mu^*(F)} \right] d\mu \geq \|g\|_\infty - \varepsilon.$$

Since $0 < \varepsilon < \|g\|_\infty$ is arbitrary, $\|F_g\| \geq \|g\|_\infty$ holds. Now, using part (a), we see that $\|F_g\| = \|g\|_\infty$ holds for all $g \in L_\infty(\mu)$. Therefore, $g \longmapsto F_g$ is a lattice isometry.

For the converse, assume that $g \longmapsto F_g$ is a lattice isometry, and let E be a measurable set with $\mu^*(E) = \infty$. Then $g = \chi_E \in L_\infty(\mu)$, and so $\|F_g\| = \|g\|_\infty = 1$. Pick some $0 \leq f \in L_1(\mu)$ with $F_g(f) = \int fg\, d\mu = \int_E f\, d\mu > \frac{1}{2}$. It is easy to see that there exists a step function $0 \leq \phi \leq f\chi_E$ with $\int \phi\, d\mu > \frac{1}{2}$. From this, it easily follows that there exists a measurable set $F \subseteq E$ with $0 < \mu^*(F) < \infty$.

Problem 31.17. *Let $(X, \mathcal{S}, \mu)$ be a measure space. Assume that there exist measurable sets $E_1, \ldots, E_n$ such that $0 < \mu(E_i) < \infty$ for $1 \leq i \leq n$, $X = \bigcup_{i=1}^n E_i$, and each E_i does not contain any proper nonempty measurable set. Then show that $L_\infty^*(\mu) = L_1(\mu)$; that is, show that $g \mapsto F_g$ from $L_1(\mu)$ to $L_\infty^*(\mu)$ is onto.*

Solution. From our assumptions, we see that $E_i \cap E_j = \emptyset$ holds whenever $i \neq j$. For each $1 \leq i \leq n$ fix some $x_i \in E_i$ and note that $\mu^*(\{x_i\}) > 0$. If f is a measurable function and $\alpha_i = f(x_i)$, then the set $f^{-1}(\{\alpha_i\}) \cap E_i$ is nonempty and measurable. Thus, by our hypothesis, $f^{-1}(\{\alpha_i\}) \cap E_i = E_i$ holds, and therefore, f must be constant on each E_i. In other words, $f = \sum_{i=1}^n f(x_i)\chi_{E_i}$ holds for each measurable function f.

To see that $g \longmapsto F_g$ from $L_1(\mu)$ to $L_\infty^*(\mu)$ is onto, let F be an arbitrary functional in $L_\infty^*(\mu)$. Put $c_i = F(\chi_{E_i})$ for $1 \leq i \leq n$, and then let $g =$

$\sum_{i=1}^{n} \left[\frac{c_i}{\mu^*(E_i)} \right] \chi_{E_i} \in L_1(\mu)$. Note that

$$F_g(\chi_{E_i}) = \int \chi_{E_i} g \, d\mu = \int \left[\frac{c_i}{\mu^*(E_i)} \right] \chi_{E_i} \, d\mu = c_i = F(\chi_{E_i}).$$

Consequently,

$$F_g(f) = F_g\left(\sum_{i=1}^{n} f(x_i)\chi_{E_i} \right) = \sum_{i=1}^{n} f(x_i) F_g(\chi_{E_i})$$

$$= \sum_{i=1}^{n} f(x_i) F(\chi_{E_i}) = F\left(\sum_{i=1}^{n} f(x_i)\chi_{E_i} \right) = F(f)$$

holds for all $f \in L_1(\mu)$, and so $F = F_g$. That is, $g \longmapsto F_g$ is onto.

Problem 31.18.   *Let $(X, \mathcal{S}, \mu)$ be a measure space, and let $0 < p < 1$.*

   a.   *Show by a counterexample that $\| \cdot \|_p$ is no longer a norm on $L_p(\mu)$.*
   b.   *For each $f, g \in L_p(\mu)$ let $d(f, g) = \int |f - g|^p \, d\mu = (\| f - g \|_p)^p$. Show that d is a metric on $L_p(\mu)$ and that $L_p(\mu)$ equipped with d is a complete metric space.*

Solution.   (a) Let $0 < p < 1$ and consider the space $L_p([0, 1])$. Take $f = \chi_{(0,\frac{1}{2})}$ and $g = \chi_{(\frac{1}{2},1)}$, and note that

$$\left\| f + g \right\|_p = 1 > 2^{1-\frac{1}{p}} = \left(\tfrac{1}{2}\right)^{\frac{1}{p}} + \left(\tfrac{1}{2}\right)^{\frac{1}{p}} = \left\| f \right\|_p + \left\| g \right\|_p.$$

That is, $\| \cdot \|_p$ does not satisfy the triangle inequality.
(b) If $a > 0$ and $b > 0$ (and $0 < p < 1$), then

$$\begin{aligned}(a + b)^p &= (a + b)(a + b)^{p-1} = a(a + b)^{p-1} + b(a + b)^{p-1} \\ &\le a \cdot a^{p-1} + b \cdot b^{p-1} = a^p + b^p.\end{aligned}$$

Thus, $(a + b)^p \le a^p + b^p$ holds for each $a \ge 0$ and each $b \ge 0$. This inequality easily implies that $d(f, g) = \int |f - g|^p \, d\mu$ is a metric on $L_p(\mu)$.

For the completeness, let $\{f_n\}$ be a Cauchy sequence in the metric space $(L_p(\mu), d)$, where $0 < p < 1$. By passing to a subsequence, we can assume that $\int |f_{n+1} - f_n|^p \, d\mu < 2^{-n}$ holds for each n. We shall establish the existence of some $f \in L_p(\mu)$ such that $\lim \| f_n - f \|_p = 0$.

Set $g_1 = 0$ and $g_n = |f_1| + |f_2 - f_1| + \cdots + |f_n - f_{n-1}|$ for $n \geq 2$. Clearly, $0 \leq g_n \uparrow$ and

$$\int (g_n)^p \, d\mu \leq \int |f_1|^p \, d\mu + \sum_{i=2}^{n} \int |f_i - f_{i-1}|^p \, d\mu \leq \int |f_1|^p \, d\mu + 1 < \infty$$

holds for each n. By Levi's Theorem 22.8, there exists some $g \in L_p(\mu)$ such that $0 \leq g_n \uparrow g$ a.e. From

$$\left| f_{n+k} - f_n \right| = \left| \sum_{i=n+1}^{n+k} (f_i - f_{i-1}) \right| \leq \sum_{i=n+1}^{n+k} \left| f_i - f_{i-1} \right| = g_{n+k} - g_n,$$

it follows that $\{f_n\}$ converges pointwise (a.e.) to some function f. Since $|f_n| = \left| f_1 + \sum_{i=2}^{n}(f_i - f_{i-1}) \right| \leq g_n \leq g$ hold a.e., we see that $|f| \leq g$ a.e. also holds. Therefore, $f \in L_p(\mu)$. Now, note that $|f_n - f| \leq 2g$ and $|f_n - f|^p \longrightarrow 0$ hold, and so by the Lebesgue Dominated Convergence Theorem, we see that

$$d(f_n, f) = \int \left| f_n - f \right|^p \, d\mu \longrightarrow 0.$$

Therefore, $\left(L_p(\mu), d \right)$ is a complete metric space.

Problem 31.19. *If $(X, \mathcal{S}, \mu)$ is a finite measure space, then show that the vector space of all step functions is norm dense in $L_\infty(\mu)$.*

Solution. Let $f \in L_\infty(\mu)$ and let $\varepsilon > 0$. Choose some $C > 0$ such that $|f(x)| < C$ holds for almost all x, and then pick a partition $-C = a_0 < a_1 < \cdots < a_n = C$ of $[-C, C]$ with $a_i - a_{i-1} < \varepsilon$ for each $1 \leq i \leq n$. Let $E_i = f^{-1}([a_{i-1}, a_i))$, and note that (since $\mu^*(X) < \infty$) the simple function $\phi = \sum_{i=1}^{n} a_i \chi_{E_i}$ is a step function satisfying $\|f - \phi\|_\infty \leq \varepsilon$.

Problem 31.20. *If K is a compact subset of a metric space X, then show that there exists a regular Borel measure μ on X such that $\operatorname{Supp} \mu = K$.*

Solution. Let K be a compact subset of a metric space X. Pick a countable dense subset $\{x_1, x_2, \ldots\}$ of K (see Problem 7.2) and then for each n consider the Dirac measure δ_{x_n} supported at the point x_n (see Example 13.4). Now, consider the measure $\mu \colon \mathcal{P}(X) \longrightarrow [0, 1]$ defined by

$$\mu(A) = \sum_{n=1}^{\infty} 2^{-n} \delta_{x_n}(A) = \sum_{n \in \hat{A}} 2^{-n},$$

where $\hat{A} = \{n \in \mathbb{N} \colon x_n \in A\}$.

Clearly, $\mu(X \setminus K) = 0$. On the other hand, if $\mathcal{O}$ is an open subset of X satisfying $\mathcal{O} \cap K \neq \emptyset$, then for some n we have $x_n \in \mathcal{O}$, and so $\mu(\mathcal{O} \cap K) \geq 2^{-n}\delta_{x_n}(\mathcal{O} \cap K) = 2^{-n} > 0$.

It remains to be shown that μ is a regular Borel measure. To this end, let $A \subseteq X$ be fixed. Note first that if $C_n = \{x_1, \ldots, x_n\} \cap A \subseteq A$, then C_n is a finite set (and hence, a compact set) and, moreover, $\mu(C_n) \uparrow \mu(A)$ holds. Therefore,

$$\mu(A) = \sup\{\mu(C): C \text{ compact and } C \subseteq A\}.$$

In the other direction, note that if for each n we consider the open set

$$\mathcal{O}_n = X \setminus \{x_i: 1 \leq i \leq n \text{ and } x_i \notin A\},$$

then $A \subseteq \mathcal{O}_n$ and $\mu(\mathcal{O}_n) \downarrow \mu(A)$ (why?). Therefore,

$$\mu(A) = \inf\{\mu(\mathcal{O}): \mathcal{O} \text{ open and } A \subseteq \mathcal{O}\}$$

also holds, proving that μ is a regular Borel measure.

Problem 31.21. *If $\{f_n\}$ is a norm bounded sequence of $L_2(\mu)$, then show that $f_n/n \to 0$ a.e.*

Solution. Assume that a sequence $\{f_n\} \subseteq L_2(\mu)$ satisfies $\int (f_n)^2 d\mu \leq C$ for all n, where $C > 0$ is a constant. Then,

$$\sum_{n=1}^{\infty} \int \left(\tfrac{f_n}{n}\right)^2 d\mu \leq C \sum_{n=1}^{\infty} \tfrac{1}{n^2} < \infty$$

holds. By the series version of Levi's Theorem 22.9, we know that the series $\sum_{n=1}^{\infty} \left(\tfrac{f_n}{n}\right)^2$ defines an integrable function. Therefore, $\tfrac{f_n}{n} \longrightarrow 0$ a.e. must hold.

Problem 31.22. *Let $(X, \mathcal{S}, \mu)$ be a measure space such that $\mu^*(X) = 1$. If $f, g \in L_1(\mu)$ are two positive functions satisfying $f(x)g(x) \geq 1$ for almost all x, then show that*

$$\left(\int f\, d\mu\right) \cdot \left(\int g\, d\mu\right) \geq 1.$$

Solution. Note that the functions $\sqrt{f}$ and $\sqrt{g}$ both belong to $L_2(\mu)$ and satisfy $\sqrt{f(x)}\sqrt{g(x)} \geq 1$ for almost all x. Applying Hölder's inequality, we see that

$$1 = \int 1 \, d\mu \le \int \sqrt{f} \sqrt{g} \, d\mu \le \left(\int (\sqrt{f})^2 \, d\mu \right)^{\frac{1}{2}} \cdot \left(\int (\sqrt{g})^2 \, d\mu \right)^{\frac{1}{2}}.$$

Squaring, we get $\left(\int f \, d\mu \right) \cdot \left(\int g \, d\mu \right) \ge 1$.

Problem 31.23. *Consider a measure space $(X, \mathcal{S}, \mu)$ with $\mu^*(X) = 1$, and let $f, g \in L_2(\mu)$. If $\int f \, d\mu = 0$, then show that*

$$\left(\int fg \, d\mu \right)^2 \le \left[\int g^2 \, d\mu - \left(\int g \, d\mu \right)^2 \right] \int f^2 \, d\mu.$$

Solution. Put $\alpha = \int g \, d\mu$. Then, using Hölder's inequality, we get

$$
\begin{aligned}
\left| \int fg \, d\mu \right| &= \left| \int (fg - \alpha f) \, d\mu \right| = \left| \int f(g - \alpha) \, d\mu \right| \\
&\le \int |f| |g - \alpha| \, d\mu \le \left(\int f^2 \, d\mu \right)^{\frac{1}{2}} \cdot \left(\int (g - \alpha)^2 \right)^{\frac{1}{2}} \\
&= \left(\int f^2 \, d\mu \right)^{\frac{1}{2}} \left(\int g^2 \, d\mu - 2\alpha \int g \, d\mu + \alpha^2 \right)^{\frac{1}{2}} \\
&= \left(\int f^2 d\mu \right)^{\frac{1}{2}} \left[\int g^2 d\mu - 2 \left(\int g d\mu \right) \left(\int g d\mu \right) + \left(\int g d\mu \right)^2 \right]^{\frac{1}{2}} \\
&= \left(\int f^2 \, d\mu \right)^{\frac{1}{2}} \left[\int g^2 \, d\mu - \left(\int g \, d\mu \right)^2 \right]^{\frac{1}{2}},
\end{aligned}
$$

and our inequality follows.

Problem 31.24. *If two functions $f, g \in L_3(\mu)$ satisfy*

$$\| f \|_3 = \| g \|_3 = \int f^2 g \, d\mu = 1,$$

then show that $g = |f|$ a.e.

Solution. Let $p = \frac{3}{2}$ and $q = 3$, and note that $\frac{1}{p} + \frac{1}{q} = 1$. Clearly, $f^2 \in L_p(\mu) = L_{\frac{3}{2}}(\mu)$, and since $g \in L_3(\mu)$, we see that $f^2 g \in L_1(\mu)$.

Now, using Hölder's inequality, we obtain

$$1 = \left| \int f^2 g \, d\mu \right| \leq \int f^2 |g| \, d\mu \leq \| f^2 \|_p \cdot \| g \|_q$$

$$= \left[\int (f^2)^{\frac{3}{2}} \, d\mu \right]^{\frac{2}{3}} \cdot \| g \|_3 = \left(\| f \|_3 \right)^2 \cdot \| g \|_3 = 1,$$

and so $\int f^2 |g| \, d\mu = \| f^2 \|_p \cdot \| g \|_q = 1$. By Problem 31.4, there exists a constant $C > 0$ such that $C |f^2|^p = |g|^q$, or $C |f|^3 = |g|^3$. From $\| f \|_3 = \| g \|_3 = 1$, we infer that $C = 1$, and so $|f|^3 = |g|^3$ holds. Therefore,

$$bigl| f | = |g| \quad a.e. \tag{$\star$}$$

From the relation

$$\int f^2 (|g| - g) \, d\mu = \int f^2 |g| \, d\mu - \int f^2 g \, d\mu = \int |f|^3 \, d\mu - 1 = 1 - 1 = 0$$

and $f^2 (|g| - g) \geq 0$ a.e., we conclude that $f^2 (|g| - g) = 0$ a.e. Taking into account $(\star)$, the latter easily implies that $g = |g| = |f|$ a.e. holds.

Problem 31.25. *For a function $f \in L_1(\mu) \cap L_2(\mu)$ establish the following properties:*

   a. $f \in L_p(\mu)$ *for each* $1 \leq p \leq 2$, *and*
   b. $\lim_{p \to 1^+} \| f \|_p = \| f \|_1$.

Solution. Let $f \in L_1(\mu) \cap L_2(\mu)$; we can assume that $f(x) \in \mathbf{R}$ for each $x \in X$. Consider the measurable set $A = \{ x \in X \colon |f(x)| \geq 1 \}$ and then define the function $g \colon X \longrightarrow \mathbf{R}$ by

$$g(x) = \begin{cases} |f(x)|^2 & \text{if } x \in A \\ |f(x)| & \text{if } x \notin A, \end{cases}$$

i.e., $g = f^2 \chi_A + f \chi_{A^c}$. From our hypothesis, we see that $g \in L_1(\mu)$.

(a) Let $1 \leq p \leq 2$. Then, the inequality

$$|f(x)|^p \leq \begin{cases} |f(x)|^2 & \text{if } x \in A \\ |f(x)| & \text{if } x \notin A \end{cases} = g(x), \tag{$\star$}$$

implies $f \in L_p(\mu)$ for each $1 \leq p \leq 2$.

(b) Let a sequence $\{p_n\}$ of the interval $[1, 2]$ satisfy $p_n \longrightarrow 1$. From $(\star)$, we see that $|f|^{p_n} \leq g$ holds for each n. Now, from $|f|^{p_n} \longrightarrow |f|$ a.e. and the Lebesgue Dominated Convergence Theorem, we infer that

$$\lim_{n \to \infty} \|f\|_{p_n} = \lim_{n \to \infty} \left(\int |f|^{p_n} \, d\mu \right)^{\frac{1}{p_n}} = \int |f| \, d\mu = \|f\|_1.$$

The preceding easily implies that $\lim_{p \to 1^+} \|f\|_p = \|f\|_1$ holds.

Problem 31.26. *Assume that the positive real numbers $\alpha_1, \ldots, \alpha_n$ satisfy $0 < \alpha_i < 1$ for each i and $\sum_{i=1}^n \alpha_i = 1$. If $f_1, \ldots, f_n$ are positive integrable functions on some measure space, then show that*

    a. $f_1^{\alpha_1} f_2^{\alpha_2} \cdots f_n^{\alpha_n} \in L_1(\mu)$, *and*
    b. $\int f_1^{\alpha_1} f_2^{\alpha_2} \cdots f_n^{\alpha_n} \, d\mu \leq (\|f_1\|_1)^{\alpha_1} (\|f_2\|_1)^{\alpha_2} \cdots (\|f_n\|_1)^{\alpha_n}$.

Solution. We shall establish the result by using induction on n. For $n = 1$ the result is trivial. For $n = 2$, note that $f_1^{\alpha_1} \in L_{\frac{1}{\alpha_1}}(\mu)$ and $f_2^{\alpha_2} \in L_{\frac{1}{\alpha_2}}(\mu)$. Since $\left(\frac{1}{\alpha_1} \right)^{-1} + \left(\frac{1}{\alpha_2} \right)^{-1} = \alpha_1 + \alpha_2 = 1$, it follows from Hölder's inequality that $f_1^{\alpha_1} f_2^{\alpha_2} \in L_1(\mu)$ and that

$$\int f_1^{\alpha_1} f_2^{\alpha_2} \, d\mu \leq \left(\int (f_1^{\alpha_1})^{\frac{1}{\alpha_1}} \, d\mu \right)^{\alpha_1} \left(\int (f_2^{\alpha_2})^{\frac{1}{\alpha_2}} \, d\mu \right)^{\alpha_2}$$

$$= \left(\|f_1\|_1 \right)^{\alpha_1} \left(\|f_2\|_1 \right)^{\alpha_2}.$$

For the inductive argument, assume that the result is true for some n. Let $f_1, \ldots, f_n, f_{n+1}$ be $n + 1$ integrable positive functions and let $\alpha_1, \ldots, \alpha_n, \alpha_{n+1}$ be positive constants such that $\sum_{i=1}^{n+1} \alpha_i = 1$. Put $\alpha = \sum_{i=1}^n \alpha_i > 0$, and note that $\sum_{i=1}^n \frac{\alpha_i}{\alpha} = 1$. Now, by our induction hypothesis, we have $f_1^{\frac{\alpha_1}{\alpha}} \cdots f_n^{\frac{\alpha_n}{\alpha}} \in L_1(\mu)$. Also, applying the case $n = 2$ for α and $1 - \alpha = \alpha_{n+1}$, we see that

$$\int f_1^{\alpha_1} \cdots f_n^{\alpha_n} f_{n+1}^{\alpha_{n+1}} \, d\mu = \int \left(f_1^{\frac{\alpha_1}{\alpha}} \cdots f_n^{\frac{\alpha_n}{\alpha}} \right)^{\alpha} f_{n+1}^{\alpha_{n+1}} \, d\mu$$

$$\leq \left(\int f_1^{\frac{\alpha_1}{\alpha}} \cdots f_n^{\frac{\alpha_n}{\alpha}} \, d\mu \right)^{\alpha} \left(\int f_{n+1} \, d\mu \right)^{\alpha_{n+1}}$$

$$\leq \left[\left(\|f_1\|_1 \right)^{\frac{\alpha_1}{\alpha}} \cdots \left(\|f_n\|_1 \right)^{\frac{\alpha_n}{\alpha}} \right]^{\alpha} \left(\|f_{n+1}\|_1 \right)^{\alpha_{n+1}}$$

$$= \left(\|f_1\|_1 \right)^{\alpha_1} \cdots \left(\|f_n\|_1 \right)^{\alpha_n} \left(\|f_{n+1}\|_1 \right)^{\alpha_{n+1}},$$

and the induction is complete.

Problem 31.27. *Let (X, S, μ) be a measure space and let $\{A_n\}$ be a sequence of measurable sets satisfying $0 < \mu^*(A_n) < \infty$ for each n and $\lim \mu^*(A_n) = 0$. Fix $1 < p < \infty$ and let $g_n = [\mu^*(A_n)]^{-\frac{1}{q}} \chi_{A_n}$ $(n = 1, 2, \ldots)$, where $\frac{1}{p} + \frac{1}{q} = 1$. Prove that $\lim \int f g_n \, d\mu = 0$ for each $f \in L_p(\mu)$.*

Solution. Pick $1 < q < \infty$ such that $\frac{1}{p} + \frac{1}{q} = 1$ and let $f \in L_p(\mu)$. Then, by Hölder's inequality, we have

$$
\left| \int f g_n \, d\mu \right| = \left| \int (f \chi_{A_n}) g_n \, d\mu \right|
$$
$$
\leq \left(\int |f \chi_{A_n}|^p \, d\mu \right)^{\frac{1}{p}} \left(\int |g_n|^q \, d\mu \right)^{\frac{1}{q}} = \left(\int_{A_n} |f|^p \, d\mu \right)^{\frac{1}{p}}.
$$

From Problem 22.6, we know that $\lim \int_{A_n} |f|^p \, d\mu = 0$, and therefore $\lim \int f g_n \, d\mu = 0$ likewise holds.

Problem 31.28. *Let (X, S, μ) be a measure space such that $\mu^*(X) = 1$. For each $1 < p < \infty$ define the set*

$$
\mathcal{E}_p = \left\{ f \in L_1(\mu) : \int |f| \, d\mu = 1 \quad and \quad \int |f|^p \, d\mu = 2 \right\}.
$$

Show that for each $0 < \epsilon < 1$ there exists some $\delta_p > 0$ such that

$$
\mu^*(\{x \in X : |f(x)| > \epsilon\}) \geq \delta_p
$$

for each $f \in \mathcal{E}_p$.

Solution. Fix $0 < \varepsilon < 1$. For each $f \in \mathcal{E}_p$ put

$$
E_f = \{x \in X : |f(x)| > \varepsilon\} \quad \text{and} \quad F_f = X \setminus E_f = \{x \in X : |f(x)| \leq \varepsilon\}.
$$

From $|f| \chi_{F_f} \leq \varepsilon \chi_{F_f}$, it follows that $\int_{F_f} |f| \, d\mu \leq \varepsilon \mu^*(F_f) \leq \varepsilon$, and so

$$
\int_{E_f} |f| \, d\mu = \int_X |f| \, d\mu - \int_{F_f} |f| \, d\mu \geq 1 - \varepsilon. \tag{$\star$}
$$

Now, if $1 < q < \infty$ satisfies $\frac{1}{p} + \frac{1}{q} = 1$, then Hölder's inequality implies

$$\int_{E_f} |f| \, d\mu \le \left(\int_{E_f} |f|^p \, d\mu \right)^{\frac{1}{p}} \cdot \left(\int_{E_f} 1^q \, d\mu \right)^{\frac{1}{q}}$$

$$= \left(\int_{E_f} |f|^p \, d\mu \right)^{\frac{1}{p}} [\mu^*(E_f)]^{\frac{1}{q}} \le 2^{\frac{1}{p}} [\mu^*(E_f)]^{\frac{1}{q}}.$$

A glance at $(\star)$ shows that $1 - \varepsilon \le 2^{\frac{1}{p}} [\mu^*(E_f)]^{\frac{1}{q}}$, or

$$\mu^*(E_f) \ge \frac{(1-\varepsilon)^q}{2^{\frac{q}{p}}} = \frac{(1-\varepsilon)^q}{2^{q-1}} = 2\left(\frac{1-\varepsilon}{2}\right)^q$$

holds for each $f \in \mathcal{E}_p$, and the desired conclusion follows.

Problem 31.29. *Let $(X, \mathcal{S}, \mu)$ be a measure space and let $1 \le p < \infty$ and $0 < \eta < p$.*

a. *Show that the nonlinear function $\psi \colon L_p(\mu) \to L_{\frac{p}{\eta}}(\mu)$, where $\psi(f) = |f|^\eta$, is norm continuous.*

b. *If $f_n \to f$ and $g_n \to g$ hold in $L_p(\mu)$, then show that*

$$\lim_{n \to \infty} \int |f_n|^{p-\eta} |g_n|^\eta \, d\mu = \int |f|^{p-\eta} |g|^\eta \, d\mu.$$

Solution. (a) It should be clear that ψ maps indeed $L_p(\mu)$ into $L_{\frac{p}{\eta}}(\mu)$, and that ψ is nonlinear. Let $f_n \to f$ in $L_p(\mu)$ (i.e., let $\|f_n - f\|_p \to 0$) and assume by way of contradiction that $\psi(f_n) \not\to \psi(f)$ in $L_{\frac{p}{\eta}}(\mu)$. So, by passing to a subsequence, we can assume that there exists some $\varepsilon > 0$ such that

$$\left\| \psi(f_n) - \psi(f) \right\|_{\frac{p}{\eta}} = \left(\int \left| |f_n|^\eta - |f|^\eta \right|^{\frac{p}{\eta}} \, d\mu \right)^{\frac{\eta}{p}} \ge \varepsilon. \qquad (\star\star)$$

Now, by passing to a subsequence again, we can assume that there exists some function $0 \le g \in L_p(\mu)$ such that $|f_n| \le g$ μ-a.e. holds for each n and $f_n \longrightarrow f$ a.e.; see Lemma 31.6. Therefore, the relations $\left| |f_n|^\eta - |f|^\eta \right|^{\frac{p}{\eta}} \le \left(|g|^\eta + |f|^\eta \right)^{\frac{p}{\eta}} \in L_1(\mu)$ and $\left| |f_n|^\eta - |f|^\eta \right|^{\frac{p}{\eta}} \longrightarrow 0$ a.e., coupled with the Lebesgue Dominated Convergence Theorem, imply $\int \left| |f_n|^\eta - |f|^\eta \right|^{\frac{p}{\eta}} \, d\mu \longrightarrow 0$, which contradicts $(\star\star)$. Consequently, the nonlinear mapping ψ is norm continuous.

(b) Notice that the two nonlinear functions $\psi_1 \colon L_p(\mu) \to L_{\frac{p}{p-\eta}}(\mu)$ and $\psi_2 \colon L_p(\mu) \to L_{\frac{p}{\eta}}(\mu)$, defined by

$$\psi_1(u) = |u|^{p-\eta} \quad \text{and} \quad \psi_2(v) = |v|^\eta.$$

are—by part (a)—both norm continuous. Therefore,

$$\left\| \, |f_n|^{p-\eta} - |f|^{p-\eta} \, \right\|_{\frac{p}{p-\eta}} \to 0 \quad \text{and} \quad \left\| \, |g_n|^{\eta} - |g|^{\eta} \, \right\|_{\frac{p}{\eta}} \to 0.$$

Now, observe that $\left(L_{\frac{p}{p-\eta}}(\mu) \right)^* = L_{\frac{p}{\eta}}(\mu)$ holds. Consequently, from the duality $\langle L_{\frac{p}{p-\eta}}(\mu), L_{\frac{p}{\eta}}(\mu) \rangle$, we see that

$$\int |f_n|^{p-\eta}|g_n|^{\eta}\, d\mu = \langle |f_n|^{p-\eta}, |g_n|^{\eta}\rangle \to \langle |f|^{p-\eta}, |g|^{\eta}\rangle = \int |f|^{p-\eta}|g|^{\eta}\, d\mu,$$

as claimed.

Problem 31.30. *Let $T: L_p(\mu) \to L_p(\mu)$ be a continuous operator, where $1 < p < \infty$, and let $0 \le \eta \le p$. Show that:*

a. *If $f \in L_p(\mu)$, then $|f|^{p-\eta}|Tf|^{\eta} \in L_1(\mu)$ and*

$$\int |f|^{p-\eta}|Tf|^{\eta}\, d\mu \le \|T\|^{\eta}\left(\|f\|_p\right)^p.$$

b. *If for some $f \in L_p(\mu)$ with $\|f\|_p \le 1$ we have $\int |f|^{p-\eta}|Tf|^{\eta}\, d\mu = \|T\|^{\eta}$, then $|Tf| = \|T\|\,|f|$.*

Solution. Assume T, η, and f are as stated in the problem.

(a) If $\eta = 0$ or $\eta = p$, then the desired inequality is obvious. So, assume $0 < \eta < p$ and consider the conjugate exponents

$$r = \frac{p}{p-\eta} \quad \text{and} \quad s = \left(1 - \frac{1}{r}\right)^{-1} = \frac{p}{\eta}.$$

Since $|f|^{p-\eta} \in L_r(\mu)$ and $|Tf|^{\eta} \in L_s(\mu)$, we see that $|f|^{p-\eta}|Tf|^{\eta}$ belongs to $L_1(\mu)$. Also, applying Hölder's inequality with exponents r and s, we obtain

$$\int |f|^{p-\eta}|Tf|^{\eta}\, d\mu \le \left[\int \left(|f|^{p-\eta}\right)^{\frac{p}{p-\eta}}\, d\mu\right]^{\frac{p-\eta}{p}} \cdot \left[\int \left(|Tf|^{\eta}\right)^{\frac{p}{\eta}}\, d\mu\right]^{\frac{\eta}{p}}$$

$$= \left(\int |f|^p\, d\mu\right)^{\frac{p-\eta}{p}} \cdot \left(\int |Tf|^p\, d\mu\right)^{\frac{\eta}{p}}$$

$$= \left(\|f\|_p\right)^{p-\eta}\left(\|Tf\|_p\right)^{\eta} \le \left(\|f\|_p\right)^{p-\eta}\|T\|^{\eta}\left(\|f\|_p\right)^{\eta}$$

$$= \|T\|^{\eta}\left(\|f\|_p\right)^p.$$

(b) Assume that some $f \in L_p(\mu)$ with $\|f\|_p \leq 1$ satisfies

$$\int |f|^{p-\eta}|Tf|^{\eta}\, d\mu = \|T\|^{\eta}.$$

From Hölder's inequality, we see that

$$\|T\|^{\eta} = \int |f|^{p-\eta}|Tf|^{\eta}\, d\mu \leq \big\||f|^{p-\eta}\big\|_r \cdot \big\||Tf|^{\eta}\big\|_s$$

$$= \big(\|f\|_p\big)^{p-\eta}\big(\|Tf\|_p\big)^{\eta} \leq \|T\|^{\eta}.$$

Thus, $\int |f|^{p-\eta}|Tf|^{\eta}\, d\mu = \big\||f|^{p-\eta}\big\|_r \cdot \big\||Tf|^{\eta}\big\|_s$. From Problem 31.4, there exists a constant $c \geq 0$ such that $\big(|Tf|^{\eta}\big)^s = c\big(|f|^{p-\eta}\big)^r$, or

$$|Tf|^p = c|f|^p.$$

Therefore, $|Tf| = \lambda|f|$ holds for some $\lambda \geq 0$. This implies $\lambda\|f\|_p = \|Tf\|_p \leq \|T\|\|f\|_p$ and so $\lambda \leq \|T\|$. Also, from

$$\|T\|^{\eta} = \int |f|^{p-\eta}|Tf|^{\eta}\, d\mu = \int |f|^{p-\eta}\lambda^{\eta}|f|^{\eta}\, d\mu \leq \lambda^{\eta},$$

we see that $\|T\| \leq \lambda$. Hence, $\lambda = \|T\|$, and so $|Tf| = \|T\|\|f\|$.

Problem 31.31. *Let $(X, \mathcal{S}, \mu)$ be a measure space and let $f \in L_p(\mu)$ for some $1 \leq p < \infty$. Show that the function $g: [0, \infty) \to [0, \infty]$ defined by*

$$g(t) = pt^{p-1}\mu^*\big(\{x \in X: |f(x)| \geq t\}\big)$$

is Lebesgue integrable over $[0, \infty)$ and that

$$\int |f|^p\, d\mu = \int_{[0,\infty)} g(t)\, d\lambda(t) = p\int_0^{\infty} t^{p-1}\mu^*(\{x \in X: |f(x)| \geq t\})\, dt.$$

Solution. Let $f \in L_p(\mu)$; we shall assume that $f(x) \in \mathbf{R}$ holds for each $x \in X$. Let $T = [0, \infty)$ and consider the product measure space $T \times X$. Also, let

$$A = \big\{(t, x) \in T \times X: 0 \leq t \leq |f(x)|\big\},$$

and note that (by Problem 26.8) the set A is $\mu \times \lambda$-measurable. Now, consider the function $h: T \times X \longrightarrow [0, \infty)$ defined by

$$h(t, x) = \begin{cases} pt^{p-1} & \text{if } 0 \leq t \leq |f(x)| \\ 0 & \text{if } t > |f(x)| \end{cases} = pt^{p-1}\chi_A(t, x).$$

In addition, we have

$$\int_X \left[\int_T h(t, x)\, d\lambda(t) \right] d\mu(x) = \int_X \left[\int_T pt^{p-1}\chi_A(t, x)\, d\lambda(t) \right] d\mu(x)$$

$$= \int_X \left[\int_0^{|f(x)|} pt^{p-1}\, dt \right] d\mu(x)$$

$$= \int_X |f(x)|^p \, d\mu(x) < \infty.$$

By Tonneli's Theorem (Theorem 26.7), the function h is integrable over $T \times X$ and

$$\int_X |f(x)|^p \, d\mu(x) = \int_T \left[\int_X h(t, x)\, d\mu(x) \right] d\lambda(t). \qquad (\star)$$

Put $E_t = \{x \in X : |f(x)| \geq t\}$ and note that

$$\int_X h(t, x)\, d\mu(x) = \int_{E_t} pt^{p-1} \, d\mu(x) = pt^{p-1}\mu^*(E_t).$$

By Fubini's Theorem (Theorem 26.6), we know that the function

$$t \longmapsto pt^{p-1}\mu^*(E_t)$$

is integrable over $[0, \infty)$ and from $(\star)$, we see that

$$\int_X |f(x)|^p \, d\mu(x) = p \int_T t^{p-1}\mu^*(E_t)\, d\lambda(t) = p \int_0^\infty t^{p-1}\mu^*(E_t)\, dt.$$

That the Lebesgue integral $\int_T t^{p-1}\mu^*(E_t)\, d\lambda(t)$ is also an improper Riemann integral follows from the fact that the function $t \longmapsto \mu^*(E_t)$ is decreasing—and hence continuous for all but at-most countably many t.

Problem 31.32. *Let (X, S, μ) be a measure space and let $f: X \to \mathbb{R}$ be a measurable function. If $\mu^*\big(\{x \in X : |f(x)| \geq t\}\big) \leq e^{-t}$ for all $t \geq 0$, then show that $f \in L_p(\mu)$ holds for each $1 \leq p < \infty$.*

Solution. Let $g(t) = \mu^*(\{x \in X: |f(x)| \geq t\})$, $t \geq 0$. In view of Problem 31.31, we must show that $\int_0^\infty t^{p-1} g(t) \, d\lambda(t) < \infty$. Since for each $t \geq 0$ we have $0 \leq g(t) \leq e^{-t}$, it suffices to establish that $\int_0^\infty t^{p-1} e^{-t} \, dt < \infty$.

To see this, start by observing that by L'Hôpital's Rule we have

$$\lim_{t \to \infty} t^{p-1} e^{-\frac{t}{2}} = \lim_{t \to \infty} \frac{t^{p-1}}{e^{\frac{t}{2}}} = 0.$$

So, there exists some $M > 0$ satisfying $0 \leq t^{p-1} e^{-\frac{t}{2}} \leq M$ for all $t \geq 0$. Hence,

$$
\begin{aligned}
0 \leq \int_0^\infty t^{p-1} e^{-t} \, dt &= \int_0^\infty t^{p-1} e^{-\frac{t}{2}} e^{-\frac{t}{2}} \, dt \\
&\leq \int_0^\infty M e^{-\frac{t}{2}} \, dt = 2M < \infty,
\end{aligned}
$$

as desired.

Problem 31.33. *Consider the vector space of functions*

$$E = \left\{ f: \mathbb{R}^n \to \mathbb{R} \,\middle|\, f \text{ is a } C^\infty\text{-function with compact support and } \int_{\mathbb{R}^n} f \, d\lambda = 0 \right\}.$$

Show that for each $1 < p < \infty$ the vector space E is dense in $L_p(\mathbb{R}^n)$. Is E dense in $L_1(\mathbb{R}^n)$?

Solution. We shall prove the result for the special case $n = 1$. The general case (whose details can be completed as in Problem 25.3) is left for the reader. The proof will be based upon the following property: If $1 < p < \infty$, $\varepsilon > 0$, $h > 0$, and a positive integer n are given, then there exists a C^∞-function $\phi: \mathbb{R} \longrightarrow \mathbb{R}$ such that

1. $\operatorname{Supp} \phi$ is compact and $\operatorname{Supp} \phi \subseteq [n, \infty)$;
2. $0 \leq \phi(x) \leq h$ for all $x \in \mathbb{R}$;
3. $\int_{\mathbb{R}} \phi \, d\lambda = 1$; and
4. $\|\phi\|_p = \left(\int_{\mathbb{R}} \phi^p \, d\lambda \right)^{\frac{1}{p}} < \varepsilon$.

To see this, assume $1 < p < \infty$, $\varepsilon > 0$, $h > 0$, and the positive integer n are given. If k is an arbitrary positive integer, then (by Problem 25.3) there exists a C^∞-function $f: \mathbb{R} \longrightarrow \mathbb{R}$ such that:

a. $\operatorname{Supp} f \subseteq [n, n+k+2]$;

b. $0 \leq f(x) \leq 1$ for each $x \in \mathbb{R}$ and $f(x) = 1$ for each $x \in [n+1, n+k+1]$.

If $c = \int_{\mathbb{R}} f \, d\lambda > 0$, then the C^∞-function $\phi = \frac{1}{c} f$ satisfies $\mathrm{Supp}\, \phi \subseteq [n, n+k+2] \subseteq [n, \infty)$, $\int_{\mathbb{R}} \phi \, d\lambda = 1$, and (in view of $c = \int_{\mathbb{R}} f \, d\lambda \geq \int_{n+1}^{n+k+1} 1 \, dx = k$) $0 \leq \phi(x) \leq \frac{1}{k}$ for each $x \in \mathbb{R}$. In addition, we have

$$\|\phi\|_p = \left(\int_{\mathbb{R}} \phi^p \, d\lambda \right)^{\frac{1}{p}} \leq \left[\int_n^{n+k+2} \left(\tfrac{1}{k} \right)^p d\lambda \right]^{\frac{1}{p}}$$
$$= \frac{(k+2)^{\frac{1}{p}}}{k} = \left(\tfrac{k+2}{k} \right)^{\frac{1}{p}} \cdot k^{\frac{1}{p}-1}.$$

In view of $1 < p < \infty$, we see that $\lim_{k \to \infty} \left(\frac{k+2}{k} \right)^{\frac{1}{p}} \cdot k^{\frac{1}{p}-1} = 0$, and so a sufficiently large k will yield a function ϕ with the desired properties.

To complete the proof, let $f \in L^p(\mathbb{R})$ and let $\varepsilon > 0$. As in Problem 25.5(b) (how?), there exists a C^∞-function g with compact support such that $\|f - g\|_p < \varepsilon$. If $m = \int_{\mathbb{R}} g \, d\lambda = 0$, then $g \in E$, and we are done. So, assume that $m \neq 0$. Pick a positive integer n such that $\mathrm{Supp}\, g \cap [n, \infty) = \emptyset$, and then (by the prior discussion) pick a C^∞-function ϕ with compact support such that:

    i. $\mathrm{Supp}\, \phi \subseteq [n, \infty)$;
   ii. $\phi(x) \geq 0$ for each $x \in \mathbb{R}$;
  iii. $\int_{\mathbb{R}} \phi \, d\lambda = 1$; and
  iv. $\|\phi\|_p = \left(\int_{\mathbb{R}} \phi^p \, d\lambda \right)^{\frac{1}{p}} < \frac{\varepsilon}{|m|}$.

Now, consider the function $\psi = g - m\phi$, and note that $\psi \in E$ and

$$\|f - \psi\|_p = \|(f - g) + m\phi\|_p$$
$$\leq \|f - g\|_p + |m| \|\phi\|_p$$
$$< \varepsilon + |m| \|\phi\|_p < \varepsilon + \varepsilon = 2\varepsilon.$$

Therefore, E is dense in $L_p(\mathbb{R})$.

The vector space E is not dense in $L_1(\mathbb{R})$. For instance, consider the function $f = \chi_{[0,1]} \in L_1(\mathbb{R})$. If $\phi \in L_1(\mathbb{R})$ satisfies $\int |f - \phi| \, d\lambda < \frac{1}{2}$, then from

$$1 - \int_{\mathbb{R}} \phi \, d\lambda = \int_{\mathbb{R}} (f - \phi) \, d\lambda \leq \int_{\mathbb{R}} |f - \phi| \, d\lambda < \tfrac{1}{2},$$

it follows that $\int_{\mathbb{R}} \phi \, d\lambda > 1 - \frac{1}{2} = \frac{1}{2}$, and so $\phi \notin E$. This shows that E is not dense in $L_1(\mathbb{R})$.

Problem 31.34. *Let* $(0, \infty)$ *be equipped with the Lebesgue measure, and let* $1 < p < \infty$. *For each* $f \in L_p(\lambda)$ *let*

$$T(f)(x) = x^{-1} \int f \chi_{(0,x)} \, d\lambda \quad \text{for } x > 0.$$

Then show that T *defines a one-to-one bounded linear operator from* $L_p((0, \infty))$ *into itself such that* $\|T\| = \frac{p}{p-1}$.

Solution. For simplicity, we shall write Tf instead of $T(f)$. Consider an arbitrary function $0 \le f \in C_c((0, \infty))$. Choose some $M > 0$ so that $0 \le f(x) \le M$ holds for all $x > 0$.

If $I = \int_0^\infty f(t) \, dt$, then the function

$$g(x) = \begin{cases} M & \text{if } 0 < x \le 1 \\ \frac{I}{x} & \text{if } x > 1 \end{cases}$$

belongs to $L_p((0, \infty))$. Since $0 \le Tf \le g$ holds, we see that Tf belongs to $L_p((0, \infty))$. Also, in view of the inequalities

$$0 \le x\Big(\tfrac{1}{x} \int_0^x f(t) \, dt\Big)^p = x[Tf(x)]^p \le x[g(x)]^p,$$

it follows that

$$x\Big(\tfrac{1}{x} \int_0^x f(t) \, dt\Big)^p \Big|_0^\infty = 0.$$

Now, integrating by parts and using Hölder's inequality, we get

$$\begin{aligned}
\big(\|Tf\|_p\big)^p &= \int_0^\infty \Big(\tfrac{1}{x} \int_0^x f(t) \, dt\Big)^p dx = \tfrac{1}{1-p} \int_0^\infty \Big(\int_0^x f(t) \, dt\Big)^p d(x^{1-p}) \\
&= \tfrac{1}{1-p}\Big[x\Big(\tfrac{1}{x}\int_0^x f(t)dt\Big)^p \Big|_0^\infty - p\int_0^\infty f(x)\Big(\tfrac{1}{x}\int_0^x f(t)dt\Big)^{p-1} dx\Big] \\
&= \tfrac{p}{p-1} \int_0^\infty f(x)[Tf(x)]^{p-1} \, dx \\
&\le \tfrac{p}{p-1} \|f\|_p \Big(\int_0^\infty [Tf(x)]^{q(p-1)} \, dx\Big)^{\frac{1}{q}} \\
&= \tfrac{p}{p-1} \|f\|_p \cdot \big(\|Tf\|_p\big)^{\frac{p}{q}}.
\end{aligned}$$

This easily implies that

$$\|Tf\|_p \le \tfrac{p}{p-1}\|f\|_p$$

holds for all $f \in C_c\big((0,\infty)\big)$. In other words,

$$T : C_c\big((0,\infty)\big) \longrightarrow L_p\big((0,\infty)\big)$$

defines a continuous operator such that $\|T\| \le \tfrac{p}{p-1}$ holds.

Since $C_c\big((0,\infty)\big)$ is norm dense in $L_p\big((0,\infty)\big)$ (Theorem 31.11), T has a unique continuous (linear) extension T^* to all of $L_p\big((0,\infty)\big)$ such that $\|T^*\| \le \tfrac{p}{p-1}$ holds. Our next objective is to show that $T^* f(x) = \tfrac{1}{x}\int_0^x f(t)\,d\lambda(t) = Tf(x)$ holds for all $f \in L_p\big((0,\infty)\big)$ and all $x > 0$.

To this end, let $0 \le \phi$ be a step function. Choose some $C > 0$ satisfying $0 \le \phi(x) \le C$ for all $x > 0$. By Theorem 31.11, there exists a sequence $\{f_n\}$ of $C_c\big((0,\infty)\big)$ with $\lim \int |f_n - \phi|^p\,d\lambda = 0$. We can assume that $\lim f_n(x) = \phi(x)$ holds for almost all x (see Lemma 31.6). In view of

$$\big|f_n \wedge C - \phi\big| = \big|f_n \wedge C - \phi \wedge C\big| \le \big|f_n - C\big|,$$

replacing $\{f_n\}$ by $\{f_n \wedge C\}$, we can assume that $0 \le f_n(x) \le C$ holds for all $x > 0$ and all n. Since $\lim \|Tf_n - T^*\phi\|_p = 0$, we can also assume (by passing to a subsequence) that $Tf_n(x) \longrightarrow T^*\phi(x)$ holds for almost all x. Next, observe that for each fixed $x > 0$ we have $\phi \in L_1\big((0,x)\big)$ and so, by the Lebesgue Dominated Convergence Theorem, we see that

$$T^*\phi(x) = \lim_{n\to\infty} Tf_n(x) = \lim_{n\to\infty} \tfrac{1}{x}\int_0^x f_n(t)\,d\lambda(t) = \tfrac{1}{x}\int_0^x \phi(t)\,d\lambda(t)$$

holds for almost all x. Now, let $0 \le f \in L_p\big((0,\infty)\big)$. Choose a sequence $\{\phi_n\}$ of step functions with $0 \le \phi_n \uparrow f$. In view of

$$\lim \|T^*\phi_n - T^*f\|_p = 0,$$

we can assume that $T^*\phi_n(x) \longrightarrow T^*f(x)$ holds for almost all x. Taking into account that for each fixed $x > 0$, we have $f \in L_p\big((0,x)\big) \subseteq L_1\big((0,x)\big)$, the Lebesgue Dominated Convergence Theorem implies

$$T^*f(x) = \lim_{n\to\infty} T^*\phi_n(x) = \lim_{n\to\infty} \tfrac{1}{x}\int_0^x \phi_n(t)\,d\lambda(t) = \tfrac{1}{x}\int_0^x f(t)\,d\lambda(t)$$

holds for almost all x. Thus, $T^* = T$ holds.

Next, we shall show that $\|T\| = \frac{p}{p-1}$ holds. We already know that $\|T\| \leq \frac{p}{p-1}$ holds. So, it must be established that $\|T\| \geq \frac{p}{p-1}$. To this end, let

$$f_n(x) = \begin{cases} x^{(n^{-1}-1)p^{-1}} & \text{if } 0 < x < 1 \\ 0 & \text{if } x \geq 1 \ . \end{cases}$$

Then, $(\|f_n\|_p)^p = \int_0^1 x^{n^{-1}-1}\, dx = n$, and moreover,

$$Tf_n(x) = \frac{np}{1+n(p-1)} \begin{cases} x^{(n^{-1}-1)p^{-1}} & \text{if } 0 < x < 1 \\ x^{-1} & \text{if } x \geq 1 \ . \end{cases}$$

Consequently, we have

$$\left(\|Tf_n\|_p\right)^p = \int_0^\infty |Tf_n(x)|^p\, d\lambda(x)$$

$$= \left[\frac{np}{1+n(p-1)}\right]^p \left[\int_0^1 x^{n^{-1}-1}\, dx + \int_1^\infty x^{-p}\, dx\right]$$

$$= \left[\frac{np}{1+n(p-1)}\right]^p \left(n + \frac{1}{p-1}\right),$$

and so

$$\frac{np}{1+n(p-1)}\left[n + \frac{1}{p-1}\right]^{\frac{1}{p}} = \|Tf_n\|_p \leq \|T\| \cdot \|f_n\|_p = \|T\| \cdot n^{\frac{1}{p}}.$$

This implies

$$\|T\| \geq \frac{p}{p-1+\frac{1}{n}} \cdot \left[1 + \frac{1}{n(p-1)}\right]^{\frac{1}{p}} \longrightarrow \frac{p}{p-1},$$

from which it follows that $\|T\| \geq \frac{p}{p-1}$ also holds.

Finally, we establish that T is one-to-one. Assume that $Tf = 0$ holds for some $f \in L_p\big((0, \infty)\big)$. Then, $\int_0^x f(t)\, d\lambda(t) = 0$ holds for all $x > 0$. Now, by Problem 22.19, we infer that $f = 0$ a.e. holds, and so the operator T is one-to-one.

CHAPTER 6

HILBERT SPACES

32. INNER PRODUCT SPACES

Problem 32.1. *Let $c_1, c_2, \ldots, c_n$ be n (strictly) positive real numbers. Show that the function of two variables $(\cdot, \cdot)$: $\mathbf{R}^n \times \mathbf{R}^n \to \mathbf{R}$, defined by $(x, y) = \sum_{i=1}^{n} c_i x_i y_i$, is an inner product on $\mathbf{R}^n$.*

Solution. Notice that for all vectors $x = (x_1, \ldots, x_n)$, $y = (y_1, \ldots, y_n)$ and $z = (z_1, \ldots, z_n)$ in $\mathbf{R}^n$ we have

$$(\alpha x + \beta y, z) = \sum_{i=1}^{n} c_i (\alpha x_i + \beta y_i) z_i = \alpha \sum_{i=1}^{n} c_i x_i z_i + \beta \sum_{i=1}^{n} c_i y_i z_i = \alpha(x, z) + \beta(y, z),$$

$$(x, y) = \sum_{i=1}^{n} c_i x_i y_i = \sum_{i=1}^{n} c_i y_i x_i = (y, x), \quad \text{and}$$

$$(x, x) = \sum_{i=1}^{n} c_i x_i^2 \geq 0.$$

Moreover, $(x, x) = \sum_{i=1}^{n} c_i x_i^2 = 0$ implies $c_i x_i^2 = 0$ for each i, and so (since $c_i > 0$ for each i) $x_i = 0$ for each i, i.e., $x = 0$. The above show the function $(\cdot, \cdot)$ is an inner product on $\mathbf{R}^n$.

Problem 32.2. *Let $(X, (\cdot, \cdot))$ be a real inner product vector space with complexification $X_{\mathbf{c}}$. Show that the function $\langle \cdot, \cdot \rangle$: $X_{\mathbf{c}} \times X_{\mathbf{c}} \to \mathbf{C}$ defined via the formula*

$$\langle x + \imath y, x_1 + \imath y_1 \rangle = (x, x_1) + (y, y_1) + \imath [(y, x_1) - (x, y_1)].$$

is an inner product on $X_{\mathbf{c}}$. Also, show that the norm induced by the inner product $\langle \cdot, \cdot \rangle$ on $X_{\mathbf{c}}$ is given by

$$\|x + \imath y\| = \sqrt{(x, x) + (y, y)} = (\|x\|^2 + \|y\|^2)^{\frac{1}{2}}.$$

Solution. Let $x_1 + \iota y_1, x_2 + \iota y_2, x_3 + \iota y_3 \in X_\mathbf{c}$. We check below the properties of the inner product.

1. (Additivity)

$$
\begin{aligned}
&\langle (x_1 + \iota y_1) + (x_2 + \iota y_2), x_3 + \iota y_3 \rangle \\
&= \langle x_1 + x_2 + \iota(y_1 + y_2), x_3 + \iota y_3 \rangle \\
&= (x_1 + x_2, x_3) + (y_1 + y_2, y_3) + \iota \big[(y_1 + y_2, x_3) - (x_1 + x_2, y_3) \big] \\
&= (x_1, x_3) + (y_1, y_3) + \iota[(y_1, x_3) - (x_1, y_3)] + ((x_2, x_3) + (y_2, y_3) \\
&\quad + \iota[(y_2, x_3) - (x_2, y_3)]) \\
&= \langle x_1 + \iota y_1, x_3 + \iota y_3 \rangle + \langle x_2 + \iota y_2, x_3 + \iota y_3 \rangle.
\end{aligned}
$$

2. (Homogeneity)

$$
\begin{aligned}
&\langle (\alpha + \iota \beta)(x_1 + \iota y_1), x_2 + \iota y_2 \rangle \\
&= \langle \alpha x_1 - \beta y_1 + \iota(\beta x_1 + \alpha y_1), x_2 + \iota y_2 \rangle \\
&= (\alpha x_1 - \beta y_1, x_2) + (\beta x_1 + \alpha y_1, y_2) + \iota[(\beta x_1 + \alpha y_1, x_2) \\
&\quad -(\alpha x_1 - \beta y_1, y_2)] \\
&= (\alpha + \iota \beta)\big[(x_1, x_2) + (y_1, y_2) + \iota[(y_1, x_2) - (x_1, y_2)] \big] \\
&= (\alpha + \iota \beta)\langle x_1 + \iota y_1, x_2 + \iota y_2 \rangle.
\end{aligned}
$$

3. (Conjugate Linearity)

$$
\begin{aligned}
\overline{\langle x_1 + \iota y_1, x_2 + \iota y_2 \rangle} &= \overline{(x_1, x_2) + (y_1, y_2) + \iota[(y_1, x_2) - (x_1, y_2)]} \\
&= (x_1, x_2) + (y_1, y_2) + \iota[(x_1, y_2) - (y_1, x_2)] \\
&= (x_2, x_1) + (y_2, y_1) + \iota[(y_2, x_1) - (x_2, y_1)] \\
&= \langle x_2 + \iota y_2, x_1 + \iota y_1 \rangle.
\end{aligned}
$$

4. (Positivity)

$$
\langle x_1 + \iota y_1, x_1 + \iota y_1 \rangle = (x_1, x_1) + (y_1, y_1) \geq 0.
$$

Moreover,

$$
\langle x_1 + \iota y_1, x_1 + \iota y_1 \rangle = (x_1, x_1) + (y_1, y_1) = 0 \iff x_1 = y_1 = 0 \iff x_1 + \iota y_1 = 0.
$$

Problem 32.3. *Let Ω be a Hausdorff compact topological space and let μ be a regular Borel measure on Ω such that $\operatorname{Supp}\mu = \Omega$. Show that the function $(\cdot,\cdot)\colon C(\Omega) \times C(\Omega) \to \mathbf{R}$, defined by*

$$(f, g) = \int_{\Omega} fg\, d\mu,$$

is an inner product. Also, describe the complexification of $C(\Omega)$ and the extension of the inner product to the complexification of $C(\Omega)$.

Solution. If $f, g, h \in C(\Omega)$ and $\alpha, \beta \in \mathbf{R}$, then note that

$$(\alpha f + \beta g, h) = \int_{\Omega} (\alpha f + \beta g) h\, d\mu = \alpha \int_{\Omega} fh\, d\mu + \beta \int_{\Omega} gh\, d\mu = \alpha(f, h) + \beta(g, h),$$

$$(f, g) = \int_{\Omega} fg\, d\mu = \int_{\Omega} gf\, d\mu = (g, f), \quad \text{and}$$

$$(f, f) = \int_{\Omega} f^2\, d\mu \geq 0.$$

Moreover, observe that (since $\operatorname{Supp}\mu = \Omega$) a function $f \in C(\Omega)$ satisfies

$$(f, f) = \int_{\Omega} f^2\, d\mu \iff f = 0.$$

The complexification $C_{\mathbf{c}}(\Omega)$ of $C(\Omega)$ consists of all complex-valued functions $f + \imath g$, where $f, g \in C(\Omega)$. The complex inner product is given by

$$\langle f, g \rangle = \int_{\Omega} f\, \overline{g}\, d\mu$$

for all $f, g \in C_{\mathbf{c}}(\Omega)$.

Problem 32.4. *Show that equality holds in the Cauchy–Schwarz inequality (i.e., $|(x, y)| = \|x\|\, \|y\|$) if and only if x and y are linearly dependent vectors.*

Solution. Assume $|(x, y)| = \|x\|\, \|y\|$. If $x = 0$, then the conclusion is obvious. So, assume $x \neq 0$. Let $(x, y) = re^{\imath\theta}$. Replacing x by $e^{-\imath\theta}x$, we can assume without loss of generality that $(x, y) = r \geq 0$, and so $(x, y) = \|x\|\, \|y\|$. Now, notice that for each real λ we have

$$
\begin{aligned}
0 \leq (\lambda x + y, \lambda x + y) &= \lambda^2 \|x\|^2 + \lambda\big[(x, y) + \overline{(x, y)}\big] + \|y\|^2 \\
&= \lambda^2 \|x\|^2 + 2\lambda(x, y) + \|y\|^2 \\
&= \lambda^2 \|x\|^2 + 2\lambda\|x\|\,\|y\| + \|y\|^2 \\
&= \big(\lambda\|x\| + \|y\|\big)^2.
\end{aligned}
$$

So, if $\lambda = -\frac{\|y\|}{\|x\|}$, then $(\lambda x + y, \lambda x + y) = 0$ or $\lambda x + y = 0$. This implies $\|y\|x - \|x\|y = 0$, which means that the vectors x and y are linearly dependent.

If x and y are linearly dependent, then the equation $|(x, y)| = \|x\|\,\|y\|$ should be obvious.

Problem 32.5. *If x is a vector in an inner product space, then show that*

$$
\|x\| = \sup_{\|y\|=1} |(x, y)|.
$$

Solution. If $x = 0$, then the conclusion is obvious. So, we consider the case $x \neq 0$. If $\|y\| = 1$, then the Cauchy–Schwarz inequality implies $|(x, y)| \leq \|x\|\,\|y\| \leq \|x\|$. Therefore, we have

$$
\sup_{\|y\|=1} |(x, y)| \leq \|x\|.
$$

For the reverse inequality, let $z = x/\|x\|$. Then, $\|z\| = 1$, and so

$$
\sup_{\|y\|=1} |(x, y)| \geq |(x, z)| = |(x, x/\|x\|)| = (x, x)/\|x\| = \|x\|.
$$

Therefore, $\|x\| = \sup_{\|y\|=1} |(x, y)|$, with the supremum being in actuality the maximum.

Problem 32.6. *Show that in a real inner product space $x \perp y$ holds if and only if $\|x + y\|^2 = \|x\|^2 + \|y\|^2$. Does $\|x + y\|^2 = \|x\|^2 + \|y\|^2$ in a complex inner product space imply $x \perp y$?*

Solution. Let x and y be two vectors in a real inner product space. If $x \perp y$, then the Pythagorean Theorem gives $\|x + y\|^2 = \|x\|^2 + \|y\|^2$. Conversely, if

$\|x + y\|^2 = \|x\|^2 + \|y\|^2$, then from

$$\begin{aligned} \|x + y\|^2 &= (x + y, x + y) = (x, x) + (x, y) + (y, x) + (y, y) \\ &= \|x\|^2 + (x, y) + (y, x) + \|y\|^2 \\ &= \|x\|^2 + 2(x, y) + \|y\|^2, \end{aligned}$$

it follows that $2(x, y) = 0$, and so $x \perp y$.

In complex inner product space the Pythagorean identity $\|x+y\|^2 = \|x\|^2 + \|y\|^2$ does not imply $x \perp y$. To see this, consider a non-zero vector x and let $y = \imath x$. Clearly, $\|y\|^2 = (\imath x, \imath x) = \|x\|^2$. Now, note that

$$\begin{aligned} \|x + y\|^2 &= \|x + \imath x\|^2 = \|(1 + \imath)x\|^2 = |1 + \imath|^2 \|x\|^2 \\ &= 2\|x\|^2 = \|x\|^2 + \|y\|^2, \end{aligned}$$

while $(x, y) = (x, \imath x) = -\imath\|x\|^2 \neq 0$.

Problem 32.7. *Assume that a sequence $\{x_n\}$ in an inner product space satisfies $(x_n, x) \to \|x\|^2$ and $\|x_n\| \to \|x\|$. Show that $x_n \to x$.*

Solution. Observe that $(x_n, x) \to \|x\|^2$ implies $(x, x_n) = \overline{(x_n, x)} \to \overline{\|x\|^2} = \|x\|^2$. So, from

$$\begin{aligned} \|x_n - x\|^2 &= (x_n - x, x_n - x) \\ &= \|x_n\|^2 - (x, x_n) - (x_n, x) + \|x\|^2 \\ &\longrightarrow \|x\|^2 - \|x\|^2 - \|x\|^2 + \|x\|^2 = 0, \end{aligned}$$

it follows that $\|x_n - x\| \to 0$, i.e., $x_n \to x$.

Problem 32.8. *Let S be an orthogonal subset of an inner product space. Show that there exists a complete orthogonal subset C such that $S \subseteq C$.*

Solution. Assume that S is an orthogonal subset of an inner product space X. Let $\mathcal{C}$ denote the collection of all orthogonal sets that contain S. That is, an orthogonal set A of vectors of X belongs to $\mathcal{C}$ if and only if $S \subseteq A$. If we consider $\mathcal{C}$ partially ordered by the inclusion relation $\subseteq$, then it is easy to see that $\mathcal{C}$ satisfies the hypotheses of Zorn's Lemma. Now, notice that any maximal element C of $\mathcal{C}$ is a complete orthogonal set satisfying $S \subseteq C$.

Problem 32.9. *Show that the norms of the following Banach spaces cannot be induced by inner products.*

a. *The norm $\|x\| = \max\{|x_1|, |x_2|, \ldots, |x_n|\}$ on $\mathbf{R}^n$.*
b. *The sup norm on $C[a, b]$.*
c. *The L_p-norm on any $L_p(\mu)$-space for each $1 \le p \le \infty$ with $p \ne 2$.*

Solution.   (a) Consider the vectors $x = (1, 0, \ldots, 0)$ and $y = (0, 1, 0, \ldots, 0)$. Clearly,

$$\|x\| = \|y\| = \|x + y\| = \|x - y\| = 1,$$

and so

$$\|x + y\|^2 + \|x - y\|^2 = 2 \quad \text{and} \quad 2\|x\|^2 + 2\|y\|^2 = 4.$$

Therefore, $\|x + y\|^2 + \|x - y\|^2 \ne 2\|x\|^2 + 2\|y\|^2$ and consequently the norm $\| \cdot \|$ does not satisfy the Parallelogram Law. This implies that the norm $\| \cdot \|$ cannot be induced by an inner product.

(b) Again, we shall show that the sup norm $\| \cdot \|_\infty$ does not satisfy the Parallelogram Law—and this will guarantee that the sup norm is not induced by an inner product. To see this, consider the two functions $\mathbf{1}$ (the constant function one) and $f : [a, b] \to \mathbf{R}$ defined by $f(x) = \frac{x-a}{b-a}$. Now, note that

$$\|\mathbf{1}\|_\infty = \|f\|_\infty = 1, \quad \|\mathbf{1} + f\|_\infty = 2, \quad \text{and} \quad \|\mathbf{1} - f\|_\infty = 1.$$

Therefore,

$$(\|\mathbf{1} + f\|_\infty)^2 + (\|\mathbf{1} - f\|_\infty)^2 = 5 \ne 4 = 2(\|\mathbf{1}\|_\infty)^2 + 2(\|f\|_\infty)^2,$$

so that the norm $\| \cdot \|_\infty$ does not satisfy the Parallelogram Law.

(c) Assume that there are two disjoint measurable sets E and F such that $0 < \mu^*(E) < \infty$ and $0 < \mu^*(F) < \infty$. First, we consider the case $p = \infty$. Then, note that

$$\|\chi_E\|_\infty = \|\chi_F\|_\infty = 1 \quad \text{and} \quad \|\chi_E + \chi_F\|_\infty = \|\chi_E - \chi_F\|_\infty = 1,$$

and consequently,

$$\left(\|\chi_E + \chi_F\|_\infty\right)^2 + (\|\chi_E - \chi_F\|_\infty)^2 = 2 \ne 4 = 2(\|\chi_E\|_\infty)^2 + 2(\|\chi_F\|_\infty)^2.$$

This shows that the norm $\| \cdot \|_\infty$ does not satisfy the Parallelogram Law and so is not induced by an inner product.

Now, consider the case $1 \le p < \infty$ with $p \ne 2$. The functions $f = [\mu^*(E)]^{-\frac{1}{p}} \chi_E$ and $g = [\mu^*(F)]^{-\frac{1}{p}} \chi_F$ satisfy $\|f\|_p = \|g\|_p = 1$, and hence,

$$2(\|f\|_p)^2 + 2(\|g\|_p)^2 = 4 = 2^2.$$

Also, from $|f + g|^p = |f - g|^p = |f|^p + |g|^p$, we see that

$$(\|f + g\|_p)^2 + (\|f - g\|_p)^2 = \left(2^{\frac{1}{p}}\right)^2 + \left(2^{\frac{1}{p}}\right)^2 = 2\left(2^{\frac{1}{p}}\right)^2 = 2^{1+\frac{2}{p}}$$

Since $p \ne 2$, we have $2^{1+\frac{2}{p}} \ne 2^2$, and so

$$(\|f + g\|_p)^2 + (\|f - g\|_p)^2 \ne 2(\|f\|_p)^2 + 2(\|g\|_p)^2.$$

This shows that the norm $\|\cdot\|_p$ does not satisfy the Parallelogram Law and so it is not induced by an inner product.

Problem 32.10. *Show that a norm $\|\cdot\|$ in a complex vector space is induced by an inner product if and only if it satisfies the Parallelogram Law, i.e., if and only if*

$$\|x + y\|^2 + \|x - y\|^2 = 2\left(\|x\|^2 + \|y\|^2\right)$$

holds for all vectors x and y. Moreover, show that if $\|\cdot\|$ satisfies the Parallelogram Law, then the inner product $(\cdot, \cdot)$ that induces $\|\cdot\|$ is given by

$$(x, y) = \frac{1}{4}\left(\|x + y\|^2 - \|x - y\|^2 + \iota\|x + \iota y\|^2 - \iota\|x - \iota y\|^2\right).$$

Solution. If $\|\cdot\|$ is induced by the inner product $(\cdot, \cdot)$, then for all vectors x and y, we have

$$\begin{aligned}
\|x + y\|^2 + \|x - y\|^2 &= (x + y, x + y) + (x - y, x - y) \\
&= [(x, x) + (y, x) + (x, y) + (y, y)] \\
&\quad + [(x, x) - (y, x) - (x, y) + (y, y)] \\
&= 2\|x\|^2 + 2\|y\|^2.
\end{aligned}$$

For the converse, assume that the norm $\|\cdot\|$ satisfies the Parallelogram Law. Consider, the complex-valued function $(\cdot, \cdot)$ defined by

$$(x, y) = \frac{1}{4}\left(\|x + y\|^2 - \|x - y\|^2 + \iota\|x + \iota y\|^2 - \iota\|x - \iota y\|^2\right).$$

Clearly, $(x, x) = \|x\|^2$ holds for all vectors x. To finish the solution, we shall verify that $(\cdot, \cdot)$ is a complex inner product. Start by observing that

$$
\begin{aligned}
(y, x) &= \frac{1}{4}\left(\|y + x\|^2 - \|y - x\|^2 + \imath\|y + \imath x\|^2 - \imath\|y - \imath x\|^2 \right) \\
&= \frac{1}{4}\left(\|x + y\|^2 - \|x - y\|^2 - \imath\|(-\imath)(x + \imath y)\|^2 + \imath\|\imath(x - \imath y)\|^2 \right) \\
&= \frac{1}{4}\left(\|x + y\|^2 - \|x - y\|^2 - \imath\|x + \imath y\|^2 + \imath\|x - \imath y\|^2 \right) \\
&= \overline{(x, y)}.
\end{aligned}
$$

Next, note that for all vectors u, v and w, we have

$$
\begin{aligned}
&4(u + v, w) + 4(u - v, w) \\
&= \left[\|u + v + w\|^2 - \|u + v - w\|^2 + \imath\|u + v + \imath w\|^2 - \imath\|u + v - \imath w\|^2 \right] \\
&\quad + \left[\|u - v + w\|^2 - \|u - v - w\|^2 + \imath\|u - v + \imath w\|^2 - \imath\|u - v - \imath w\|^2 \right] \\
&= \left[\|u + w + v\|^2 + \|u + w - v\|^2 \right] - \left[\|u - w + v\|^2 + \|u - w - v\|^2 \right] \\
&\quad + \imath\left[\|u + \imath w + v\|^2 + \|u + \imath w - v\|^2 \right] - \imath\left[\|u - \imath w + v\|^2 + \|u - \imath w - v\|^2 \right] \\
&= 2\|u + w\|^2 + 2\|v\|^2 - 2\|u - w\|^2 - 2\|v\|^2 \\
&\quad + \imath\left[2\|u + \imath w\|^2 + 2\|v\|^2 - 2\|u - \imath w\|^2 - 2\|v\|^2 \right] \\
&= 2\left[\|u + w\|^2 - \|u - w\|^2 + \imath\|u + \imath w\|^2 - \imath\|u - \imath w\|^2 \right] \\
&= 8(u, w).
\end{aligned}
$$

Thus, for all vectors u, v, and w we have

$$
(u + v, w) + (u - v, w) = 2(u, w). \tag{$\star$}
$$

When $v = u$, $(\star)$ yields $(2u, w) = 2(u, w)$. Now, letting $u = \frac{1}{2}(x + y)$, $v = \frac{1}{2}(x - y)$ and $w = z$ in $(\star)$, we get

$$
(x, z) + (y, z) = (u + v, z) + (u - v, z) = 2(u, z) = (2u, z) = (x + y, z),
$$

which is the additivity of $(\cdot, \cdot)$ in the first variable.

For the homogeneity, note first that

$$
\begin{aligned}
(\imath x, y) &= \frac{1}{4}\left(\|\imath x + y\|^2 - \|\imath x - y\|^2 + \imath\|\imath x + \imath y\|^2 - \imath\|\imath x - \imath y\|^2 \right) \\
&= \frac{1}{4}\left(\|\imath x + y\|^2 - \|\imath x - y\|^2 + \imath\|x + y\|^2 - \imath\|x - y\|^2 \right) \\
&= \imath\left[\frac{1}{4}\left(\|x + y\|^2 - \|x - y\|^2 - \imath\|\imath x + y\|^2 + \imath\|\imath x - y\|^2 \right) \right]
\end{aligned}
$$

$$= \imath \left[\frac{1}{4} \left(\|x+y\|^2 - \|x-y\|^2 - \imath \|\imath(x-\imath y)\|^2 + \imath \|\imath(x+\imath y)\|^2 \right) \right]$$

$$= \imath \left[\frac{1}{4} \left(\|x+y\|^2 - \|x-y\|^2 + \imath \|x+\imath y\|^2 - \imath \|x-\imath y\|^2 \right) \right]$$

$$= \imath(x,y).$$

Now, as in the proof of Lemma 18.7, we can establish that $(rx, y) = r(x, y)$ holds for each "real" rational number r and all $x, y \in X$. Since $(\cdot, \cdot)$, as defined above, is a jointly continuous function (relative to the norm $\| \cdot \|$), it easily follows that $(\alpha x, y) = \alpha(x, y)$ holds for all $\alpha \in \mathbf{R}$ and all $x, y \in X$. Finally, for an arbitrary complex number $\alpha + \imath \beta$ and arbitrary vectors x and y, note that

$$\begin{aligned} \big((\alpha + \imath \beta)x, y\big) &= (\alpha x + \imath \beta x, y) = (\alpha x, y) + \big(\beta(\imath x), y\big) \\ &= \alpha(x, y) + \beta(\imath x, y) = \alpha(x, y) + \beta \imath(x, y) \\ &= (\alpha + \imath \beta)(x, y). \end{aligned}$$

This establishes that $(\cdot, \cdot)$ is an inner product that induces the norm $\| \cdot \|$.

Problem 32.11. *Let X be a complex inner product space and let $T: X \to X$ be a linear operator. Show that $T = 0$ if and only if $(Tx, x) = 0$ for each $x \in X$. Is this result true for real inner product spaces?*

Solution. Assume that $(Tx, x) = 0$ holds for all $x \in X$. From the identity

$$\big(T(x+y), x+y\big) = (Tx, x) + (Tx, y) + (Ty, x) + (Ty, y)$$

and our hypothesis, it follows that

$$(Tx, y) + (Ty, x) = 0 \qquad\qquad (\star\star)$$

for all $x, y \subset X$. Replacing y by $\imath y$ in $(\star\star)$ yields $(Tx, \imath y) + (T(\imath y), x) - \imath\big[-(Tx, y) + (Ty, x)\big] = 0$. So,

$$-(Tx, y) + (Ty, x) = 0 \qquad\qquad (\star\star\star)$$

holds for all $x, y \in X$. Adding $(\star\star)$ and $(\star\star\star)$, we get $2(Ty, x) = 0$ or $(Ty, x) = 0$ for all $x, y \in X$. Letting $x = Ty$, we get $(Ty, Ty) = 0$ and so $Ty = 0$ for all $y \in X$, i.e. $T = 0$.

For real inner product spaces the preceding conclusion is false. Here is an example. Consider the Euclidean space $\mathbf{R}^2$ equipped with its standard inner product and define the linear operator $T: \mathbf{R}^2 \to \mathbf{R}^2$ by $T(x) = (-x_2, x_1)$ for all

$x = (x_1, x_2) \in \mathbb{R}^2$. Clearly, $T \neq 0$, and

$$(Tx, x) = (-x_2, x_1) \cdot (x_1, x_2) = -x_2 x_1 + x_1 x_2 = 0$$

holds for all $x \in \mathbb{R}^2$.

Problem 32.12. *If $\{x_n\}$ is an orthonormal sequence in an inner product space, then show that $\lim(x_n, y) = 0$ for each vector y.*

Solution. Let $\{x_n\}$ be an orthonormal sequence in an inner product space, and let y be an arbitrary vector. Then, from Bessel's Inequality, we have

$$\sum_{n=1}^{\infty} |(x_n, y)|^2 \leq \|y\|^2 < \infty.$$

This implies $|(x_n, y)|^2 \to 0$, and so $(x_n, y) \to 0$.

Problem 32.13. *The **orthogonal complement** of a nonempty subset A of an inner product space X is defined by*

$$A^{\perp} = \{x \in X : x \perp y \text{ for all } y \in A\}.$$

We shall denote $(A^{\perp})^{\perp}$ by $A^{\perp\perp}$. Establish the following properties regarding orthogonal complements:

a. *$A^{\perp}$ is a closed subspace of X, $A \subseteq A^{\perp\perp}$ and $A \cap A^{\perp} = \{0\}$.*
b. *If $A \subseteq B$, then $B^{\perp} \subseteq A^{\perp}$.*
c. *$A^{\perp} = \overline{A}^{\perp} = [\mathcal{L}(A)]^{\perp} = [\overline{\mathcal{L}(A)}]^{\perp}$, where $\mathcal{L}(A)$ denotes the vector subspace generated by A in X.*
d. *If M and N are two vector subspaces of X, then $M^{\perp\perp} + N^{\perp\perp} \subseteq (M+N)^{\perp\perp}$.*
e. *If M is a finite dimensional subspace, then $X = M \oplus M^{\perp}$.*

Solution. (a) If $x, y \in A^{\perp}$ and α, β are arbitrary scalars, then for each $z \in A$ we have

$$(\alpha x + \beta y, z) = \alpha(x, z) + \beta(y, z) = \alpha 0 + \beta 0 = 0,$$

and so $\alpha x + \beta y \in A^{\perp}$. Therefore, $A^{\perp}$ is a vector subspace of X. Since $x \in A$ implies $x \perp y$ for all $y \in A^{\perp}$, it follows that $x \in A^{\perp\perp}$, i.e., $A \subseteq A^{\perp\perp}$. Now if $x \in A \cap A^{\perp}$, then $(x, x) = 0$ or $x = 0$, and thus $A \cap A^{\perp} = \{0\}$.

(b) Assume $A \subseteq B$ and $x \in B^{\perp}$. If $y \in A$, then $y \in B$, and so $y \perp x$. This implies $x \in A^{\perp}$, and so $B^{\perp} \subseteq A^{\perp}$.

(c) From $A \subseteq \overline{A}$ and Part (b), it follows that $\overline{A}^{\perp} \subseteq A^{\perp}$. Now, let $x \in A^{\perp}$ and let $y \in \overline{A}$. Pick a sequence $\{y_n\} \subseteq A$ satisfying $y_n \to y$ and note that

$$(y, x) = \lim_{n \to \infty} (y_n, x) = 0.$$

Therefore, $x \in \overline{A}^{\perp}$. Hence, $A^{\perp} \subseteq \overline{A}^{\perp}$, and thus $A^{\perp} = \overline{A}^{\perp}$.

For the other equalities, note first that $A \subseteq \mathcal{L}(A)$ implies $[\mathcal{L}(A)]^{\perp} \subseteq A^{\perp}$. Now, fix $x \in A^{\perp}$, and let $y \in \mathcal{L}(A)$. Pick $y_1, \ldots, y_k \in A$ and scalars $\lambda_1, \ldots, \lambda_k$ such that $y = \sum_{i=1}^{k} \lambda_i y_i$. Then,

$$(y, x) = \left(\sum_{i=1}^{k} \lambda_i y_i, x \right) = \sum_{i=1}^{k} \lambda_i (y_i, x) = 0.$$

This shows that $x \in [\mathcal{L}(A)]^{\perp}$. Thus, $A^{\perp} \subseteq [\mathcal{L}(A)]^{\perp}$, and so $A^{\perp} = [\mathcal{L}(A)]^{\perp}$.

(d) From $M \subseteq M + N$, it follows that $M^{\perp\perp} \subseteq (M + N)^{\perp\perp}$ holds. Likewise, $N \subseteq M + N$ implies $N^{\perp\perp} \subseteq (M+N)^{\perp\perp}$. Therefore, $M^{\perp\perp} + N^{\perp\perp} \subseteq (M+N)^{\perp\perp}$.

(e) Let M be a finite dimensional subspace of dimension n. In order to establish that $X = M \oplus M^{\perp}$, we must show that every vector can be written in the form $y + z$ with $y \in M$ and $y \in M^{\perp}$. (The uniqueness of the decomposition should be obvious.)

Start by fixing a Hamel basis $\{x_1, x_2, \ldots, x_n\}$ of M. Replacing (if necessary) $\{x_1, x_2, \ldots, x_n\}$ by the normalized set of vectors that can be obtained by applying the Gram–Schmidt orthogonalization process (Theorem 32.11) to $\{x_1, x_2, \ldots, x_n\}$, we can assume that the set $\{x_1, x_2, \ldots, x_n\}$ is also an orthonormal set.

Now, fix $x \in X$ and consider the vectors

$$z = \sum_{k=1}^{n} (x, x_k) x_k \quad \text{and} \quad y = x - \sum_{k=1}^{n} (x, x_k) x_k.$$

Clearly, $z \in M$ and since $(y, x_k) = 0$ for each k, it easily follows that $y \in M^{\perp}$. Now, note that $x = y + z \in M \oplus M^{\perp}$.

Problem 32.14. *Let V be a vector subspace of a real inner product space X. A linear operator $L: V \to X$ is said to be* **symmetric** *if $(Lx, y) = (x, Ly)$ holds for all $x, y \in V$.*

  a. *Consider the real inner product space $C[a, b]$ and let $V = \{f \in C^2[a, b]:$ $f(a) = f(b) = 0\}$. Also, let $p \in C^1[a, b]$ and $q \in C[a, b]$ be two fixed functions. Show that the linear operator $L: V \to C[a, b]$, defined by*

$$L(f) = (pf')' + qf,$$

is a symmetric operator.

b.   *Consider $\mathbf{R}^n$ equipped with its standard inner product and let $A: \mathbf{R}^n \to \mathbf{R}^n$ be a linear operator. As usual, we identify the operator with the matrix $A = [a_{ij}]$ representing it, where the jth column of the matrix A is the column vector Ae_j. Show that A is a symmetric operator if and only if A is a symmetric matrix. (Recall that an $n \times n$ matrix $B = [b_{ij}]$ is said to be **symmetric** if $b_{ij} = b_{ji}$ holds for all i and j.)*

c.   *Let $L: V \to X$ be a symmetric operator. Then L extends naturally to a linear operator $L: V_c = \{x + \iota y: x, y \in V\} \to X_c$ via the formula $L(x + \iota y) = Lx + \iota Ly$. Show that L also satisfies $(Lu, v) = (u, Lv)$ for all $u, v \in V_c$ and that the eigenvalues of L are all real numbers.*

d.   *Show that eigenvectors of a symmetric operator corresponding to distinct eigenvalues are orthogonal.*

Solution.   (a) If $f, g \in V$, then note that

$$
\begin{aligned}
(Lf, g) &= \int_a^b \left([p(x)f'(x)]' + q(x)f(x) \right) g(x) \, dx \\
&= \int_a^b [p(x)f'(x)]' g(x) \, dx + \int_a^b q(x)f(x)g(x) \, dx \\
&= p(x)f'(x)g(x) \Big|_a^b - \int_a^b p(x)f'(x)g'(x) \, dx + \int_a^b q(x)f(x)g(x) \, dx \\
&= \int_a^b q(x)f(x)g(x) \, dx - \int_a^b p(x)f'(x)g'(x) \, dx \\
&= (f, Lg).
\end{aligned}
$$

(b) Recall that the transpose of a matrix $B = [b_{ij}]$ is the matrix $B^t = [b_{ji}]$. In terms of the transpose, a matrix A is symmetric if and only if $A^t = A$. Now, our conclusion follows immediately from the following two identities:

$$(Ax, y) = (x, A^t y) \text{ for all } x, y \in \mathbf{R}^n, \quad \text{and}$$
$$a_{ij} = (e_i, Ae_j).$$

(c) If $u = x + \iota y$ and $v = x_1 + \iota y_1$ are vectors of V_c, then note that

$$
\begin{aligned}
(Lu, v) &= \left(L(x + \iota y), x_1 + \iota y_1 \right) = \left(Lx + \iota Ly, x_1 + \iota y_1 \right) \\
&= (Lx, x_1) + (Ly, y_1) + \iota\left[(Ly, x_1) - (Lx, y_1) \right] \\
&= (x, Lx_1) + (y, Ly_1) + \iota\left[(y, Lx_1) - (x, Ly_1) \right] \\
&= (x + \iota y, Lx_1 + \iota Ly_1) = (u, Lv).
\end{aligned}
$$

Now, assume that $\lambda \in \mathbf{C}$ is an eigenvalue of $L \colon V_{\mathbf{c}} \to X_{\mathbf{c}}$. Fix a unit vector $u \in X_{\mathbf{c}}$ satisfying $Lu = \lambda u$, and note that

$$\lambda = \lambda(u, u) = (\lambda u, u) = (Lu, u) = (u, Lu) = (u, \lambda u) = \overline{\lambda}(u, u) = \overline{\lambda}.$$

This shows that λ is a real number.

(d) Assume that $L \colon V \to X$ is a symmetric operator and let two nonzero vectors $u, v \in V_{\mathbf{c}}$ satisfy $Lu = \lambda u$ and $Lv = \mu v$ with $\lambda \neq \mu$. By part (c), we know that λ and μ are real numbers. Therefore,

$$(\lambda - \mu)(u, v) = \lambda(u, v) - \mu(u, v) = (\lambda u, v) - (u, \mu v) = (Lu, v) - (u, Lv) = 0,$$

and so $(u, v) = 0$.

Problem 32.15. *Let $(\cdot, \cdot)$ denote the standard inner product on $\mathbf{R}^n$, i.e., $(x, y) = \sum_{i=1}^{n} x_i y_i$ for all $x, y \in \mathbf{R}^n$. Recall that an $n \times n$ matrix A is said to be* **positive definite** *if $(x, Ax) > 0$ holds for all nonzero vectors $x \in \mathbf{R}^n$.*

Show that a function of two variables $\langle \cdot, \cdot \rangle \colon \mathbf{R}^n \times \mathbf{R}^n \to \mathbf{R}$ is an inner product on $\mathbf{R}^n$ if and only if there exists a unique real symmetric positive definite matrix A such that

$$\langle x, y \rangle = (x, Ay)$$

holds for all $x, y \in \mathbf{R}^n$. (It is known that a symmetric matrix is positive definite if and only if its eigenvalues are all positive.)

Solution. Let $\langle \cdot, \cdot \rangle \colon \mathbf{R}^n \times \mathbf{R}^n \to \mathbf{R}$ be a function of two variables. Assume first that there exists a real symmetric positive definite matrix A such that

$$\langle x, y \rangle = (x, Ay)$$

holds for all $x, y, z \in \mathbf{R}^n$. Then, for all $x, y \in \mathbf{R}^n$ and all $\alpha, \beta \in \mathbf{R}$, we have

$$\langle x, y \rangle = (x, Ay) = (Ax, y) = (y, Ax) = \langle y, x \rangle,$$
$$\langle \alpha x + \beta y, z \rangle = (\alpha x + \beta y, Az) = \alpha(x, Az) + \beta(y, Az) = \alpha \langle x, z \rangle + \beta \langle y, z \rangle, \quad \text{and}$$
$$\langle x, x \rangle = (x, Ax) \geq 0 \text{ for all } x \in \mathbf{R}^n \text{ and } \langle x, x \rangle = 0 \iff x = 0.$$

This shows that $\langle \cdot, \cdot \rangle$ is an inner product.

For the converse assume that the function of two variables $\langle \cdot, \cdot \rangle \colon \mathbf{R}^n \times \mathbf{R}^n \to \mathbf{R}$ is a real inner product. Let $e_1, e_2, \ldots, e_n$ denote the standard unit vectors, and so each vector $x \in \mathbf{R}^n$ is written as $x = \sum_{i=1}^{n} x_i e_i$. It follows that

$$\langle x, y \rangle = \left\langle \sum_{i=1}^{n} x_i e_i, \sum_{j=1}^{n} y_j e_j \right\rangle = \sum_{i=1}^{n} \sum_{j=1}^{n} x_i y_j \langle e_i, e_j \rangle = (x, Ay)$$

for all $x, y \in \mathbf{R}^n$, where A is the $n \times n$ matrix $A = \left[\langle e_i, e_j \rangle \right]$. Clearly, A is a real symmetric matrix and in view of $(x, Ax) = \langle x, x \rangle$, we see that A is also a positive definite matrix. The uniqueness of A should be obvious.

33. HILBERT SPACES

Problem 33.1. *Let (X, S, μ) be a measure space and let $\rho \colon X \to (0, \infty)$ be a measurable function—called a **weight function**. Show that the collection of measurable functions*

$$L_2(\rho) = \left\{ f \in \mathcal{M} \colon \int \rho |f|^2 \, d\mu < \infty \right\}$$

under the inner product $(\cdot, \cdot) \colon L_2(\rho) \times L_2(\rho) \to \mathbf{R}$, defined by

$$(f, g) = \int \rho f g \, d\mu,$$

is a real Hilbert space.

Solution. It should be clear that $L_2(\rho)$ is a vector space. Moreover, since $f \in L_2(\rho)$ is equivalent to $\sqrt{\rho} f \in L_2(\mu)$, it follows from Hölder's inequality that

$$\left| \int \rho f g \, d\mu \right| \leq \left(\int \rho |f|^2 \, d\mu \right)^{\frac{1}{2}} \left(\int \rho |g|^2 \, d\mu \right)^{\frac{1}{2}} < \infty,$$

and so $(\cdot, \cdot)$ is well-defined. We leave it as an exercise for the reader to verify that $(\cdot, \cdot)$ is indeed a real inner product. We shall prove that $L_2(\rho)$ is a Hilbert space by establishing that it is complete.

To this end, let $\{f_n\} \subseteq L_2(\rho)$ be a Cauchy sequence. That is, for each $\epsilon > 0$ there exists some n_0 such that

$$\|f_n - f_m\|^2 = \int \rho |f_n - f_m|^2 \, d\mu = \int \left| \sqrt{\rho} f_n - \sqrt{\rho} f_m \right|^2 \, d\mu < \epsilon^2$$

holds for all $n, m \geq n_0$. This means that the sequence of functions $\{\sqrt{\rho} f_n\} \subseteq$ $L_2(\mu)$ is a norm Cauchy sequence of the Hilbert space $L_2(\mu)$. Since $L_2(\mu)$ is a Banach space (Theorem 31.5), it follows that there exists some function $g \in L_2(\mu)$ such that

$$\int \left| \sqrt{\rho} f_n - g \right|^2 d\mu \longrightarrow 0.$$

Now, note that if $f = g/\sqrt{\rho}$, then $f \in L_\rho(\mu)$ and

$$\| f_n - f \|^2 = \int \rho \left| f_n - f \right|^2 d\mu = \int \left| \sqrt{\rho} f_n - g \right|^2 d\mu \longrightarrow 0.$$

This shows that $L_2(\rho)$ is norm complete and hence, it is a Hilbert space.

The reader should also notice that $L_2(\rho)$ is exactly the Hilbert space $L_2(\nu)$ for the measure $\nu \colon \Lambda_\mu \to [0, \infty]$ defined by

$$\nu(A) = \int_A \rho(x) \, d\mu(x)$$

for each $A \in \Lambda_\mu$.

Problem 33.2.   *Show that the Hilbert space $L_2[0, \infty)$ is separable.*

Solution.   Consider the countable set of functions $\{ f_{k,n} \colon k, n = 1, 2, \ldots \}$, where

$$f_{k,n}(x) = \begin{cases} x^n & \text{if } 0 \leq x \leq k \\ 0 & \text{if } k < x. \end{cases}$$

We know that the continuous functions with compact support are dense in $L^2[0, \infty)$ (Theorem 31.11) and so, we need only prove that the linear span of $\{ f_{k,n} \}$ is dense in the vector space of continuous functions with compact support. Observe that if this is established, then the linear span of $\{ f_{k,n} \}$ with rational coefficients would be a countable dense set.

Let $f \in L_2[0, \infty)$ be a continuous function with compact support, and let $\epsilon > 0$. Fix an integer k such that $f(x) = 0$ for all $x \geq k$. By the Stone-Weierstrass approximation theorem, there exists a polynomial $P(x) = \sum_{n=0}^m c_n x^n$ satisfying $|f(x) - P(x)| < \epsilon/\sqrt{k}$ for each $x \in [0, k]$. Now, notice that if we consider the function $g \in L_2[0, \infty)$ defined by $g(x) = \sum_{n=0}^m c_n f_{k,n}(x)$, then

$$\| f - g \|_2 = \left(\int_0^\infty |f(x) - g(x)|^2 \, dx \right)^{\frac{1}{2}} = \left(\int_0^k |f(x) - P(x)|^2 \, dx \right)^{\frac{1}{2}}$$

$$< \left(\int_0^k \frac{\epsilon^2}{k} \, dx \right)^{\frac{1}{2}} = \epsilon$$

holds. This shows that the linear span of the countable set $\{f_{k,n}\colon k, n = 1, 2, \ldots\}$ is dense in $L_2[0, \infty)$, and so $L_2[0, \infty)$ is separable.

Problem 33.3. *Let $\{\psi_n\}$ be an orthonormal sequence of functions in the Hilbert space $L_2[a, b]$ which is also uniformly bounded. If $\{\alpha_n\}$ is a sequence of scalars such that $\alpha_n \psi_n \to 0$ a.e., then show that $\lim \alpha_n = 0$.*

Solution. Fix some constant C such that $|\psi_n(x)| \le C$ hold for all n and for all $x \in [a, b]$. Also, let $\{\alpha_n\}$ be a sequence of scalars such that $\alpha_n \psi_n(x) \to 0$ holds for almost all x.

Next, fix $\epsilon > 0$ so that $\epsilon C^2 < \frac{1}{2}$. Now, by Egorov's Theorem 16.7, there exists a measurable set $E \subseteq [a, b]$ with $\lambda(E^c) < \epsilon$ such that the sequence of functions $\{\alpha_n \psi_n\}$ converges uniformly to zero on E. So, there exists an integer m such that $|\alpha_n \psi_n(x)| \le \epsilon$ for all $n \ge m$ and all $x \in E$. Then, we have

$$
\begin{aligned}
|\alpha_n|^2 &= \int_a^b |\alpha_n \psi_n(t)|^2 \, dt = \int_E |\alpha_n \psi_n(t)|^2 \, dt + \int_{E^c} |\alpha_n \psi_n(t)|^2 \, dt \\
&\le \int_E \epsilon^2 \, dt + |\alpha_n|^2 \int_{E^c} |\psi_n(t)|^2 \, dt \\
&\le \epsilon^2(b - a) + |\alpha_n|^2 \int_{E^c} C^2 \, dt \\
&\le \epsilon^2(b - a) + \epsilon |\alpha_n|^2 C^2.
\end{aligned}
$$

This implies $\frac{1}{2}|\alpha_n|^2 < (1 - \epsilon C^2)|\alpha_n|^2 \le \epsilon^2(b - a)$ for all $n \ge m$, or

$$
|\alpha_n| \le \epsilon \sqrt{2(b - a)}
$$

for all $n \ge m$. Since $0 < \epsilon < \frac{1}{2C^2}$ is arbitrary, we have established that $\alpha_n \to 0$.

Problem 33.4. *Let $\{\phi_n\}$ be an orthonormal sequence of functions in the Hilbert space $L_2[-1, 1]$. Show that the sequence of functions $\{\psi_n\}$, where*

$$
\psi_n(x) = \left(\tfrac{2}{b-a}\right)^{\frac{1}{2}} \phi_n\left(\tfrac{2}{b-a}\left(x - \tfrac{b+a}{2}\right)\right),
$$

is an orthonormal sequence in the Hilbert space $L_2[a, b]$.

Solution. Observe that the inner product satisfies

$$
\begin{aligned}
(\psi_n, \psi_m) &= \int_a^b \psi_n(x) \overline{\psi_m}(x) \, dx \\
&= \tfrac{2}{b-a} \int_a^b \phi_n\left(\tfrac{2}{b-a}\left(x - \tfrac{b+a}{2}\right)\right) \overline{\phi_m}\left(\tfrac{2}{b-a}\left(x - \tfrac{b+a}{2}\right)\right) dx.
\end{aligned}
$$

Making the substitution $t = \frac{2}{b-a}\left(x - \frac{b+a}{2}\right)$, we have $dt = \frac{2}{b-a}dx$, and so

$$(\psi_n, \psi_m) = \int_{-1}^{1} \phi_n(t)\,\overline{\phi_m(t)}\,dt = \delta_{mn}.$$

That is, $\{\psi_n\}$ is an orthonormal sequence in the Hilbert space $L_2[a, b]$.

Problem 33.5. *Show that the norm completion $\hat{X}$ of an inner product space X is a Hilbert space. Moreover, if $x, y \in \hat{X}$ and two sequences $\{x_n\}$ and $\{y_n\}$ of X satisfy $x_n \to x$ and $y_n \to y$ in $\hat{X}$, then establish that the inner product of $\hat{X}$ is given by*

$$\langle x, y \rangle = \lim_{n \to \infty} (x_n, y_n).$$

Solution. Assume that $\hat{X}$ is the norm completion of an inner product space X and let $x, y \in \hat{X}$. Pick two sequences $\{x_n\}$ and $\{y_n\}$ of X such that $\|x_n - x\| \to 0$ and $\|y_n - y\| \to 0$, where $\| \cdot \|$ is the norm of $\hat{X}$ (which is the unique continuous extension of the norm of X to $\hat{X}$). Fix some constant $M > 0$ such that $\|x_n\| \leq M$ and $\|y_n\| \leq M$ hold for each n. Then, using the Cauchy–Schwarz inequality, we have

$$\begin{aligned}
\left| (x_n, y_n) - (x_m, y_m) \right| &= \left| (x_n, y_n) - (x_n, y_m) + (x_n, y_m) - (x_m, y_m) \right| \\
&= \left| (x_n, y_n - y_m) + (x_n - x_m, y_m) \right| \\
&\leq \left| (x_n, y_n - y_m) \right| + \left| (x_n - x_m, y_m) \right| \\
&\leq \|x_n\|\,\|y_n - y_m\| + \|x_n - x_m\|\,\|y_m\| \\
&\leq M\left(\|x_n - x_m\| + \|y_n - y_m\| \right).
\end{aligned}$$

This shows that the sequence of scalars $\{(x_n, y_n)\}$ is a Cauchy sequence and hence, convergent.

Next, assume two other sequences $\{x_n'\}$ and $\{y_n'\}$ of X satisfy $\|x_n' - x\| \to 0$ and $\|y_n' - y\| \to 0$. We can assume without loss of generality that $\|x_n'\| \leq M$ and $\|y_n'\| \leq M$ holds for each n. By the preceding $\lim(x_n', y_n')$ exists, and since

$$\begin{aligned}
\left| (x_n, y_n) - (x_n', y_n') \right| &= \left| (x_n, y_n) - (x_n, y_n') + (x_n, y_n') - (x_n', y_n') \right| \\
&= \left| (x_n, y_n - y_n') + (x_n - x_n', y_n') \right| \\
&\leq \left| (x_n, y_n - y_n') \right| + \left| (x_n - x_n', y_n') \right| \\
&\leq \|x_n\|\,\|y_n - y_n'\| + \|x_n - x_n'\|\,\|y_n'\| \\
&\leq M\left(\|x_n - x_n'\| + \|y_n - y_n'\| \right) \longrightarrow 0,
\end{aligned}$$

it follows that $\lim(x_n, y_n) = \lim(x'_n, y'_n)$. In other words, the formula

$$\langle x, y \rangle = \lim_{n\to\infty}(x_n, y_n) \qquad\qquad (\star)$$

gives rise to well-defined scalar-valued function on $\hat{X} \times \hat{X}$.

Now, it should be clear that the properties of the inner product are transferred via $(\star)$ from the inner product of X to the function $\langle \cdot, \cdot \rangle : \hat{X} \times \hat{X} \to \mathbf{C}$. In other words, the formula of $(\star)$ is an inner product on $\hat{X}$. Moreover, from

$$\|x\|^2 = \lim_{n\to\infty}\|x_n\|^2 = \lim_{n\to\infty}(x_n, x_n) = \langle x, x \rangle,$$

we see that the inner product given by $(\star)$ induces the norm of $\hat{X}$.

Problem 33.6. *Show that the closed unit ball of ℓ_2 is not a norm compact set.*

Solution. Let $\mathcal{U} = \{x \in \ell_2 \colon \|x\| \le 1\}$ be the closed unit ball of ℓ_2. Now, for each n let $e_n = (0, 0, \ldots, 0, 1, 0, 0, \ldots)$, the sequence with 1 in its nth coordinate and zero elsewhere. Note that $\|e_n\| = 1$ for each n and thus $\{e_n\}$ is a sequence of the unit ball of ℓ^2. (In fact $\{e_n\}$ is an orthonormal sequence of ℓ_2.) Now, notice that for $n \ne m$ we have $\|e_n - e_m\| = \sqrt{2}$. This implies that $\{e_n\}$ does not have any Cauchy subsequences—and hence, it does not have any convergent subsequences either. Now a glance at Theorem 7.3 guarantees that $\mathcal{U}$ is not a norm compact subset of ℓ_2.

Problem 33.7. *Show that the **Hilbert cube** (the set of all $x = (x_1, x_2, \ldots) \in \ell_2$ such that $|x_n| \le \frac{1}{n}$ holds for all n) is a compact subset of ℓ_2.*

Solution. Let $C = \{(x_1, x_2, \ldots) \in \ell_2 \colon |x_n| \le \frac{1}{n}$ for each $n = 1, 2, \ldots\}$. Clearly, C is a closed subset of ℓ_2. Thus, in order to establish the compactness of C, it suffices to prove (by Theorem 7.8) that C is totally bounded.

To this end, let $\varepsilon > 0$. Fix some n such that $\sum_{k=n+1}^{\infty} \frac{1}{k^2} < \varepsilon$. Since the set $A = \{(x_1, \ldots, x_n) \in \mathbf{R}^n \colon |x_i| \le \frac{1}{i}$ for $1 \le i \le n\}$ is closed and bounded, it must be a compact subset of $\mathbf{R}^n$. Pick $x^1, \ldots, x^m \in A$ $\big($where $x^i = (x_1^i, \ldots, x_n^i)\big)$ so that $A \subseteq \bigcup_{i=1}^{m} B(x^i, \varepsilon)$ holds. (We consider, of course, $\mathbf{R}^n$ equipped with the Euclidean distance.) Now, for each $1 \le i \le m$ let $y_i = (x_1^i, \ldots, x_n^i, 0, 0, \ldots)$. Then, it is easy to see that $C \subseteq \bigcup_{i=1}^{m} B(y_i, 2\varepsilon)$ holds in ℓ_2. This shows that C is totally bounded, as required.

Problem 33.8. *Show that every subspace M of a Hilbert space satisfies $\overline{M} = M^{\perp\perp}$.*

Solution. It should be clear that $(\overline{M})^\perp = M^\perp$; see Problem 32.13. Therefore, by Theorem 33.7, $H = \overline{M} \oplus M^\perp$. Also, it should be noticed that $M \subseteq \overline{M} \subseteq M^{\perp\perp}$.

Now, let $x \in M^{\perp\perp}$. From $H = \overline{M} \oplus M^\perp$, it follows that we can write $x = u + v$ with $u \in \overline{M}$ and $v \in M^\perp$. This implies $x - u = v \in M^\perp$ and since $u \in \overline{M} \subseteq M^{\perp\perp}$, we have $x - u \in M^{\perp\perp}$. Hence, $x - u \in M^\perp \cap M^{\perp\perp} = \{0\}$ or $x = u \in \overline{M}$. Therefore, $M^{\perp\perp} \subseteq \overline{M}$ also holds true, and so $M^{\perp\perp} = \overline{M}$, as desired.

Problem 33.9. *For two arbitrary vector subspaces M and N of a Hilbert space establish the following:*

   a. $(M + N)^\perp = M^\perp \cap N^\perp$, *and*
   b. *if M and N are both closed, then $(M \cap N)^\perp = \overline{M^\perp + N^\perp}$.*

Solution. (a) Let $x \perp M + N$. Then, $x \perp y$ holds for all $y \in M$ and $x \perp z$ holds for all $z \in N$. That is, $x \in M^\perp$ and $x \in N^\perp$. Therefore, $(M + N)^\perp \subseteq M^\perp \cap N^\perp$.

For the reverse inclusion, let $x \in M^\perp \cap N^\perp$ and let $y \in M + N$. Write $y = u + v$ with $u \in M$ and $v \in N$ and note that $(x, y) = (x, u) + (x, v) = 0$ holds. Hence, $x \in (M + N)^\perp$, and therefore $M^\perp \cap N^\perp \subseteq (M + N)^\perp$. Thus, $(M + N)^\perp = M^\perp \cap N^\perp$.

(b) Now, suppose that M and N are closed subspaces. By the preceding problem we know that $M = M^{\perp\perp}$ and $N = N^{\perp\perp}$. Now, use part (a) to get

$$\left(M^\perp + N^\perp\right)^\perp = M \cap N.$$

Therefore,

$$(M \cap N)^\perp = \left[(M^\perp + N^\perp)^\perp\right]^\perp = \overline{M^\perp + N^\perp}.$$

Problem 33.10. *Let X be an inner product space such that $M = M^{\perp\perp}$ holds for every closed subspace M. Show that X is a Hilbert space.*

Solution. We need to show that X is complete in the induced norm. For this, it suffices to establish that $X = \hat{X}$, where $\hat{X}$ denotes the norm completion of X. (We already know that $\hat{X}$ is a Hilbert space; see Problem 33.5.) To this end, let $\hat{u} \in \hat{X}$ be a nonzero vector.

The linear functional $f : \hat{X} \to \mathbf{C}$ defined by $f(x) = (x, \hat{u})$ is nonzero and continuous. So, f restricted to X is also continuous and since X is norm dense in $\hat{X}$, it follows that $f : X \to \mathbf{C}$ is a nonzero continuous linear functional. In particular, its kernel $M = \{x \in X : f(x) = 0\}$ is a proper closed subspace of X. We claim that $M^\perp \neq \{0\}$. Indeed, if $M^\perp = \{0\}$, then it follows from our hypothesis that $M = M^{\perp\perp} = \{0\}^\perp = X$, which is a contradiction.

Next, fix a vector $u \in M^{\perp}$ with $\|u\| = 1$ and let $v = \overline{f(u)}u \in X$. Now, taking into account that $f(x)u - f(u)x \in M$ holds for each $x \in X$, it follows that

$$f(x) = f(x)(u, u) = f(u)(x, u) = (x, v).$$

for each $x \in X$. That is, $(x, \hat{u}) = (x, v)$ for all $x \in X$. Since X is dense in $\hat{X}$, we get $(x, v) = (x, \hat{u})$ for all $x \in \hat{X}$. That is, $(x, v - \hat{u}) = 0$ for all $x \in \hat{X}$, and from this we conclude that $\hat{u} = v \in X$. So, $X = \hat{X}$, and thus X is a Hilbert space.

Problem 33.11. *Consider the linear operator* $V : L_2[a, b] \rightarrow L_2[a, b]$ *defined by*

$$V f(x) = \int_a^x f(t) \, dt.$$

Show that the norm of the operator satisfies $\|V\| \leq b - a$.

Solution. By Hölder's Inequality, we get

$$
\begin{aligned}
|V f(x)| &\leq \int_a^x |f(t)| \, dt \leq \int_a^b |f(t)| \, dt \\
&\leq \left[\int_a^b |f(t)|^2 \, dt \right]^{\frac{1}{2}} \left[\int_a^b 1^2 \, dt \right]^{\frac{1}{2}} \\
&\leq (b - a)^{\frac{1}{2}} \|f\| \, .
\end{aligned}
$$

Therefore, the norm of V satisfies

$$
\begin{aligned}
\|V f\|^2 &= \int_a^b |V f(t)|^2 \, dt \leq (b - a) \int_a^b \|f\|^2 \, dt \\
&\leq (b - a)^2 \|f\|^2.
\end{aligned}
$$

This implies $\|V\| \leq b - a$.

Problem 33.12. *Let* $\{x_n\}$ *be a norm bounded sequence of vectors in the Hilbert space* ℓ_2, *where* $x_n = (x_1^n, x_2^n, x_3^n, \ldots)$. *If for each fixed coordinate* k *we have* $\lim_{n \to \infty} x_k^n = 0$, *then show that*

$$\lim_{n \to \infty} (x_n, y) = 0$$

holds for each vector $y \in \ell_2$.

Solution. Choose some $\lambda > 0$ such that $\|x_n\| \leq \lambda$ holds for all n. Now, fix a vector $y = (y_1, y_2, \ldots) \in \ell_2$ and let $\epsilon > 0$. Pick some m satisfying $\left(\sum_{k=m}^{\infty} |y_k|^2 \right)^{\frac{1}{2}} < \epsilon$.

Since for each fixed k we have $\lim_{n \to \infty} x_k^n = 0$, there exists an integer n_0 satisfying $\left| \sum_{k=1}^{m} x_k^n \, \overline{y_k} \right| < \epsilon$ for all $n \geq n_0$. Now, using the Cauchy–Schwarz inequality, we see that for each $n \geq n_0$ we have

$$|(x_n, y)| = \left| \sum_{k=1}^{\infty} x_k^n \, \overline{y_k} \right| \leq \left| \sum_{k=1}^{m} x_k^n \, \overline{y_k} \right| + \sum_{k=m+1}^{\infty} |x_k^n \, \overline{y_k}|$$

$$< \epsilon + \left(\sum_{k=m+1}^{\infty} |x_k^n|^2 \right)^{\frac{1}{2}} \left(\sum_{k=m+1}^{\infty} |y_k|^2 \right)^{\frac{1}{2}}$$

$$\leq \epsilon + \lambda \epsilon = (1 + \lambda) \epsilon.$$

Since $\epsilon > 0$ is arbitrary, we have shown that $\lim_{n \to \infty} (x_n, y) = 0$.

Problem 33.13. *Let H be a Hilbert space and let $\{x_n\}$ be a sequence satisfying*

$$\lim_{n \to \infty} (x_n, y) = (x, y)$$

for each $y \in H$. Show that there exists a subsequence $\{x_{k_n}\}$ of $\{x_n\}$ such that

$$\lim_{n \to \infty} \left\| \frac{x_{k_1} + x_{k_2} + x_{k_3} + \cdots + x_{k_n}}{n} - x \right\| = 0.$$

Solution. Start by noticing that we can assume without loss of generality that $x = 0$. Therefore, suppose

$$\lim_{n \to \infty} (x_n, y) = 0$$

for each $y \subset X$. We claim that the sequence $\{x_n\}$ is norm bounded. To see this, for each n consider the continuous linear functional $f_n \colon H \to \mathbf{C}$ defined by $f_n(y) = (y, x_n)$ for each $y \in H$. By our condition, the sequence of bounded linear functionals $\{f_n\}$ is pointwise bounded. So, by the Principle of Uniform Boundedness (Theorem 28.8), there exists some $C > 0$ such that $\|f_n\| \leq C$ for each n. Now, notice that (by Theorem 33.9) $\|x_n\| = \|f_n\|$ holds for each n.

Now, let $k_1 = 1$ and then choose $k_2 > k_1$ with $|(x_{k_1}, x_{k_2})| < 1$. Next, an inductive argument shows that there exist integers $k_1 < k_2 < k_3 < \cdots < k_n < k_{n+1} < \cdots$ satisfying

$$\left| (x_{k_i}, x_{k_{n+1}}) \right| < 2^{-n} \quad \text{for each } 1 \leq i \leq n.$$

To finish the solution, we shall show that

$$\left\| \frac{x_{k_1} + x_{k_2} + x_{k_3} + \cdots + x_{k_n}}{n} \right\|^2 \le \frac{2 + C^2}{n}$$

holds for each n. To see this, take inner products to get

$$\left\| \frac{\sum_{i=1}^{n} x_{k_i}}{n} \right\|^2 = \frac{\sum_{i=1}^{n}(x_{k_1}, x_{k_i}) + \sum_{i=1}^{n}(x_{k_2}, x_{k_i}) + \cdots + \sum_{i=1}^{n}(x_{k_n}, x_{k_i})}{n^2}$$

$$\le \frac{n\left[C^2 + \left(1 + 2^{-1} + 2^{-2} + 2^{-3} + \cdots + 2^{-n}\right)\right]}{n^2} \le \frac{2 + C^2}{n}.$$

This implies

$$\lim_{n \to \infty} \left\| \frac{x_{k_1} + x_{k_2} + x_{k_3} + \cdots + x_{k_n}}{n} \right\| = 0,$$

as required.

Problem 33.14. *Let $\rho: [a, b] \to (0, \infty)$ be a measurable essentially bounded function and for each $n = 0, 1, 2, \ldots$ let P_n be a nonzero polynomial of degree n. Assume that*

$$\int_a^b \rho(x) P_n(x) \overline{P_m(x)} \, dx = 0 \text{ for } n \ne m.$$

Show that each P_n has n distinct real roots all lying in the open interval (a, b).

Solution. By Theorem 33.12, we know that the sequence of orthogonal polynomials $P_0, P_1, P_2, \ldots$ is complete and coincides (aside of scalar factors) with the sequence of orthogonal functions of $L_2(\rho)$ that is obtained by applying the Gram–Schmidt orthogonalization process to the sequence of linearly independent functions $\{1, x, x^2, x^3, \ldots\}$. In particular, we have $\int_a^b \rho(x) x^m P_n(x) \, dx = 0$ for all $m = 0, 1, \ldots, n - 1$. Also, by multiplying each P_n by an appropriate scalar, we can assume that each P_n has real coefficients and leading coefficient 1.

Now, fix one of these polynomials P_n, where $n \ge 1$. First, we shall show that P_n cannot have any complex roots. If P_n has a complex root, then P_n has a factorization of the form $P_n(x) = [(x + \alpha)^2 + \beta^2] Q(x)$, where α and β are real numbers and Q is a polynomial of degree $n - 2$. Hence $Q(x) P_n(x) = [(x + \alpha)^2 + \beta^2][Q(x)]^2 \ge 0$, and from $\int_a^b \rho(x)x^m P_n(x)\,dx = 0$ for all $m = 0, 1, \ldots, n-1$, it follows that

$$0 < \int_a^b \rho(x)Q(x)P_n(x)\,dx = 0,$$

which is impossible. Hence, each P_n has only real roots. This means that P_n has a factorization of the form

$$P_n(x) = (x - r_1)^{m_1}(x - r_2)^{m_2} \cdots (x - r_k)^{m_k},$$

where $r_1, r_2, \ldots, r_k$ are real number and $m_1, m_2, \ldots, m_k$ are natural numbers such that $m_1 + m_2 + \cdots + m_k = n$.

Next, we claim that P_n does not have any root outside of the open interval (a, b). To see this, assume that one root lies outside of (a, b), say $r_1 \leq a$. Then the polynomial $Q(x) = (x - r_2)^{m_2} \cdots (x - r_k)^{m_k}$ has degree less than n and satisfies $Q(x)P_n(x) \geq 0$ for each $a \leq x \leq b$. But then, we have

$$0 < \int_a^b \rho(x)Q(x)P_n(x)\,dx = 0,$$

which is a contradiction.

Finally, to see that each root appears with multiplicity one, assume by way of contradiction that one root has multiplicity more than one, say $m_1 > 1$. If, again

$$Q(x) = \begin{cases} (x - r_2)^{m_2} \cdots (x - r_k)^{m_k} & \text{if } m_1 \text{ is even} \\ (x - r_1)(x - r_2)^{m_2} \cdots (x - r_k)^{m_k} & \text{if } m_1 \text{ is odd,} \end{cases}$$

then Q is a polynomial of degree strictly less than n and satisfies $Q(x)P_n(x) \geq 0$ for all $a \leq x \leq b$. But then, as previously,

$$0 = \int_a^h \rho(x)Q(x)P_n(x)\,dx > 0,$$

which is absurd. Hence, each polynomial P_n has n distinct real roots all lying in the open interval (a, b).

Problem 33.15. *In Example 33.13 we defined the sequence $P_0, P_1, P_2, \ldots$ of Legendre polynomials by the formulas*

$$P_n(x) = \frac{1}{2^n n!} \frac{d^n}{dx^n} (x^2 - 1)^n.$$

We also proved that these are (aside of scalar factors) the polynomials obtained by applying the Gram–Schmidt orthogonalization process to the sequence of linearly independent functions $\{1, x, x^2, \ldots\}$ in the Hilbert space $L_2([-1, 1])$. Show that for each n we have

$$P_n(1) = 1 \quad and \quad \|P_n\| = \sqrt{\tfrac{2}{2n+1}}.$$

Solution. The proof of the formula $P_n(1) = 1$ is by induction. Notice that for $n = 0$ and $n = 1$ the formula is trivially true. So, for the induction argument, assume that $P_n(1) = 1$ holds true for some n. To complete the proof, we must show that $P_{n+1}(1) = 1$. To see this, note that

$$
\begin{aligned}
P_{n+1}(x) &= \frac{1}{2^{n+1}(n+1)!} \frac{d^{n+1}}{dx^{n+1}} (x^2 - 1)^{n+1} \\
&= \frac{1}{2^{n+1}(n+1)!} \frac{d^n}{dx^n} \Big[\big((x^2 - 1)^{n+1}\big)' \Big] \\
&= \frac{1}{2^{n+1}(n+1)!} \frac{d^n}{dx^n} \big[2(n+1)x(x^2 - 1)^n \big] \\
&= \frac{1}{2^n n!} \frac{d^n}{dx^n} \big[(x-1)(x^2 - 1)^n \big] + \frac{1}{2^n n!} \frac{d^n}{dx^n} (x^2 - 1)^n \\
&= (x-1)Q(x) + P_n(x),
\end{aligned}
$$

where the term $(x - 1)Q(x)$ designates the form of the expression

$$\frac{1}{2^n n!} \frac{d^n}{dx^n} (x-1)(x^2 - 1)^n = \frac{1}{2^n n!} \frac{d^n}{dx^n} (x-1)^{n+1}(x+1)^n.$$

So, $P_{n+1}(1) = P_n(1) = 1$.

Next, we shall compute the norm of P_n. Clearly,

$$
\begin{aligned}
\|P_n\|^2 &= \int_{-1}^{1} P_n(x)\, P_n(x)\, dx \\
&= \frac{1}{(2^n n!)^2} \int_{-1}^{1} \frac{d^n}{dx^n}(x^2 - 1)^n \frac{d^n}{dx^n}(x^2 - 1)^n \, dx.
\end{aligned}
$$

Next, observe that the function $(x^2 - 1)^n = (x-1)^n(x+1)^n$ and all of its derivatives of order less than or equal to $n - 1$ vanish at the points ± 1. So, integrating by parts n-times and using the previous observation, we obtain

$$\|P_n\|^2 = \frac{(-1)^n}{(2^n n!)^2} \int_{-1}^{1} (x^2 - 1)^n \frac{d^{2n}}{dx^{2n}} (x^2 - 1)^n \, dx = \frac{(-1)^n}{(2^n n!)^2} \int_{-1}^{1} (x^2 - 1)^n (2n)! \, dx$$

$$= \frac{(-1)^n (2n)!}{(2^n n!)^2} \int_{-1}^{1} (x^2 - 1)^n \, dx = \frac{(2n)!}{(2^n n!)^2} \int_{-1}^{1} (1 - x^2)^n \, dx.$$

Using integration by parts to evaluate this last integral gives

$$\int_{-1}^{1} (1 - x^2)^n \, dx = \int_{-1}^{1} (1 - x)^n (1 + x)^n \, dx$$

$$= \frac{(1-x)^n (1+x)^{n+1}}{n+1} \Big|_{-1}^{1} + \int_{-1}^{1} n(1 - x)^{n-1} \frac{(1+x)^{n+1}}{n+1} \, dx$$

$$= \frac{n}{n+1} \int_{-1}^{1} (1 - x)^{n-1} (1 + x)^{n+1} \, dx = \cdots$$

$$= \frac{n!}{(n+1) \cdots (2n)} \int_{-1}^{1} (1 + x)^{2n} \, dx$$

$$= \frac{(n!)^2 \, 2^{2n+1}}{(2n)! \, (2n + 1)}.$$

Consequently, the norm of P_n is given by

$$\|P_n\|^2 = \left[\frac{(2n)!}{(2^n n!)^2} \right] \left[\frac{(n!)^2 \, 2^{2n+1}}{(2n)! \, (2n + 1)} \right] = \frac{2}{2n + 1},$$

as claimed.

Problem 33.16. *Let $\{T_u\}_{u \in A}$ be a family of linear continuous operators from a complex Hilbert space X into another complex Hilbert space Y. Assume that for each $x \in X$ and each $y \in Y$ the set of complex numbers $\{(T_\alpha(x), y) : \alpha \in A\}$ is bounded. Show that the family of operators $\{T_\alpha\}_{\alpha \in A}$ is uniformly norm bounded, i.e., show that there exists some constant $M > 0$ satisfying $\|T_\alpha\| \leq M$ for all $\alpha \in A$.*

Solution. Observe that if Z is a Banach space over the field of complex numbers, then we may also consider Z as a Banach space over the field of real numbers. Therefore, the Principle of Uniform Boundedness (Theorem 28.8) can be applied to any Banach space over the field of complex numbers.

Fix a vector $x \in X$. For each $\alpha \in A$ define the complex valued continuous linear operator $B_\alpha : Y \to \mathbf{C}$ by

$$B_\alpha(y) = \big(y, T_\alpha(x)\big).$$

Thus, $\{B_\alpha\}_{\alpha \in A}$ is family of bounded linear operators from the Banach space Y to the Banach space of complex numbers C. From Theorem 33.9, we know that

$$\|B_\alpha\| = \|T_\alpha(x)\|.$$

Next, notice that for each fixed $y \in Y$, it follows from

$$|B_\alpha(y)| = |(y, T_\alpha(x))| = |(T_\alpha(x), y)|$$

and our hypothesis that the family of continuous linear operators $\{B_\alpha\}_{\alpha \in A}$ is point-wise bounded. Hence, by the Principle of Uniform Boundedness (Theorem 28.8), the family $\{B_\alpha\}_{\alpha \in A}$ is norm bounded. This means that there exists a constant $M_x > 0$ (that depends upon x) such that $\|B_\alpha\| \leq M_x$ for all $\alpha \in A$. Thus, we have $\|T_\alpha(x)\| \leq M_x$ for all α.

Therefore, the family $\{T_\alpha\}_{\alpha \in A}$ of continuous linear operators $T_\alpha : X \to Y$ is pointwise bounded. Invoking the Principle of Uniform Boundedness once more, we conclude that there exists a constant $M > 0$ satisfying $\|T_\alpha\| \leq M$ for all $\alpha \in A$.

Problem 33.17. *Let $\{\phi_n\}$ be an orthonormal sequence in a Hilbert space H and consider the operator $T : H \to H$ defined by*

$$T(x) = \sum_{n=1}^{\infty} \alpha_n(x, \phi_n)\phi_n,$$

where $\{\alpha_n\}$ is a sequence of scalars satisfying $\lim \alpha_n = 0$. Show that T is a compact operator.

Solution. Let B be the open unit ball of H. We need to show that $\overline{T(B)}$ is a compact set. For this, it suffices to show that $T(B)$ is totally bounded. To this end, fix $\epsilon > 0$ and observe that there exists an integer m such that $|\alpha_n| < \epsilon$ holds for all $n > m$.

Next, define the operator $T_m : H \to H$ by

$$T_m(x) = \sum_{i=1}^{m} \alpha_i(x, \phi_i)\phi_i.$$

Clearly, the range of T_m is a finite dimensional subspace of H and thus, T_m is a compact operator. Therefore, $T_m(B)$ is a totally bounded set. Thus, there exists a finite set $\{y_1, y_2, \ldots, y_n\}$ such that for each $x \in B$ there exists some $1 \leq k \leq n$ such that $\|T_m(x) - y_k\| < \epsilon$. Now, Parseval's Inequality $\sum_{i=1}^{\infty} |(x, \phi_i)|^2 \leq \|x\|^2 < 1$ implies

$$\left\| T(x) - T_m(x) \right\|^2 = \left\| \sum_{i=m+1}^{\infty} \alpha_i (x, \phi_i) \phi_i \right\|^2 = \sum_{i=m+1}^{\infty} |\alpha_i|^2 |(x, \phi_i)|^2$$

$$\leq \epsilon^2 \sum_{i=m+1}^{\infty} |(x, \phi_i)|^2 \leq \epsilon^2 \|x\|^2 < \epsilon^2.$$

Therefore, for each $x \in B$ there exists some $1 \leq k \leq n$ such that

$$\|T(x) - y_k\| \leq \|T(x) - T_m(x)\| + \|T_m(x) - y_k\| \leq \epsilon + \epsilon = 2\epsilon.$$

This shows that $T(B)$ is totally bounded, and hence T is a compact operator.

Problem 33.18. *Assume that $T, T^*: H \to H$ are two functions on a Hilbert space satisfying*

$$(Tx, y) = (x, T^*y)$$

for all $x, y \in H$. Show that T and T^* are both bounded linear operators satisfying

$$\|T\| = \|T^*\| \quad and \quad \|TT^*\| = \|T\|^2.$$

Solution. By the symmetry of the situation, it suffices to show that T is a bounded linear operator. We shall show first that T is a linear operator. To this end, fix $x, y \in H$ and two scalars α and β. Then, for each $z \in H$ we have

$$\begin{aligned}\left(T(\alpha x + \beta y), z\right) &= \left(\alpha x + \beta y, T^*z\right) = \alpha(x, T^*z) + \beta(y, T^*z) \\ &= \alpha(Tx, z) + \beta(Ty, z) = \left(\alpha Tx + \beta Ty, z\right),\end{aligned}$$

or $\left(T(\alpha x + \beta y) - \alpha Tx - \beta Ty, z\right) = 0$ for all $z \in H$. This implies (by letting $z = T(\alpha x + \beta y) - \alpha Tx - \beta Ty$) that

$$T(\alpha x + \beta y) = \alpha T(x) + \beta T(y),$$

i.e., that T is a linear operator.

Next, for each $y \in H$ with $\|y\| \leq 1$ consider the linear functional $f_y \colon H \to H$ defined by $f_y(x) = (x, T^*y)$. Then, using the Cauchy–Schwarz Inequality, we see that

$$|f_y(x)| = |(x, T^*(y)| = |(T(x), y)| \leq \|T(x)\| \, \|y\| \leq \|T(x)\|$$

for all $y \in H$ with $\|y\| \leq 1$. This implies that the set of linear functionals $\{f_y \colon \|y\| \leq 1\}$ is pointwise bounded and therefore, by the Principle of Uniform Boundedness, the set of linear functionals $\{f_y \colon \|y\| \leq 1\}$ is norm bounded. Thus, there exists some constant $C > 0$ such that $\|f_y\| \leq C$ for all $y \in H$ with $\|y\| \leq 1$.

Now, assume that $y \in H$ satisfies $\|y\| \leq 1$ and $T^*(y) \neq 0$. Letting $x = T^*(y)/\|T^*(y)\|$, we obtain

$$\|T^*(y)\| = \frac{1}{\|T^*(y)\|} \left|(T^*(y), T^*(y))\right| = \left|(x, T^*(y))\right| = |f_y(x)| \leq C.$$

This implies

$$\|T^*\| = \sup\{\|T^*(y)\| \colon \|y\| \leq 1\} \leq C,$$

and thus, T^* is a bounded linear operator. By the symmetry of the situation, T is likewise a bounded linear operator.

Now, note that for all $x, y \in H$ with $\|x\| = \|y\| = 1$, we have

$$|(Tx, y)| = |(x, T^*y)| \leq \|x\| \, \|T^*y\| \leq \|x\| \, \|T^*\| \, \|y\| = \|T^*\|,$$

and hence, $\|Tx\| = \sup\{|(Tx, y)| \colon \|y\| \leq 1\} \leq \|T^*\|$ for all unit vectors $x \in H$. This implies

$$\|T\| = \sup_{\|x\|=1} \|T(x)\| \leq \|T^*\|.$$

Using the symmetry once more, we get $\|T^*\| \leq \|T\|$, and so $\|T\| = \|T^*\|$.

Finally, for each $x \in H$ with $\|x\| = 1$, we have

$$\|TT^*x\| \leq \|T\| \, \|T^*\| \, \|x\| = \|T\|^2, \text{ and}$$
$$\|T^*x\|^2 = (T^*x, T^*x) = (TT^*x, x) \leq \|TT^*\| \, \|x\| \, \|x\| = \|TT^*\|,$$

and so by taking suprema, we get $\|TT^*\| = \|T\|^2$.

Problem 33.19. *Show that if $T \colon H \to H$ is a bounded linear operator on a Hilbert space, then there exists a unique bounded operator $T^* \colon H \to H$ (called* *the* **adjoint operator** *of T) satisfying*

$$(Tx, y) = (x, T^*y)$$

for all $x, y \in H$. *Moreover, show that* $\|T\| = \|T^*\|$.

Solution. Assume that $T: H \to H$ is a bounded linear operator on a Hilbert space. For each fixed $y \in H$, the formula $f_y(x) = (Tx, y)$ defines a bounded linear functional on H. By Theorem 33.9, there exists a unique vector $T^*y \in H$ satisfying

$$f_y(x) = (Tx, y) = (x, T^*y)$$

for all $x \in H$. Now, use the preceding problem to conclude that the unique function $T^*: H \to H$ defined above is, in fact, a bounded linear operator satisfying $\|T^*\| = \|T\|$.

34.   ORTHONORMAL BASES

Problem 34.1.   *Let* $\{e_i\}_{i \in I}$ *and* $\{f_j\}_{j \in J}$ *be two orthonormal bases of a Hilbert space. Show that* I *and* J *have the same cardinality.*

Solution.   Assume that $\{e_i\}_{i \in I}$ and $\{f_j\}_{j \in J}$ are two orthonormal bases of a Hilbert space. First, suppose that I is a finite set. Then, from Theorem 34.2, it follows that $\{e_i\}_{i \in I}$ is also a Hamel basis and so H is finite dimensional. Since the f_j (as being mutually orthogonal vectors) are also linearly independent, we conclude that J must also be a finite set. This implies that $\{f_j\}_{j \in J}$ is itself a Hamel basis for H, and so I and J must have the same number of elements.

Now, suppose that I and J are infinite sets. For each $i \in I$, we define the set of indices $J_i = \{j \in J: (e_i, f_j) \neq 0\}$. By Theorem 34.2, we know that each J_i is nonempty and at-most countable. Next, we claim that

$$J = \bigcup_{i \in I} J_i. \qquad (\star)$$

To see this, let $j \in J$. Since $\{e_i\}_{i \in I}$ is an orthonormal basis, it follows from Parseval's Identity that $\sum_{i \in I} |(f_j, e_i)|^2 = \|f_j\|^2 = 1$, and so $(e_i, f_j) \neq 0$ holds true for some $i \in I$. Thus, $j \in J_i$ holds true for at least one $i \in I$, and thus $J = \bigcup_{i \in I} J_i$.

Finally, to see that I and J have the same cardinality use $(\star)$ together with the standard "cardinality" arithmetic; see, for instance, P. R. Halmos, *Naive Set Theory*, Springer–Verlag, 1974, pp. 94–98.

Problem 34.2. *Let $\{e_i\}_{i\in I}$ be an orthonormal basis in a Hilbert space H. If D is a dense subset of H, then show that the cardinality of D is at least as large as that of I. Use this conclusion to provide an alternate proof of Theorem 34.4 by proving that for an infinite dimensional Hilbert space H the following statements are equivalent:*

1. *H has a countable orthonormal basis.*
2. *H is separable.*
3. *H is linearly isometric to ℓ_2.*

Solution. Let $\{e_i\}_{i\in I}$ be an orthonormal basis in a Hilbert space H and let D be a dense subset of H. Consider the family of open balls $\left\{B\left(e_i, \frac{1}{2}\right)\right\}_{i\in I}$, where

$$B\left(e_i, \tfrac{1}{2}\right) = \left\{x \in H\colon \|e_i - x\| < \tfrac{1}{2}\right\}.$$

Since $\|e_i - e_j\| = \sqrt{2}$ for $i \neq j$, it follows that $\{B(e_i, \frac{1}{2})\}_{i\in I}$ is a pairwise disjoint family of open sets. Since D is dense in H for each $i \in I$, there exists some $d_i \in D \cap B(e_i, \frac{1}{2})$. Clearly, the mapping $i \mapsto d_i$, from I into D, is one-to-one and this shows that D has cardinality greater than or equal of the cardinality of I.

Next, we shall prove the equivalent statements. To this end, assume that H is an infinite dimensional Hilbert space.

(1) $\iff$ (2) Let $\{e_1, e_2, \ldots\}$ be a countable orthonormal basis for H. Then, the finite linear combinations of the e_n with "rational" coefficients is a countable dense set.

Now, assume that H is separable, and let D be a countable dense subset of H. If $\{e_i\}_{i\in I}$ is an orthonormal basis, it follows form the first part that I has cardinality at most that of D, and hence, I is at-most countable. Since H is infinite dimensional, I must be countable, and so H has a countable orthonormal basis.

(2) $\implies$ (3) If H has a countable orthonormal basis, then H is linearly isometric (by Theorem 34.9) to $\ell_2(\mathbb{N}) = \ell_2$.

(3) $\implies$ (1) Obvious.

Problem 34.3. *Let I be an arbitrary nonempty set and for each $i \in I$ let $e_i = \chi_{\{i\}}$. Show that the family of functions $\{e_i\}_{i\in I}$ is an orthonormal basis for the Hilbert space $\ell_2(I)$.*

Solution. For each i, let $e_i = \chi_{\{i\}}$ and note that the family of functions $\{e_i\}_{i\in I}$ is an orthonormal family. Now, notice that if $x = \{x_i\}_{i\in I} \in \ell_2(I)$, then

$$\|x\|^2 = \sum_{i\in I} |x_i|^2 = \sum_{i\in I} |(x, e_i)|^2.$$

This shows that $\{e_i\}_{i\in I}$ is an orthonormal basis for the Hilbert space $\ell_2(I)$.

Problem 34.4. *Let $\{e_i\}_{i \in i}$ be an orthonormal basis in a Hilbert space and let x be a unit vector, i.e, $\|x\| = 1$. Show that for each $k \in \mathbb{N}$ the set $\left\{i \in I: |(x, e_i)| \geq \frac{1}{k}\right\}$ has at most k^2 elements.*

Solution. From Parseval's Identity we know that

$$1 = \|x\|^2 = \sum_{i \in I} |(x, e_i)|^2.$$

Let $A = \left\{i \in I: |(x, e_i)| \geq \frac{1}{k}\right\}$. If A has more than k^2 elements, then by choosing $k^2 + 1$ indices from A, we see that

$$1 = \|x\|^2 = \sum_{i \in I} |(x, e_i)|^2 \geq (k^2 + 1)\frac{1}{k^2} > 1,$$

which is impossible. Therefore, A has at most k^2 elements.

Problem 34.5. *Let M be a closed vector subspace of a Hilbert space H and let $\{e_i\}_{i \in I}$ be an orthonormal basis of M; where M is now considered as a Hilbert space in its own right under the induced operations. If $x \in H$, then show that the unique vector of M closest to x (which is guaranteed by Theorem 33.6) is the vector $y = \sum_{i \in I}(x, e_i)e_i$.*

Solution. Assume that $\{e_i\}_{i \in I}$ is an orthonormal basis for a closed subspace M of a Hilbert space H and let $x \in H$ be a fixed vector. Note first that Parseval's Inequality guarantees that $\sum_{i \in I} |(x, e_i)|^2 \leq \|x\|^2$, and so $y = \sum_{i \in I}(x, e_i)e_i$ is a well-defined vector of M.

We claim that $x - y \perp M$. To see this, let z be an arbitrary vector of M, and let $z = \sum_{j \in I}(z, e_j)e_j$ be its Fourier series expansion as a vector of M. Then, we have

$$
\begin{aligned}
(z, x - y) &= \left(\sum_{j \in I}(z, e_j)e_j, x - \sum_{i \in I}(x, e_i)e_i\right) \\
&= \sum_{j \in I}(z, e_j)(e_j, x) - \sum_{j \in I}\sum_{i \in I}((z, e_j)e_j, (x, e_i)e_i) \\
&= \sum_{j \in I}(z, e_j)(e_j, x) - \sum_{j \in I}(z, e_j)(e_j, x) = 0.
\end{aligned}
$$

Now, if z is an arbitrary vector of M, then $y - z \in M$ and so $x - y \perp y - z$. Hence, by the Pythagorean Theorem,

$$\|x - z\|^2 = \|(x - y) + (y - z)\|^2 = \|x - y\|^2 + \|y - z\|^2 \geq \|x - y\|^2.$$

This shows that y is the vector in M closest to x.

Problem 34.6. *Let $\{e_n\}$ be an orthonormal basis of a separable Hilbert space. For each n let $f_n = e_{n+1} - e_n$. Show that the vector subspace generated by the sequence $\{f_n\}$ is dense.*

Solution. We need to show that if $x \perp f_n$ for all n, then $x = 0$. So, let x be a vector satisfying $x \perp (e_{n+1} - e_n)$ for each n. That is,

$$0 = (x, e_{n+1} - e_n) = (x, e_{n+1}) - (x, e_n).$$

This implies $(x, e_{n+1}) = (x, e_n)$ for each n. If we let $\delta = (x, e_1)$, then $\delta = (x, e_n)$ for all n, and by Parseval's Identity

$$\|x\|^2 = \sum_{n=1}^{\infty} |(x, e_n)|^2 = \sum_{n=1}^{\infty} \delta^2$$

Therefore, $\delta = 0$, and hence, $\|x\| = 0$, or $x = 0$.

Problem 34.7. *Show that a linear operator $L: H_1 \to H_2$ between two Hilbert spaces is norm preserving if and only if it is inner product preserving.*

Solution. Let $L: H_1 \to H_2$ be a linear operator between two Hilbert spaces. If L is inner product preserving, then

$$\|Lx\|^2 = (Lx, Lx) = (x, x) = \|x\|^2$$

holds or $\|Lx\| = \|x\|$ for each $x \in H_1$, i.e., L is norm preserving. For the converse, assume that L is norm preserving. Then, from Theorem 32.6, it follows that

$$
\begin{aligned}
(Lx, Ly) &= \tfrac{1}{4}\big(\|Lx + Ly\|^2 - \|Lx - Ly\|^2 + \imath\|Lx + \imath Ly\|^2 - \imath\|Lx - \imath Ly\|^2\big) \\
&= \tfrac{1}{4}\big(\|L(x + y)\|^2 - \|L(x - y)\|^2 + \imath\|L(x + \imath y)\|^2 - \imath\|L(x - \imath y)\|^2\big) \\
&= \tfrac{1}{4}\big(\|x + y\|^2 - \|x - y\|^2 + \imath\|x + \imath y\|^2 - \imath\|x - \imath y\|^2\big) \\
&= (x, y).
\end{aligned}
$$

That is, L is inner product preserving.

Problem 34.8. *Show that the vector space $\ell_2(Q)$ of all square summable complex-valued functions defined on a nonempty set Q under the inner product*

$$(x, y) = \sum_{q \in Q} x(q)\overline{y(q)}$$

is a Hilbert space.

Solution. The verification of the inner product properties of the function $(\cdot, \cdot)$ are straightforward. We shall show that $\ell_2(Q)$ is a Hilbert space, i.e., complete under its induced normed.

To see this, assume that Q is an infinite set, and let $\{x_n\} \subseteq \ell_2(Q)$ be a Cauchy sequence. Since for each n we have $x_n(q) \neq 0$ for at-most countably many $q \in Q$, there exists an at-most countable subset C of Q such that $x_n(q) = 0$ for all $q \in Q \setminus C$ and all n. We consider only the case when C is a countable set, say $C = \{q_1, q_2, \ldots\}$. For each n, let

$$y_n = \big(x_n(q_1), x_n(q_2), \ldots\big).$$

Then, it is easy to see that we can consider $\{y_n\}$ as a Cauchy sequence in ℓ_2. The completeness of ℓ_2 implies that $\{y_n\}$ converges to some sequence $y = (y_1, y_2, \ldots)$ in ℓ_2. If $x: Q \to \mathbf{C}$ is defined by $x(q_i) = y_i$ and $x(q) = 0$ whenever $q \in Q \setminus C$, then $x \in \ell_2(Q)$ and $\|x_n - x\| = \|y_n - y\| \to 0$ holds in $\ell_2(Q)$. This shows that $\ell_2(Q)$ is a Hilbert space.

Problem 34.9. *Let $\{e_i\}_{i \in I}$ be an orthonormal basis of a Hilbert space H. Show that the linear operator $L: H \to \ell_2(I)$, defined by*

$$L(x) = \{(x, e_i)\}_{i \in I},$$

is a surjective linear isometry.

Solution. Clearly, L is linear and by Parseval's Identity (Theorem 34.2(5)), it is also an isometry. We shall verify next that L is also surjective. To this end, let $\{\lambda_i\}_{i \in I} \in \ell_2(I)$. From $\sum_{i \in I} |\lambda_i|^2 < \infty$, it follows that $\lambda_i \neq 0$ for at-most countably many indices i. Assume that $\{i \in I: \lambda_i \neq 0\} = \{\lambda_{i_1}, \lambda_{i_2}, \ldots\}$. (We consider only the countable case; the finite case is trivial.) Clearly, $\sum_{n=1}^{\infty} |\lambda_{i_n}|^2 < \infty$.

From the Pythagorean Theorem, we have

$$\left\| \sum_{k=n}^{m} \lambda_{i_k} e_{i_k} \right\|^2 = \sum_{k=n}^{m} |\lambda_{i_k}|^2,$$

and from this it follows that the series $\sum_{n=1}^{\infty} \lambda_{i_n} e_{i_n}$ is norm convergent in H. Let $x = \sum_{n=1}^{\infty} \lambda_{i_n} e_{i_n} = \sum_{i \in I} \lambda_i e_i$. Then, $(x, e_i) = \lambda_i$ for each, i and so $L(x) = \{\lambda_i\}_{i \in I}$. This shows that L is also surjective, as required.

Problem 34.10. *Let $\{e_n\}$ be an orthonormal sequence of vectors in the Hilbert space $L_2[0, 2\pi]$. Suppose that for each continuous function f in $L_2[0, 2\pi]$ we have $f = \sum_{n=1}^{\infty} (f, e_n)e_n$. Show that $\{e_n\}$ is an orthonormal basis.*

Solution. We need only show that the linear span of the set $\{e_n\}$ is dense. Let $\epsilon > 0$ and let $g \in L_2[0, 2\pi]$. Since the continuous functions are dense in $L_2[0, 2\pi]$ (see Theorem 31.10), there exists a continuous function $f \in L_2[0, 2\pi]$ with $\|f - g\| \le \epsilon$. By our assumption, we have $f = \sum_{n=1}^{\infty}(f, e_n)e_n$ and, by Bessel's Inequality, we know that $\sum_{n=1}^{\infty}|(f, e_n)|^2 < \infty$. Next, choose an integer m such that $\left[\sum_{k=m}^{\infty}|(f, e_k)|^2\right]^{\frac{1}{2}} < \epsilon$, and then let $h = \sum_{k=1}^{m}(f, e_k)e_k$. Then, h is in the linear span of the sequence $\{e_n\}$, and moreover

$$\|g - h\| = \left\| g - \sum_{k=1}^{m}(f, e_k)e_k \right\| \le \|g - f\| + \left\| \sum_{k=m+1}^{\infty}(f, e_k)e_k \right\|$$

$$\le \|g - f\| + \left[\sum_{k=m+1}^{\infty}|(f, e_k)|^2 \right]^{\frac{1}{2}} < \epsilon + \epsilon = 2\epsilon.$$

Therefore, the linear span of $\{e_n\}$ is dense and hence, $\{e_n\}$ is a complete orthonormal set, i.e., it is an orthonormal basis.

Problem 34.11. *Let $\{\phi_n\}$ be an orthonormal sequence of vectors in the Hilbert space $L_2[0, 2\pi]$. Suppose that for each continuous function f in $L_2[0, 2\pi]$ we have $\|f\|^2 = \sum_{n=1}^{\infty}|(f, \phi_n)|^2$. Show that $\{\phi_n\}$ is an orthonormal basis.*

Solution. It suffices to show that the linear span of the set $\{\phi_n\}$ is dense. Let $\epsilon > 0$ and let $g \in L_2[0, 2\pi]$. Since the continuous functions are dense in $L_2[0, 2\pi]$, there exists a continuous function $f \in L_2[0, 2\pi]$ with $\|f - g\| < \epsilon$.

Now, by our hypothesis, we have $\|f\|^2 = \sum_{n=1}^{\infty}|(f, \phi_n)|^2$. Choose an integer m such that $\left[\sum_{k=m}^{\infty}|(f, \phi_k)|^2\right]^{\frac{1}{2}} < \epsilon$, and note that

$$\left\| g - \sum_{k=1}^{m}(f, \phi_k)\phi_k \right\| \le \|g - f\| + \left\| f - \sum_{k=1}^{m}(f, \phi_k)\phi_k \right\|.$$

Using once more our hypothesis, we see that

$$\left\| g - \sum_{k=1}^{m}(f, \phi_k)\phi_k \right\| \le \|g - f\| + \left[\sum_{k=1}^{\infty}\left|\left(f - \sum_{i=1}^{m}(f, \phi_i)\phi_i, \phi_k\right)\right|^2 \right]^{\frac{1}{2}}$$

$$\le \|g - f\| + \left[\sum_{k=m+1}^{\infty}|(f, \phi_k)|^2 \right]^{\frac{1}{2}} < \epsilon + \epsilon = 2\epsilon.$$

Therefore, the linear span of $\{\phi_n\}$ is dense and hence, $\{\phi_n\}$ is an orthonormal basis.

Problem 34.12. *Let $\{\phi_1, \phi_2, \ldots\}$ be an orthonormal basis of the Hilbert space $L_2(\mu)$, where μ is a finite measure. Fix a function $f \in L_2(\mu)$ and let $\{\alpha_1, \alpha_2, \ldots\}$ be its sequence of Fourier coefficients relative to $\{\phi_n\}$, i.e., $\alpha_n = \int f \overline{\phi_n} \, d\mu$. Show that (although the series $\sum_{n=1}^{\infty} \alpha_n \phi_n$ need not converge pointwise almost everywhere to f) the Fourier series $\sum_{n=1}^{\infty} \alpha_n \phi_n$ can be integrated term-by-term in the sense that for every measurable set E we have*

$$\int_E f \, d\mu = \sum_{n=1}^{\infty} \alpha_n \int_E \phi_n \, d\mu.$$

Solution. Let $s_n = \sum_{k=1}^{n} \alpha_n \phi_n$, and note that $\| f - s_n \| \to 0$. Now, using the Cauchy–Schwarz inequality, we see that

$$\left| \int_E f \, d\mu - \int_E s_n \, d\mu \right|^2 = \left| \int_E (f - s_n) \, d\mu \right|^2 \le \left(\int_E |f - s_n| \, d\mu \right)^2$$

$$\le \int_E |f - s_n|^2 \, d\mu \cdot \int_E 1^2 \, d\mu$$

$$\le \| f - s_n \|^2 \, \mu^*(E) \longrightarrow 0.$$

Hence, $\int_E f \, d\mu = \sum_{n=1}^{\infty} \alpha_n \int_E \phi_n \, d\mu$.

Problem 34.13. *Establish the following "perturbation" property of orthonormal bases. If $\{e_i\}_{i \in I}$ is an orthonormal basis and $\{f_i\}_{i \in I}$ is another orthonormal family satisfying*

$$\sum_{i \in I} \| e_i - f_i \|^2 < \infty,$$

then $\{f_i\}_{i \in I}$ is also an orthonormal basis.

Solution. Let $\{e_i\}_{i \in I}$ be an orthonormal basis in a Hilbert space H, and let $\{f_i\}_{i \in I}$ be another orthonormal family satisfying $\sum_{i \in I} \| e_i - f_i \|^2 < \infty$. To establish that the orthonormal family $\{f_i\}_{i \in I}$ is an orthonormal basis, it suffices to show that if a vector u satisfies $u \perp f_i$ for each $i \in I$, then $u = 0$. So, fix a vector $u \in H$ such that $u \perp f_i$ for all $i \in I$.

From $\sum_{i \in I} \| e_i - f_i \|^2 < \infty$, we know that the set $J = \{i \in I : f_i \ne e_i\}$ is at-most countable. We distinguish two cases.

CASE I: *J is finite, say $J = \{k_1, k_2, \ldots, k_\ell\}$.*

Let $M = \{y \in H : y \perp f_i \text{ for all } i \notin J\}$. Then, M is a closed vector subspace of H satisfying $\{f_{k_1}, \ldots, f_{k_\ell}\} \subseteq M$ and $\{e_{k_1}, \ldots, e_{k_\ell}\} \subseteq M$. Moreover, we claim that $\{e_{k_1}, \ldots, e_{k_\ell}\}$ must be an orthonormal basis for M. Indeed, if $x \in M$ satisfies

$z \perp e_{k_r}$ for $r = 1, \ldots, \ell$, then (in view $x \perp f_i = e_i$ for each $i \notin J$) we have $x \perp e_i$ for each $i \in I$. Since $\{e_i\}_{i \in I}$ is an orthonormal basis of H, it follows that $x = 0$. Thus, $\{e_{k_1}, \ldots, e_{k_\ell}\}$ is (as being an orthonormal basis) also a Hamel basis for M, and so M is ℓ-dimensional. This implies that $\{f_{k_1}, \ldots, f_{k_\ell}\}$ is also a Hamel basis. The latter implies that every e_{k_r} is a linear combination of the vectors $f_{k_1}, \ldots, f_{k_\ell}$. Consequently, $u \perp e_{k_r}$ for each $r = 1, \ldots, \ell$, and hence $u \perp e_i$ for all $i \in I$. This implies $u = 0$, and thus, in this case, $\{f_i\}_{i \in I}$ is an orthonormal basis.

CASE II: *J is countable, say $J = \{k_1, k_2, k_3, \ldots, \}$.*

In this case, choose a natural number ℓ such that

$$\sum_{j=\ell+1}^{\infty} \|e_{k_j} - f_{k_j}\|^2 = \delta < 1,$$

and let $J_1 = \{k_1, k_2, \ldots, k_\ell\}$. Next, define the vectors

$$g_r = e_{k_r} - \sum_{j=\ell+1}^{\infty} (e_{k_r}, f_{k_j}) f_{k_j}, \quad r = 1, 2, \ldots, \ell.$$

We claim the following:

- *If a vector $x \in H$ satisfies $x \perp g_r$ for $r = 1, 2, \ldots, \ell$ and $x \perp f_j$ for $j \notin J_1$, then $x = 0$.*

To see this, assume that vector $x \in H$ is orthogonal to g_r for $r = 1, 2, \ldots, \ell$ and to each f_j for $j \notin J_1$. Then, for $j \notin J_1$, we have $(x, e_j) = (x, e_j - f_j)$ and for each $1 \leq r \leq \ell$, we have

$$(x, e_{k_r}) = (x, g_r) + \sum_{j=\ell+1}^{\infty} (e_{k_r}, f_{k_j})(x, f_{k_j}) = 0.$$

Now, from Parseval's Identity, we have

$$\|x\|^2 = \sum_{i \in I} |(x, e_i)|^2 = \sum_{j=\ell+1}^{\infty} |(x, e_{k_j} - f_{k_j})|^2$$

$$\leq \left[\sum_{j=\ell+1}^{\infty} \|e_{k_j} - f_{k_j}\|^2 \right] \|x\|^2 = \delta \|x\|^2.$$

This implies $0 \leq (1 - \delta)\|x\|^2 \leq 0$, or $x = 0$, as claimed.

Next, we consider the closed vector subspace

$$M = \{y \in H: y \perp f_i \text{ for all } i \notin J_1\}.$$

Clearly, $\{g_1, g_2, \ldots, g_\ell\} \subseteq M$. Moreover, if some vector $x \in M$ is orthogonal to $g_1, g_2, \ldots, g_\ell$, then by property ($\bullet$) we have $x = 0$. This means that the vector space generated by $g_1, g_2, \ldots, g_\ell$ coincides with M. Since the orthogonal vectors $f_{k_1}, f_{k_2}, \ldots, f_{k_\ell}$ belong to M, M is ℓ-dimensional and so $\{f_{k_1}, f_{k_2}, \ldots, f_{k_\ell}\}$ is a Hamel basis of M. In particular, for each $1 \leq r \leq \ell$ the vector g_r is a linear combination of the vectors $f_{k_1}, f_{k_2}, \ldots, f_{k_\ell}$. This implies $u \perp g_r$ for each $1 \leq r \leq \ell$ and $u \perp f_j$ for $j \notin J_1$. Using ($\bullet$) once more, we conclude that $u = 0$. Therefore, the orthonormal family $\{f_i\}_{i \in I}$ is an orthonormal basis.

35. FOURIER ANALYSIS

Problem 35.1. *Show that* $\sin^n x$ *is a linear combination of*

$$\left\{ 1, \sin x, \cos x, \sin 2x, \cos 2x, \sin 3x, \cos 3x, \ldots, \sin nx, \cos nx \right\}.$$

Furthermore, show that the coefficients of the cosine terms are zero when n is an odd integer, and the coefficients of the sine terms are zero when n is an even integer.

Solution. Observe that

$$\sin x = \frac{e^{\imath x} - e^{-\imath x}}{2\imath}.$$

Then, using the binomial theorem, we get

$$\sin^n x = \left(\frac{e^{\imath x} - e^{-\imath x}}{2\imath}\right)^n = \frac{1}{(2\imath)^n} \sum_{k=0}^{n} \left[\frac{n!(-1)^k}{k!(n-k)!} e^{\imath(n-k)x} e^{-\imath kx} \right]$$

$$= \frac{1}{(2\imath)^n} \sum_{k=0}^{n} \left[\frac{n!(-1)^k}{k!(n-k)!} e^{\imath(n-2k)x} \right]$$

$$= \frac{1}{(2\imath)^n} \sum_{k=0}^{n} \left[\frac{n!(-1)^k}{k!(n-k)!} \left(\cos(n-2k)x + \imath \sin(n-2k)x \right) \right]$$

$$= \frac{1}{(2\imath)^n} \sum_{k=0}^{n} \left[\frac{n!(-1)^k}{k!(n-k)!} \cos(n-2k)x \right] + \frac{\imath}{(2\imath)^n} \sum_{k=0}^{n} \left[\frac{n!(-1)^k}{k!(n-k)!} \sin(n-2k)x \right].$$

Now, observe that if n is odd, then $\imath^n = \pm\imath$ and if n is even, then $\imath^n = \pm 1$. Since

$\sin^n x$ is equal to the real part of the preceding expression, we have the following two cases:

$$\sin^n x = \frac{1}{(2\iota)^n} \sum_{k=0}^{n} \left[\frac{n!(-1)^k}{k!(n-k)!} \cos(n-2k)x \right] \quad \text{for } n \text{ even, and}$$

$$\sin^n x = \frac{\iota}{(2\iota)^n} \sum_{k=0}^{n} \left[\frac{n!(-1)^k}{k!(n-k)!} \sin(n-2k)x \right] \quad \text{for } n \text{ odd.}$$

Problem 35.2. *Show that the Dirichlet kernel D_n and the Fejér kernel K_n satisfy*

$$\frac{1}{\pi} \int_{-\pi}^{\pi} D_n(t)\, dt = \frac{1}{\pi} \int_{-\pi}^{\pi} K_n(t)\, dt = 1.$$

Solution. The Dirichlet kernel is given by

$$D_n(t) = \frac{1}{2} + \sum_{k=1}^{n} \cos kt.$$

Integrating gives

$$\frac{1}{\pi} \int_{-\pi}^{\pi} D_n(t)\, dt = 1 + \frac{1}{\pi} \sum_{k=1}^{n} \int_{-\pi}^{\pi} \cos kt\, dt = 1.$$

Likewise, the Fejér kernel is defined by

$$K_n(t) = \frac{1}{n+1} \sum_{k=0}^{n} D_k(t).$$

So, integrating and using the previous result on the Dirichlet kernel, we get

$$\frac{1}{\pi} \int_{-\pi}^{\pi} K_n(t)\, dt = \frac{1}{n+1} \sum_{k=0}^{n} \frac{1}{\pi} \int_{-\pi}^{\pi} D_n(t)\, dt = \frac{1}{n+1} \sum_{k=0}^{n} 1 = 1.$$

Problem 35.3. *Let X denote the Banach space of all continuous periodic real-valued functions defined on $[0, 2\pi]$. Fix some $x \in [0, 2\pi]$ and define the linear* *functional* $S_n: X \to \mathbf{R}$ *by the formula*

$$S_n(f) = \frac{1}{\pi} \int_0^{2\pi} f(t) D_n(x - t)\, dt.$$

Show that the norm of the linear functional S_n satisfies

$$\|S_n\| = \frac{1}{\pi} \int_0^{2\pi} |D_n(x - t)|\, dt.$$

Solution. The norm of S_n is defined by

$$\|S_n\| = \sup_{\|f\|_\infty \le 1} \left| \frac{1}{\pi} \int_0^{2\pi} f(t) D_n(x - t)\, dt \right|.$$

Since $\|f\|_\infty \le 1$ implies $|f(t) D(x - t)| \le |D(x - t)|$ for each t, we see that

$$\|S_n\| \le \frac{1}{\pi} \int_0^{2\pi} |D_n(x - t)|\, dt.$$

Next, we shall establish the reverse inequality. Since the Dirichlet kernel D_n has period 2π, it follows that the continuous function

$$f(\epsilon, t) = \frac{D_n(x - t)}{|D_n(x - t)| + \epsilon}$$

also has period 2π with respect to t, and clearly $|f(\epsilon, t)| \le 1$ for each t and all $\epsilon > 0$. This implies

$$\|S_n\| \ge S_n(f) = \frac{1}{\pi} \int_0^{2\pi} \frac{|D_n(x - t)|^2}{|D_n(x - t)| + \epsilon}\, dt.$$

Taking into account Theorem 24.4 and letting $\epsilon \to 0^+$ yields

$$\|S_n\| \ge \frac{1}{\pi} \int_0^{2\pi} |D_n(x - t)|\, dt.$$

Therefore, $\|S_n\| = \frac{1}{\pi} \int_0^{2\pi} |D_n(x - t)|\, dt$ holds true.

Problem 35.4.   *Show that the sequence of functions*

$$\left\{ \left(\tfrac{1}{\pi}\right)^{\frac{1}{2}},\ \left(\tfrac{2}{\pi}\right)^{\frac{1}{2}}\cos x,\ \left(\tfrac{2}{\pi}\right)^{\frac{1}{2}}\cos 2x,\ \left(\tfrac{2}{\pi}\right)^{\frac{1}{2}}\cos 3x,\ \left(\tfrac{2}{\pi}\right)^{\frac{1}{2}}\cos 4x, \dots \right\}$$

is an orthonormal basis in $L_2[0, \pi]$. *Also show that the preceding sequence is an orthogonal sequence of functions in* $L_2[0, 2\pi]$ *which is not complete.*

Solution.   It is easy to verify that the functions

$$\left(\tfrac{1}{\pi}\right)^{\frac{1}{2}},\ \left(\tfrac{2}{\pi}\right)^{\frac{1}{2}}\cos x,\ \left(\tfrac{2}{\pi}\right)^{\frac{1}{2}}\cos 2x,\ \left(\tfrac{2}{\pi}\right)^{\frac{1}{2}}\cos 3x, \dots$$

are mutually orthogonal and of norm one in $L_2[0, \pi]$.

To show that the preceding orthonormal sequence is complete (i.e., that it is an orthonormal basis), we need to show that if $f \in L_2[0, \pi]$ is perpendicular to the functions $1, \cos x, \cos 2x, \cos 3x, \dots$, then $f = 0$. To this end, suppose that a function $f \in L_2[0, \pi]$ satisfies

$$\int_0^\pi f(x)\cos nx\, dx = 0$$

for all $n = 0, 1, 2, \dots$. Define the function $g: [0, 2\pi] \to \mathbb{R}$ by

$$g(x) = \begin{cases} f(x) & \text{if } 0 \le x \le \pi \\ f(2\pi - x) & \text{if } \pi < x \le 2\pi. \end{cases}$$

Then, in $L_2[0, 2\pi]$, for each n we have

$$(g, \cos nx) = \int_0^{2\pi} g(x)\cos nx\, dx$$

$$= \int_0^\pi f(x)\cos nx\, dx + \int_\pi^{2\pi} f(2\pi - x)\cos nx\, dx.$$

The change of variable $t = 2\pi - x$ gives

$$(g, \cos nx) = \int_0^\pi f(t)\cos nt\, dt + \int_0^\pi f(t)\cos nt\, dt = 0.$$

Next, observe that

$$(g, \sin nx) = \int_0^{2\pi} g(x)\sin nx\, dx = \int_0^\pi f(x)\sin nx\, dx + \int_\pi^{2\pi} f(2\pi - x)\sin nx\, dx$$

$$= \int_0^\pi f(t)\sin nt\, dt - \int_0^\pi f(t)\sin nt\, dt = 0.$$

The preceding show that g is perpendicular to every vector of a complete orthogonal sequence of $L_2[0, 2\pi]$. Therefore, $g = 0$ and hence, $f = 0$. Thus, the sequence

$$\left(\tfrac{1}{\pi}\right)^{\frac{1}{2}}, \ \left(\tfrac{2}{\pi}\right)^{\frac{1}{2}} \cos x, \ \left(\tfrac{2}{\pi}\right)^{\frac{1}{2}} \cos 2x, \ \left(\tfrac{2}{\pi}\right)^{\frac{1}{2}} \cos 3x, \dots$$

is an orthonormal basis of $L_2[0, \pi]$.

To see that the orthogonal set

$$\left(\tfrac{1}{\pi}\right)^{\frac{1}{2}}, \ \left(\tfrac{2}{\pi}\right)^{\frac{1}{2}} \cos x, \ \left(\tfrac{2}{\pi}\right)^{\frac{1}{2}} \cos 2x, \ \left(\tfrac{2}{\pi}\right)^{\frac{1}{2}} \cos 3x, \dots$$

is not complete in $L_2[0, 2\pi]$ notice that the nonzero function $\sin x$ is perpendicular to each of these functions in $L_2[0, 2\pi]$.

Problem 35.5. *Show that the sequence of functions*

$$\left\{ \left(\tfrac{2}{\pi}\right)^{\frac{1}{2}} \sin x, \ \left(\tfrac{2}{\pi}\right)^{\frac{1}{2}} \sin 2x, \ \left(\tfrac{2}{\pi}\right)^{\frac{1}{2}} \sin 3x, \ \left(\tfrac{2}{\pi}\right)^{\frac{1}{2}} \sin 4x, \ \dots \right\}$$

is an orthonormal basis of $L_2[0, \pi]$. Also prove that this set of functions is an orthogonal set of functions in $L_2[0, 2\pi]$ which is not complete.

Solution.   It is easy to verify that the collection of functions

$$\left\{ \left(\tfrac{2}{\pi}\right)^{\frac{1}{2}} \sin x, \ \left(\tfrac{2}{\pi}\right)^{\frac{1}{2}} \sin 2x, \ \left(\tfrac{2}{\pi}\right)^{\frac{1}{2}} \sin 3x, \ \left(\tfrac{2}{\pi}\right)^{\frac{1}{2}} \sin 4x, \ \dots \right\}$$

is an orthonormal set of functions of $L_2[0, \pi]$.

Thus, in order to show that it is an orthonormal basis, we need to show that if a function $f \in L_2[0, \pi]$ is perpendicular to each function of the preceding set in $L_2[0, \pi]$, then $f = 0$. To this end, assume that a function $f \in L_2[0, \pi]$ satisfies

$$\int_0^\pi f(x) \sin nx \, dx = 0$$

for all $n = 1, 2, 3, \dots$. Define the function $g: [0, 2\pi] \to \mathbf{R}$ by

$$g(x) = \begin{cases} f(x) & \text{if } 0 \le x \le \pi \\ -f(2\pi - x) & \text{if } \pi < x \le 2\pi. \end{cases}$$

Then, in $L_2[0, 2\pi]$, for each n we have

$$
\begin{aligned}
(g, \sin nx) &= \int_0^{2\pi} g(x) \sin nx \, dx \\
&= \int_0^{\pi} f(x) \sin nx \, dx - \int_{\pi}^{2\pi} f(2\pi - x) \sin nx \, dx \\
&= \int_0^{\pi} f(t) \sin nt \, dt + \int_0^{\pi} f(t) \sin nt \, dt = 0 + 0 = 0.
\end{aligned}
$$

Next, observe that

$$
\begin{aligned}
(g, \cos nx) &= \int_0^{2\pi} g(x) \cos nx \, dx \\
&= \int_0^{\pi} f(x) \cos nx \, dx - \int_{\pi}^{2\pi} f(2\pi - x) \cos nx \, dx \\
&= \int_0^{\pi} f(t) \cos nt \, dt - \int_0^{\pi} f(t) \cos nt \, dt = 0.
\end{aligned}
$$

Therefore, g is perpendicular to every vector of a complete orthogonal set of functions in $L_2[0, 2\pi]$. Therefore, $g = 0$ and hence, $f = 0$. Thus, the orthonormal set of functions

$$
\left\{ \left(\tfrac{2}{\pi}\right)^{\frac{1}{2}} \sin x, \; \left(\tfrac{2}{\pi}\right)^{\frac{1}{2}} \sin 2x, \; \left(\tfrac{2}{\pi}\right)^{\frac{1}{2}} \sin 3x, \; \left(\tfrac{2}{\pi}\right)^{\frac{1}{2}} \sin 4x, \; \dots \right\}
$$

is an orthonormal basis of $L_2[0, \pi]$.

To see that the orthogonal set of functions

$$
\left\{ \left(\tfrac{2}{\pi}\right)^{\frac{1}{2}} \sin x, \; \left(\tfrac{2}{\pi}\right)^{\frac{1}{2}} \sin 2x, \; \left(\tfrac{2}{\pi}\right)^{\frac{1}{2}} \sin 3x, \; \left(\tfrac{2}{\pi}\right)^{\frac{1}{2}} \sin 4x, \; \dots \right\}
$$

is not complete in $L_2[0, 2\pi]$, observe that $\cos x$ is perpendicular to each of these functions.

Problem 35.6. *The original Weierstrass approximation theorem showed that every continuous function of period 2π can be uniformly approximated by trigonometric polynomials. Establish this result.*

Solution. Weierstrass originally gave a direct proof, however, the result can be derived directly from Fejér's Theorem 35.8. Let f be a continuous function of period 2π defined on the entire real line. Let $\epsilon > 0$ be fixed. Let $\{s_n\}$ be the sequence of partial sums of the Fourier series of f and let $\{\sigma_n\}$ be the sequence arithmetic means.

From Theorem 35.8 we know that the sequence $\{\sigma_n\}$ converges uniformly to f on $[0, 2\pi]$. Now, notice that each σ_n is a trigonometric polynomial, and the claim is established.

Problem 35.7. *Find the Fourier coefficients of the function*

$$f(x) = \begin{cases} 1 & \text{if } 0 \le x < \frac{\pi}{2} \\ 0 & \text{if } \frac{\pi}{2} \le x < 2\pi. \end{cases}$$

Solution. The Fourier coefficients are given by the formulas

$$a_0 = \frac{1}{\pi} \int_0^{2\pi} f(x)\,dx = \frac{1}{\pi} \int_0^{\frac{\pi}{2}} dx = \frac{1}{2},$$

$$a_n = \frac{1}{\pi} \int_0^{2\pi} f(x) \cos nx\,dx = \frac{1}{\pi} \int_0^{\frac{\pi}{2}} \cos nx\,dx = \frac{1}{n\pi} \sin\left(n\frac{\pi}{2}\right), \text{ and}$$

$$b_n = \frac{1}{\pi} \int_0^{2\pi} f(x) \sin nx\,dx = \frac{1}{\pi} \int_0^{\frac{\pi}{2}} \sin nx\,dx = -\frac{1}{n\pi}\left[\cos\left(n\frac{\pi}{2}\right) - 1\right].$$

Simplifying yields

$$a_0 = \tfrac{1}{2},$$

$$a_n = \begin{cases} 0 & \text{if } n = 2, 4, 6, \ldots \\ 1/n\pi & \text{if } n = 1, 5, 9, \ldots \\ -1/n\pi & \text{if } n = 3, 7, 11, \ldots, \end{cases}$$

$$b_n = \begin{cases} 1/n\pi & \text{if } n = 1, 3, 5, \ldots \\ 2/n\pi & \text{if } n = 2, 6, 10, \ldots \\ 0 & \text{if } n = 4, 8, 12, \ldots. \end{cases}$$

Problem 35.8. *Find the Fourier series of the function*

$$f(x) = \begin{cases} \sin x & \text{if } 0 < x < \pi \\ -\sin x & \text{if } \pi \le x < 2\pi. \end{cases}$$

Solution. The function f is continuous, even, and periodic. Its Fourier coefficients are given by

$$a_0 = \tfrac{2}{\pi} \int_0^{\pi} \sin x\,dx = \tfrac{4}{\pi},$$

$$a_n = \tfrac{2}{\pi} \int_0^{\pi} \sin x \cos nx\,dx = \begin{cases} -\frac{4}{\pi(n^2-1)} & \text{if } n \text{ is even} \\ 0 & \text{if } n \text{ is odd,} \end{cases} \text{ and}$$

$$b_n = 0.$$

So, the Fourier series of f is given by $\frac{2}{\pi} - \frac{4}{\pi} \sum_{n=1}^{\infty} \frac{\cos 2nx}{4n^2-1}$. Since this series converges at every x and the periodic function f is continuous everywhere, it follows from Corollary 35.9 that the series converges to $f(x)$ for each x. That is, we have

$$f(x) = \frac{2}{\pi} - \frac{4}{\pi} \sum_{n=1}^{\infty} \frac{\cos 2nx}{4n^2 - 1}$$

for each real number x.

Problem 35.9. *Show that for each $0 < x < 2\pi$ we have*

$$x = \pi - 2 \sum_{n=1}^{\infty} \frac{\sin nx}{n}.$$

Solution. We consider the periodic function $f: [0, 2\pi] \to \mathbf{R}$ defined by

$$f(x) = \begin{cases} x & \text{if } 0 \le x < 2\pi \\ 0 & \text{if } x = 2\pi . \end{cases}$$

Computing the Fourier coefficients of f, we obtain

$$a_0 = \frac{1}{\pi} \int_0^{2\pi} x \, dx = \frac{x^2}{2\pi} \Big|_0^{2\pi} = 2\pi,$$

$$a_n = \frac{1}{\pi} \int_0^{2\pi} x \cos nx \, dx = \frac{1}{n\pi} \int_0^{2\pi} x \, d(\sin nx)$$

$$= \frac{1}{n\pi} \left[x \sin nx \Big|_0^{2\pi} - \int_0^{2\pi} \sin nx \, dx \right] = 0, \text{ and}$$

$$b_n = \frac{1}{\pi} \int_0^{2\pi} x \sin nx \, dx = -\frac{1}{n\pi} \int_0^{2\pi} x \, d(\cos nx)$$

$$= -\frac{1}{n\pi} \left[x \cos nx \Big|_0^{2\pi} - \int_0^{2\pi} \cos nx \, dx \right] = -\frac{1}{n\pi} \left[2\pi - \frac{1}{n} \sin nx \Big|_0^{2\pi} \right] = -\frac{2}{n}.$$

So, the Fourier series of the function f is $\pi - 2 \sum_{n=1}^{\infty} \frac{\sin nx}{n}$. Given that the function f is continuous at every $0 < x < 2\pi$ and that the preceding Fourier series converges for each $0 < x < 2\pi$ (see Example 9.7), it follows from Corollary 35.9 that

$$x = \pi - 2 \sum_{n=1}^{\infty} \frac{\sin nx}{n}$$

holds for each $0 < x < 2\pi$.

Problem 35.10. *Show that*

$$\frac{x^2}{2} = \pi x - \frac{\pi^2}{3} + 2 \sum_{n=1}^{\infty} \frac{\cos nx}{n^2}$$

holds for all $0 \le x \le 2\pi$. *Letting* $x = 0$ *we obtain the formula* $\sum_{n=1}^{\infty} \frac{1}{n^2} = \frac{\pi^2}{6}$.

Solution. Consider the periodic function $f: [0, 2\pi] \to \mathbb{R}$ defined by $f(x) = \frac{x^2}{2} - \pi x$. Computing its Fourier coefficients, we get

$$a_0 = \frac{1}{\pi} \int_0^{2\pi} \left(\frac{x^2}{2} - \pi x \right) dx = \frac{1}{\pi} \left(\frac{x^3}{6} - \frac{\pi x^2}{2} \right) \Big|_0^{2\pi} = -\frac{2\pi^2}{3},$$

$$a_n = \frac{1}{\pi} \int_0^{2\pi} \left(\frac{x^2}{2} - \pi x \right) \cos nx \, dx = \frac{2}{n^2}, \quad \text{and}$$

$$b_n = \frac{1}{\pi} \int_0^{2\pi} \left(\frac{x^2}{2} - \pi x \right) \sin nx \, dx = 0.$$

Therefore, the Fourier series of the function f is $\frac{\pi^2}{3} + 2 \sum_{n=1}^{\infty} \frac{\cos nx}{n^2}$. Since this series converges for each x and f is a continuous function, it follows from Corollary 35.9 that

$$\frac{x^2}{2} - \pi x = -\frac{\pi^2}{3} + 2 \sum_{n=1}^{\infty} \frac{\cos nx}{n^2},$$

and the desired identity follows.

Problem 35.11. *Show that*

$$x^2 = \frac{4}{3} \pi^2 + 4 \sum_{n=1}^{\infty} \left(\frac{\cos nx}{n^2} - \frac{\pi \sin nx}{n} \right)$$

holds for each $0 < x < 2\pi$.

Solution. Consider the periodic function $f: [0, 2\pi] \to \mathbb{R}$ defined by

$$f(x) = \begin{cases} x^2 & \text{if } 0 \le x < 2\pi \\ 0 & \text{if } x = 2\pi \, . \end{cases}$$

Computing the Fourier coefficients of f, we obtain

$$a_0 = \frac{1}{\pi} \int_0^{2\pi} x^2 \, dx = \frac{x^3}{3\pi} \Big|_0^{2\pi} = \frac{8}{3} \pi^2,$$

$$a_n = \frac{1}{\pi} \int_0^{2\pi} x^2 \cos nx\, dx = \frac{4}{n^2}, \text{ and}$$

$$b_n = \frac{1}{\pi} \int_0^{2\pi} x^2 \sin nx\, dx = -\frac{4\pi}{n}.$$

Therefore, the Fourier series of the function f is $\frac{4}{3}\pi^2 + 4\sum_{n=1}^{\infty}\left(\frac{\cos nx}{n^2} - \frac{\pi \sin nx}{n}\right)$. Since this series converges for each x (see Example 9.7) and f is continuous at each $0 < x < 2\pi$, it follows from Corollary 35.9 that

$$x^2 = \frac{4}{3}\pi^2 + 4\sum_{n=1}^{\infty}\left(\frac{\cos nx}{n^2} - \frac{\pi \sin nx}{n}\right),$$

for each $0 < x < 2\pi$.

Problem 35.12. *Consider the "integral" operator $T: L_2[0, \pi] \to L_2[0, \pi]$ defined by*

$$Tf(x) = \int_0^{\pi} K(x, t)f(t)\, dt,$$

where the kernel $K: [0, \pi] \times [0, \pi] \to \mathbf{R}$ is given by

$$K(x, t) = \sum_{n=1}^{\infty} \frac{[\sin(n + 1)x] \sin nt}{n^2}.$$

Show that the norm of the operator T satisfies $\|T\| = \pi/2$.

Solution. By Problem 35.5, we know that the sequence of functions

$$\left\{\left(\tfrac{2}{\pi}\right)^{\frac{1}{2}}\sin nx\colon n = 1, 2, \ldots\right\}$$

is an orthonormal basis for $L_2[0, \pi]$. Also, as usual, the norm of the operator is given by

$$\|T\| = \sup\{\|T(f)\|\colon f \in L_2[0, \pi] \text{ and } \|f\| = 1\}.$$

Now, fix a function $f \in L_2[0, \pi]$ with $\|f\| = 1$, and write

$$f = \sum_{n=1}^{\infty} c_n\left(\tfrac{2}{\pi}\right)^{\frac{1}{2}}\sin nx$$

in its Fourier expansion relative to the above basis. By Parseval's Identity, we have

$$\|f\|^2 = \sum_{n=1}^{\infty} |c_n|^2.$$

Next, notice that the operator satisfies

$$
\begin{aligned}
Tf(x) &= \int_0^{\pi} \left[\sum_{n=1}^{\infty} \frac{\sin(n+1)x \, \sin nt}{n^2} \right] \left[\sum_{m=1}^{\infty} c_m \left(\tfrac{2}{\pi}\right)^{\frac{1}{2}} \sin mt \right] dt \\
&= \sum_{n=1}^{\infty} \left[\frac{\sin(n+1)x}{n^2} \left(\int_0^{\pi} \sin nt \sum_{m=1}^{\infty} c_m \left(\tfrac{2}{\pi}\right)^{\frac{1}{2}} \sin mt \, dt \right) \right] \\
&= \frac{\pi}{2} \sum_{n=1}^{\infty} \left[\frac{c_n}{n^2} \left(\tfrac{2}{\pi}\right)^{\frac{1}{2}} \sin(n+1)x \right].
\end{aligned}
$$

Now, notice that the latter expression is the Fourier expansion of $T(f)$ with respect to the orthonormal basis described at the beginning of the solution. Thus, by Parseval's Identity, we have

$$\|T(f)\|^2 = \frac{\pi^2}{4} \sum_{n=1}^{\infty} \frac{|c_n|^2}{n^4} \leq \frac{\pi^2}{4} \sum_{n=1}^{\infty} |c_n|^2 = \frac{\pi^2}{4} \|f\|^2,$$

from which it follows that $\|T\| \leq \pi/2$ holds.

Finally, if $f_0(x) = \left(\tfrac{2}{\pi}\right)^{\frac{1}{2}} \sin x$, then $\|f_0\| = 1$ and by Parseval's Identity we have $c_1 = 1$ and $c_n = 0$ for $n \neq 1$, and so $\|T(f_0)\|^2 = \frac{\pi^2}{4}$. Therefore, $\|T\| \geq \pi/2$, and hence, $\|T\| = \pi/2$.

CHAPTER 7 _____

SPECIAL TOPICS IN INTEGRATION

36. SIGNED MEASURES

Problem 36.1. *Give an example of a signed measure and two Hahn decompositions (A, B) and (A_1, B_1) of X with respect to the signed measure such that $A \neq A_1$ and $B \neq B_1$.*

Solution. Let $X = \mathbf{R}$ and let Σ be the σ-algebra of all Lebesgue measurable sets. Consider the measures μ_1, $\mu_2 \in M(\Sigma)$ defined by $\mu_1(E) = \lambda(E \cap [0, 1])$ and $\mu_2(E) = \lambda(E \cap [1, 2])$ for each $E \in \Sigma$ (where λ denotes the Lebesgue measure on $\mathbf{R}$). Now, consider the signed measure $\mu = \mu_1 - \mu_2$, and note that $\big((-\infty, 1), [1, \infty)\big)$ and $\big([0, 1), (-\infty, 0) \cup [1, \infty)\big)$ are two Hahn decompositions of X with respect to the signed measure μ.

Problem 36.2. *If μ is a signed measure, then show that $\mu^+ \wedge \mu^- = 0$.*

Solution. Let (A, B) be a Hahn decomposition of X with respect to μ. If $E \in \Sigma$, then note that

$$
\begin{aligned}
0 \leq \mu^+ \wedge \mu^-(E) &= \mu^+ \wedge \mu^-(E \cap B) + \mu^+ \wedge \mu^-(E \cap A) \\
&\leq \mu^+(E \cap B) + \mu^-(E \cap A) \\
&= \mu(E \cap B \cap A) - \mu(E \cap A \cap B) = 0.
\end{aligned}
$$

Problem 36.3. *If μ is a signed measure, then show that for each $A \in \Sigma$ we have*

$$
|\mu|(A) = \sup\left\{ \sum_{n=1}^{\infty} |\mu(A_n)| \colon \{A_n\} \text{ is a disjoint sequence of } \Sigma \text{ with } \bigcup_{n=1}^{\infty} A_n = A \right\}.
$$

Solution.   Fix $A \in \Sigma$. From Theorem 36.9, we know that

$$|\mu|(A) = \sup\left\{\sum_{n=1}^{k}|\mu(A_n)|\colon \{A_1, \ldots, A_k\} \subseteq \Sigma \text{ is disjoint and } \bigcup_{n=1}^{k} A_n \subseteq A \right\}.$$

Also, let

$$s = \sup\left\{\sum_{n=1}^{\infty}|\mu(A_n)|\colon \{A_n\} \subseteq \Sigma \text{ is disjoint and } A = \bigcup_{n=1}^{\infty} A_n \right\}.$$

Now, let $\{A_n\}$ be a pairwise disjoint sequence of Σ such that $\bigcup_{n=1}^{\infty} A_n = A$. Clearly, $\sum_{n=1}^{k} |\mu(A_n)| \le |\mu|(A)$ holds for each k, and so

$$\sum_{n=1}^{\infty} |\mu(A_n)| = \lim_{k \to \infty} \sum_{n=1}^{k} |\mu(A_n)| \le |\mu|(A).$$

Therefore, $s \le |\mu|(A)$. On the other hand, if $\{A_1, \ldots, A_k\}$ is a finite pairwise disjoint collection of Σ satisfying $\bigcup_{n=1}^{k} A_n \subseteq A$, then

$$A = A_1 \cup \cdots \cup A_k \cup \left(A \setminus \bigcup_{n=1}^{k} A_k\right) \cup \emptyset \cup \emptyset \cdots,$$

and so

$$\sum_{n=1}^{k}|\mu(A_n)| \le \sum_{n=1}^{k}|\mu(A_n)| + \left|\mu\left(A \setminus \bigcup_{n=1}^{k} A_n\right)\right| + \mu(\emptyset) + \mu(\emptyset) + \cdots \le s.$$

Consequently, $|\mu|(A) \le s$ also holds. Thus, $|\mu|(A) = s$, as claimed.

Problem 36.4.   *Verify that if μ and ν are two finite signed measures, then the least upper bound $\mu \vee \nu$ and the greatest lower bound $\mu \wedge \nu$ holds in $M(\Sigma)$ are given by*

$$\mu \vee \nu(A) = \sup\{\mu(B) + \nu(A \setminus B)\colon B \in \Sigma \text{ and } B \subseteq A\}, \text{ and}$$
$$\mu \wedge \nu(A) = \inf\{\mu(B) + \nu(A \setminus B)\colon B \in \Sigma \text{ and } B \subseteq A\}$$

for each $A \in \Sigma$.

Solution. The proof parallels the one of Theorem 36.1. We shall verify that if $\mu, \nu \in M(\Sigma)$, then the formula

$$\omega(A) = \inf\{\mu(B) + \nu(A \setminus B)\colon B \in \Sigma \text{ and } B \subseteq A\}, \quad A \in \Sigma,$$

defines a finite signed measure (i.e., $\omega \in M(\Sigma)$), and that ω is the greatest lower bound of μ and ν in $M(\Sigma)$.

Since μ and ν are both bounded from below (and also both bounded from above), it follows that $\omega(A) \in \mathbb{R}$ for each $A \in \Sigma$. Clearly, $\omega(\emptyset) = 0$ holds. Next, we shall establish that ω is σ-additive. To this end, let $\{A_n\}$ be a pairwise disjoint sequence of Σ and let $A = \bigcup_{n=1}^{\infty} A_n$. If $B \in \Sigma$ satisfies $B \subseteq A$, then

$$\mu(B) + \nu(A \setminus B) = \mu\left(\bigcup_{n=1}^{\infty} A_n \cap B\right) + \nu\left(\bigcup_{n=1}^{\infty}(A_n \setminus A_n \cap B)\right)$$

$$= \sum_{n=1}^{\infty}\left[\mu(A_n \cap B) + \nu(A_n \setminus A_n \cap B)\right]$$

$$\geq \sum_{n=1}^{\infty} \omega(A_n),$$

and so $\omega(A) \geq \sum_{n=1}^{\infty} \omega(A_n)$ holds. For the reverse inequality, let $\varepsilon > 0$. Then, for each n pick some $B_n \in \Sigma$ with $B_n \subseteq A_n$ and

$$\mu(B_n) + \nu(A_n \setminus B_n) < \omega(A_n) + \frac{\varepsilon}{2^n}.$$

Obviously, $\{B_n\}$ is a pairwise disjoint sequence of Σ. Put $B = \bigcup_{n=1}^{\infty} B_n \subseteq A$, and note that $\bigcup_{n=1}^{\infty}(A_n \setminus B_n) = A \setminus B$ holds. Moreover,

$$\omega(A) \leq \mu(B) + \nu(A \setminus B) = \mu\left(\bigcup_{n=1}^{\infty} B_n\right) + \nu\left(\bigcup_{n=1}^{\infty}(A_n \setminus B_n)\right)$$

$$= \sum_{n=1}^{\infty}\left[\mu(B_n) + \nu(A_n \setminus B_n)\right]$$

$$\leq \sum_{n=1}^{\infty}\left[\omega(A_n) + \tfrac{\varepsilon}{2^n}\right] = \sum_{n=1}^{\infty} \omega(A_n) + \varepsilon.$$

Since $\varepsilon > 0$ is arbitrary, we infer that $\omega(A) \leq \sum_{n=1}^{\infty} \omega(A_n)$ also holds, and so $\omega \in M(\Sigma)$.

Finally, we shall establish that ω is the greatest lower bound of μ and ν in $M(\Sigma)$. Note first that ω is a lower bound for both μ and ν. Indeed, if $A \in \Sigma$, then (by letting $B = A$), we see that

$$\omega(A) \le \mu(A) + \nu(\emptyset) = \mu(A) \quad \text{and} \quad \omega(A) \le \mu(\emptyset) + \nu(A) = \nu(A).$$

On the other hand, if $\pi \in M(\Sigma)$ satisfies $\pi \le \mu$ and $\pi \le \nu$ and $A \in \Sigma$, then for each $B \in \Sigma$ with $B \subseteq A$ we have

$$\pi(A) = \pi(B) + \pi(A \setminus B) \le \mu(B) + \nu(A \setminus B),$$

from which it follows that $\pi(A) \le \omega(A)$, i.e., $\pi \le \omega$. This shows that $\omega = \mu \wedge \nu$ holds in $M(\Sigma)$.

Problem 36.5. *Let λ be the Lebesgue measure on the Lebesgue measurable subsets of $\mathbf{R}$. If μ is the Dirac measure, defined by $\mu(A) = 0$ if $0 \notin A$ and $\mu(A) = 1$ if $0 \in A$, describe $\lambda \vee \mu$ and $\lambda \wedge \mu$.*

Solution. If $A = \mathbf{R} \setminus \{0\}$, $B = \{0\}$, and E is an arbitrary Lebesgue measurable set, then

$$0 \le \lambda \wedge \mu(E) \le \lambda(E \cap B) + \mu(E \cap A) = 0$$

holds. That is, $\lambda \wedge \mu = 0$. Moreover, we have

$$\lambda \vee \mu = \lambda \vee \mu + \lambda \wedge \mu = \lambda + \mu.$$

Problem 36.6. *Show that the collection of all σ-finite measures forms a distributive lattice. That is, show that if μ, ν, and ω are three σ-finite measures, then*

$$(\mu \vee \nu) \wedge \omega = (\mu \wedge \omega) \vee (\nu \wedge \omega) \quad \text{and} \quad (\mu \wedge \nu) \vee \omega = (\mu \vee \omega) \wedge (\nu \vee \omega).$$

Solution. We shall show first that every vector lattice satisfies the distributive law. To do this, we shall use the identity (a) of Problem 9.1.

Let x, y, and z be elements in a vector lattice. Since $x \vee y \ge x$, it follows that $(x \vee y) \wedge z \ge x \wedge z$, and similarly $(x \vee y) \wedge z \ge y \wedge z$. Thus,

$$(x \vee y) \wedge z \ge (x \wedge z) \vee (y \wedge z).$$

On the other hand, if $u = (x \wedge z) \vee (y \wedge z)$, then $u \ge x \wedge z = x + z - x \vee z$ holds.

Hence, $x \le u - z + x \vee z \le u - z + (x \vee y) \vee z$, and similarly $y \le u - z + (x \vee y) \vee z$. It follows that $x \vee y \le u - z + (x \vee y) \vee z$, and so

$$(x \wedge z) \vee (y \wedge z) = u \ge x \vee y + z - (x \vee y) \vee z = (x \vee y) \wedge z.$$

Therefore, $(x \vee y) \wedge z = (x \wedge z) \vee (y \wedge z)$ holds. The other identity can be established in a similar manner.

Now, let $\{X_n\} \subseteq \Sigma$ satisfy $\mu(X_n) < \infty$, $\nu(X_n) < \infty$, $\omega(X_n) < \infty$ for all n, and $X_n \uparrow X$. If $\Sigma_n = \{A \cap X_n \colon A \in \Sigma\}$, then clearly μ, ν, and ω (restricted to Σ_n) belong to the vector lattice $M(\Sigma_n)$. Thus, if $E \in \Sigma$, then

$$(\mu \vee \nu) \wedge \omega(E \cap X_n) = (\mu \wedge \omega) \vee (\nu \wedge \omega)(E \cap X_n)$$

and

$$(\mu \wedge \nu) \vee \omega(E \cap X_n) = (\mu \vee \omega) \wedge (\nu \vee \omega)(E \cap X_n)$$

hold. To finish the proof note that $E \cap X_n \uparrow E$, and then use the "order continuity" of the measure (Theorem 15.4).

Problem 36.7. *If Σ is a σ-algebra of subsets of a set X and $\mu \colon \Sigma \to \mathbf{R}^*$ is a signed measure, then show that*

$$\Lambda_{\mu^+} \cap \Lambda_{\mu^-} = \Lambda_{|\mu|}.$$

Solution. Assume that $\mu \colon \Sigma \longrightarrow \mathbf{R}^*$ is an arbitrary signed measure. Let E be in $\Lambda_{\mu^+} \cap \Lambda_{\mu^-}$ and let $A \in \Sigma$ be an arbitrary set. Then,

$$
\begin{aligned}
|\mu|(A) &= \mu^+(A) + \mu^-(A) \\
&= \left[\mu^+(A \cap E) + \mu^+(A \cap E^c)\right] + \left[\mu^-(A \cap E) + \mu^-(A \cap E^c)\right] \\
&= \left[\mu^+(A \cap E) + \mu^-(A \cap E)\right] + \left[\mu^+(A \cap E^c) + \mu^-(A \cap E^c)\right] \\
&= |\mu|(A \cap E) + |\mu|(A \cap E^c),
\end{aligned}
$$

and so $E \in \Lambda_{|\mu|}$, i.e., $\Lambda_{\mu^+} \cap \Lambda_{\mu^-} \subseteq \Lambda_{|\mu|}$.

For the reverse inclusion, let $E \in \Lambda_{|\mu|}$. If $A \in \Sigma$ is arbitrary, then note that

$$
\begin{aligned}
\mu^+(A) + \mu^-(A) = |\mu|(A) &= |\mu|(A \cap E) + |\mu|(A \cap E^c) \\
&= \left[\mu^+(A \cap E) + \mu^+(A \cap E^c)\right] + \left[\mu^-(A \cap E) + \mu^-(A \cap E^c)\right].
\end{aligned}
$$

Since $\mu^+(A) = \mu^+\big((A \cap E) \cup (A \cap E^c)\big) \le \mu^+(A \cap E) + \mu^+(A \cap E^c)$ and

$\mu^-(A) \leq \mu^-(A \cap E) + \mu^-(A \cap E^c)$ both hold, it follows from the preceding equality that

$$\mu^+(A) = \mu^+(A \cap E) + \mu^+(A \cap E^c) \text{ and } \mu^-(A) = \mu^-(A \cap E) + \mu^-(A \cap E^c),$$

which shows that $E \in \Lambda_{\mu^+} \cap \Lambda_{\mu^-}$. Thus, $\Lambda_{|\mu|} \subseteq \Lambda_{\mu^+} \cap \Lambda_{\mu^-}$, and consequently, $\Lambda_{|\mu|} = \Lambda_{\mu^+} \cap \Lambda_{\mu^-}$ holds, as desired.

Problem 36.8. *Let μ and ν be two measures on a σ-algebra Σ with at least one of them finite. Assume also that S is a semiring such that $S \subseteq \Sigma$, $X \in S$, and that the σ-algebra generated by S equals Σ. Then show that $\mu = \nu$ on Σ if and only if $\mu = \nu$ on S.*

Solution. Assume that μ is finite and that $\mu = \nu$ on S. If we consider the measure space (X, S, μ), then it is easy to see that $S \subseteq \Sigma \subseteq \Lambda_\mu$ holds. Now, apply Theorem 15.10 to get that $\mu = \mu^* = \nu$ holds on Σ.

Problem 36.9. *Let (X, S, μ) be a measure space, and let $f \in L_1(\mu)$. Then show that*

$$\nu(A) = \int_A f \, d\mu$$

for each $A \in \Lambda_\mu$ defines a finite signed measure on Λ_μ. Also, show that

$$\nu^+(A) = \int_A f^+ \, d\mu, \quad \nu^-(A) = \int_A f^- \, d\mu \quad and \quad |\nu|(A) = \int_A |f| \, d\mu$$

holds for each $A \in \Lambda_\mu$.

Solution. If $\{A_n\}$ is a pairwise disjoint sequence of Λ_μ satisfying $A = \bigcup_{n=1}^{\infty} A_n$, then $\lim \sum_{i=1}^{n} f \chi_{A_i} = f \chi_A$ and $\left| \sum_{i=1}^{n} f \chi_{A_i} \right| \leq |f|$ holds for each n. Thus, from the Lebesgue Dominated Convergence Theorem, it follows that

$$\nu(A) = \int f \chi_A \, d\mu = \lim_{n \to \infty} \int \left(\sum_{i=1}^{n} f \chi_{A_i} \right) d\mu = \lim_{n \to \infty} \sum_{i=1}^{n} \nu(A_i) = \sum_{i=1}^{\infty} \nu(A_i).$$

Therefore, ν is a finite signed measure.

Now, note that if

$$A = \{x \in X \colon f(x) \geq 0\} \quad \text{and} \quad B = \{x \in X \colon f(x) < 0\},$$

then it is easy to see that (A, B) is a Hahn decomposition of X with respect to v. Since $f\chi_{E\cap A} = f^+\chi_E$ holds, we see that

$$v^+(E) = v(E \cap A) = \int_{E\cap A} f\, d\mu = \int_E f^+\, d\mu$$

for each $E \in \Lambda_\mu$. The proof for v^- is similar. The absolute value formula follows from the identity $|v| = v^+ + v^-$.

Problem 36.10. *Let v be a signed measure on Σ. A function $f: X \to \mathbb{R}$ is said to be v-integrable if f is simultaneously v^+- and v^--integrable (in this case, we write $\int f\, dv = \int f\, dv^+ - \int f\, dv^-$). Show that a function f is v-integrable if and only if $f \in L_1(|v|)$.*

Solution. Assume that f is simultaneously v^+- and v^--integrable. We can assume that $f(x) \geq 0$ holds for all $x \in X$. Since each set of the form $\{x \in X: a \leq f(x) < b\}$ belongs to $\Lambda_{\mu^+} \cap \Lambda_{\mu^-} = \Lambda_{|\mu|}$ (for this identity see Problem 36.7), we see that there exists a sequence $\{\phi_n\}$ of simultaneously v^+- and v^--step functions such that $\phi_n(x) \uparrow f(x)$ holds for all $x \in X$; see the proof of Theorem 17.7. Clearly, each ϕ_n is a $|\mu|$-step function and from

$$\int \phi_n\, d|\mu| = \int \phi_n\, d\mu^+ + \int \phi_n\, d\mu^- \uparrow \int f\, d\mu^+ + \int f\, d\mu^- < \infty,$$

we see that f is $|\mu|$-integrable and that $\int f\, d|\mu| = \int f\, d\mu^+ + \int f\, d\mu^-$ holds.

For the converse, assume that f belongs to $L_1(|v|)$. We can assume that $f(x) \geq 0$ holds for each x. Note first that if $f = \chi_A$ for a $|v|$-measurable set A with $|v|^*(A) < \infty$, then there exists (by Theorem 15.11) some $B \in \Sigma$ with $A \subseteq B$ and $|v|^*(A) = |v|^*(B)$. It follows that $|v|^*(B \setminus A) = 0$, and in view of $0 \leq v^+ \leq |v|$, we have $(v^+)^*(B \setminus A) = 0$. Thus, $B \setminus A$ is a v^+-measurable set, and consequently, $A = B \setminus (B \setminus A)$ is also v^+-measurable. This shows that χ_A is v^+-integrable.

Now, choose a sequence $\{\phi_n\}$ of $|v|$-step functions with $0 \leq \phi_n(x) \uparrow f(x)$ for each x. By the previous discussion, $\{\phi_n\}$ is a sequence of v^+-step functions. Moreover,

$$\int \phi_n\, dv^+ \leq \int \phi_n\, d|v| \leq \int f\, d|v| < \infty$$

holds for all n. Thus, f is v^+-integrable.

The v^--integrability of f can be established in a similar manner.

Problem 36.11. *Show that the Jordan decomposition is unique in the following sense. If ν is a signed measure, and μ_1 and μ_2 are two measures such that $\nu = \mu_1 - \mu_2$ and $\mu_1 \wedge \mu_2 = 0$, then $\mu_1 = \nu^+$ and $\mu_2 = \nu^-$.*

Solution. First, we shall establish that $\nu^+ = \mu_1$ holds. Start by observing that $\nu \le \mu_1$ implies $\nu^+ \le \mu_1$.

Now, let $E \in \Sigma$. If $\nu^+(E) = \infty$, then $\nu^+(E) = \mu_1(E) = \infty$ holds trivially. Thus, we can suppose $\nu^+(E) < \infty$. Since $\nu(E) = \mu_1(E) - \mu_2(E) \le \nu^+(E) < \infty$, it follows that $\mu_1(E) < \infty$. Let $\varepsilon > 0$. Then, in view of

$$0 = \mu_1 \wedge \mu_2(E) = \inf\{\mu_1(E \setminus B) + \mu_2(B) \colon B \in \Sigma \text{ and } B \subseteq E\},$$

there exists some $B \in \Sigma$ with $B \subseteq E$ and $\mu_1(E \setminus B) + \mu_2(B) < \varepsilon$. Thus,

$$\nu^+(E) = \sup\{\nu(F) \colon F \in \Sigma \text{ and } F \subseteq E\} \ge \nu(B) = \mu_1(B) - \mu_2(B)$$
$$\ge \mu_1(B) - \varepsilon = \mu_1(E) - \mu_1(E \setminus B) - \varepsilon \ge \mu_1(E) - 2\varepsilon$$

holds for all $\varepsilon > 0$. That is, $\nu^+(E) \ge \mu_1(E)$ for each $E \in \Sigma$, and therefore $\nu^+ = \mu_1$ holds. For the other identity note that

$$\nu^- = (-\nu)^+ = (\mu_2 - \mu_1)^+ = \mu_2.$$

Problem 36.12. *In a vector lattice $x_n \downarrow x$ means that $x_{n+1} \le x_n$ for each n and that x is the greatest lower bound of the sequence $\{x_n\}$. A normed vector lattice is said to have σ-**order continuous norm** if $x_n \downarrow 0$ implies $\lim \|x_n\| = 0$.*

   a. *Show that every $L_p(\mu)$ with $1 \le p < \infty$ has σ-order continuous norm.*
   b. *Show that $L_\infty([0,1])$ does not have σ-order continuous norm.*
   c. *Let Σ be a σ-algebra of sets, and let $\{\mu_n\}$ be a sequence of $M(\Sigma)$ such that $\mu_n \downarrow \mu$. Show that $\lim \mu_n(A) = \mu(A)$ holds for all $A \in \Sigma$.*
   d. *Show that the Banach lattice $M(\Sigma)$ has σ-order continuous norm.*

Solution. (a) Note first that $f_n \downarrow f$ in $L_p(\mu)$ is equivalent to $f_n \downarrow f$ a.e. (why?). If for some $1 \le p < \infty$ a sequence $\{f_n\}$ of $L_p(\mu)$ satisfies $f_n \downarrow 0$ a.e., then

$$\|f_n\|_p = \left(\int |f_n|^p \, d\mu\right)^{\frac{1}{p}} \downarrow 0$$

holds by virtue of the Lebesgue Dominated Convergence Theorem.

(b) If $f_n = \chi_{(0,\frac{1}{n})}$, then $f_n \downarrow 0$ holds in $L_\infty([0,1])$. However, note that $\|f_n\|_\infty = 1$ holds for each n.

(c) Let $\mu_n \downarrow \mu$ in $M(\Sigma)$. Then $0 \leq \mu_1 - \mu_n \uparrow \mu_1 - \mu$ in $M(\Sigma)$. By Theorem 36.2, it follows that $\mu_1(A) - \mu_n(A) \uparrow \mu_1(A) - \mu(A)$ holds for each $A \in \Sigma$. Thus, $\mu_n(A) \downarrow \mu(A)$ holds for each $A \in \Sigma$.

(d) If $\mu_n \downarrow 0$ in $M(\Sigma)$, then from part (c) it follows that $\|\mu_n\| = \mu_n(X) \downarrow 0$.

Problem 36.13.   *Prove the following additivity property of the Banach lattice $M(\Sigma)$: If $\mu, \nu \in M(\Sigma)$ are disjoint (i.e., $|\mu| \wedge |\nu| = 0$), then $\|\mu + \nu\| = \|\mu\| + \|\nu\|$ holds.*

Solution.   If $|\mu| \wedge |\nu| = 0$ holds in $M(\Sigma)$, then $|\mu + \nu| = |\mu| + |\nu|$ holds (see Problems 9.2 and 9.3). Thus, $\left\| \mu + \nu \right\| = \left| \mu + \nu \right|(X) = \left| \mu \right|(X) + \left| \nu \right|(X) = \|\mu\| + \|\nu\|$.

Problem 36.14.   *Let Σ be a σ-algebra of subsets of a set X and let $\{\mu_n\}$ be a disjoint sequence of $M(\Sigma)$. If the sequence of signed measures $\{\mu_n\}$ is order bounded, then show that $\lim \|\mu_n\| = 0$.*

Solution.   Let $\{\mu_n\}$ be a disjoint sequence of the Banach lattice $M(\Sigma)$ such that for some $0 \leq \mu \in M(\Sigma)$ we have $|\mu_n| \leq \mu$ for each n. From $|\mu_n| \wedge |\mu_m| = 0$ for $n \neq m$, we see that

$$\sum_{n=1}^{k} |\mu_n| = \bigvee_{n=1}^{k} |\mu_n| \leq \mu$$

holds for each k. In particular, we have

$$\sum_{n=1}^{k} \|\mu_n\| = \sum_{n=1}^{k} |\mu_n|(X) \leq \left[\bigvee_{n=1}^{k} |\mu_n| \right](X) \leq \mu(X) < \infty$$

holds for each n, and so $\sum_{n=1}^{\infty} \|\mu_n\| < \infty$. The latter easily implies $\lim \|\mu_n\| = 0$.

37.   COMPARING MEASURES AND THE RADON–NIKODYM THEOREM

Problem 37.1.   *Verify the following properties of signed measures:*

a. $\mu \ll \mu$.

b. $\nu \ll \mu$ and $\mu \ll \omega$ imply $\nu \ll \omega$.

c.   *If $0 \leq \nu \leq \mu$, then $\nu \ll \mu$.*

d.   *If $\mu \ll 0$, then $\mu = 0$.*

Solution. (a) From Theorem 36.9, we have $|\mu(A)| \leq |\mu|(A)$, and so if $|\mu|(A) = 0$ holds, then $\mu(A) = 0$ likewise holds. That is, $\mu \ll \mu$.

(b) Assume $\nu \ll \mu$ and $\mu \ll \omega$ and $|\omega|(A) = 0$. Theorem 37.2 applied twice shows that $|\mu|(A) = 0$ and $|\nu|(A) = 0$. Hence, $\nu \ll \omega$ holds.

(c) Let $0 \leq \nu \leq \mu$. If $|\mu|(A) = \mu(A) = 0$, then clearly $\nu(A) = 0$, and so $\nu \ll \mu$ holds.

(d) Let $\mu \ll 0$. Since the zero measure assumes the zero value at every $A \in \Sigma$, it follows that $\mu(A) = 0$ holds for every $A \in \Sigma$. This means that $\mu = 0$.

Problem 37.2. *Verify the following statements about signed measures on a σ-algebra Σ of sets:*

1. *If $\mu \ll \omega$ and $\nu \ll \omega$, then $|\mu| + |\nu| \ll \omega$.*
2. *If $\mu \perp \omega$ and $\nu \perp \omega$, then $|\mu| + |\nu| \perp \omega$.*
3. *If $\mu \ll \omega$ and $|\nu| \leq |\mu|$, then $\nu \ll \omega$.*
4. *If $\mu \perp \omega$ and $|\nu| \leq |\mu|$, then $\nu \perp \omega$.*
5. *If $\nu \ll \mu$ and $\nu \perp \mu$, then $\nu = 0$.*

Solution. (1) This follows immediately from Theorem 37.2.

(2) Since $\mu \perp \omega$, there exists (by Theorem 37.5) some $A_1 \in \Sigma$ with $|\omega|(A_1) = |\mu|(A_1^c) = 0$. Similarly, there exists some $A_2 \in \Sigma$ with $|\omega|(A_2) = |\nu|(A_2^c) = 0$. Put $A = A_1 \cup A_2$ and $B = (A_1 \cup A_2)^c = A_1^c \cap A_2^c$. Then A, $B \in \Sigma$, $A \cup B = X$, $A \cap B = \emptyset$, $|\omega|(A) = 0$, and $(|\mu| + |\nu|)(B) = 0$. By Theorem 37.5 we infer that $\omega \perp |\mu| + |\nu|$ holds.

(3) This follows easily from Theorem 37.2.

(4) This follows immediately from Theorem 37.5.

(5) Since $\nu \perp \mu$, there exists some $A \in \Sigma$ such that $|\nu|(A) = |\mu|(A^c) = 0$. By $\nu \ll \mu$ and Theorem 37.2, $|\nu|(A^c) = 0$, and so $|\nu|(X) = |\nu|(A) + |\nu|(A^c) = 0$. That is, $|\nu| = 0$, so that $\nu = 0$.

Problem 37.3. *Let μ and ν be two measures on a σ-algebra Σ. If ν is a finite measure, then show that the following statements are equivalent.*

a. *$\nu \ll \mu$ holds.*
b. *For each sequence $\{A_n\}$ of Σ with $\lim \mu(A_n) = 0$, we have $\lim \nu(A_n) = 0$.*
c. *For each $\epsilon > 0$ there exists some $\delta > 0$ (depending on ϵ) such that whenever $A \in \Sigma$ satisfies $\mu(A) < \delta$, then $\nu(A) < \epsilon$ holds.*

Solution. (a) $\Longrightarrow$ (b) If (b) is not true, then there exists some $\varepsilon > 0$ and some sequence $\{A_n\}$ of Σ such that $\mu(A_n) < 2^{-n}$ and $\nu(A_n) > \varepsilon$ for each n. Set

$A = \bigcap_{n=1}^{\infty} \bigcup_{i=n}^{\infty} A_i \in \Sigma$. From $A \subseteq \bigcup_{i=n}^{\infty} A_i$, we see that

$$\mu(A) \leq \sum_{i=n}^{\infty} \mu(A_i) \leq \sum_{i=n}^{\infty} \tfrac{1}{2^n} = 2^{1-n}$$

holds for each n, and so $\mu(A) = 0$. However, from Theorem 15.4(2), we see that $v(A) \geq \varepsilon$, contrary to $v \ll \mu$. Hence, (a) implies (b).

(b) $\Longrightarrow$ (c) If (c) is not true, then there exist some $\varepsilon > 0$ and a sequence $\{A_n\}$ of Σ such that $\mu(A_n) < \frac{1}{n}$ and $v(A_n) \geq \varepsilon$ hold for all n. Clearly, this contradicts (b).

(c) $\Longrightarrow$ (a) Let $A \in \Sigma$ satisfy $\mu(A) = 0$. Given $\varepsilon > 0$, choose some $\delta > 0$ so that (c) is satisfied. In view of $\mu(A) < \delta$, it follows that $v(A) < \varepsilon$. Since $\varepsilon > 0$ is arbitrary, $v(A) = 0$, and so $v \ll \mu$ holds.

Problem 37.4.   *Let μ be a finite measure, and let $\{v_n\}$ be a sequence of finite measures (all on Σ) such that $v_n \ll \mu$ holds for each n. Furthermore, assume that $\lim v_n(A)$ exists in $\mathbb{R}$ for each $A \in \Sigma$. Then, show that:*

  a.   *For each $\epsilon > 0$ there exists some $\delta > 0$ such that whenever $A \in \Sigma$ satisfies $\mu(A) < \delta$, then $v_n(A) < \epsilon$ holds for each n.*
  b.   *The set function $v \colon \Sigma \to [0, \infty]$, defined by $v(A) = \lim v_n(A)$ for each $A \in \Sigma$, is a measure such that $v \ll \mu$.*

Solution.   (a) From Problem 31.3, we know that Σ under the distance $d(A, B) = \mu(A \triangle B)$ is a complete metric space. From $v_k \ll \mu$ and the inequality

$$\left| v_k(A) - v_k(B) \right| \leq v_k(A \triangle B),$$

it easily follows that the function $v_k \colon \Sigma \longrightarrow \mathbb{R}$ is well defined (i.e., $v_k(A) = v_k(B)$ holds whenever $\mu(A \triangle B) = 0$) and is continuous.

Now, let $\varepsilon > 0$. Define

$$\mathcal{C}_k = \left\{ A \in \Sigma \colon |v_n(A) - v_m(A)| \leq \varepsilon \text{ for all } n, m \geq k \right\}.$$

Note that each $\mathcal{C}_k$ is closed and that $\Sigma = \bigcup_{k=1}^{\infty} \mathcal{C}_k$ holds. By Baire's Category Theorem 6.18), we have $\mathcal{C}_k^{\circ} \neq \emptyset$ for some k. Thus, there exist $A_0 \in \mathcal{C}_k$ and $\delta_1 > 0$ such that $A \in \Sigma$ and $\mu(A \triangle A_0) < \delta_1$ imply $A \in \mathcal{C}_k$.

From $v_i \ll \mu$ ($1 \leq i \leq k$) and the preceding problem, there exists some $0 < \delta < \delta_1$ such that $A \in \Sigma$ and $\mu(A) < \delta$ imply $v_i(A) < \varepsilon$ for all $1 \leq i \leq k$.

Now, if $A \in \Sigma$ satisfies $\mu(A) < \delta$, then $A \cup (A_0 \setminus A) = A \cup A_0$ satisfies $\mu\big((A \cup A_0)\Delta A_0\big) \leq \mu(A) < \delta_1$, and so

$$
\begin{aligned}
\big|v_n(A) - v_k(A)\big| &= \big|(v_n - v_k)(A \cup A_0) - (v_n - v_k)(A_0 \setminus A)\big| \\
&\leq \big|(v_n - v_k)(A \cup A_0)\big| + \big|(v_n - v_k)(A_0 \setminus A)\big| \leq 2\varepsilon
\end{aligned}
$$

holds for all $n > k$. Thus, $A \in \Sigma$ and $\mu(A) < \delta$ imply

$$
\big|v_n(A)\big| \leq 2\varepsilon + v_k(A) < 3\varepsilon
$$

for all $n > k$ (and all $1 \leq n \leq k$).

(b) Let $A = \bigcup_{n=1}^{\infty} A_n$ with the sequence $\{A_n\}$ of Σ pairwise disjoint, and let $\varepsilon > 0$. Choose some $\delta > 0$ so that statement (a) is satisfied. Next, choose some m so that $\mu\big(A \setminus \bigcup_{i=1}^{n} A_i\big) < \delta$ holds for all $n \geq m$. Then,

$$
\left|v_k(A) - \sum_{i=1}^{n} v_k(A_i)\right| = v_k\left(A \setminus \bigcup_{i=1}^{n} A_i\right) < \varepsilon
$$

holds for all k and all $n \geq m$. Thus, $\big|v(A) - \sum_{i=1}^{n} v(A_i)\big| \leq \varepsilon$ holds for all $n \geq m$, and so $\big|v(A) - \sum_{i=1}^{\infty} v(A_i)\big| \leq \varepsilon$. Since $\varepsilon > 0$ is arbitrary, we see that $v(A) = \sum_{n=1}^{\infty} v(A_n)$. Thus, v is a measure, and from part (a) and the preceding problem it follows immediately that $v \ll \mu$ holds.

Problem 37.5. *Let $\{v_n\}$ be a sequence of nonzero finite measures such that $\lim v_n(A)$ exists in $\mathbf{R}$ for each $A \in \Sigma$. Show that $v(A) = \lim v_n(A)$ for $A \in \Sigma$ is a finite measure.*

Solution. Consider the set function $\mu \colon \Sigma \longrightarrow [0, \infty)$ defined by

$$
\mu(A) = \sum_{n=1}^{\infty} \frac{v_n(A)}{v_n(X)} 2^{-n},
$$

and note that μ is in fact a measure. In addition, note that $v_n \ll \mu$ holds for each n. Now, invoke part (b) of the preceding problem to conclude that the set function v is also a measure.

Problem 37.6. *Verify the uniqueness of the Radon–Nikodym derivative by proving the following statement: If $(X, \mathcal{S}, \mu)$ is a measure space and $f \in L_1(\mu)$ satisfies $\int_A f \, d\mu = 0$ for all $A \in \mathcal{S}$, then $f = 0$ a.e.*

Solution.   From the given condition, it is easy to see that $\int_A f \, d\mu = 0$ must hold for each σ-set A. Now, consider the measurable sets

$$A = \left\{ x \in X \colon \, f(x) > 0 \right\} \quad \text{and} \quad B = \left\{ x \in X \colon \, f(x) < 0 \right\}.$$

By Problem 22.7 we know that A and B are both σ-finite sets. Now, in view of $\int_A f \, d\mu = \int_B f \, d\mu = 0$, it follows from Problem 22.13 that $\mu^*(A) = \mu^*(B) = 0$. Therefore, $f = 0$ a.e. holds.

Problem 37.7.   *This problem shows that the hypothesis of σ-finiteness of μ in the Radon–Nikodym Theorem cannot be omitted. Consider $X = [0, 1]$, Σ the σ-algebra of all Lebesgue measurable subsets of $[0, 1]$, ν the Lebesgue measure on Σ and μ the measure defined by $\mu(\emptyset) = 0$ and $\mu(A) = \infty$ if $A \neq \emptyset$. (Incidentally, μ is the largest measure on Σ.) Show that:*

a.   *ν is a finite measure, μ is not σ-finite, and $\nu \ll \mu$.*
b.   *There is no function $f \in L_1(\mu)$ such that $\nu(A) = \int_A f \, d\mu$ holds for all $A \in \Sigma$.*

Solution.   (a) Note that $\mu(A) = 0$ means $A = \emptyset$, and so $\nu \ll \mu$ holds.
(b) Observe that $L_1(\mu) = \{0\}$.

Problem 37.8.   *Let μ be a finite signed measure on Σ. Show that there exists a unique function $f \in L_1(|\mu|)$ such that*

$$\mu(A) = \int_A f \, d|\mu|$$

holds for all $A \in \Sigma$.

Solution.   The conclusion follows from the Radon–Nikodym Theorem by observing that $\mu \ll |\mu|$ holds.

Problem 37.9.   *Assume that ν is a finite measure and μ is a σ-finite measure such that $\nu \ll \mu$. Let $g = d\nu/d\mu \in L_1(\mu)$ be the Radon–Nikodym derivative of ν with respect to μ. Then show that:*

a.   *If $Y = \{x \in X \colon g(x) > 0\}$, then $Y \cap A$ is a μ-measurable set for each ν-measurable set A.*
b.   *If $f \in L_1(\nu)$, then $fg \in L_1(\mu)$ and $\int f \, d\nu = \int fg \, d\mu$ holds.*

Solution.   (a) Note first that by Theorem 37.3, $\Sigma \subseteq \Lambda_\mu \subseteq \Lambda_\nu$ holds, and that $Y \in \Lambda_\mu$.

First consider the case when $A \in \Lambda_\nu$ satisfies $A \subseteq Y$ and $\nu^*(A) = 0$. By Theorem 15.11 there exists some $B \in \Sigma$ with $A \subseteq B$ and $\nu^*(B) = 0$. Now, if $\mu^*(B \cap Y) > 0$, then we have the contradiction $0 = \nu^*(B \cap Y) = \int_{B \cap Y} g \, d\mu > 0$ (see Problem 22.13). Consequently, $\mu^*(B \cap Y) = \mu^*(A) = 0$ holds, and so $A \in \Lambda_\mu$.

Now, let $A \in \Lambda_\nu$. Choose some $B \in \Sigma$ with $A \subseteq B$ and $\nu^*(A) = \nu^*(B)$. Thus, $\nu^*(B \setminus A) = 0$, and so $(B \setminus A) \cap Y \in \Lambda_\mu$. Now, note that

$$A \cap Y = B \cap Y \setminus (B \setminus A) \cap Y \in \Lambda_\mu.$$

(b) It follows immediately from Problem 22.15.

Problem 37.10. *Establish the **chain rule** for Radon–Nikodym derivatives: If ω is a σ-finite measure and ν and μ are two finite measures (all on Σ) such that $\nu \ll \mu$ and $\mu \ll \omega$, then $\nu \ll \omega$ and*

$$\frac{d\nu}{d\omega} = \frac{d\nu}{d\mu} \cdot \frac{d\mu}{d\omega} \quad (\omega\text{-}a.e.)$$

holds.

Solution. Clearly, $\nu \ll \mu$ and $\Sigma \subseteq \Lambda_\omega \subseteq \Lambda_\mu \subseteq \Lambda_\nu$. Put $f = \frac{d\nu}{d\mu} \in L_1(\mu)$ and $g = \frac{d\mu}{d\omega} \in L_1(\omega)$. If $A \in \Sigma$, then by part (b) of the preceding problem, we infer that

$$\nu(A) = \int_A f \, d\mu = \int f \chi_A \, d\mu = \int f \chi_A g \, d\omega = \int_A fg \, d\omega.$$

This combined with the Radon–Nikodym Theorem shows that

$$\frac{d\nu}{d\mu} = fg, \quad \omega\text{-}a.e.$$

Problem 37.11. *All measures considered here will be assumed defined on a fixed σ-algebra Σ.*

a. *Call two measures μ and ν equivalent (in symbols, $\mu \equiv \nu$) if $\mu \ll \nu$ and $\nu \ll \mu$ both hold. Show that $\equiv$ is an equivalence relation among the measures on Σ.*

b. *If μ and ν are two equivalent σ-finite measures, then show that $\Lambda_\mu = \Lambda_\nu$.*

c. *Show that if μ and ν are two equivalent finite measures, then*

$$\frac{d\mu}{d\nu} \cdot \frac{d\nu}{d\mu} = 1 \quad a.e. \ holds.$$

d. *If μ and ν are two equivalent finite measures, then show that*

$$f \mapsto f \cdot \frac{d\mu}{d\nu},$$

from $L_1(\mu)$ to $L_1(\nu)$, is an onto lattice isometry. Thus, under this identification $L_1(\mu) = L_1(\nu)$ holds.

e. *Generalize (d) to equivalent σ-finite measures. That is, if μ and ν are two equivalent σ-finite measures, then show that the Banach lattices $L_1(\mu)$ and $L_1(\nu)$ are lattice isometric.*

f. *Show that if μ and ν are two equivalent σ-finite measures, then the Banach lattices $L_p(\mu)$ and $L_p(\nu)$ are lattice isometric for each $1 \le p \le \infty$.*

Solution.   (a) Straightforward.

(b) It follows immediately from Theorem 37.3.

(c) Use the relation $\nu \ll \mu \ll \nu$ and the preceding problem.

(d) Let $f \mapsto f \cdot \frac{d\mu}{d\nu} = T(f)$. Since $\frac{d\mu}{d\nu} \in L_1(\nu)$, it follows from Problem 37.9(b) that $T(f) = f \cdot \frac{d\mu}{d\nu} \in L_1(\nu)$ and $\int f \, d\mu = \int f \cdot \frac{d\mu}{d\nu} \, d\nu$ hold for each $f \in L_1(\mu)$. Thus, T defines a mapping from $L_1(\mu)$ to $L_1(\nu)$ which is clearly linear. Since $\frac{d\mu}{d\nu} \ge 0$ holds, it follows that

$$T(|f|) = |f| \cdot \frac{d\mu}{d\nu} = \left| f \cdot \frac{d\mu}{d\nu} \right| = |T(f)|$$

and

$$\|T(f)\|_1 = \int \left| f \cdot \frac{d\mu}{d\nu} \right| d\nu = \int |f| \cdot \frac{d\mu}{d\nu} \, d\nu = \int |f| \, d\mu = \|f\|_1$$

hold for each $f \in L_1(\mu)$. Thus, $T: L_1(\mu) \longrightarrow L_1(\nu)$ is a lattice isometry. To see that T is also onto, note that if $g \in L_1(\nu)$, then $g \cdot \frac{d\nu}{d\mu} \in L_1(\mu)$ and by part (c), we see that

$$T\left(g \cdot \frac{d\nu}{d\mu} \right) = g \cdot \frac{d\nu}{d\mu} \cdot \frac{d\mu}{d\nu} = g.$$

(e) Let $\{E_n\}$ be a pairwise disjoint sequence of Σ such that $\bigcup_{n=1}^{\infty} E_n = X$, $\mu(E_n) < \infty$, and $\nu(E_n) < \infty$ for each n. Let

$$T_n: L_1(E_n, \mu) \longrightarrow L_1(E_n, \nu)$$

be the onto lattice isometry determined by part (d) previously. Now, it is a routine matter to verify that $T: L_1(\mu) \longrightarrow L_1(\nu)$ defined by

$$T(f) = \sum_{n=1}^{\infty} T_n\left(f \chi_{E_n} \right)$$

for each $f \in L_1(\mu)$ is an onto lattice isometry.

(f) Suppose first that μ and ν are finite. Then,

$$f \longmapsto f \cdot \left(\tfrac{d\mu}{d\nu}\right)^{\frac{1}{p}}$$

is a lattice isometry from $L_p(\mu)$ onto $L_p(\nu)$ for each $1 \leq p < \infty$. Now, if μ and ν are σ-finite, then use the arguments of part (e) to establish that $L_p(\mu)$ and $L_p(\nu)$ are lattice isometric.

If $p = \infty$, then from part (b) it follows that $L_\infty(\mu) = L_\infty(\nu)$ holds, and so in this case the identity operator is a lattice isometry.

Problem 37.12. *Let μ be a σ-finite measure, and let $AC(\mu)$ be the collection of all finite signed measures that are absolutely continuous with respect to μ; that is,*

$$AC(\mu) = \{\nu \in M(\Sigma): \ \nu \ll \mu\}.$$

a. *Show that $AC(\mu)$ is a norm closed ideal of $M(\Sigma)$ (and hence $AC(\mu)$, with the norm $\|\nu\| = |\nu|(X)$, is a Banach lattice in its own right).*

b. *For each $f \in L_1(\mu)$, let μ_f be the finite signed measure defined by $\mu_f(A) = \int_A f \, d\mu$ for each $A \in \Sigma$. Then show that $f \mapsto \mu_f$ is a lattice isometry from $L_1(\mu)$ onto $AC(\mu)$.*

Solution. (a) Clearly, $AC(\mu)$ is an ideal of $M(\Sigma)$. If $\{\nu_n\}$ is a sequence of $AC(\mu)$ satisfying $\nu_n \longrightarrow \nu$ in $M(\Sigma)$, then $\nu_n(A) \longrightarrow \nu(A)$ holds for each $A \in \Sigma$. Problem 37.4 shows that $\nu \in AC(\mu)$. Thus, $AC(\mu)$ is a closed vector sublattice of $M(\Sigma)$, and hence, a Banach lattice in its own right.

(b) Clearly, $f \longmapsto \mu_f$ is a linear operator. By Problem 37.6 this operator is one-to-one. From Problem 36.9, it follows that $f \longmapsto \mu_f$ is a lattice isometry, and the Radon–Nikodym Theorem implies that it is also onto.

Problem 37.13. *Let Σ be a σ-algebra of subsets of a set X and μ a measure on Σ. Assume also that Σ^* is a σ-algebra of subsets of a set Y and that $T: X \to Y$ has the property that $T^{-1}(A) \in \Sigma$ for each $A \in \Sigma^*$.*

a. *Show that $\nu(A) = \mu(T^{-1}(A))$ for each $A \in \Sigma^*$ is a measure on Σ^*.*

b. *If $f \in L_1(\nu)$, then show that $f \circ T \in L_1(\mu)$ and*

$$\int_Y f \, d\nu = \int_X f \circ T \, d\mu.$$

c. *If μ is finite and ω is a σ-finite measure on Σ^* such that $\nu \ll \omega$, then show that there exists a function $g \in L_1(\omega)$ such that*

$$\int_X f \circ T \, d\mu = \int_Y fg \, d\omega$$

holds for each $f \in L_1(\nu)$.

Solution. (a) Straightforward.

(b) Note first that if A is a ν-null set, then $T^{-1}(A)$ is a μ-null set. Indeed, if $\nu^*(A) = 0$ holds, then there exists (by Theorem 15.11) some $B \in \Sigma^*$ with $A \subseteq B$ and $\nu(B) = 0$. Therefore,

$$0 \le \mu^*\big(T^{-1}(A)\big) \le \mu^*\big(T^{-1}(B)\big) = \mu\big(T^{-1}(B)\big) = \nu(B) = 0.$$

Now, let A be a ν-measurable set with $\nu^*(A) < \infty$. Choose some $B \in \Sigma^*$ with $A \subseteq B$ and $\nu^*(B) = \nu^*(A)$. Since $\nu^*(B \setminus A) = 0$, it follows from the preceding discussion that $\mu^*\big(T^{-1}(B \setminus A)\big) = 0$. Thus, $T^{-1}(A) = T^{-1}(B) \setminus T^{-1}(B \setminus A)$ is μ-measurable, and moreover,

$$\int_Y \chi_A \, d\nu = \nu^*(A) = \mu^*\big(T^{-1}(A)\big) = \int_X \chi_{T^{-1}(A)} \, d\mu = \int_X \chi_A \circ T \, d\mu.$$

It follows that for every ν-step function ϕ we have $\phi \circ T \in L_1(\mu)$, and $\int_Y \phi \, d\nu = \int_X \phi \circ T \, d\mu$. An easy continuity argument can complete the proof.

(c) It follows immediately from part (b) and Problem 37.9.

Problem 37.14. *Let $(X, \mathcal{S}, \mu)$ be a σ-finite measure space, and let g be a measurable function. Show that if for some $1 \le p < \infty$ we have $fg \in L_1(\mu)$ for all $f \in L_p(\mu)$, then $g \in L_q(\mu)$, where $\frac{1}{p} + \frac{1}{q} = 1$.*

Also, show by a counterexample that for $1 < p < \infty$ the σ-finiteness of μ cannot be dropped.

Solution. We can assume that $g \ge 0$ holds (why?). Then the formula $F(f) = \int fg \, d\mu$ for $f \in L_p(\mu)$ defines a positive linear functional on $L_p(\mu)$. By Theorem 40.10, F is continuous. Now, by Theorems 37.9 and 37.10 there exists some $h \in L_q(\mu)$ such that $\int fg \, d\mu = \int fh \, d\mu$ for each $f \in L_p(\mu)$. This implies (how?) $\int_A (g - h) \, d\mu = 0$ for each measurable subset A. Now, a glance at Problem 22.13 guarantees that $g = h$ a.e. holds.

The σ-finiteness of μ cannot be dropped. Consider $X = (0, \infty)$ with the measure μ defined on the σ-algebra $\mathcal{P}(X)$ by $\mu(A) = \infty$ if $A \ne \emptyset$ and $\mu(\emptyset) = 0$. Then for $1 < p < \infty$ we have $L_1(\mu) = L_p(\mu) = L_q(\mu) = \{0\}$, and $L_\infty(\mu) = B(X)$, the bounded real-valued functions on X. On the other hand, if $g(x) = x$, then $fg = 0 \in L_1(\mu)$ holds for all $f \in L_p(\mu)$ ($1 \le p < \infty$), while $g \notin L_q(\mu)$.

Problem 37.15. *Let $(X, \mathcal{S}, \mu)$ be a σ-finite measure space, g a measurable function, and $1 \le p < \infty$. Assume that there exists some real number $M > 0$ such that $\phi g \in L_1(\mu)$ and $\int \phi g \, d\mu \le M \|\phi\|_p$ holds for every step function ϕ. Then, show that:*

a. $g \in L_q(\mu)$, where $\frac{1}{p} + \frac{1}{q} = 1$, and
b. $\int fg \, d\mu \leq M\|f\|_q$ holds for all $f \in L_p(\mu)$.

Solution. Let L denote the vector space of all step functions. The given conditions show that the function $F: L \longrightarrow \mathbf{R}$, defined by $F(\phi) = \int \phi g \, d\mu$, is a continuous linear functional. Since L is dense in $L_p(\mu)$ (Theorem 31.10), it follows that F has a continuous extension (which we shall denote by F again) to all of $L_p(\mu)$. By Theorems 37.9 and 37.10 there exists some $h \in L_q(\mu)$ such that $F(f) = \int fh \, d\mu$ holds for all $f \in L_p(\mu)$. Clearly,

$$\left| F(f) \right| = \left| \int fh \, d\mu \right| \leq M\|f\|_p$$

holds for all $f \in L_p(\mu)$.

To complete the proof, it suffices to show that $g = h$ a.e. holds. To see this, let $E \in \Lambda_\mu$ satisfy $\mu^*(E) < \infty$. Then, consider the step function $\phi = \chi_E \operatorname{Sgn}(g - h) \in L$, and note that $\int \phi(g - h) \, d\mu = 0$ implies $\int_E |g - h| \, d\mu = 0$. That is, $g = h$ a.e. holds on E; see Problem 22.13. Since μ is σ-finite, we see that $g = h$ a.e. holds on X.

Problem 37.16. *Let μ be a Borel measure on $\mathbf{R}^k$ and suppose that there exists a constant $c > 0$ such that whenever a Borel set E satisfies $\lambda(E) = c$, then $\mu(E) = c$. Show that μ coincides with λ, i.e., show that $\mu = \lambda$.*

Solution. Assume that the Borel measure μ and the constant $c > 0$ satisfy the properties of the problem. Clearly, μ is a σ-finite Borel measure. By Theorem 37.7, we can write

$$\mu = \mu_1 + \mu_2, \quad \text{where } \mu_1 \ll \lambda \text{ and } \mu_2 \perp \lambda.$$

First, we shall establish that $\mu_2 = 0$. From $\mu_2 \perp \lambda$, there exist two disjoint Borel sets A and B with $A \cup B = \mathbf{R}^k$ and $\mu_2(A) = \lambda(B) = 0$. Since $\lambda(A) = \infty$, there exists (by Problem 18.19) a Borel subset C of A with $\lambda(C) = c$. From $\lambda(C \cup B) = \lambda(C) + \lambda(B) = \lambda(C) = c$ and our hypothesis, we see that $\mu(C \cup B) = c$. Now, note that

$$c \leq c + \mu_2(B) \leq c + \mu(B) = \mu(C) + \mu(B) = \mu(C \cup B) = c,$$

and so $\mu_2(B) = 0$. This shows that $\mu_2 = 0$, and consequently $\mu = \mu_1$ is absolutely continuous with respect to λ.

Next, fix a compact set K with $\lambda(K) \geq c$ and consider both μ and λ restricted to K. By the Radon–Nikodym Theorem, there exists a non-negative function $f \in L_1(K, \mathcal{B}, \lambda)$ such that

$$\mu(E) = \int_E f \, d\lambda$$

holds for each Borel subset E of K (see Problem 12.13). We claim that $f = 1$ a.e. To establish this, assume by way of contradiction that the Lebesgue measurable set $D = \{x \in K \colon f(x) < 1\}$ satisfies $\lambda(D) > 0$; we can assume (why?) that D is a Borel set. If $\lambda(D) \geq c$ holds, then pick a Borel subset D_1 of D with $\lambda(D_1) = c$; if $\lambda(D) < c$, then pick a Borel set D_1 with $D \subseteq D_1 \subseteq K$ and $\lambda(D_1) = c$; (see Problem 18.19). Now, note that in either case, we have $\mu(D_1) < c$, which contradicts our hypothesis. Hence, $\lambda(D) = 0$. Similarly, $\lambda(\{x \in K \colon f(x) > 1\}) = 0$, and so $f = 1$ a.e. Therefore, $\mu(E) = \lambda(E)$ holds for each Borel subset E of K. Now, pick a sequence $\{K_n\}$ of compact subsets of $\mathbf{R}^k$ with $\lambda(K_n) \geq c$ and $K_n \uparrow \mathbf{R}^k$. If E is an arbitrary Borel subset of $\mathbf{R}^k$, then note that

$$\mu(E) = \lim_{n \to \infty} \mu(E \cap K_n) = \lim_{n \to \infty} \lambda(E \cap K_n) = \lambda(E).$$

Problem 37.17. *Let μ and ν be two σ-finite measures on a σ-algebra Σ of subsets of a set X such that $\nu \ll \mu$ and $\nu \neq 0$. Show that there exist a set $E \in \Sigma$ and an integer n such that*

   a. $\nu(E) > 0$; and

   b. $A \in \Sigma$ and $A \subseteq E$ imply $\frac{1}{n}\mu(A) \leq \nu(A) \leq n\mu(A)$.

Solution. Pick a sequence $\{X_n\}$ of Σ with $X = \bigcup_{n=1}^{\infty} X_n$, $\nu(X_n) < \infty$, and $\mu(X_n) < \infty$ for each n. Since $\nu \neq 0$, there exists some n such that $\nu(X_n) > 0$. From $\nu \ll \mu$, it follows that $\mu(X_n) > 0$ also holds. Thus, replacing X by X_n, we can assume from the outset that both ν and μ are finite measures.

Now, by the Radon–Nikodym Theorem, there exists a function $0 \leq f \in L_1(\mu)$ such that

$$\nu(A) = \int_A f \, d\mu$$

holds for each $A \in \Sigma$. From $\nu \neq 0$, we see that $f \neq 0$, and so the μ-measurable set $F = \{x \in X \colon f(x) > 0\}$ satisfies $\mu^*(F) > 0$. Next, put

$$E_n = \left\{x \in X \colon \tfrac{1}{n} \leq f(x) \leq n\right\}$$

and note that $E_n \uparrow E$ a.e. Thus, for some n, we have $\mu^*(E_n) > 0$. By Theorem 15.11, there exists some $E \in \Sigma$ with $E_n \subseteq E$ and $\mu(E) = \mu^*(E_n)$. We claim that the set E satisfies the desired properties.

To see this, note first that

$$\nu(E) = \int_E f \, d\mu = \int_{E_n} f \, d\mu \geq \tfrac{1}{n}\mu^*(E_n) > 0.$$

Now, if $A \in \Sigma$ satisfies $A \subseteq E$, then note that $\frac{1}{n}\chi_A \leq f \leq n\chi_A$ μ – a.e., and consequently

$$\tfrac{1}{n}\mu(A) = \tfrac{1}{n}\int_A \chi_A \, d\mu \leq \int_A f \, d\mu = \nu(A) \leq \int_A n\chi_A \, d\mu = n\mu(A).$$

Problem 37.18. *Let μ be a finite Borel measure on $[1, \infty)$ such that*

a. $\mu \ll \lambda$, *and*
b. $\mu(B) = a\mu(aB)$ *for each $a \geq 1$ and each Borel subset B of $[1, \infty)$, where $aB = \{ab\colon b \in B\}$.*

If the Radon–Nikodym derivative $d\mu/d\lambda$ is a continuous function, then show that there exists a constant $c \geq 0$ such that $[d\mu/d\lambda](x) = \frac{c}{x^2}$ for each $x \geq 1$.

Solution. For simplicity, let us write $\frac{d\mu}{d\lambda} = f$. Then, the given identity $\mu(B) = a\mu(aB)$ can be written in the form

$$\int_B f \, d\lambda = a \int_{aB} f \, d\lambda.$$

For $B = [1, x]$, we get

$$\int_1^x f(t)\,dt = a \int_a^{ax} f(t)\,dt$$

for each $a \geq 1$ and each $x \geq 1$. Differentiating with respect to x (and taking into account the Fundamental Theorem of Calculus), we see that

$$f(x) = a^2 f(ax)$$

holds for each $x \geq 1$ and each $a \geq 1$. Letting $x = 1$, we obtain

$$f(a) = \frac{f(1)}{a^2}$$

for all $a \geq 1$, and our conclusion follows.

Problem 37.19. *Let μ be a finite Borel measure on $(0, \infty)$ such that*

a. $\mu \ll \lambda$, *and*

b. $\mu(aB) = \mu(B)$ *for each $a > 0$ and each Borel subset B of $(0, \infty)$.*

If the Radon–Nikodym derivative is a continuous function, then show that there exists a constant $c \geq 0$ such that $[d\mu/d\lambda](x) = \frac{c}{x}$ for each $x > 0$.

Solution. Let $\frac{d\mu}{d\lambda} = f$. Then, (by The Radon–Nikodym Theorem) the given identity $\mu(B) = \mu(aB)$ can be written in the form

$$\int_B f \, d\lambda = \int_{aB} f \, d\lambda.$$

For $B = [1, x]$ (put $B = [x, 1]$ if $0 < x < 1$), we get

$$\int_1^x f(t) \, dt = \int_a^{ax} f(t) \, dt$$

for each $a > 0$ and each $x > 0$. Differentiating with respect to x (and taking into account the Fundamental Theorem of Calculus), we see that

$$f(x) = af(ax)$$

holds for each $x > 0$ and each $a > 0$. Letting $x = 1$, we obtain

$$f(a) = \frac{f(1)}{a}$$

for all $a > 0$, as desired.

38. THE RIESZ REPRESENTATION THEOREM

Problem 38.1. *If X is a compact topological space, then show that a continuous linear functional F on $C(X)$ is positive if and only if $F(\mathbf{1}) = \|F\|$ holds.*

Solution. Let F be a continuous linear functional on $C(X)$, where X is compact. Note first that

$$\{f \in C(X): \|f\|_\infty \leq 1\} = \{f \in C(X): |f| \leq 1\}.$$

Thus, if F is also positive, then

$$\begin{aligned}
\|F\| &= \sup\{F(f): f \in C(X) \text{ and } \|f\|_\infty \leq 1\} \\
&= \sup\{F(f): f \in C(X) \text{ and } |f| \leq 1\} = F(\mathbf{1}).
\end{aligned}$$

On the other hand, assume $F(1) = \|F\|$. Let $0 \le f \in C(X)$ be nonzero, and put $g = \frac{f}{\|f\|_\infty}$. Clearly, $\|1 - g\|_\infty \le 1$. Thus,

$$F(1) - F(g) = F(1 - g) \le \|F\| = F(1)$$

holds, which implies $F(g) \ge 0$. Therefore, $F(f) = \|f\|_\infty F(g) \ge 0$ holds, and so F is a positive linear functional.

Problem 38.2. *Let X be a compact topological space, and let F and G be two positive linear functionals on $C(X)$. If $F(1) + G(1) \le \|F - G\|$, then show that $F \wedge G = 0$.*

Solution. Since $F, G \ge 0$, it follows that $F - G \le F \vee G$ and $G - F \le F \vee G$, and so $|F - G| \le F \vee G$. Thus, by the preceding problem

$$\begin{aligned}
\|F - G\| &\le \|F \vee G\| = F \vee G(1) \le \|F + G\| \le \|F\| + \|G\| \\
&= F(1) + G(1) \le \|F - G\|,
\end{aligned}$$

and hence, $F \vee G(1) = F(1) + G(1)$ holds. From $F + G = F \vee G + F \wedge G$, it follows that

$$\|F \wedge G\| = F \wedge G(1) = F(1) + G(1) - F \vee G(1) = 0,$$

and so $F \wedge G = 0$.

Problem 38.3. *Let X be a Hausdorff locally compact topological space and let*

$$c_0(X) = \left\{ f \in C(X) \colon \forall \epsilon > 0 \; \exists K \text{ compact with } |f(x)| < \epsilon \; \forall x \notin K \right\}.$$

Show that:

a. *$c_0(X)$ equipped with the sup norm is a Banach lattice.*
b. *The norm completion of $C_c(X)$ is the Banach lattice $c_0(X)$.*

Solution. (a) Clearly, $c_0(X)$ with the sup norm is a normed vector lattice. For the completeness, let $\{f_n\}$ be a Cauchy sequence of $c_0(X)$. Then $\{f_n\}$ converges uniformly on X to some function f. By Theorem 9.2 we infer that $f \in C(X)$. Now, if $\varepsilon > 0$ is given, pick some n with $\|f_n - f\|_\infty < \varepsilon$, and then choose some compact set K with $|f_n(x)| < \varepsilon$ for $x \notin K$. Thus,

$$|f(x)| \le |f_n(x) - f(x)| + |f_n(x)| < \varepsilon + \varepsilon = 2\varepsilon$$

holds for all $x \notin K$, so that $f \in c_0(X)$.

(b) Obviously, $C_c(X)$ is a vector sublattice of $c_0(X)$. We have to show that $C_c(X)$ is dense in $c_0(X)$.

To this end, let $f \in c_0(X)$ and let $\varepsilon > 0$. Pick some compact set K with $|f(x)| < \varepsilon$ for $x \notin K$, and then choose some open set V with compact closure such that $K \subseteq V$. By Theorem 10.8 there exists a function $g: X \longrightarrow \mathbf{R}$ with $K \prec g \prec V$. Then, $fg \in C_c(X)$ and $\|f - fg\|_\infty \leq 2\varepsilon$ holds, proving that $C_c(X)$ is dense in the Banach lattice $c_0(X)$.

Problem 38.4. *Let F be a positive linear functional on $C_c(X)$, where X is Hausdorff and locally compact, and let μ be the outer measure induced by F on X. Show that if μ^* is the outer measure generated by the measure space $(X, \mathcal{B}, \mu)$, then $\mu^*(A) = \mu(A)$ holds for every subset A of X.*

Solution. Let $A \subseteq X$. We know that

$$\mu(A) = \inf\{\mu(V): V \text{ open and } A \subseteq V\}.$$

So, if $A \subseteq V$ holds with V open, then $\mu^*(A) \leq \mu^*(V) = \mu(V)$ also holds, and thus $\mu^*(A) \leq \mu(A)$. On the other hand, by Theorem 15.11 there exists some $B \in \mathcal{B}$ with $A \subseteq B$ and $\mu^*(A) = \mu(B)$. Thus, $\mu(A) \leq \mu(B) = \mu^*(A)$, proving that $\mu^*(A) = \mu(A)$ holds.

Problem 38.5. *Let μ and ν be two regular Borel measures on a Hausdorff locally compact topological space X. Then show that $\mu \geq \nu$ holds if and only if $\int f \, d\mu \geq \int f \, d\nu$ for each $f \in C_c(X)^+$.*

Solution. Let μ and ν be two regular Borel measures on a Hausdorff locally compact topological space X.

Assume first that $\mu \geq \nu$ holds (i.e., assume that $\mu(A) \geq \nu(A)$ holds for each $A \in \mathcal{B}$). Clearly, if ϕ is a μ-step function of the form $\phi = \sum_{i=1}^n a_i \chi_{A_i}$ with each $a_i \geq 0$ and each $A_i \in \mathcal{B}$, then ϕ is a ν-step function and $\int \phi \, d\mu \geq \int \phi \, d\nu$ holds. Now, let $0 \leq f \in C_c(X)$. Since

$$f^{-1}\big([a, b)\big) = \big[f^{-1}\big((-\infty, a)\big)\big]^c \cap f^{-1}\big((-\infty, b)\big) \in \mathcal{B},$$

it follows from Theorem 17.7 that there exists a sequence $\{\phi_n\}$ of μ-step functions of the preceding type satisfying $\phi_n(x) \uparrow f(x)$ for each $x \in X$. This implies

$$\int f \, d\mu = \lim_{n \to \infty} \int \phi_n \, d\mu \geq \lim_{n \to \infty} \int \phi_n \, d\nu = \int f \, d\nu.$$

For the converse, assume that $\int f \, d\mu \geq \int f \, d\nu$ holds for each $0 \leq f \in C_c(X)$. In view of the regularity of the measures, in order to establish that $\mu \geq \nu$ holds it suffices to show that $\mu(K) \geq \nu(K)$ holds for each compact set K. To this end, let K be a compact set. Given $\varepsilon > 0$, choose an open set V such that $K \subseteq V$ and $\mu(V) < \mu(K) + \varepsilon$. By Theorem 10.8 there exists a function $f \in C_c(X)$ such that $K \prec f \prec V$. Now note that from $\chi_K \leq f \leq \chi_V$, it follows that

$$v(K) = \int \chi_K \, dv \leq \int f \, dv \leq \int f \, d\mu \leq \int \chi_V \, d\mu = \mu(V) < \mu(K) + \varepsilon$$

for all $\varepsilon > 0$. That is, $v(K) \leq \mu(K)$ holds, as desired.

Problem 38.6. *Fix a point x in a Hausdorff locally compact topological space X, and define $F(f) = f(x)$ for each $f \in C_c(X)$. Show that F is a positive linear functional on $C_c(X)$ and then describe the unique regular Borel measure μ that satisfies $F(f) = \int f \, d\mu$ for each $f \in C_c(X)$. What is the support of μ?*

Solution.   Clearly, F is a positive linear functional. The regular Borel measure representing F is the Dirac measure with "base point" at x. Its support is, of course, the set $\{x\}$.

Problem 38.7.   *Let X be a compact Hausdorff topological space. If μ and v are regular Borel measures, then show that the regular Borel measures $\mu \vee v$ and $\mu \wedge v$ satisfy*
  a. $\mathrm{Supp}(\mu \vee v) = \mathrm{Supp}\,\mu \cup \mathrm{Supp}\,v$, *and*
  b. $\mathrm{Supp}(\mu \wedge v) \subseteq \mathrm{Supp}\,\mu \cap \mathrm{Supp}\,v$.
Use (b) to show that if $\mathrm{Supp}\,\mu \cap \mathrm{Supp}\,v = \emptyset$, then $\mu \perp v$ holds. Also, give an example for which $\mathrm{Supp}(\mu \wedge v) \neq \mathrm{Supp}\,\mu \cap \mathrm{Supp}\,v$.

Solution.   (a) Let $A = \mathrm{Supp}(\mu \vee v)$, $B = \mathrm{Supp}\,\mu$, and $C = \mathrm{Supp}\,v$. From $\mu \leq \mu \vee v$, $v \leq \mu \vee v$, and $\mu \vee v(A^c) = 0$, it follows that $\mu(A^c) = v(A^c) = 0$, and so $B \subseteq A$ and $C \subseteq A$. That is, $B \cup C \subseteq A$. On the other hand, the inequality $\mu \vee v \leq \mu + v$ implies

$$\mu \vee v \left(B^c \cap C^c\right) \leq \left(\mu + v\right)\left(B^c \cap C^c\right) \leq \mu\left(B^c\right) + v\left(C^c\right) = 0,$$

and so $A \subseteq \left(B^c \cap C^c\right)^c = B \cup C$.
(b) The inclusion follows easily from the inequalities

$$\mu \wedge v \leq \mu \quad \text{and} \quad \mu \wedge v \leq v.$$

If $\mathrm{Supp}\,\mu \cap \mathrm{Supp}\,v = \emptyset$, then by part (b) $\mathrm{Supp}(\mu \wedge v) = \emptyset$ holds, and so $\mu \wedge v = 0$. For an example showing that equality need not hold in (b), let $X = \mathbf{R}$, $\mu = $ the Lebesgue measure, and $v = $ the Dirac measure with "base point" at 0.

By Problem 36.5, we have $\mu \wedge \nu = 0$. Therefore, $\mathrm{Supp}(\mu \wedge \nu) = \emptyset$ holds, while $\mathrm{Supp}\,\mu \cap \mathrm{Supp}\,\nu = \mathbb{R} \cap \{0\} = \{0\}$.

Problem 38.8. *Let X be a Hausdorff locally compact topological space X. Characterize the positive linear functionals F on $C_c(X)$ that are also lattice homomorphisms; that is, $F(f \vee g) = \max\{F(f), F(g)\}$ holds for each pair $f, g \in C_c(X)$.*

Solution. Let F be a positive linear functional on $C_c(X)$. Then we shall show that F is a lattice homomorphism if and only if there exist some $c \geq 0$ and some $a \in X$ such that $F(f) = cf(a)$ holds for all $f \in C_c(X)$.

Clearly, if for some $c \geq 0$ and some $a \in X$ we have $F(f) = cf(a)$ for each $f \in C_c(X)$, then F is a lattice homomorphism. For the converse, assume that F is a non-zero lattice homomorphism. Let μ be the regular Borel measure that represents F. If $x, y \in \mathrm{Supp}\,\mu$ satisfy $x \neq y$, then it is not difficult to see that there exist f, g in $C_c(X)$ with $f \wedge g = 0$ and $f(x) = g(y) = 1$. Therefore,

$$F(f \vee g) = F(f + g) = F(f) + F(g) > \max\{F(f), F(g)\}$$

must hold, which is a contradiction. Thus, $\mathrm{Supp}\,\mu$ consists precisely of one point; let $\mathrm{Supp}\,\mu = \{a\}$. Set $c = \mu(\{a\}) > 0$, and note that for every $f \in C_c(X)$, we have

$$F(f) = \int f \, d\mu = f(a) \cdot \mu(\{a\}) = cf(a).$$

Problem 38.9. *Let X be a Hausdorff locally compact topological space such that X is an uncountable set. Then show that*

    a. $C_c^*(X)$ *is not separable, and*

    b. $C[0, 1]$ *(with the sup norm) is not a reflexive Banach space.*

Solution. (a) For each $x \in X$ define the positive linear functional $F_x \colon C_c(X) \longrightarrow \mathbb{R}$ by $F_x(f) = f(x)$ and note that $\|F_x - F_y\| = 2$ holds for $x \neq y$. Clearly, the set $\{F_x \colon x \in X\}$ is an uncountable subset of $C_c^*(X)$. Therefore, $\{B(F_x, 1) \colon x \in X\}$ is an uncountable collection of pairwise disjoint open balls. From this, it easily follows that no countable subset of $C_c^*(X)$ can be dense in $C_c^*(X)$.

(b) By Problem 11.12, we know that $C[0, 1]$ is separable. If $C[0, 1]$ is reflexive, then its second dual is likewise separable. But then (by Problem 29.8) its first dual must be separable, contradicting part (a). Thus, $C[0, 1]$ is not a reflexive Banach lattice.

Problem 38.10. *Let X be a Hausdorff locally compact topological space. For a finite signed measure μ on $\mathcal{B}$ show that the following statements are equivalent:*

   a.   *μ belongs to $M_b(X)$.*

   b.   *μ^+ and μ^- are both finite regular Borel measures.*

   c.   *For each $A \in \mathcal{B}$ and $\epsilon > 0$, there exist a compact set K and an open set V with $K \subseteq A \subseteq V$ such that $|\mu(B)| < \epsilon$ holds for all $B \in \mathcal{B}$ with $B \subseteq V \setminus K$.*

Solution. (a) $\Longrightarrow$ (b) Pick two finite regular Borel measures μ_1 and μ_2 such that $\mu = \mu_1 - \mu_2$. Then, $\mu^+ = (\mu_1 - \mu_2)^+ = \mu_1 \vee \mu_2 - \mu_2$ holds. By Theorem 38.5, $\mu_1 \vee \mu_2$ is a finite regular Borel measure, and from this it follows that μ^+ is a finite regular Borel measure. Similarly, μ^- is a finite regular Borel measure.

(b) $\Longrightarrow$ (c) Note that $|\mu| = \mu^+ + \mu^-$ is a finite regular Borel measure. Now, let $A \in \mathcal{B}$ and let $\varepsilon > 0$ be given. Then, there exists a compact set K and an open set V with $K \subseteq A \subseteq V$ and $|\mu|(V \setminus K) < \varepsilon$. Therefore, if $B \in \mathcal{B}$ satisfies $B \subseteq V \setminus K$, then $|\mu(B)| \leq |\mu|(B) \leq |\mu|(V \setminus K) < \varepsilon$ holds.

(c) $\Longrightarrow$ (a) Let $A \in \mathcal{B}$ and let $\varepsilon > 0$. Choose a compact set K and an open set V so that (c) is satisfied. Then, by Theorem 36.9, we have

$$0 \leq \mu^+(A) - \mu^+(K) = \mu^+(A \setminus K) = \sup\{\mu(B) : B \in \mathcal{B} \text{ and } B \subseteq A \setminus K\}$$

and

$$0 \leq \mu^+(V) - \mu^+(A) = \mu^+(V \setminus A) = \sup\{\mu(B) : B \in \mathcal{B} \text{ and } B \subseteq V \setminus A\}.$$

Thus, $\mu^+(A) - \mu^+(K) \leq \varepsilon$ and $\mu^+(V) - \mu^+(A) \leq \varepsilon$ both hold. Hence, μ^+ is a finite regular Borel measure. Similarly, μ^- is a finite regular Borel measure, and so $\mu = \mu^+ - \mu^- \in M_b(X)$.

Problem 38.11. *A sequence $\{x_n\}$ in a normed space is said to **converge weakly** to some vector x if $\lim f(x_n) = f(x)$ holds for every continuous linear functional f.*

   a.   *Show that a sequence in a normed space can have at most one weak limit.*

   b.   *Let X be a Hausdorff compact topological space. Then show that a sequence $\{f_n\}$ of $C(X)$ converges weakly to some function f if and only if $\{f_n\}$ is norm bounded and $\lim f_n(x) = f(x)$ holds for each $x \in X$.*

Solution. (a) Assume that a sequence $\{x_n\}$ in a normed vector space Y satisfies $\lim f(x_n) = f(x)$ and $\lim f(x_n) = f(y)$ for every $f \in Y^*$. Then, $f(x - y) = 0$ holds for all $f \in Y^*$. By Theorem 29.4, we see that $x - y = 0$, and so $\{x_n\}$ can have at most one weak limit.

(b) Assume first that the sequence $\{f_n\}$ of $C(X)$ converges weakly to some function $f \in C(X)$. By Theorem 29.8, $\{f_n\}$ is norm bounded. If $x \in X$, let μ_x denote the Dirac measure with support $\{x\}$, and note that

$$f_n(x) = \int f_n \, d\mu_x \longrightarrow \int f \, d\mu_x = f(x).$$

Conversely, if $\{f_n\}$ is norm bounded and $\lim f_n(x) = f(x)$ holds for each $x \in X$ (where, of course, $f \in C(X)$), then the Lebesgue Dominated Convergence Theorem implies that $\lim \int f_n \, d\mu = \int f \, d\mu$ holds for every Borel measure μ. This, coupled with the Riesz Representation Theorem, shows that $\{f_n\}$ converges weakly to f.

Problem 38.12. *Let μ be a regular Borel measure on a Hausdorff locally compact topological space X, and let $f \in L_1(\mu)$. Show that the finite signed measure ν, defined by*

$$\nu(E) = \int_E f \, d\mu$$

for each Borel set E, is a (finite) regular Borel signed measure. In other words, show that $\nu \in M_b(X)$.

Solution. We can assume that $f(x) \geq 0$ holds for all x. By Problem 22.7, the set $A = \{x \in X \colon f(x) > 0\}$ is a σ-finite set with respect to μ. Choose a sequence $\{X_n\}$ of μ-measurable sets with $\mu(X_n) < \infty$ for each n and $X_n \uparrow A$.

Now, let E be a Borel set and let $\varepsilon > 0$; clearly, $\nu(E) = \nu(E \cap A)$. Select some n with $\nu(E) - \nu(X_n \cap E) < \varepsilon$. Also, using the regularity of μ and Problem 22.6, we see that there exists a compact set $K \subseteq X_n \cap E$ with

$$\nu(X_n \cap E) - \nu(K) = \int_{X_n \cap E} f \, d\mu - \int_K f \, d\mu < \varepsilon.$$

Thus, the compact set $K \subseteq E$ satisfies

$$0 \leq \nu(E) - \nu(K) = \big[\nu(E) - \nu(X_n \cap E)\big] + \big[\nu(X_n \cap E) - \nu(K)\big] < 2\varepsilon.$$

Next, use Problem 22.6 and the regularity of μ to see that for each n there exists an open set V_n satisfying $X_n \cap E \subseteq V_n$ and $\nu(V_n) - \nu(X_n \cap E) < \frac{\varepsilon}{2^n}$. Then, the open set $V = \bigcup_{n=1}^{\infty} V_n$ satisfies $E \subseteq V$, and in view of $V \setminus E \subseteq \bigcup_{n=1}^{\infty}(V_n \setminus X_n \cap E)$, we see that

$$0 \leq \nu(V) - \nu(E) = \nu(V \setminus E) \leq \sum_{n=1}^{\infty} \nu(V_n \setminus X_n \cap E) < \varepsilon.$$

Altogether, the preceding show that ν is a regular Borel measure.

Problem 38.13.   *Generalize part (3) of Theorem 38.5 as follows: If μ and ν are two regular Borel measures on a Hausdorff locally compact topological space and one of then is σ-finite, then show that $\mu \wedge \nu$ is also a regular Borel measure.*

Solution.   Let μ and ν be two regular Borel measures on a locally compact Hausdorff topological space X and assume that μ is σ-finite. Also, let $\omega = \mu \wedge \nu$ and note (in view of $\omega \le \mu$) that ω is a σ-finite Borel measure which is absolutely continuous with respect to μ.

Now, let E be a Borel subset of X satisfying $\mu(E) < \infty$ and let $\varepsilon > 0$. Consider ω and μ restricted to the Borel sets $\mathcal{B}_E$ of E (from Problem 12.13 we know that $\mathcal{B}_E = \{B \cap E \colon B \in \mathcal{B}\}$). Now, by the Radon–Nikodym Theorem there exists a (unique) non-negative function $f \in L_1(E, \mathcal{B}_E, \mu)$ satisfying

$$\omega(B \cap E) = \int_{B \cap E} f \, d\mu, \quad \text{for each} \quad B \in \mathcal{B}.$$

Since μ is a regular Borel measure, it follows from Problem 22.6 that there exists a compact subset K of E such that

$$0 \le \omega(E) - \omega(K) = \int_E f \, d\mu - \int_K f \, d\mu = \int_{E \setminus K} f \, d\mu < \varepsilon.$$

Therefore, we infer that

$$\omega(E) = \sup\{\omega(K) \colon K \text{ compact and } K \subseteq E\}. \qquad (\star)$$

Now, use the σ-finiteness of ω to show that $(\star)$ holds true for each Borel subset of E.

It remains to be shown that the measure of every Borel set can be approximated from above by the measures of the open sets. To this end, let E be an arbitrary Borel set, and recall that

$$\omega(E) = \mu \wedge \nu(E) = \inf\{\mu(B) + \nu(E \setminus B) \colon B \in \mathcal{B} \text{ and } B \subseteq E\}.$$

Let $c = \inf\{\omega(O) \colon O \text{ open and } E \subseteq O\}$ and let $\varepsilon > 0$. Given $B \in \mathcal{B}$ with $B \subseteq E$, choose open sets V and W such that $B \subseteq V$, $E \setminus B \subseteq W$, $\mu(V) \le \mu(B) + \varepsilon$, and $\nu(W) \le \nu(E \setminus B) + \varepsilon$. Then, we have

$$\omega(E) \le c \le \omega(V \cup W) \le \omega(V) + \omega(W) \le \mu(V) + \nu(W)$$
$$\le \mu(B) + \varepsilon + \nu(E \setminus B) + \varepsilon = \mu(B) + \nu(E \setminus B) + 2\varepsilon.$$

Thus, $\omega(E) \le c \le \omega(E) + 2\varepsilon$ holds for each $\varepsilon > 0$, and so $\omega(E) = c$, and we are finished.

Problem 38.14. *Show that every finite Borel measure on a complete separable metric space is a regular Borel measure. Use this conclusion to present an alternate proof of the fact that the Lebesgue measure is a regular Borel measure.*

Solution. Let X be a complete separable metric space, let $\mathcal{B}$ be the σ-algebra of all Borel sets of X, let $\{x_1, x_2, \ldots\}$ be a dense countable subset of X, and let $\mu \colon \mathcal{B} \longrightarrow [0, \infty)$ be a measure.

Consider the collection $\mathcal{A}$ of subsets of X defined by

$$\mathcal{A} = \Big\{ A \in \mathcal{B} \colon \ \mu(A) = \inf\{\mu(O) \colon A \subseteq O \ \text{and} \ O \ \text{open}\}$$
$$= \sup\{\mu(K) \colon K \subseteq A \ \text{and} \ K \ \text{compact}\} \Big\}.$$

The collection $\mathcal{A}$ has the following properties:

1. $\mathcal{A}$ contains the open and closed sets.

To see this, assume first that V is an open set, and let $\varepsilon > 0$. For each n let $\mathcal{F}_n$ be the collection of all open balls of the form $B(x_i, r)$ with r a rational number less than or equal to $\frac{1}{n}$ and $\overline{B(x_i, r)} \subseteq V$. Clearly, each $\mathcal{F}_n$ is at most countable and $V = \bigcup_{B \in \mathcal{F}_n} B$ holds. For each n pick $B_1^n, \ldots, B_{k_n}^n \in \mathcal{F}_n$ such that

$$\mu\Big(V \setminus \bigcup_{i=1}^{k_n} B_i^n \Big) < \frac{\varepsilon}{2^n}.$$

Next, put $C = \bigcap_{n=1}^{\infty} \bigcup_{i=1}^{k_n} B_i^n$, and note that C is a totally bounded set. Hence, its closure $\overline{C}$ is also a totally bounded set (why?). Since (by Theorem 6.13) $\overline{C}$ is a complete metric space in its own right, it follows from Theorem 7.8 that $\overline{C}$ is a compact set. Now, note that $\overline{C} \subseteq V$ holds, and that

$$0 \le \mu(V) - \mu(\overline{C}) \le \mu(V) - \mu(C) = \mu(V \setminus C)$$
$$= \mu\Big(\bigcup_{n=1}^{\infty} \Big(V \setminus \bigcup_{i=1}^{k_n} B_i^n \Big) \Big) \le \sum_{n=1}^{\infty} \mu\Big(V \setminus \bigcup_{i=1}^{k_n} B_i^n \Big)$$
$$< \sum_{n=1}^{\infty} \frac{\varepsilon}{2^n} = \varepsilon.$$

Therefore,

$$\mu(V) = \inf\{\mu(O) \colon V \subseteq O \ \text{and} \ O \ \text{open}\}$$
$$= \sup\{\mu(K) \colon K \subseteq V \ \text{and} \ K \ \text{compact}\}$$

holds, and so $V \in \mathcal{A}$.

Now, let C be a closed set, and let $\varepsilon > 0$. By the preceding, there exists a compact subset K with $\mu(X) - \mu(K) < \varepsilon$. Then, the compact subset $C \cap K$ of C satisfies

$$\mu(C) - \mu(C \cap K) = \mu(C \setminus C \cap K)$$
$$= \mu(C \setminus K) \leq \mu(X \setminus K) = \mu(X) - \mu(K) < \varepsilon.$$

Also, by the previous part, there exists a compact set K_1 with $K_1 \subseteq X \setminus C$ and $\mu(X \setminus C) - \mu(K_1) < \varepsilon$. Now, the open set $O = X \setminus K_1$ satisfies $C \subseteq O$ and

$$\mu(O) - \mu(C) = \mu(X \setminus K_1) - \mu(C) = \mu(X) - \mu(K_1) - \mu(C)$$
$$= \mu(X \setminus C) - \mu(K_1) < \varepsilon.$$

Thus,

$$\mu(C) = \inf\{\mu(O)\colon C \subseteq O \text{ and } O \text{ open}\}$$
$$= \sup\{\mu(K)\colon K \subseteq C \text{ and } K \text{ compact}\}$$

also holds, and so $C \in \mathcal{A}$.

2. If $A \in \mathcal{A}$, then $A^c \in \mathcal{A}$.

From $\mu(A) = \sup\{\mu(K)\colon K \subseteq A \text{ and } K \text{ compact}\}$, it follows that

$$\mu(A^c) = \mu(X) - \mu(A)$$
$$= \inf\{\mu(X) - \mu(K)\colon K \subseteq A \text{ and } K \text{ compact}\}$$
$$= \inf\{\mu(K^c)\colon K \subseteq A \text{ and } K \text{ compact}\}$$
$$= \inf\{\mu(O)\colon A^c \subseteq O \text{ and } O \text{ open}\}.$$

Similarly, $\mu(A) = \inf\{\mu(O)\colon A \subseteq O \text{ and } O \text{ open}\}$ implies

$$\mu(A^c) = \sup\{\mu(C)\colon C \subseteq A^c \text{ and } C \text{ closed}\}.$$

Since, by part (1), $\mu(C) = \sup\{\mu(K)\colon K \subseteq C \text{ and } K \text{ compact}\}$ holds for each closed set C, we see that

$$\mu(A^c) = \sup\{\mu(K)\colon K \subseteq A^c \text{ and } K \text{ compact}\}.$$

3. If $\{A_n\}$ is a sequence of $\mathcal{A}$, then $\bigcup_{n=1}^{\infty} A_n \in \mathcal{A}$.

Let $\{A_n\} \subseteq \mathcal{A}$, let $A = \bigcup_{n=1}^{\infty} A_n$, and let $\varepsilon > 0$. For each n pick some open set O_n with $A_n \subseteq O_n$ and $\mu(O_n \setminus A_n) < \varepsilon 2^{-n}$. Then, the open set $O = \bigcup_{n=1}^{\infty} O_n$

satisfies $A \subseteq O$ and from $O \setminus A = \bigcup_{n=1}^{\infty} O_n \setminus \bigcup_{n=1}^{\infty} A_n \subseteq \bigcup_{n=1}^{\infty} (O_n \setminus A_n)$, we get

$$\mu(O) - \mu(A) = \mu(O \setminus A) \leq \mu\left(\bigcup_{n=1}^{\infty} (O_n \setminus A_n)\right) \leq \sum_{n=1}^{\infty} \mu(O_n \setminus A_n) < \varepsilon.$$

On the other hand, fix some k with $\mu(A \setminus \bigcup_{i=1}^{k} A_i) < \varepsilon$, and then for each $1 \leq i \leq k$ pick a compact set $K_i \subseteq A_i$ with $\mu(A_i \setminus K_i) < \varepsilon 2^{-i}$. Then, the compact set $K = \bigcup_{i=1}^{k} K_i$ satisfies $K \subseteq \bigcup_{i=1}^{k} A_i \subseteq A$ and

$$\mu(A) - \mu(K) = \mu(A \setminus K) = \mu\left(A \setminus \bigcup_{i=1}^{k} A_i\right) + \mu\left(\left(\bigcup_{i=1}^{k} A_i\right) \setminus K\right)$$

$$< \varepsilon + \mu\left(\left(\bigcup_{i=1}^{k} A_i\right) \setminus K\right) = \varepsilon + \mu\left(\bigcup_{i=1}^{k} A_i \setminus \bigcup_{i=1}^{k} K_i\right)$$

$$\leq \varepsilon + \sum_{i=1}^{k} \mu(A_i \setminus K_i) < \varepsilon + \varepsilon = 2\varepsilon.$$

The validity of statement (3) has been established.

Now, from the preceding statements, we see that $\mathcal{A}$ is a σ-algebra that contains the open sets. Consequently, every Borel set belongs to $\mathcal{A}$ (i.e., $\mathcal{A} = \mathcal{B}$), and so μ is a regular Borel measure.

Now, let us use the previous conclusion to establish that the Lebesgue measure λ on $\mathbb{R}^n$ is a regular Borel measure. To this end, let A be an arbitrary Borel set. Also, let V_n (resp. C_n) denote the open (resp. the closed) ball of $\mathbb{R}^n$ with center at zero and radius n. Since each C_n is a complete separable metric space in its own right, it follows from the previous result that λ restricted to each C_n is a regular Borel measure. Therefore, we have

$$\lambda(A \cap C_n) = \sup\{\lambda(K): \; K \subseteq A \cap C_n \text{ and } K \text{ compact}\}.$$

From $A \cap C_n \uparrow A$, it follows that $\lambda(A \cap C_n) \uparrow \lambda(A)$, and an easy argument shows that

$$\lambda(A) = \sup\{\lambda(K): \; K \subseteq A \text{ and } K \text{ compact}\}.$$

Next, note that if $\lambda(A) = \infty$, then

$$\lambda(A) = \inf\{\lambda(O): \; A \subseteq O \text{ and } O \text{ open}\}$$

is trivially true. So, assume that $\lambda(A) < \infty$, and let $\varepsilon > 0$. By the regularity of λ on C_n (and the fact that V_n is an open set), we see that

$$
\begin{aligned}
\lambda(A \cap V_n) &= \inf\{\lambda(O \cap C_n):\ V_n \cap A \subseteq O \cap C_n \text{ and } O \text{ open}\} \\
&= \inf\{\lambda(O \cap V_n):\ V_n \cap A \subseteq O \cap V_n \text{ and } O \text{ open}\} \\
&= \inf\{\lambda(O):\ A \cap V_n \subseteq O \text{ and } O \text{ open}\}.
\end{aligned}
$$

Therefore, for each n there exists an open set O_n with $A \cap V_n \subseteq O_n$ and $\lambda(O_n \setminus A \cap V_n) < \varepsilon 2^{-n}$. Now, the set $O = \bigcup_{n=1}^{\infty} O_n$ is open and satisfies $A \subseteq O$. From

$$
O \setminus A = \bigcup_{n=1}^{\infty} O_n \setminus \bigcup_{n=1}^{\infty} A \cap V_n \subseteq \bigcup_{n=1}^{\infty}(O_n \setminus A \cap V_n),
$$

we see that

$$
0 \leq \lambda(O) - \lambda(A) = \lambda(O \setminus A) \leq \sum_{n=1}^{\infty} \lambda(O_n \setminus A \cap V_n) < \varepsilon.
$$

Hence, $\lambda(A) = \inf\{\lambda(O):\ A \subseteq O \text{ and } O \text{ open}\}$ also holds, and so the Lebesgue measure λ is a regular Borel measure.

Problem 38.15. *Let X be a Hausdorff compact topological space. If $\phi: X \to X$ is a continuous function, then show that there exists a regular Borel measure on X such that*

$$
\int f \circ \phi \, d\mu = \int f \, d\mu
$$

holds for each $f \in C(X)$.

Solution. Let X be a Hausdorff compact topological space and let $\phi: X \to X$ be a continuous function. Fix some $\omega \in X$ and let $\mathcal{L}im: \ell_\infty \to \ell_\infty$ is a Banach–Mazur limit (see Problem 29.7). Now, consider the positive linear functional $F: C(X) \to \mathbb{R}$ defined by

$$
F(f) = \mathcal{L}im(f(\phi(\omega)),\ f(\phi^2(\omega)),\ f(\phi^3(\omega)), \ldots),
$$

and let μ be the regular Borel measure on X representing F, i.e., $F(f) = \int f \, d\mu$

holds for each $f \in C(X)$. The identity

$$\mathcal{L}im(x_1, x_2, \ldots) = \mathcal{L}im(x_2, x_3, \ldots)$$

for all $(x_1, x_2, \ldots) \in \ell_\infty$ easily implies $F(f) = F(f \circ \phi)$ for each $f \in C(X)$. Consequently, the regular Borel measure μ satisfies $\int f \, d\mu = \int f \circ \phi \, d\mu$ for each $f \in C(X)$.

Problem 38.16. *This exercise gives an identification of the order dual $C_c^\sim(X)$ of $C_c(X)$. Consider the collection $\mathcal{M}(X)$ of all formal expressions $\mu_1 - \mu_2$ with μ_1 and μ_2 regular Borel measures. That is,*

$$\mathcal{M}(X) = \{\mu_1 - \mu_2 \colon \mu_1 \text{ and } \mu_2 \text{ are regular Borel measures on } X\}.$$

a. *Define $\mu_1 - \mu_2 \equiv \nu_1 - \nu_2$ in $\mathcal{M}(X)$ to mean $\mu_1(A) + \nu_2(A) = \nu_1(A) + \mu_2(A)$ for all $A \in \mathcal{B}$. Show that $\equiv$ is an equivalence relation.*
b. *Denote the collection of all equivalence classes by $\mathcal{M}(X)$ again. That is, $\mu_1 - \mu_2$ and $\nu_1 - \nu_2$ are considered to be identical if $\mu_1 + \nu_2 = \nu_1 + \mu_2$ holds. In $\mathcal{M}(X)$ define the algebraic operations*

$$(\mu_1 - \mu_2) + (\nu_1 - \nu_2) = (\mu_1 + \nu_1) - (\mu_2 + \nu_2),$$

$$\alpha(\mu_1 - \mu_2) = \begin{cases} \alpha\mu_1 - \alpha\mu_2 & \text{if } \alpha \geq 0 \\ (-\alpha)\mu_2 - (-\alpha)\mu_1 & \text{if } \alpha < 0. \end{cases}$$

Show that these operations are well defined (i.e., show that they depend only upon the equivalence classes) and that they make $\mathcal{M}(X)$ a vector space.
c. *Define an ordering in $\mathcal{M}(X)$ by $\mu_1 - \mu_2 \geq \nu_1 - \nu_2$ whenever*

$$\mu_1(A) + \nu_2(A) \geq \nu_1(A) + \mu_2(A)$$

holds for each $A \in \mathcal{B}$. Show that $\geq$ is well defined and that it is an order relation on $\mathcal{M}(X)$ under which $\mathcal{M}(X)$ is a vector lattice.
d. *Consider the mapping $\mu = \mu_1 - \mu_2 \mapsto F_\mu$ from $\mathcal{M}(X)$ to $C_c^\sim(X)$ defined by $F_\mu(f) = \int f \, d\mu_1 - \int f \, d\mu_2$ for each $f \in C_c(X)$. Show that F_μ is well defined and that $\mu \mapsto F_\mu$ is a lattice isomorphism (Lemma 38.6 may be helpful here) from $\mathcal{M}(X)$ onto $C_c^\sim(X)$. That is, show that $C_c^\sim(X) = \mathcal{M}(X)$ holds.*

Solution. (a) Clearly, $\mu_1 - \mu_2 \equiv \mu_1 - \mu_2$ and $\mu_1 - \mu_2 \equiv \nu_1 - \nu_2$ implies $\nu_1 - \nu_2 \equiv \mu_1 - \mu_2$. For the transitivity, let $\mu_1 - \mu_2 \equiv \nu_1 - \nu_2$ and $\nu_1 - \nu_2 \equiv \omega_1 - \omega_2$. That is, assume that $\mu_1 + \nu_2 = \nu_1 + \mu_2$ and $\nu_1 + \omega_2 = \omega_1 + \nu_2$. Adding the last two equalities, we see that

$$\mu_1 + \omega_2 + (\nu_1 + \nu_2) = \omega_1 + \mu_2 + (\nu_1 + \nu_2). \qquad (\star)$$

Since all measures involved are regular Borel measures, it follows from $(\star)$ that $\mu_1(K) + \omega_2(K) = \omega_1(K) + \mu_2(K)$ holds for each compact subset K of X. The regularity of the measures implies

$$\mu_1(A) + \omega_2(A) = \omega_1(A) + \mu_2(A)$$

for each $A \in \mathcal{B}$, and so $\mu_1 - \mu_2 = \omega_1 - \omega_2$ holds.

(b) To see that the addition is well defined, assume that $\mu_1 - \mu_2 \equiv \nu_1 - \nu_2$ and $\omega_1 - \omega_2 \equiv \pi_1 - \pi_2$. That is, $\mu_1 + \nu_2 = \nu_1 + \mu_2$ and $\omega_1 + \pi_2 = \pi_1 + \omega_2$, and so $(\mu_1 + \omega_1) + (\nu_2 + \pi_2) = (\nu_1 + \pi_1) + (\mu_2 + \omega_2)$. That is,

$$(\mu_1 - \mu_2) + (\omega_1 - \omega_2) = (\mu_1 + \omega_1) - (\mu_2 + \omega_2)$$
$$\equiv (\nu_1 + \pi_1) - (\nu_2 + \pi_2) = (\nu_1 - \nu_2) + (\pi_1 - \pi_2).$$

Similarly, the multiplication is well defined. Now, it is a routine matter to verify that under these algebraic operations $\mathcal{M}(X)$ is a vector space.

(c) To verify that $\geq$ is well defined, proceed as in part (b) above. It is a routine matter to check that $\geq$ makes $\mathcal{M}(X)$ a partially ordered vector space.

Next, we shall show that $\mathcal{M}(X)$ is a vector lattice. It suffices to verify that $(\mu_1 - \mu_2)^+$ exists in $\mathcal{M}(X)$ for each $\mu_1 - \mu_2 \in \mathcal{M}(X)$. To this end, let $\mu_1 - \mu_2$ in $\mathcal{M}(X)$. By Theorem 38.5, $\mu_1 \vee \mu_2$ is a regular Borel measure, and we claim that $(\mu_1 - \mu_2)^+ = \mu_1 \vee \mu_2 - \mu_2$ holds in $\mathcal{M}(X)$. Clearly,

$$\mu_1 - \mu_2 \leq \mu_1 \vee \mu_2 - \mu_2 \quad \text{and} \quad 0 \leq \mu_1 \vee \mu_2 - \mu_2$$

both hold. To see that $\mu_1 \vee \mu_2 - \mu_2$ is the least upper bound of $\mu_1 - \mu_2$ and 0, assume $\mu_1 - \mu_2 \leq \nu_1 - \nu_2 = \nu$ and $\nu \geq 0$. Then, $\nu + \mu_2$ is a regular Borel measure such that $\nu + \mu_2 \geq \mu_1$ and $\nu + \mu_2 \geq \mu_2$ both hold. By Theorem 38.5, $\nu + \mu_2 \geq \mu_1 \vee \mu_2$, and hence $\nu \geq \mu_1 \vee \mu_2 - \mu_2$ holds in $\mathcal{M}(X)$. This shows that $\mu_1 \vee \mu_2 - \mu_2$ is the least upper bound of $\mu_1 - \mu_2$ and 0.

(d) It is a routine matter to verify that $\mu \longmapsto F_\mu$ from $\mathcal{M}(X)$ into $C_c^\sim(X)$ is well defined and linear. Moreover, since every $F \in C_c^\sim(X)$ can be written as a difference of two positive linear functionals, the Riesz Representation Theorem guarantees that $\mu \longmapsto F_\mu$ is onto. To see that $\mu \longmapsto F_\mu$ is one-to-one, assume that $F_\mu = 0$. Then, $\int f \, d\mu_1 = \int f \, d\mu_2$ holds for each $f \in C_c(X)$, and so by (the Riesz Representation Theorem) $\mu_1 = \mu_2$. Therefore, $\mu = \mu_1 - \mu_2 = 0$ and so $\mu \longmapsto F_\mu$ is one-to-one.

Finally, observe that $\mu \geq 0$ holds in $\mathcal{M}(X)$ if and only if $F_\mu \geq 0$ holds in $C_c^\sim(X)$, and then invoke Lemma 38.6 to see that $\mu \longmapsto F_\mu$ is a lattice isomorphism from $\mathcal{M}(X)$ onto $C_c^\sim(X)$. Thus, under this lattice isomorphism, we can say that $C_c^\sim(X) = \mathcal{M}(X)$.

Problem 38.17. *This problem shows that for a noncompact space X, in general $C_c^*(X)$ is a proper ideal of $C_c^\sim(X)$. Let X be a Hausdorff locally compact topological space having a sequence $\{\mathcal{O}_n\}$ of open sets such that $\mathcal{O}_n \subseteq \mathcal{O}_{n+1}$ and $\mathcal{O}_n \neq \mathcal{O}_{n+1}$ for each n, and with $X = \bigcup_{n=1}^\infty \mathcal{O}_n$.*

   a. *Show that if X is σ-compact but not a compact space, then X admits a sequence $\{\mathcal{O}_n\}$ of open sets with the preceding properties.*

   b. *Choose $x_1 \in \mathcal{O}_1$ and $x_n \in \mathcal{O}_n \setminus \mathcal{O}_{n-1}$ for $n \geq 2$. Then, show that*

$$F(f) = \sum_{n=1}^\infty f(x_n) \ \ for \ \ f \in C_c(X)$$

defines a positive linear functional on $C_c(X)$ that is not continuous.

   c. *Determine the (unique) regular Borel measure μ on X that represents F. What is the support of μ?*

Solution.   (a) Let $\{K_n\}$ be a sequence of compact sets with $K_n \uparrow X$. For each n pick an open set V_n with compact closure such that $K_n \subseteq V_n$. Put $O_n = \bigcup_{i=1}^n V_i$ and note that $O_n \uparrow X$. Since each O_n has compact closure and X is not compact, $O_n \neq X$ holds for each n. By passing to a subsequence of $\{O_n\}$, we can assume that $O_n \neq O_{n+1}$ also holds for each n.

(b) Let $f \in C_c(X)$. Since Supp $f \subseteq \bigcup_{n=1}^\infty O_n$ holds and Supp f is compact, there exists some k with Supp $f \subseteq O_k$. Thus, $f(x_n) = 0$ for $n > k$, and so F clearly defines a positive linear functional on $C_c(X)$.

Next, we shall show that F is not continuous. By Theorem 10.8 there exists some $g_n \in C_c(X)$ with $\{x_1, \ldots, x_n\} \prec g_n \prec O_n$. Therefore, $\|F\| \geq F(g_n) = n$ holds for each n, and so $\|F\| = \infty$.

(c) The regular Borel measure μ that represents F is defined on the Borel set B by

$$\mu(B) = \text{The number of elements of } \{x_1, x_2, \ldots\} \cap B.$$

(If $\{x_1, x_2, \ldots\} \cap B$ is countable, then $\mu(B) = \infty$, and if $\{x_1, x_2, \ldots\} \cap B = \emptyset$, then $\mu(B) = 0$.) Also, note that

$$\text{Supp } \mu = \{x_1, x_2, \ldots\}.$$

39. DIFFERENTIATION AND INTEGRATION

Problem 39.1. *If μ is a Borel measure on $\mathbf{R}^k$, then show that $\mu \perp \lambda$ holds if and only if $D\mu(x) = 0$ for almost all x.*

Solution. Assume $\mu \perp \lambda$. Choose two disjoint Borel sets A and B with $A \cup B = \mathbf{R}^k$ and $\mu(A) = \lambda(B) = 0$. By Lemma 39.3, $D\mu(x) = 0$ holds for almost all x in A, and so $D\mu(x) = 0$ holds for almost all x in $\mathbf{R}^k$.

Now, suppose that $D\mu(x) = 0$ holds for almost all $x \in \mathbf{R}^k$. Use the Lebesgue Decomposition Theorem 37.7 to write $\mu = \mu_1 + \mu_2$ with $\mu_1 \ll \lambda$ and $\mu_2 \perp \lambda$. By the preceding, $D\mu_2(x) = 0$ holds for almost all x in $\mathbf{R}^k$. Thus, from Theorem 39.4, it follows that

$$\tfrac{d\mu_1}{d\lambda} = D\mu_1 = D\mu = 0,$$

and so $\mu_1 = 0$. Therefore, $\mu = \mu_2 \perp \lambda$ holds.

Problem 39.2. *Show that if E is a Lebesgue measurable subset of $\mathbf{R}^k$, then almost all points of E are density points.*

Solution. For each $x = (x_1, \ldots, x_k) \in \mathbf{R}^k$ and each $\varepsilon > 0$, consider the open interval $I_\varepsilon = \prod_{i=1}^{k}(x_i - \varepsilon, x_i + \varepsilon)$. If E is a Lebesgue measurable set, then

$$\lim_{\varepsilon \to 0^+} \tfrac{\lambda(E \cap I_\varepsilon)}{(2\varepsilon)^k} = \chi_E(x) \qquad\qquad (\star)$$

holds for almost all x. To see this, note first that we can assume without loss of generality that $\lambda(E) < \infty$ holds (why?). Now, consider the finite Borel measure μ on $\mathbf{R}^k$ defined by

$$\mu(A) = \lambda(E \cap A) = \int_A \chi_E \, d\lambda.$$

Clearly, $\mu \ll \lambda$ and $\tfrac{d\mu}{d\lambda} = \chi_E$. By Theorem 39.4, we have $D\mu = \chi_E$ a.e., and the validity of $(\star)$ follows.

Problem 39.3. *Write $B_r(a)$ for the open ball with center at $a \in \mathbf{R}^k$ and radius r. If f is a Lebesgue integrable function on $\mathbf{R}^k$, then a point $a \in \mathbf{R}^k$ is called a* **Lebesgue point** *for f if*

$$\lim_{r \to 0^+} \frac{1}{\lambda(B_r(a))} \int_{B_r(a)} |f(x) - f(a)| \, d\lambda(x) = 0.$$

Show that if f is a Lebesgue integrable function on $\mathbf{R}^k$, then almost all points of $\mathbf{R}^k$ are Lebesgue points.

Solution. Denote by Q the set of all rational numbers of $\mathbb{R}$. Fix some $a \in Q$, and let $B_n = B(0, n)$. Now, define the finite Borel measure μ by

$$\mu(E) = \int_{E \cap B_n} |f(x) - a| \, d\lambda(x).$$

Since $\mu \ll \lambda$, it follows from Theorem 39.4 that

$$D\mu = \tfrac{d\mu}{d\lambda} = |f - a|\chi_{B_n} \quad a.e.$$

Consequently,

$$\lim_{r \to 0^+} \tfrac{1}{\lambda(B_r(x))} \int_{B_r(x)} |f(t) - a| \, d\lambda(t) = |f(x) - a| \qquad (\star)$$

holds for almost all x in B_n, and therefore (since n is arbitrary) $(\star)$ holds for almost all x in $\mathbb{R}^k$. Let E_a be a Lebesgue null set for which $(\star)$ holds for all $x \notin E_a$. Set $E = \bigcup_{a \in Q} E_a$, and note that $\lambda(E) = 0$.

Now, let $y \notin E$ and let $\varepsilon > 0$. Choose some rational number $s \in Q$ with $|s - f(y)| < \varepsilon$ (we shall assume that f is real-valued everywhere). In view of $|f(x) - f(y)| \le |f(x) - s| + |s - f(y)|$, we see that

$$\limsup_{r \to 0^+} \tfrac{1}{\lambda(B_r(y))} \int_{B_r(y)} |f(x) - f(y)| \, d\lambda(x)$$

$$\le \lim_{r \to 0^+} \tfrac{1}{\lambda(B_r(y))} \int_{B_r(y)} |f(x) - s| \, d\lambda(x)$$

$$+ \lim_{r \to 0^+} \tfrac{1}{\lambda(B_r(y))} \int_{B_r(y)} |s - f(y)| \, d\lambda(x)$$

$$= |f(y) - s| + |s - f(y)| < 2\varepsilon,$$

and from this the desired conclusion follows.

Problem 39.4. *Let $f \colon \mathbb{R} \to \mathbb{R}$ be an increasing, left continuous function. Show directly (i.e., without using Theorem 38.4) that the Lebesgue–Stieltjes measure μ_f is a regular Borel measure.*

Solution. Let (a, b) be an open interval. Then, there exists a sequence $\{[a_n, b_n]\}$ of closed intervals with $[a_n, b_n] \uparrow (a, b)$. It follows that $\mu_f([a_n, b_n]) \uparrow \mu_f((a, b))$. Since every open subset of $\mathbb{R}$ can be written as an at most countable union of pairwise disjoint open intervals, it follows that

$$\mu_f(O) = \sup\{\mu_f(K) \colon K \subseteq O \text{ and } K \text{ compact}\}$$

holds for all open sets O.

Now, let $[a, b)$ be a finite interval. Then, for each point $c < a$ of continuity of f, we have $[a, b) \subseteq (c, b)$ and $\mu_f((c, b)) - \mu_f([a, b)) = f(a) - f(c)$. By the left continuity of f, we see that $\mu_f([a, b)) = \inf\{\mu_f((c, b)) : c < a\}$. Next, consider a σ-set A with $\mu_f(A) < \infty$. Choose a pairwise disjoint sequence $\{[a_n, b_n)\}$ with $A = \bigcup_{n=1}^{\infty} [a_n, b_n)$. Given $\varepsilon > 0$, for each n choose some real number $c_n < a_n$ with $\mu_f((c_n, b_n) \setminus [a_n, b_n)) < \frac{\varepsilon}{2^n}$, and then set $V = \bigcup_{n=1}^{\infty} (c_n, b_n)$. Clearly, V is an open set, $A \subseteq V$, and

$$\mu_f(V) - \mu_f(A) = \mu_f(V \setminus A) \leq \mu_f\left(\bigcup_{n=1}^{\infty} [(c_n, b_n) \setminus [a_n, b_n)]\right)$$

$$\leq \sum_{n=1}^{\infty} \mu_f((c_n, b_n) \setminus [a_n, b_n)) < \sum_{n=1}^{\infty} \frac{\varepsilon}{2^n} = \varepsilon.$$

Thus, $\mu_f(A) = \inf\{\mu_f(V) : A \subseteq V \text{ and } V \text{ open}\}$.

Now, to complete the proof, use Problem 15.2. (For a general result about regular Borel measures, see also Problem 38.14.)

Problem 39.5 (Fubini).   *Let $\{f_n\}$ be a sequence of increasing functions defined on $[a, b]$ such that $\sum_{n=1}^{\infty} f_n(x) = f(x)$ converges in $\mathbb{R}$ for each $x \in [a, b]$. Then, show that f is differentiable almost everywhere and that $f'(x) = \sum_{n=1}^{\infty} f_n'(x)$ holds for almost all x.*

Solution.   Replacing each f_n by $f_n - f_n(a)$, we can assume that $f_n \geq 0$ holds for each n. Set $s_n = f_1 + \cdots + f_n$, and note that each s_n is increasing and $s_n(x) \uparrow f(x)$ holds for each x. Clearly, f is also an increasing function. By Theorem 39.9, f and all the f_n are differentiable almost everywhere. Since $f_{n+1} = s_{n+1} - s_n$ is an increasing function, we see that $s_{n+1}'(x) \geq s_n'(x)$ must hold for almost all x. Similarly, since $f(x) - s_n(x) = \sum_{i=n+1}^{\infty} s_i(x)$ is an increasing function, it follows that $f'(x) \geq s_n'(x)$ holds for almost all x. Thus,

$$\lim_{n \to \infty} s_n'(x) = \sum_{n=1}^{\infty} f_n'(x)$$

exists for almost all x.

Now, for each n let

$$t_n(x) = f(x) - s_n(x) = \sum_{i=n+1}^{\infty} f_i(x) \geq 0.$$

Clearly, each t_n is an increasing function. Pick a subsequence $\{s_{k_n}\}$ of $\{s_n\}$ such that

$$\sum_{n=1}^{\infty} t_{k_n}(x) \leq \sum_{n=1}^{\infty} \left[f(b) - s_{k_n}(b) \right] < \infty.$$

The same arguments applied to $\{t_{k_n}\}$ instead of $\{s_n\}$ show that

$$\sum_{n=1}^{\infty} t'_{k_n}(x) = \sum_{n=1}^{\infty} \left[f'(x) - s'_{k_n}(x) \right]$$

converges for almost all x. In particular, $s'_{k_n}(x) \longrightarrow f'(x)$ holds for almost all x, and so

$$\sum_{n=1}^{\infty} f'_n(x) = f'(x)$$

holds for almost all x.

Problem 39.6. *Suppose $\{f_n\}$ is a sequence of increasing functions on $[a, b]$ and that f is an increasing function on $[a, b]$ such that $\mu_{f_n} \uparrow \mu_f$. Establish that $f'(x) = \lim f'_n(x)$ holds for almost all x.*

Solution. We shall present a solution of this problem based upon the following general continuity property of the Differential Operator D: *If $\{\mu_n\}$ is a sequence of Borel measures in $\mathbf{R}^k$ and $\mu_n \uparrow \mu$ holds for some Borel measure μ, then $D\mu_n \uparrow D\mu$ a.e. also holds.*

If this property is established, then using Theorem 39.8, we see that

$$f'_n(x) = D\mu_{f_n}(x) \uparrow D\mu_f(x) = f'(x)$$

must hold for almost all x.

To establish the validity of the continuity property start by observing that if two Borel measures μ and ν satisfy $\mu \leq \nu$, then $D\mu \leq D\nu$ a.e. holds. Indeed, by Theorem 39.6 both μ and ν are differentiable almost everywhere. If $x \in \mathbf{R}^k$ is a point for which $D\mu(x)$ and $D\nu(x)$ exist and $B_n = B(x, \frac{1}{n})$, then

$$D\mu(x) = \lim_{n \to \infty} \frac{\mu(B_n)}{\lambda(B_n)} \leq \lim_{n \to \infty} \frac{\nu(B_n)}{\lambda(B_n)} = D\nu(x).$$

Now, let $\mu_n \uparrow \mu$. Restricting ourselves to the open balls $\{x \in \mathbf{R}^k : \|x\| < n\}$, we can assume without loss of generality that all measures are finite.

By the Lebesgue Decomposition Theorem 37.7, we can write $\mu_n = \nu_n + \omega_n$ with $\nu_n \ll \lambda$ and $\omega_n \perp \lambda$. It follows from the proof of Theorem 37.7 that $\mu_n \wedge m\lambda \uparrow_m \nu_n$. Clearly, this implies $\nu_n \leq \nu_{n+1}$ for each n. From formula (c) of Problem 9.1, we get

$$\mu_n - \mu_n \wedge m\lambda = 0 \vee (\mu_n - m\lambda) \leq 0 \vee (\mu_{n+1} - m\lambda) = \mu_{n+1} - \mu_{n+1} \wedge m\lambda$$

for each m. Letting $m \longrightarrow \infty$, we obtain

$$\omega_n = \mu_n - \nu_n \leq \mu_{n+1} - \nu_{n+1} = \omega_{n+1}$$

for each n. Let $\nu_n \uparrow \nu$ and $\omega_n \uparrow \omega$. Since $\mu_n = \nu_n + \omega_n \uparrow \mu$, it follows that $\mu = \nu + \omega$. The relation $\nu_n \ll \lambda$ for each n easily implies $\nu \ll \lambda$. In view of $\omega_n \perp \lambda = 0$ for each n, it follows from Lemma 37.6 that $\omega \perp \lambda = 0$. That is, $\nu \ll \lambda$ and $\omega \perp \lambda$ both hold, and so $\mu = \nu + \omega$ is the Lebesgue decomposition of μ with respect to λ.

From Problem 39.1 (or by repeating the proof of Theorem 39.6), we see that $D\mu_n(x) = D\nu_n(x)$ and $D\mu(x) = D\nu(x)$ both hold for almost all x. Let $D\mu_n = \frac{d\mu_n}{d\lambda} = f_n$ for each n. In view of $\mu_n(\mathbf{R}^k) = \int f_n \, d\lambda \leq \mu(\mathbf{R}^k) < \infty$, Levi's Theorem 22.8 shows that there exists some $f \in L_1(\mathbf{R}^k)$ with $f_n \uparrow f$. Now note that $\nu_n(E) = \int_E f_n \, d\lambda$ implies $\nu(E) = \int_E f \, d\lambda$ for each Borel set E. This implies $f = D\nu$ a.e., and so

$$D\mu_n(x) = D\nu_n(x) = f_n(x) \uparrow f(x) = D\nu(x) = D\mu(x)$$

holds for almost all x, as desired.

Problem 39.7. *This problem reveals some basic properties of functions of bounded variation on an interval* $[a, b]$.

a. *If f is differentiable at every point and $|f'(x)| \leq M < \infty$ holds for all $x \in [a, b]$, then show that f is absolutely continuous (and hence, of bounded variation).*

b. *Show that the function $f : [0, 1] \longrightarrow \mathbf{R}$ defined by*

$$f(x) = \begin{cases} 0 & \text{if } x = 0 \\ x^2 \cos(\frac{1}{x^2}) & \text{if } 0 < x \leq 1 \end{cases}$$

is differentiable at each x, but is not of bounded variation (and hence, f is continuous but not absolutely continuous).

c. *If f is a function of bounded variation and $|f(x)| \geq M > 0$ holds for each $x \in [a, b]$, then show that $g(x) = \frac{1}{f(x)}$ is a function of bounded variation.*

d.   *If a function* $f:[a, b] \longrightarrow \mathbb{R}$ *satisfies a Lipschitz condition (i.e., if there exists a real number* M *such that* $|f(x) - f(y)| \leq M|x - y|$ *holds for all* $x, y \in [a, b]$), *then show that* f *is absolutely continuous.*

Solution.   (a) If $(a_1, b_1), \ldots, (a_n, b_n)$ are pairwise disjoint open subintervals of $[a, b]$, then by the Mean Value Theorem we have

$$\sum_{i=1}^{n} |f(b_i) - f(a_i)| \leq M \sum_{i=1}^{n} (b_i - a_i).$$

The preceding easily implies that f is an absolutely continuous function.

(b) Only the differentiability of f at zero needs verification. The inequality

$$\left| \frac{f(x) - f(0)}{x - 0} \right| = \left| x \cos\left(\frac{1}{x^2}\right) \right| \leq x$$

for $0 < x \leq 1$ yields $f'(0) = 0$. Now if

$$P_n = \left\{ 0, \ \sqrt{\frac{2}{2n\pi}}, \ \sqrt{\frac{2}{(2n-1)\pi}}, \ \cdots, \ \sqrt{\frac{2}{3\pi}}, \ \sqrt{\frac{2}{2\pi}}, \ 1 \right\},$$

then an easy computation shows that the variation of f with respect to the partition P_n is

$$\cos 1 + \frac{2}{\pi} \cdot \sum_{k=1}^{n} \frac{1}{k} \leq V_f.$$

This implies $V_f = \infty$.

(c) Note that for each $a \leq x < y \leq b$, we have

$$|g(x) - g(y)| = \frac{|f(x) - f(y)|}{|f(x)f(y)|} \leq \frac{1}{M^2} |f(x) - f(y)|.$$

Therefore, $V_g \leq \frac{1}{M^2} V_f < \infty$ holds.

(d) Let a function $f:[a, b] \longrightarrow \mathbb{R}$ satisfy a Lipschitz condition as stated in the problem and let $\varepsilon > 0$. Put $\delta = \frac{\varepsilon}{M} > 0$ and note that if $(a_1, b_1), \ldots, (a_n, b_n)$ are pairwise disjoint open subintervals of $[a, b]$ satisfying $\sum_{i=1}^{n}(b_i - a_i) < \delta$, then

$$\sum_{i=1}^{n} |f(b_i) - f(a_i)| \leq \sum_{i=1}^{n} M(b_i - a_i)$$

$$= M \sum_{i=1}^{n} (b_i - a_i) < M\delta = \varepsilon.$$

Problem 39.8.   *This problem presents an example of a continuous increasing function (and hence, of bounded variation) that is not absolutely continuous.*

Consider the Cantor set C as constructed in Example 6.15 of the text. Recall that C was obtained from [0, 1] *by removing certain open intervals by steps. In the first step we removed the open middle third interval. At the nth step there were* 2^{n-1} *closed intervals, all of the same length, and we removed the open middle third interval from each one of them. Let us denote by* $I_1^n, \ldots, I_{2^{n-1}}^n$ *(counted from left to right) the removed open intervals at the nth step. Now, define the function* $f: [0, 1] \to [0, 1]$ *as follows:*

   i. $f(0) = 0$;
   ii. *if* $x \in I_i^n$ *for some* $1 \le i \le 2^{n-1}$, *then* $f(x) = (2i - 1)/2^n$; *and*
   iii. *if* $x \in C$ *with* $x \ne 0$, *then* $f(x) = \sup\{f(t): t < x \text{ and } t \in [0, 1] \setminus C\}$.

Part of the graph of f is shown in Figure 7.1.

   a. *Show that f is an increasing continuous function from* [0, 1] *to* [0, 1].
   b. *Show that* $f'(x) = 0$ *for almost all x.*
   c. *Show that f is not absolutely continuous.*
   d. *Show that* $\mu_f \perp \lambda$ *holds.*

Solution.   Notice again that parts of the graph of the function f are shown in Figure 7.1.

(a) Straightforward.

(b) Observe that f is constant on each I_i^n. This implies that $f'(x) = 0$ holds

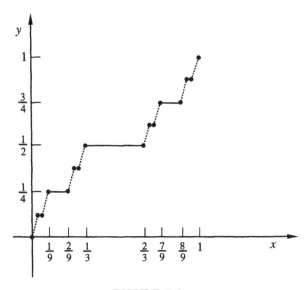

FIGURE 7.1.

for all $x \in [0, 1] \setminus C$. Since $\lambda(C) = 0$, it follows that $f'(x) = 0$ holds for almost all x.

(c) If f is absolutely continuous, then by Theorem 39.15 we should have

$$1 = f(1) - f(0) = \int_0^1 f'(x) \, d\lambda(x) = 0,$$

which is impossible.

(d) Note that if $B = [0, 1] \setminus C$, then $B \cup C = [0, 1]$ and $\mu_f(B) = \lambda(C) = 0$. Hence, $\mu_f \perp \lambda$ holds.

Problem 39.9. *Let $f: [a, b] \to \mathbf{R}$ be an absolutely continuous function. Then show that f is a constant function if and only if $f'(x) = 0$ holds for almost all x.*

Solution. Assume that $f: [a, b] \longrightarrow \mathbf{R}$ is an absolutely continuous function such that $f'(x) = 0$ holds for almost all x. By Theorem 39.15, we have

$$f(x) - f(a) = \int_a^x f'(t) \, d\lambda(t) = 0$$

for each $x \in [a, b]$. Hence, $f(x) = f(a)$ holds for each $x \in [a, b]$, so that f is a constant function.

Problem 39.10. *Let f and g be two left continuous functions (on $\mathbf{R}$). Show that $\mu_f = \mu_g$ holds if and only if $f - g$ is a constant function.*

Solution. If $f - g$ is a constant, then it is easy to see that $\mu_f = \mu_g$ holds. For the converse, assume $\mu_f = \mu_g$. If $x > 0$, then

$$f(x) - f(0) = \mu_f([0, x)) = \mu_g([0, x)) = g(x) - g(0)$$

implies $f(x) - g(x) = f(0) - g(0)$. Similarly, if $x < 0$, then

$$f(0) - f(x) = \mu_f([x, 0)) = \mu_g([x, 0)) = g(0) - g(x),$$

and so $f(x) - g(x) = f(0) - g(0)$ holds in this case too.

Problem 39.11. *This problem presents another characterization of the norm dual of $C[a, b]$. Start by letting L denote the collection of all functions of bounded variation on $[a, b]$ that are left continuous and vanish at a.*

    a. *Show that L under the usual algebraic operations is a vector space, and that $f \mapsto \mu_f$, from L to $M_b([a, b])$, is linear, one-to-one, and onto.*

b. *Define $f \succeq g$ to mean that $f - g$ is an increasing function. (Note that $f \succeq g$ does not imply $f \geq g$.) Show that L under $\succeq$ is a partially ordered vector space such that $f \succeq g$ holds in L if and only if $\mu_f \geq \mu_g$ in $M_b([a, b])$.*

c. *Establish that L with the norm $\|f\| = V_{|f|}$ is a Banach lattice.*

d. *Show, with an appropriate interpretation, that $C^*[a, b] = L$.*

Solution.   (a) Clearly, L under the usual algebraic operations is a vector space. Also, it should be clear that $f \longmapsto \mu_f$ from L to $M_b([a, b])$ is a linear mapping.

To see that $f \longmapsto \mu_f$ is one-to-one, assume $\mu_f = 0$. Then,

$$f(x) = f(x) - f(a) = \mu_f([a, x)) = 0$$

holds for all $a < x \leq b$, and so $f = 0$.

Next, we shall show that $\mu \longmapsto \mu_f$ is onto. Assume at the beginning that $0 \leq \mu \in M_b([a, b])$. Define the function

$$f(x) = \begin{cases} 0 & \text{if } x \leq a \\ \mu([a, x)) & \text{if } a < x < b \\ \mu([a, b]) & \text{if } x \geq b \ , \end{cases}$$

and note that f is increasing, left continuous, and satisfies $f(a) = 0$. Thus, $f \in L$. Now, an easy argument shows that $\mu = \mu_f$. Finally, if $\mu \in M_b([a, b])$, then pick two increasing functions $f, g \in L$ with $\mu^+ = \mu_f$ and $\mu^- = \mu_g$. Note that the function $h = f - g \in L$ satisfies $\mu = \mu^+ - \mu^- = \mu_f - \mu_g = \mu_{f-g} = \mu_h$.

(b) Straightforward.

(c) Since $f \succeq g$ holds in L if and only if $\mu_f \geq \mu_g$ holds in $M_b([a, b])$ and $M_b([a, b])$ is a vector lattice, it is easy to see that L must likewise be a vector lattice. Moreover, by Lemma 38.6 the mapping $f \longmapsto \mu_f$ is a lattice isomorphism from L onto $M_b([a, b])$.

Now, note that if $f \succeq 0$ holds in L (i.e., if f is an increasing function), then

$$V_f = f(b) - f(a) = \mu_f([a, b]) = \|\mu_f\|$$

holds. Thus, for each $f \in L$ we have

$$\|\mu_f\| = \||\mu_f|\| = \|\mu_{|f|}\| = V_{|f|}.$$

This implies that $\|f\| = V_{|f|}$ defines a lattice norm on L, and that $f \longmapsto \mu_f$ from L onto $M_b([a, b])$ is a lattice isometry. In particular, L with the norm $\|f\| = V_{|f|}$ is a Banach lattice.

(d) Using the notation of Theorem 38.7, we see that the composition of the two operators

$$f \longmapsto \mu_f \longmapsto F_{\mu_f}$$

is a lattice isometry from L onto $C^*[a, b]$.

Problem 39.12.   *If* $f: [a, b] \to \mathbf{R}$ *is an increasing function, then show that* f' *is Lebesgue integrable and that* $\int_a^b f'(x)\,dx \le f(b) - f(a)$ *holds. Give an example of an increasing function* f *for which* $\int_a^b f'(x)\,dx < f(b) - f(a)$ *holds.*

Solution.   Let $g_n(x) = n[f(x + \frac{1}{n}) - f(x)]$ for each $x \in [a, b]$ (where, of course, $f(x) = f(b)$ for $x > b$.) Clearly, $g_n(x) \longrightarrow f'(x)$ holds for almost all x; see Theorem 39.9. On the other hand, the relation $g_n(x) \ge 0$ for each x, and

$$
\begin{aligned}
\int_a^b g_n(x)\,dx &= n\left[\int_a^b f\left(x + \tfrac{1}{n}\right) dx - \int_a^b f(x)\,dx\right] \\
&= n\left[\int_{a+\frac{1}{n}}^{b+\frac{1}{n}} f(x)\,dx - \int_a^b f(x)\,dx\right] \\
&= n\left[\int_b^{b+\frac{1}{n}} f(x)\,dx - \int_a^{a+\frac{1}{n}} f(x)\,dx\right] \\
&= n\left[\tfrac{1}{n} f(b) - \int_a^{a+\frac{1}{n}} f(x)\,dx\right] \\
&\le n\left[\tfrac{1}{n} f(b) - \tfrac{1}{n} f(a)\right] = f(b) - f(a),
\end{aligned}
$$

coupled with Fatou's Lemma show that $f' \in L_1([a, b])$ and that

$$
\int_a^b f'(x)\,dx = \int_a^b \lim_{n\to\infty} g_n(x)\,dx \le \liminf_{n\to\infty} \int_a^b g_n(x)\,dx \le f(b) - f(a).
$$

Finally, an example of a function that yields strict inequality is provided by the function described in Problem 39.8.

Problem 39.13.   *If* $f: [a, b] \to \mathbf{R}$ *is an absolutely continuous function, then show that*

$$
V_f = \int_a^b |f'(x)|\,dx
$$

holds.

Solution.   By Theorem 39.15 we have $f' \in L_1([a, b])$ and

$$
\mu_f(E) = \int_E f'(x)\,dx
$$

holds for each Borel subset E of $[a, b]$.

Start by observing that if $a = t_0 < t_1 < \cdots < t_n = b$ is a partition of $[a, b]$, then

$$\sum_{i=1}^{n} \left| f(t_i) - f(t_{i-1}) \right| = \sum_{i=1}^{n} \left| \int_{t_{i-1}}^{t_i} f'(x)\, dx \right|$$

$$\leq \sum_{i=1}^{n} \int_{t_{i-1}}^{t_i} \left| f'(x) \right| dx = \int_a^b \left| f'(x) \right| dx.$$

Therefore, $V_f \leq \int_a^b |f'(x)|\, dx$ holds. Now, since the continuous functions are dense (in the L_1-norm) in $L_1([a, b])$ (Theorem 25.3) and the functions of the form

$$\phi = \sum_{i=1}^{n} a_i \chi_{(t_{i-1}, t_i)},$$

where $a = t_0 < t_1 < \cdots < t_n = b$ is a partition of $[a, b]$, are dense (in the L_1-norm) in $C[a, b]$, these functions are also dense in $L_1([a, b])$. Thus, given $\varepsilon > 0$, there exist a partition $a = t_0 < t_1 < \cdots < t_n = b$ and real numbers $a_1, \ldots, a_n$ so that $\phi = \sum_{i=1}^{n} a_i \chi_{(t_{i-1}, t_i)}$ satisfies $\|\phi - \operatorname{Sgn} f'\|_1 < \varepsilon$. In view of $\left| (-1 \vee \phi) \wedge 1 - \operatorname{Sgn} f' \right| \leq |\phi - \operatorname{Sgn} f'|$, we can assume that $|\phi(x)| \leq 1$ holds for all $x \in [a, b]$. Moreover, we have

$$\int_a^b \phi(x) f'(x)\, dx = \sum_{i=1}^{n} a_i \int_{t_{i-1}}^{t_i} f'(x)\, dx = \sum_{i=1}^{n} a_i \left[f(t_i) - f(t_{i-1}) \right]$$

$$\leq \sum_{i=1}^{n} \left| f(t_i) - f(t_{i-1}) \right| \leq V_f.$$

Next, choose a sequence $\{\phi_n\}$ of step functions of the previous type satisfying $\phi_n \longrightarrow \operatorname{Sgn} f'$ a.e. (see Lemma 31.6 of the text). In view of $|\phi_n f'| \leq |f'|$, the Lebesgue Dominated Convergence Theorem implies

$$\int_a^b |f'(x)|\, dx = \int_a^b f'(x) \cdot \operatorname{Sgn} f'(x)\, dx$$

$$= \lim_{n \to \infty} \int_a^b \phi_n(x) f'(x)\, dx \leq V_f.$$

Thus, $V_f = \int_a^b |f'(x)|\, dx$ holds.

It is interesting to observe that $V_f = |\mu_f|([a, b])$ also holds. To see this, let $a = t_0 < t_1 < \cdots < t_n = b$ be an arbitrary partition of $[a, b]$. Then, we have

$$\sum_{i=1}^{n} |f(t_i) - f(t_{i-1})| = \sum_{i=1}^{n} |\mu_f([t_{i-1}, t_i))|$$

$$\leq \sum_{i=1}^{n} |\mu_f|([t_{i-1}, t_i)) = |\mu_f|([a, b]),$$

and so $V_f \leq |\mu_f|([a, b])$. On the other hand, if $E_1, \ldots, E_n$ are pairwise disjoint Borel subsets of $[a, b]$, then

$$\sum_{i=1}^{n} |\mu_f(E_i)| = \sum_{i=1}^{n} \left| \int_{E_i} f'(x)\, dx \right| \leq \sum_{i=1}^{n} \int_{E_i} |f'(x)|\, dx$$

$$\leq \int_a^b |f'(x)|\, dx = V_f$$

holds, which (by Theorem 36.9) implies that $|\mu_f|([a, b]) \leq V_f$. Consequently, $|\mu_f|([a, b]) = V_f$ holds.

Problem 39.14. *For a continuously differentiable function $f : [a, b] \to \mathbb{R}$ establish the following properties:*

a. *The signed measure μ_f is absolutely continuous with respect to the Lebesgue measure and $d\mu_f/d\lambda = f'$ a.e.*
b. *If $g : [a, b] \to \mathbb{R}$ is Riemann integrable, then gf' is also Riemann integrable and*

$$\int g\, d\mu_f = \int_a^b g(x) f'(x)\, dx.$$

Solution. (a) By Problem 39.7, we know that f is absolutely continuous and so (by Theorem 39.12) μ_f is absolutely continuous with respect to the Lebesgue measure. Now, combining Theorems 39.14(2), 39.8, and 39.4, we see that

$$\frac{d\mu_f}{d\lambda} = D\mu_f = f' \text{ a.e.}$$

(b) Since f' is a continuous function and g is Riemann integrable, it follows that gf' is also Riemann (and hence Lebesgue) integrable over $[a, b]$. From

$\mu_f(A) = \int_A f' \, d\lambda$ for every Borel subset A of $[a, b]$ and Problem 22.15, we see that

$$\int_{[a,b]} g \, d\mu_f = \int_{[a,b]} gf' \, d\lambda = \int_a^b g(x) f'(x) \, dx.$$

Problem 39.15. *For each n consider the increasing continuous function* $f_n \colon \mathbf{R} \to \mathbf{R}$ *defined by*

$$f_n(x) = \begin{cases} 1 & \text{if } x > 0, \\ n(x-1) + 1 & \text{if } 1 - \frac{1}{n} < x < 1, \\ 0 & \text{if } x \le 1 - \frac{1}{n}. \end{cases}$$

If $f \colon \mathbf{R} \to \mathbf{R}$ *is a continuous function, then show that*
  a. *f is μ_{f_n}-integrable for each n, and*
  b. *$\lim \int f \, d\mu_{f_n} = f(1)$.*

Solution. Note that $\operatorname{Supp} \mu_{f_n} = [1 - \frac{1}{n}, 1]$. This easily implies that f is μ_{f_n}-integrable for each n. In addition, note that $f_n'(x) = n$ holds for each $1 - \frac{1}{n} \le x \le 1$. By the preceding problem, we see that

$$\int f \, d\mu_{f_n} = \int_{1-\frac{1}{n}}^1 f(x) f_n'(x) \, dx = \int_{1-\frac{1}{n}}^1 n f(x) \, dx = \frac{\int_{1-\frac{1}{n}}^1 f(x) \, dx}{\frac{1}{n}}.$$

Therefore, by the Fundamental Theorem of Calculus, we infer that $\lim \int f \, d\mu_{f_n} = f(1)$.

Problem 39.16. *Let* $f \colon \mathbf{R} \to \mathbf{R}$ *be a (uniformly) bounded function and let*

$$E = \{ x \in \mathbf{R} \colon \ f'(x) \ \text{exists in } \mathbf{R} \}.$$

If $\lambda(E) = 0$, *then show that* $\lambda\big(f(E)\big) = 0$.

Solution. For each natural number n, let

$$E_n = \{ a \in E \colon \ |f(x) - f(a)| \le n|x - a| \ \text{for all } x \in \mathbf{R} \}.$$

Since f is bounded, it is easy to see that $E = \bigcup_{n=1}^\infty E_n$, and so $f(E) = \bigcup_{n=1}^\infty f(E_n)$ (see Problem 1.1(6)). Thus, in order to establish that $\lambda\big(f(E)\big) = 0$, it suffices to show that $\lambda\big(f(E_n)\big) = 0$ holds for each n. To this end, fix n and $\varepsilon > 0$.

From $\lambda(E) = 0$, we obtain $\lambda(E_n) = 0$, and so there exists a sequence of open intervals $\{(b_k - r_k, b_k + r_k)\}$ satisfying

$$E_n \subseteq \bigcup_{k=1}^{\infty}(b_k - r_k, b_k + r_k) \quad \text{and} \quad 2\sum_{k=1}^{\infty} r_k < \varepsilon.$$

Now, note that if $a \in E_n$, then there exists some m with $|b_m - a| < r_m$, and hence $|f(b_m) - f(a)| \leq n|b_m - a| < nr_m$ holds. It follows that $f(E_n) \subseteq \bigcup_{k=1}^{\infty}(f(b_k) - nr_k, f(b_k) + nr_k)$. Therefore,

$$\sum_{k=1}^{\infty} \lambda\big((f(b_k) - nr_k, f(b_k) + nr_k)\big) = 2n\sum_{k=1}^{\infty} r_k < 2n\varepsilon.$$

Since $\varepsilon > 0$ is arbitrary, we infer that $\lambda\big(f(E_n)\big) = 0$, as desired. (Compare this problem with Problem 18.9.)

Problem 39.17. *This problem presents an example of a continuous function $f: \mathbb{R} \to \mathbb{R}$ which is nowhere differentiable; this example should be compared with Problem 9.28. Consider the function $\phi: [0, 2] \to \mathbb{R}$ defined by $\phi(x) = x$ if $0 \leq x \leq 1$ and $\phi(x) = 2 - x$ if $1 < x \leq 2$. Extend ϕ to all of $\mathbb{R}$ (periodically) so that $\phi(x) = \phi(x + 2)$ holds for all $x \in \mathbb{R}$. Now, define the function $f: \mathbb{R} \to \mathbb{R}$ by*

$$f(x) = \sum_{n=0}^{\infty}\left(\tfrac{3}{4}\right)^n \phi\big(4^n x\big).$$

Show that f is a continuous nowhere differentiable function.

Solution. Since the series $\sum_{n=0}^{\infty}\left(\tfrac{3}{4}\right)^n$ converges and $0 < \phi(x) < 1$ holds for all x, it is easy to see that the sequence of partial sums of the series $f(x) = \sum_{n=0}^{\infty}\left(\tfrac{3}{4}\right)^n \phi\big(4^n x\big)$ converges uniformly to f on $\mathbb{R}$. So, by Theorem 9.2, f is a well-defined continuous function.

Now, fix $x_0 \in \mathbb{R}$. The proof of the nondifferentiability of f at x_0 will be based upon the following property of differentiable functions.

- *If $h: (a, b) \to \mathbb{R}$ is differentiable at some $x_0 \in (a, b)$ and $\mu = h'(x_0)$, then for each $\epsilon > 0$ there exists some $\delta > 0$ such that whenever $x, y \in (a, b)$ satisfy $x < x_0 < y$ and $y - x < \delta$, then $\left|\frac{h(y)-h(x)}{y-x} - \mu\right| < \epsilon$.*

This conclusion follows easily from the inequalities

$$
\left| \frac{h(y)-h(x)}{y-x} - \mu \right| = \left| \frac{[h(y)-h(x_0)-\mu(y-x_0)]+[h(x_0)-h(x)-\mu(x_0-x)]}{y-x} \right|
$$

$$
\leq \left| \frac{h(y)-h(x_0)-\mu(y-x_0)}{y-x_0} \right| \cdot \left| \frac{y-x_0}{y-x} \right| + \left| \frac{h(x_0)-h(x)-\mu(x_0-x))}{x_0-x} \right| \cdot \left| \frac{x_0-x}{y-x} \right|
$$

$$
\leq \left| \frac{h(y)-h(x_0)}{y-x_0} - \mu \right| + \left| \frac{h(x_0)-h(x)}{x_0-x} - \mu \right|.
$$

Now, for each natural number m, then there exists a unique integer k_m such that $k_m \leq 4^m x_0 < k_m + 1$. Let

$$
s_m = 4^{-m} k_m \quad \text{and} \quad t_m = 4^{-m}(k_m + 1),
$$

and note that $s_m \leq x_0 < t_m$ holds for each m. From $t_m - s_m = 4^{-m}$, we see that $\lim t_m = \lim s_m = x_0$.

The reader should keep in mind that if p and q are two integers, then $\phi(p) - \phi(q) = 0$ if $p - q$ is an even integer and $|\phi(p) - \phi(q)| = 1$ if $p - q$ is an odd integer. Next, observe that if n is a non-negative integer, then $4^n t_m - 4^n s_m = 4^{n-m}$. So, from the definition of ϕ, we have:

    a.   if $n > m$, then $\phi(4^n t_m) - \phi(4^n s_m) = 0$,
    b.   if $n = m$, then $\phi(4^n t_m) - \phi(4^n s_m) = 1$, and
    c.   if $0 \leq n < m$, then $\phi(4^n t_m) - \phi(4^n s_m) = 4^{n-m}$.

Therefore, for each m we have

$$
\left| f(t_m) - f(s_m) \right| = \left| \sum_{n=0}^{\infty} \left(\tfrac{3}{4} \right)^n [\phi(4^n t_m) - \phi(4^n s_m)] \right|
$$

$$
= \left| \sum_{n=0}^{m} \left(\tfrac{3}{4} \right)^n [\phi(4^n t_m) - \phi(4^n s_m)] \right|
$$

$$
\geq \left(\tfrac{3}{4} \right)^m - \sum_{n=0}^{m-1} \left(\tfrac{3}{4} \right)^n 4^{n-m} > \tfrac{1}{2} \left(\tfrac{3}{4} \right)^m.
$$

This implies $\left| \frac{f(t_m)-f(s_m)}{t_m-s_m} \right| > \frac{3^m}{2}$ for each m. Now, a glance at ($\bullet$) shows that f cannot be differentiable at x_0. Since x_0 is arbitrary, f is differentiable at no point of $\mathbf{R}$.

40. THE CHANGE OF VARIABLES FORMULA

Problem 40.1. *Show that an open ball in a Banach space is a connected set. That is, show that if B is an open ball in a Banach space such that $B = \mathcal{O}_1 \cup \mathcal{O}_2$ holds with both $\mathcal{O}_1$ and $\mathcal{O}_2$ open and disjoint, then either $\mathcal{O}_1 = \emptyset$ or $\mathcal{O}_2 = \emptyset$.*

Solution. Let B be an open ball in a Banach space. Assume by way of contradiction that there exist two nonempty open sets O_1 and O_2 such that $B = O_1 \cup O_2$ and $O_1 \cap O_2 = \emptyset$. Fix two elements $a \in O_1$ and $b \in O_2$, and then define the function $f : [0, 1] \longrightarrow B$ by $f(t) = ta + (1 - t)b$. Clearly, $f(0) = b$ and $f(1) = a$. Moreover, in view of the inequality

$$\big\| f(t) - f(s) \big\| = \big\| (t - s)a + (s - t)b \big\| \leq \big(\|a\| + \|b\| \big) |t - s|,$$

we see that f is a (uniformly) continuous function.

Let $\alpha = \inf\{t \in [0, 1] : f(t) \in O_1\}$. Choose a sequence $\{\alpha_n\}$ of $[0, 1]$ with $\alpha_n \to \alpha$ and $f(\alpha_n) \in O_1$ for each n. By the continuity of f we have $f(\alpha_n) \to f(\alpha)$. Since O_2 is open and disjoint from O_1, it follows that $f(\alpha) \notin O_2$. In particular, $\alpha > 0$ must hold. Thus, there exists a sequence $\{\beta_n\}$ of real numbers with $0 < \beta_n < \alpha$ for each n and $\beta_n \longrightarrow \alpha$. By the definition of α, we see that $f(\beta_n) \in O_2$ holds for each n, and hence, as above $f(\alpha) \notin O_1$. Now, note that

$$f(\alpha) \notin O_1 \cup O_2 = B$$

holds, which is impossible.

Problem 40.2. *Let $T : V \to \mathbf{R}^k$ be C^1-differentiable. Show that the mapping $x \mapsto T'(x)$ from V into $L(\mathbf{R}^k, \mathbf{R}^k)$ is a continuous function.*

Solution. We know that

$$T'(x) = \begin{bmatrix} \frac{\partial T_1}{\partial x_1}(x) & \cdots & \frac{\partial T_1}{\partial x_k}(x) \\ \vdots & \ddots & \vdots \\ \frac{\partial T_k}{\partial x_1}(x) & \cdots & \frac{\partial T_k}{\partial x_k}(x) \end{bmatrix}.$$

So, if $a = (a_1, \ldots, a_k) \in \mathbf{R}^k$ satisfies $\|a\|_2 = \left(\sum_{i=1}^{k} a_i^2 \right)^{\frac{1}{2}} = 1$, then using the

Cauchy–Schwarz inequality, we see that

$$\big\| [T'(x) - T'(y)]a \big\|_2 = \Big(\sum_{i=1}^{k} \Big[\sum_{j=1}^{k} \big(\tfrac{\partial T_i}{\partial x_j}(x) - \tfrac{\partial T_i}{\partial x_j}(y) \big) \cdot a_j \Big]^2 \Big)^{\frac{1}{2}}$$

$$\leq \Big[\sum_{i=1}^{k} \Big(\sum_{j=1}^{k} [\tfrac{\partial T_i}{\partial x_j}(x) - \tfrac{\partial T_i}{\partial x_j}(y)]^2 \Big) \cdot \Big(\sum_{j=1}^{k} a_j^2 \Big) \Big]^{\frac{1}{2}}$$

$$= \Big(\sum_{i=1}^{k} \sum_{j=1}^{k} [\tfrac{\partial T_i}{\partial x_j}(x) - \tfrac{\partial T_i}{\partial x_j}(y)]^2 \Big)^{\frac{1}{2}} .$$

Consequently,

$$\big\| T'(x) - T'(y) \big\| = \sup \big\{ \| [T'(x) - T'(y)]a \|_2 : \ \|a\|_2 = 1 \big\}$$

$$\leq \Big(\sum_{i=1}^{k} \sum_{j=1}^{k} [\tfrac{\partial T_i}{\partial x_j}(x) - \tfrac{\partial T_i}{\partial x_j}(y)]^2 \Big)^{\frac{1}{2}}$$

for each pair x, $y \in V$. This inequality, coupled with the fact that $T : V \longrightarrow \mathbf{R}^k$ is C^1-differentiable, implies that $x \longmapsto T'(x)$ from V into $L(\mathbf{R}^k, \mathbf{R}^k)$ is a continuous function.

Problem 40.3.   *Show that the Lebesgue measure on $\mathbf{R}^2$ is "rotation" invariant.*

Solution.   A *"rotation"* of the plane is a linear operator $T : \mathbf{R}^2 \longrightarrow \mathbf{R}^2$ whose representing matrix A is orthogonal (i.e., it satisfies $AA^t = A^t A = I$). Any such orthogonal matrix is of the form

$$A = \begin{bmatrix} \cos\theta & \sin\theta \\ -\sin\theta & \cos\theta \end{bmatrix},$$

where θ represents the angle of rotation; see Figure 7.2.

In particular, note that $\det A = 1$. Thus, by Lemma 40.4, we see that

$$\lambda\big(A(E) \big) = \big| \det A \big| \lambda(E) = \lambda(E)$$

holds for each Lebesgue measurable subset E of $\mathbf{R}^2$.

Problem 40.4 (Polar Coordinates).   *Let*

$$E = \{ (r, \theta) \in \mathbf{R}^2 : \ r \geq 0 \ \text{and} \ 0 \leq \theta \leq 2\pi \}.$$

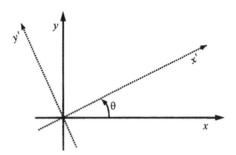

FIGURE 7.2.   Rotation by an Angle θ

The transformation $T: E \to \mathbb{R}^2$ defined by $T(r, \theta) = (r \cos \theta, r \sin \theta)$, or as it is usually written

$$x = r \cos \theta \quad and \quad y = r \sin \theta,$$

is called the **polar coordinate transformation** *on $\mathbb{R}^2$, shown graphically in Figure 7.3.*

a. *Show that $\lambda(E \setminus E^\circ) = 0$.*

b. *If $A = \{(x, 0): x \geq 0\}$, then show that A is a closed subset of $\mathbb{R}^2$ whose (2-dimensional) Lebesgue measure is zero.*

c. *Show that $T: E^\circ \to \mathbb{R}^2 \setminus A$ is a diffeomorphism whose Jacobian determinant satisfies $J_T(r, \theta) = r$ for each $(r, \theta) \in E^\circ$.*

d. *Show that if G is a Lebesgue measurable subset of E with $\lambda(G \setminus G^\circ) = 0$, then $T(G)$ is a Lebesgue measurable subset of $\mathbb{R}^2$. Moreover, show that if $f \in L_1\big(T(G)\big)$, then*

$$\int_{T(G)} f \, d\lambda = \int\int_G f(r \cos \theta, r \sin \theta) r \, dr \, d\theta$$

holds.

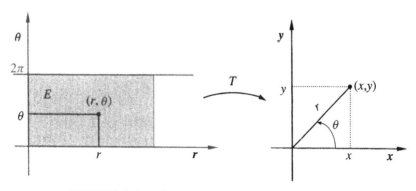

FIGURE 7.3.   The Polar Coordinate Transformation

Solution. (a) If we consider the sets $X = \{(r, 0): r \geq 0\}$, $Y = \{(r, 2\pi): r \geq 0\}$, and $Z = \{(0, \theta): 0 \leq \theta \leq 2\pi\}$, then $E \setminus E^\circ = X \cup Y \cup Z$. To show that $\lambda(E \setminus E^\circ) = 0$, it suffices to establish that $\lambda(X) = \lambda(Y) = \lambda(Z) = 0$.

Let $X_n = \{(r, 0): 0 \leq r \leq n\}$ and $Y_n = \{(r, 2\pi): 0 \leq r \leq n\}$. In view of $X_n \subseteq [0, n] \times [-\varepsilon, \varepsilon]$ and $Y_n \subseteq [0, n] \times [2\pi - \varepsilon, 2\pi + \varepsilon]$, we see that $\lambda(X_n) = \lambda(Y_n) = 0$ holds for each n. Since $X_n \uparrow X$ and $Y_n \uparrow Y$, it follows that $\lambda(X) = \lambda(Y) = 0$.

Also, the inclusion $Z \subseteq [-\varepsilon, \varepsilon] \times [0, 2\pi]$ implies $\lambda(Z) \leq 4\pi\varepsilon$ for each $\varepsilon > 0$, and thus $\lambda(Z) = 0$.

(b) This is proven in part (a) previously.

(c) Clearly, $T: E^\circ \longrightarrow \mathbf{R}^2 \setminus A$ is one-to-one, onto, and C^1-differentiable. The Jacobian determinant is

$$J_T(r, \theta) = \det \begin{bmatrix} \frac{\partial x}{\partial r} & \frac{\partial y}{\partial r} \\ \frac{\partial x}{\partial \theta} & \frac{\partial y}{\partial \theta} \end{bmatrix} = \det \begin{bmatrix} \cos\theta & \sin\theta \\ -r\sin\theta & r\cos\theta \end{bmatrix} = r,$$

which implies that $J_T(r, \theta) = r \neq 0$ holds for each $(r, \theta) \in E^\circ$. The preceding are enough to guarantee that $T: E^\circ \longrightarrow \mathbf{R}^2 \setminus A$ is a diffeomorphism.

(d) Clearly, $G^\circ \subseteq E^\circ$. Thus, by part (c), $T(G^\circ)$ is an open subset of $\mathbf{R}^2 \setminus A$ and $T: G^\circ \longrightarrow T(G^\circ)$ is a diffeomorphism.

Since $T: \mathbf{R}^2 \longrightarrow \mathbf{R}^2$ (defined by $T(r, \theta) = (r\cos\theta, r\sin\theta)$) is a C^1-diffeomorphism, it follows from Lemma 40.1 that $\lambda(T(G \setminus G^\circ)) = 0$. Now, if we consider the sets $A = G$, $B = T(G)$, $V = G^\circ$, and $W = T(G^\circ)$, then $\lambda(A \setminus V) = \lambda(G \setminus G^\circ) = 0$ and $\lambda(T(G) \setminus T(G^\circ)) \leq \lambda(T(G \setminus G^\circ)) = 0$ both hold. Thus, Theorem 40.8 applies and gives us the desired formula.

Problem 40.5. *This problem uses polar coordinates (introduced in the preceding problem) to present an alternate proof of Euler's formula $\int_0^\infty e^{-x^2}\, dx = \sqrt{\pi}/2$.*

a. *For each $r > 0$, let $C_r = \{(x, y) \in \mathbf{R}^2: x^2 + y^2 \leq r^2,\ x \geq 0,\ y \geq 0\}$ and $S_r = [0, r] \times [0, r]$. Show that $C_r \subseteq S_r \subseteq C_{r\sqrt{2}}$.*

b. *If $f(x, y) = e^{-(x^2+y^2)}$, then show that*

$$\int_{C_r} f\, d\lambda \leq \int_{S_r} f\, d\lambda \leq \int_{C_{r\sqrt{2}}} f\, d\lambda,$$

   where λ is the two-dimensional Lebesgue measure.

c. *Use the change of variables to polar coordinates and Fubini's Theorem to show that*

$$\int_{C_r} f,\, d\lambda = \int_0^{\frac{\pi}{2}} \int_0^r e^{-t^2} t\, dt\, d\theta = \frac{\pi}{4}(1 - e^{-r^2}).$$

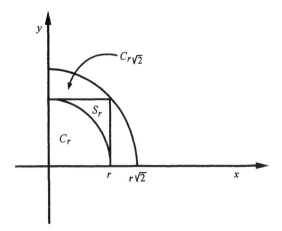

FIGURE 7.4.

d. *Use (b) to establish that*

$$\frac{\pi}{4}(1 - e^{-r^2}) \le \left(\int_0^r e^{-x^2}\,dx\right)^2 \le \frac{\pi}{4}(1 - e^{-2r^2}),$$

and then let $r \to \infty$ to obtain the desired formula.

Solution. (a) Geometrically the three sets are as shown in Figure 7.4.
(b) Since $f(x, y) = e^{-(x^2+y^2)} \ge 0$ holds for all (x, y), we see that

$$f\chi_{C_r} \le f\chi_{S_r} \le f\chi_{C_{r\sqrt{2}}},$$

and the desired inequality follows.
(c) Consider the polar coordinates transformation described in the preceding problem. For the set $G = \{(t, \theta): 0 \le t \le r \text{ and } 0 \le \theta \le \frac{\pi}{2}\}$, we have

$$\int_{C_r} f\,d\lambda = \int_{T(G)} f\,d\lambda = \int\int_G f(t\cos\theta, t\sin\theta)t\,dt\,d\theta$$

$$= \int_0^{\frac{\pi}{2}} \int_0^r e^{-t^2}t\,dt\,d\theta = \tfrac{\pi}{4}\left(1 - e^{-r^2}\right).$$

(d) Note that

$$\int_{C_r} f\,d\lambda = \int_0^r \int_0^r e^{-(x^2+y^2)}\,dx\,dy = \left(\int_0^r e^{-x^2}\,dx\right) \cdot \left(\int_0^r e^{-y^2}\,dy\right)$$

$$= \left(\int_0^r e^{-x^2}\,dx\right)^2.$$

Thus, using (b) and (c), we see that

$$\tfrac{\pi}{4}\bigl(1 - e^{-r^2}\bigr) \le \left(\int_0^r e^{-x^2}\, dx\right)^2 \le \tfrac{\pi}{4}\bigl(1 - e^{-2r^2}\bigr),$$

and by letting $r \longrightarrow \infty$ we get $\left(\int_0^\infty e^{-x^2}\, dx\right)^2 = \tfrac{\pi}{4}$.

Problem 40.6. *In $\mathbf{R}^4$, "double" polar coordinates are defined by*

$$x = r\cos\theta, \quad y = r\sin\theta, \quad z = \rho\cos\phi, \quad w = \rho\sin\phi.$$

State the change of variables formula for this transformation, and use it to show that the "volume" of the open ball in $\mathbf{R}^4$ with center at zero and radius a is $\tfrac{1}{2}\pi^2 a^4$.

Solution. The transformation $T\colon \mathbf{R}^4 \longrightarrow \mathbf{R}^4$ is given by

$$T(r, \rho, \theta, \phi) = (r\cos\theta, r\sin\theta, \rho\cos\phi, \rho\sin\phi)$$

for each $(r, \rho, \theta, \phi) \in \mathbf{R}^4$. Its Jacobian determinant is

$$J_T(r, \rho, \theta, \phi) = \det \begin{bmatrix} \cos\theta & \sin\theta & 0 & 0 \\ 0 & 0 & \cos\phi & \sin\phi \\ -r\sin\theta & r\cos\theta & 0 & 0 \\ 0 & 0 & -\rho\sin\phi & \rho\cos\phi \end{bmatrix} = -r\rho.$$

Write $\mathbf{R}^4 = \mathbf{R}^2 \times \mathbf{R}^2$, and consider the Lebesgue measure on $\mathbf{R}^4$ as the product measure of the corresponding Lebesgue measures on the two factors. Fix $a > 0$, and let

$$E = \bigl\{(r, \rho)\colon r \ge 0, \ \rho \ge 0, \ \text{and} \ r^2 + \rho^2 < a^2\bigr\}$$

and

$$F = [0, 2\pi] \times [0, 2\pi] \subseteq \mathbf{R}^2.$$

Put $G = E \times F \subseteq \mathbf{R}^2 \times \mathbf{R}^2 = \mathbf{R}^4$, and note that $T(G) = B$, the open ball of $\mathbf{R}^4$ with center at zero and radius a. Now, if $C = \{(r, \rho)\colon r\rho = 0\} \subseteq \mathbf{R}^2$ and $D = \{(r, \rho, 0, 0)\colon r \ge 0, \ \rho \ge 0\} \subseteq \mathbf{R}^4$, then both sets are closed in their corresponding spaces and their corresponding Lebesgue measures are zero. Thus, if

$$V = (E \setminus C) \times \left[(0, 2\pi) \times (0, 2\pi)\right] \text{ and } W = B \setminus D,$$

then both V and W are open subsets of $\mathbb{R}^4$ and $T: V \longrightarrow W$ is a diffeomorphism (onto). Since $\lambda(G \setminus V) = \lambda(B \setminus W) = 0$, Theorem 40.8 combined with Fubini's Theorem shows that

$$
\begin{aligned}
\text{Volume of } B = \lambda(B) &= \int_B d\lambda = \int_{T(G)} d\lambda = \iiiint_G r\rho \, dr \, d\rho \, d\theta \, d\phi \\
&= \int_E \left(\int_F r\rho \, d\lambda \right) d\lambda = 4\pi^2 \int_E r\rho \, dr \, d\rho \\
&= 4\pi^2 \cdot \tfrac{a^4}{8} = \tfrac{1}{2}\pi^2 a^4.
\end{aligned}
$$

Problem 40.7 (Cylindrical Coordinates).   *Let*

$$E = \{(r, \theta, z) \in \mathbb{R}^3 \colon r \geq 0, \ 0 \leq \theta \leq 2\pi, \ z \in \mathbb{R}\}.$$

The transformation $T: E \to \mathbb{R}^3$ defined by $T(r, \theta, z) = (r \cos\theta, r \sin\theta, z)$ or as it is usually written

$$x = r\cos\theta, \quad y = r\sin\theta, \quad z = z,$$

is called the **cylindrical coordinate transformation**, *shown graphically in Figure 7.5.*

   a.   *Show that $\lambda(E \setminus E^\circ) = 0$.*
   b.   *If $A = \{(x, 0, z) \in \mathbb{R}^3 \colon x \geq 0, \ z \in \mathbb{R}\}$, then show that A is a closed subset of $\mathbb{R}^3$ whose (three-dimensional) Lebesgue measure is zero.*

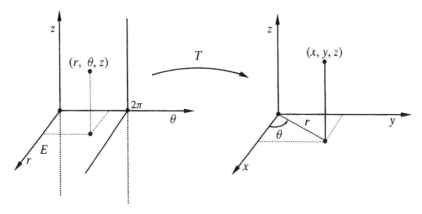

FIGURE 7.5.   The Cylindrical Coordinate Transformation

c. Show that $T: E^\circ \to \mathbf{R}^3 \setminus A$ is a diffeomorphism whose Jacobian determinant satisfies $J_T(r, \theta, z) = r$ for each $(r, \theta, z) \in E^\circ$.

d. Show that if G is a Lebesgue measurable subset of E with $\lambda(G \setminus G^\circ) = 0$, then $T(G)$ is a Lebesgue measurable subset of $\mathbf{R}^3$. Moreover, show that if $f \in L_1(T(G))$, then

$$\int_{T(G)} f \, d\lambda = \int \int \int_G f(r \cos \theta, r \sin \theta, z) r \, dr \, d\theta \, dz$$

holds.

Solution. Repeat the solution of Problem 40.4.

Problem 40.8 (Spherical Coordinates). *Let*

$$E = \{(r, \theta, \phi) \in \mathbf{R}^3 : r \geq 0, \ 0 \leq \theta \leq 2\pi, \ 0 \leq \phi \leq \pi\}.$$

The transformation $T: E \to \mathbf{R}^3$ defined by

$$T(r, \theta, \phi) = (r \cos \theta \sin \phi, r \sin \theta \sin \phi, r \cos \phi),$$

or as it is usually written

$$x = r \cos \theta \sin \phi, \quad y = r \sin \theta \sin \phi, \quad z = r \cos \phi,$$

is called the **spherical coordinate transformation,** *shown graphically in Figure 7.6.*

a. Show that $\lambda(E \setminus E^\circ) = 0$.

b. If $A = \{(x, 0, z): x \geq 0 \text{ and } z \in \mathbf{R}\}$, then show that A is a closed subset of $\mathbf{R}^3$ whose (3-dimensional) Lebesgue measure is zero.

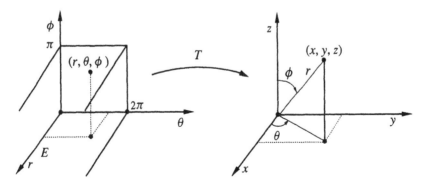

FIGURE 7.6. The Spherical Coordinate Transformation

c. Show that $T: E^{\circ} \rightarrow \mathbf{R}^3 \setminus A$ is a diffeomorphism whose Jacobian determinant satisfies $J_T(r, \theta, \phi) = -r^2 \sin \phi$.

d. Show that if G is a Lebesgue measurable subset of E with $\lambda(G \setminus G^{\circ}) = 0$, then $T(G)$ is a measurable subset of $\mathbf{R}^3$. In addition, show that if $f \in L_1(T(G))$, then

$$\int_{T(G)} f \, d\lambda = \iiint_G f(r \cos \theta \sin \phi, r \sin \theta \sin \phi, r \cos \phi) r^2 \sin \phi \, dr \, d\theta \, d\phi$$

holds.

Solution. Repeat the solution of Problem 40.4.